geoENV I – GEOSTATISTICS FOR ENVIRONMENTAL APPLICATIONS

Quantitative Geology and Geostatistics

VOLUME 9

The titles published in this series are listed at the end of this volume.

geoENV I – GEOSTATISTICS FOR ENVIRONMENTAL APPLICATIONS

Proceedings of the Geostatistics for Environmental Applications Workshop, Lisbon, Portugal, 18–19 November 1996

Edited by

AMÍLCAR SOARES

Department of Geo-Resources, IST,
University of Lisbon,
Lisbon, Portugal

JAIME GÓMEZ-HERNANDEZ

Department of Hydraulics and Environmental Engineering,
Polytechnic University of Valencia,
Valencia, Spain

and

ROLAND FROIDEVAUX

FSS Consultants SA,
Geneva, Switzerland

SPRINGER-SCIENCE+BUSINESS MEDIA, B.V.

A C.I.P. Catalogue record for this book is available from the Library of Congress.

ISBN 978-90-481-4861-5 ISBN 978-94-017-1675-8 (eBook)
DOI 10.1007/978-94-017-1675-8

Printed on acid-free paper

Scientific Committee

José António Almeida
Amvrossios Bagtzoglou
András Bárdossy
Peter Burrough
Bruce Buxton
Jesus Carrera
Andrew Cliffe
Noel Cressie
Fernando Durão
Roland Froidevaux
Henrique Garcia Pereira
Michel Garcia
Jaime Gómes-Hernández
Pierre Goovaerts
Peter Jackson
André Journel
Carlos López
Dennis Marcotte
Ghislain Marsily
Juan Mayor
Pascal Monestiez
Henning Omre
Maria João Pereira
Donato Posa
Christopher Rautman
Javier Samper
Xavier Sánchez-Vila
Júlia Seixas
Amílcar Soares
Jorge Sousa
Georg Teutsch
Marc Van Meirvenne
Hans Wackernagel

Organizing Committee

Peter Burrough
Vincenzo Capasso
Jesus Carrera
Sevket Durucan
Roland Froidevaux
Jaime Gómez-Hernández
Pierre Goovaerts
Ghislain Marsily
Amílcar Soares
Georg Teutsch

FOREWORD

GeoENV96, the First European Conference on Geostatistics for Environmental Applications held in Lisbon, was conceived to bring together researchers, mostly from, but not limited to Europe, working on environmental issues approached by geostatistical methods. Papers were attracted from fields as diverse as hydrogeology, biology, soil sciences, air pollution or ecology. It is clear that there is a lot of activity on geostatistics for environmental applications as the collection of papers in this book reveals.

GeoENV96 was successful in the number and quality of the papers presented which surpassed the initial expectations. There is still a large dispersion on the level of application of geostatistics in the different areas. To help in spreading the most novel applications of geostatistics across disciplines and to discuss the specific problems related to the application of geostatistics to environmental applications, geoENV96 is intended to set the pace and to be the first of a series of biennial meetings. The pace is set, now let us wait for geoENV98.

Lisbon, November 1996

The Executive Committee:

Jaime Gómez-Hernandez
Roland Froidevaux
Amílcar Soares

TABLE OF CONTENTS

Air Pollution. Ecology and Other Applications

EQUIVALENT TRANSMISSIVITIES IN HETEROGENEOUS POROUS MEDIA UNDER RADIALLY CONVERGENT FLOW

X. SÁNCHEZ-VILA, C.L. AXNESS, and J. CARRERA
Departament d'Enginyeria del Terreny i Cartogràfica
Universitat Politècnica de Catalunya, 08034 Barcelona, Spain

Abstract

Most field methods used to estimate transmissivity values rely on the analysis of radial flow towards a single point (pump tests). Using a perturbation approach, we present an analytical solution to the problem of equivalent transmissivity (T_{eq}) under radially convergent steady-state flow conditions produced by constant pumping from a well of radius r_w. In this context T_{eq} is defined as the value that best fits Thiem's equation, and would be the transmissivity assigned to the well location in a pump test. Using a Green's function technique, we derive an expression for T_{eq}, up to second-order in the expansion, which is given as a weighted average of the fluctuations in log-T throughout the domain. Previous work (Desbarats, 1992) hypothesized, based on empirical evidence, that the weighting function for log-T_{eq} could be written as a function of the log-T fluctuations normalized by the square of their distance from the well. We find that this is indeed the case, although our second-order terms differ from that of Desbarats. We conclude with a highly heterogeneous example that illustrates the relative ranges over which the empirical formula and our second order expansion are accurate.

1. Introduction

Most field methods to obtain hydraulic parameter values rely on flow towards a fully penetrating well (pump tests). Despite so, non-uniform flow in heterogeneous media has been addressed in the literature less frequently than uniform flow. Most of the early work was devoted to find effective conductivity values, defined as the average between radial specific discharge

A. Soares et al. (eds.), geoENV I–Geostatistics for Environmental Applications, 1–12.

and head gradient in an ensemble of T field realizations (Schvidler, 1964; Matheron, 1967; Dagan, 1989).

The problem of upscaling the transmissivity values into blocks of certain size (equivalent transmissivities) under radial convergent flow conditions in heterogeneous domains has received less attention. Most of the analytical work has concentrated on units composed of two bounded regions of different (fixed) T values, separated by some sharp discontinuity: either radial (Butler, 1988), or elliptical (Trederman et al., 1995). Butler (1991) considers a fully heterogeneous domain, characterized by some random function and studies numerically the impact of azimuthal variations in the transmissive properties obtained from conventional pump test analysis.

The work most related to ours is that of Desbarats (1992), who uses a combined numerical-empirical approach to obtain transmissivities in a block; these values are then compared with the ones obtained by a weighted spatial average of point support scale transmissivities in a block V, T_b, which is given by

$$T_b = exp\left(\frac{1}{W}\int_V \frac{Y(\boldsymbol{x})}{r^2} dV\right) \tag{1}$$

where $Y(\boldsymbol{x}) = \ln T(\boldsymbol{x})$ are the point log-transmissivity values, r is the radial distance from the center of the well, $dV = r dr d\theta$, and W is given by

$$W = \int_V \frac{d\boldsymbol{x}}{r^2(\boldsymbol{x})} \tag{2}$$

In this paper, we analyze the validity of the averaging formula (1) in a heterogeneous domain under saturated, steady-state flow conditions. The analysis can be looked at in a different way. When performing a field test interpretation, what we find is not a real point transmissivity value, but a weighted average of all the values surrounding the well. It is important to know the relationship between the average value obtained from the interpretation of the pump test results (upscaled value), and the transmissivity at the well (the point support value).

2. Previous Existing Averaging Formulae

Consider a two-dimensional annular domain, V, defined by an internal radius, r_w, and an external one, r_e. Cardwell and Parsons (1945) proved that the equivalent transmissivity T_{eq} is bounded by the weighted harmonic and the weighted arithmetic averages of T over V; that is

$$W\left(\int_V \frac{d\boldsymbol{x}}{T(\boldsymbol{x})r^2(\boldsymbol{x})}\right)^{-1} \leq T_{eq} \leq \frac{1}{W}\int_V \frac{T(\boldsymbol{x})d\boldsymbol{x}}{r^2(\boldsymbol{x})} \tag{3}$$

where T_{eq} is defined as the value that suits the following relationship

$$Q = 2\pi \frac{h_e - h_w}{\ln r_e/r_w} T_{eq} \quad (4)$$

where Q is pumping rate; h_w, and h_e are the heads at the inner and outer radii, respectively. In short, T_{eq} is the value that best suits Thiem's formula.

Only a few simplistic cases of heterogeneity exist where we can provide an exact formula for T_{eq}. One is the case where the transmissivity values are distributed in circular, radially symmetric annuli, which gives the lower bound in (3). Another one is the case where T is constant in sectors extending from the inner to the outer radius; in this case, T_{eq} is equal to the arithmetic mean. An exact solution has not been found for other cases.

In this work we develop a second-order analytical expression for T_{eq} based on a small perturbation expansion in $\ln T$, and prove that both (1) and the two bounds in (3) provide exactly the same expression up to first order, but lead to different second order terms. Futhermore, we find that the expansion of (1) up to second order misses some terms from the analytical expression and, thus, is not completely valid. The rest of the work is devoted to analyzing this additional term and seeing under which circumstances it may or may not be relevant.

3. Derivation of Apparent Equivalent Transmissivity

In the annular domain, V, fixing h_w and h_e at the boundaries, and measuring Q, we may use (4) to obtain the apparent equivalent transmissivity. At this point we do not need to impose any condition to the structure of $Y(\mathbf{x})$ other than being an integrable function; that is, our solution is valid under very general conditions. Our objective is to develop Q in a consistent expansion, and to write the different terms as a function of the spatial distribution of Y. This is done by first writing Darcy's law and then substituting it into the groundwater flow equation.

3.1. DEVELOPMENT OF THE PERTURBED FLOW EQUATION

Under steady-state conditions, the total flow rate Q can be obtained by integration of the specific discharge through any surface containing the well. In particular, we can use the surface of the well itself and write

$$Q = Q(r_w) = -\int_0^{2\pi} q_r(r_w, \theta)\, r_w d\theta \quad (5)$$

where q_r is the radial component of the specific discharge (normal to the surface of the well at every point). Darcy's law, at the local scale, becomes

$$q_r(r_w, \theta) = -T(r_w, \theta) \frac{\partial}{\partial r} h(r, \theta)|_{r=r_w} = -T_w \frac{\partial}{\partial r} h(r, \theta)|_{r=r_w} \qquad (6)$$

so that

$$Q = \int_0^{2\pi} r_w T_w \left[\frac{\partial}{\partial r} h(r, \theta) \right]_{r=r_w} d\theta \qquad (7)$$

Equation (6) is valid assuming that T is constant at the well (equal to T_w), since the integral scale of T is large compared to the well radius. Our next step is to compute $\frac{\partial h}{\partial r}(r_w, \theta)$ using a perturbation approach.

The flow equation under steady-state conditions, written in terms of Y is

$$\nabla^2 h(\mathbf{x}) + \nabla Y(\mathbf{x}) \nabla h(\mathbf{x}) = 0 \qquad (8)$$

After normalizing the radial distances, so that the inner radius is equal to one and the outer radius is $R = r_e/r_w$, the boundary conditions may be written as

$$h = h_w \quad \text{at} \quad r = 1; \qquad h = h_e \quad \text{at} \quad r = R \qquad (9)$$

As $Y(\mathbf{x})$ is heterogeneous, (8) cannot be solved analytically. A classical way to proceed is to decompose Y into a deterministic part plus a random component. As the transmissivity at the well, T_w, should have a very strong impact on the final formula, we choose its logarithm as the deterministic value in the decomposition. This is an important difference with the traditional stochastic approach, where the deterministic value chosen is equal to the expected value of the random variable. Thus, in the traditional approach the random component has a zero mean, while in our approach it may not. Then, finally

$$Y = Y_w + Y' \qquad (10)$$

where $Y_w = \ln T_w$, so that $< Y' >= C$, a constant ($<>$ standing for expected value). Note that the value of C depends upon Y_w, and it is usually different from zero, except in the particular case where $Y_w =< Y >$. Now, we formally expand h in an asymptotic sequence: $h = h^{(0)} + h^{(1)} + h^{(2)} + \ldots$.

By substituting the expansions of h and Y into (8) and retaining terms up to second order in the expansion, we have (remember $\nabla Y_w = 0$)

$$\nabla^2 h^{(0)} + \nabla^2 h^{(1)} + \nabla^2 h^{(2)} + \nabla Y' \nabla h^{(0)} + \nabla Y' \nabla h^{(1)} + \ldots = 0 \qquad (11)$$

If we now expand $e^{-Y'} = 1 - Y' + \frac{1}{2} Y'^2 + \ldots$ (considering that close to the well we have $Y' \simeq 0$, if Y is assumed a continuous random function), and substituting in (11), we finally obtain a sequence of equations that can be solved iteratively:

$$\nabla^2 h^{(0)} = 0 \tag{12}$$

$$\nabla^2 h^{(i)} + \nabla Y' \nabla h^{(i-1)} = 0 \text{ for } i = 1, \ldots, n \tag{13}$$

and we can choose the $h^{(i)}$ functions to satisfy the following boundary conditions

$$h^{(0)} = h_w \quad \text{at} \quad r = 1; \qquad h^{(0)} = h_e \quad \text{at} \quad r = R \tag{14}$$

$$h^{(i)} = 0 \quad \text{at} \quad r = 1; \qquad h^{(i)} = 0 \quad \text{at} \quad r = R, \; i \geq 1 \tag{15}$$

and we now also expand Q as $Q = Q_0 + Q_1 + Q_2 + \ldots$, where, from (7), each of the Q_i terms is given by

$$Q_i = \int_0^{2\pi} T_w \left[\frac{\partial}{\partial r} h^{(i)}(r,\theta) \right]_{r=1} d\theta \tag{16}$$

Our goal is to evaluate the first three terms in the expansion of Q.

3.2. EVALUATION OF Q_0

From (13) and (15), $h^{(0)}$ can be easily computed, as it is the only equation from the set that does not depend on the Y' function

$$h^{(0)}(r) = h_w + A \ln r \tag{17}$$

with $A = \frac{h_e - h_w}{ln\ r_e/r_w}$; due to the radial symmetry of the problem, $h^{(0)}$ is just a function of r (normalized radial distance to the pumping well) and not of θ. So, from (16)

$$Q_0 = \int_0^{2\pi} T_w \frac{h_e - h_w}{ln \frac{r_e}{r_w}} d\theta = 2\pi \frac{h_e - h_w}{ln \frac{r_e}{r_w}} T_w \tag{18}$$

which is just Thiem's formula for a homogeneous medium of transmissivity T_w.

3.3. EVALUATION OF Q_1

First, we must find $\frac{\partial}{\partial r} h^{(1)}(r,\theta)$. From (13) and (17), we have

$$\nabla^2 h^{(1)} = -\frac{A}{r} \frac{\partial Y'}{\partial r} \tag{19}$$

with homogeneous boundary conditions. The solution of this stochastic partial differential equation can be obtained with the help of Green's functions

$$h^{(1)}(r,\theta) = A\int\int \frac{1}{\rho}\frac{\partial Y'}{\partial\rho}(\rho,\phi)G(r,\theta,\rho,\phi)\rho d\rho\, d\phi \tag{20}$$

where $G(r,\theta,\rho,\phi)$ is the Green's function corresponding to the following problem:

$$\nabla^2 G(r,\theta,\rho,\phi) = -\,\delta(r-\rho,\theta-\phi) \tag{21}$$

with homogeneous boundary conditions at $r = 1$ and $r = R$. The solution for this problem is given, for example, in Weinberger (1965)

$$G(r,\theta,\rho,\phi) = \frac{lnr\; ln(R/\rho)}{2\pi lnR} + \sum_{n=1}^{\infty}\frac{r^n - r^{-n}}{R^n - R^{-n}}\left[\left(\frac{R}{\rho}\right)^n - \left(\frac{\rho}{R}\right)^n\right]\frac{cosn(\theta-\phi)}{2\pi n} \tag{22}$$

valid for $\rho > r$, where $r = 1$ in our case. From (16), after integration by parts, Q_1 is equal to

$$Q_1 = -T_w A\int_0^{2\pi}\int\int Y'(\rho,\phi)\frac{\partial^2 G}{\partial r\partial\rho}|_{r=1} d\rho d\phi d\theta \tag{23}$$

Integrating with respect to θ and taking into account that the integration of the cosine is carried out over $[0, 2\pi]$, the term including the infinite series is identically zero except for $\rho = 1$, where the infinite series is not uniformly convergent. In this case however, $Y'(\rho = 1, \theta) = 0$, and so the term drops cancels out. Then, we get

$$Q_1 = \frac{h_e - h_w}{(ln\frac{r_e}{r_w})^2}T_w\int_V \frac{Y'(\rho,\phi)}{\rho^2}dV = Q_0\frac{\int_V \frac{Y'(\rho,\phi)}{\rho^2}dV}{\int_V dV/\rho^2} \tag{24}$$

where we have used the notation $\int\int \rho d\rho d\phi = \int_V dV$ and $\int_V dV/\rho^2 = 2\pi ln\frac{r_e}{r_w}$ (easily derived). This is our first important result. Notice that if the expansion of Y had been performed around a value $\bar{Y}$ different than Y_w, the discussion after (23) would not apply, and the infinite series in (22) would give a correction term.

3.4. EVALUATION OF Q_2

Q_2 can be obtained similarly to Q_1. From (13)

$$\nabla^2 h^{(2)} = -\nabla Y'\nabla h^{(1)} \tag{25}$$

with homogeneous boundary conditions. The solution is again given as

$$h^{(2)}(r,\theta) = \int_V \nabla_\rho Y' \nabla_\rho h^{(1)} G(r,\theta,\rho,\phi) dV \tag{26}$$

Then substituting $h^{(1)}$ from (20) we get

$$\frac{\partial h^{(2)}}{\partial r}(r,\theta) = \frac{\partial}{\partial r} \int_V \nabla_\rho Y' \nabla_\rho h^{(1)} G(r,\theta,\rho,\phi) \rho d\rho d\phi =$$

$$A\frac{\partial}{\partial r} \int_V \int_V \nabla_\rho Y'(\rho,\phi) \frac{\partial Y'}{\partial \rho^*}(\rho^*,\phi^*) \nabla_\rho G(\rho,\phi,\rho^*,\phi^*) G(r,\theta,\rho,\phi) \frac{1}{\rho^*} dV dV^* \tag{27}$$

and after a very involved manipulation we can write

$$Q_2 = Q_{2a} + Q_{2b} + Q_{2c} = Q_0 \left\{ \frac{\frac{-1}{2} \int_V \frac{Y'^2(\rho,\phi)dV}{\rho^2}}{\int_V dV/\rho^2} + \right.$$

$$\left. \left(\frac{\int_V Y'(\rho,\phi)/\rho^2 dV}{\int_V dV/\rho^2} \right)^2 + \frac{\int_V \int_V \frac{Y'(\rho,\phi)Y'(\rho^*,\phi^*)}{\rho^2 \rho^{*2}} H(\rho,\phi,\rho^*,\phi^*) dV dV^*}{2\pi \int_V dV/\rho^2} \right\} \tag{28}$$

where

$$H = \begin{cases} \sum_{n=1}^{\infty} n\left(\frac{\rho^n + \rho^{-n}}{R^n - R^{-n}}\right)\left[\left(\frac{R}{\rho^*}\right)^n + \left(\frac{\rho^*}{R}\right)^n\right] \cos n(\phi - \phi^*) & \text{for}(\rho^* > \rho) \\ \sum_{n=1}^{\infty} n\left(\frac{\rho^{*n} + \rho^{*-n}}{R^n - R^{-n}}\right)\left[\left(\frac{R}{\rho}\right)^n + \left(\frac{\rho}{R}\right)^n\right] \cos n(\phi - \phi^*) & \text{for}(\rho^* \leq \rho) \end{cases} \tag{29}$$

and, finally, we can express Q by its approximation up to second order, and from (4)

$$2\pi \frac{h_e - h_w}{lnR} T_{eq} = Q_0 + Q_1 + Q_{2a} + Q_{2b} + Q_{2c} \tag{30}$$

4. Analysis of the Solution

Using the previous notations it is very easy to write the second order expansions of both the formula by Desbarats (1992),

$$2\pi \frac{h_e - h_w}{lnR} T_{eq} = Q_0 + Q_1 + \frac{1}{2} Q_{2b} \tag{31}$$

that corresponding to the weighted harmonic mean,

$$2\pi \frac{h_e - h_w}{lnR} T_{eq} = Q_0 + Q_1 + Q_{2a} + Q_{2b} \tag{32}$$

and that corresponding to the weighted arithmetic mean,

$$2\pi \frac{h_e - h_w}{lnR} T_{eq} = Q_0 + Q_1 - Q_{2a} \tag{33}$$

where we note that $Q_{2b} = Q_1^2/Q_0$. It is immediately apparent from the above formulae that the perturbed solution agrees with the weighted harmonic mean when Q_{2c} is equal to zero. That is the case, for example, when the transmissivity field does not depend on θ since the cosine term is integrated over its full range. Thus (32) provides a simple verification of the terms Q_{2a} and Q_{2b} in one limiting case. We note that up to first order in the expansion there are no differences in the terms for any of these formulae. Examining the second-order terms we note that neither Desbarats' formula nor the generalized means are correct since they miss some of the terms in the expansion. We will see in the next section a synthetic example where the term Q_{2c} is important. Precisely in this case the term Q_1 becomes identically zero and Desbarats formula gives only a constant Q_0. In other more realistic cases the importance of the Q_{2c} term should be established before using any of the simplified formulae for upscaling purposes.

5. Verification of the Perturbed Solution

To verify the different Q_i terms in the perturbed solution, we have written a finite element code that solves the flow equation in an annular domain. This code is novel in that it solves the flow equation in $(\ln r, \theta)$ space. In this space the discretization error is proportional to $(\Delta\theta)^2$ and $(1/r^2)\Delta r^2$. Since r is normalized with respect to the well radius this second error decreases rapidly with radius, allowing coarser discretization when the principal flow component is radially aligned as is the case for most transmissivity fields. Furthermore, under this transformation the flow domain becomes rectangular, Thiem's equation becomes Darcy's law, and a homogeneous flow becomes linear with respect to ln (r). The numerical simulations discussed in this section were performed using a 80x26 mesh in $(\theta$, ln$r)$ space.

One often verifies an analytic solution by looking at limiting conditions. We have verified our solution for the limiting case where the transmissivity is only a function of the radius. In this case, however, Q_{2c} is equal to zero, and so, another transmissivity perturbation function is needed to verify Q_{2c}. We selected $Y' = \beta cos(m\theta)$ for $r \geq R_I > 1$, where β is real and $m > 1$ is an integer. For this test function the integrals of the perturbed solution are not difficult to calculate, Q_{2c} is non-zero, and Q_1 is zero. The comparison between the numerically calculated Q and the non-zero terms used to calculate Q_0 and Q_2 as a function of the parameter β is shown in Figure 1 for the case m=3. The figure clearly shows a range of $\beta \leq 1$ in which the perturbations are small and the numerically calculated Q is approximately constant and well-represented by Q_0 (which in this particular case corresponds to Desbarats' solution). When $1 \leq \beta \leq 10$ the flow at the well is dominated by the high-T regions, and the addition of Q_2 is necessary in order for the perturbed solution to accurately calculate the numerical solution. The flat region for $\beta \geq 10$ corresponds to the range over which the flow to the well is controlled by the area near the well bore where the transmissivity is small compared to the amplitude of the cosine function. It is apparent for $\beta \geq 4$ that higher-order terms are necessary to accurately calculate Q. Notice that $\beta = 4$ corresponds to a very high variance as $\sigma_Y^2 = \beta^2/2 = 8$. This is an important point to make, as perturbation approaches are demonstrably valid only for $\sigma_Y^2 \leq 1$.

6. Extension of the Perturbed Solution and Implications on the Equivalent Transmissivity

We might postulate from the form of the second-order solution for Q that the exact solution of the problem may be written as an infinite series of integrated multiples of Y' along with more complex terms involving a transformed form of the Green's function. In fact, we might examine the case where Y' has no dependence on θ to try to guess the higher order terms that are not null in this case. The exact solution of the flow from a well enclosed by a series of annuli, each of constant transmissivity, was first published by Craft and Hawkins (1959)

$$Q = \frac{2\pi\Delta h}{\int_{r_w}^{r_e} \frac{dr}{T(r)r}} \tag{34}$$

Now, writing $T^{-1}(r)$ as $exp(-Y'(r))/T_w$, expanding the exponential, and dividing, we get $Q = Q_0 \sum_{i=0}^{\infty} Z_i$, where $Z_0 = 1$; $Z_1 = \frac{\gamma_1}{W}$; $Z_2 = -\frac{\gamma_2}{W} + \frac{\gamma_1^2}{W^2}$; $Z_3 = \frac{\gamma_3}{W} - 2\frac{\gamma_1\gamma_2}{W^2} + \frac{\gamma_1^3}{W^3}$; and so on, and $\gamma_i = \int_V \frac{(Y'^i/i!)}{r^2} dV$. The first three terms in the series agree exactly with Q_0, Q_1, Q_{2a} and Q_{2b}. One might think of extending the range of our solution by the addition of succesive Z_i terms, keeping in mind that the Z_i terms provide an inconsistent expansion, as there are some missing terms, related to the infinite series in the Green's function, that may or may not be important. Obviously, in the case of the transmissivity rings arranged in series the addition of each Z_i improves the accuracy of the perturbed solution, but in the previous example, where the perturbations vary significantly with θ, adding the Z_3 and Z_4 terms did not improve the accuracy of the solution.

If we consider $T(r)$ to be a one-dimensional Random Space Process, the solution implies that the equivalent transmissivity is the harmonic mean weighted by a $1/r^2$ factor, and it is independent of the distribution function assumed for the process. The analogous problem in planar flow has an exact solution (Terzaghi and Peck, 1967), the equivalent transmissivity being the unweighted harmonic mean. In fact, using a change of variable of $z = ln\, r$ in (34) we arrive at the planar flow solution scaled by a factor of 2π. The similar form of the solutions in these one-dimensional flow problems suggests a limit for T_{eq} equal to the weighted harmonic mean, independently of the transmissivity distribution, when flow is normal to the direction in which transmissivity is well-correlated.

Transmissivity is less restrictive to flow when it is arranged in n constant transmissivity sectors of width $\Delta\theta$. Due to constant head boundary conditions, the gradient with respect to $\ln(r)$ is the same in each sector, and the contribution to the well flow for any particular sector is just the solution of the Thiem equation multiplied by the factor $\Delta\theta/2\pi$. Thus the overall solution in this case for an infinite number of infinitesimal "sectors" is just the arithmetic mean of the transmissivities. In this case the transmissivity is again a unidimensional function but the flow is aligned in the direction in which the transmissivity is perfectly-correlated (radially). The analogous problem in planar flow is a perfectly stratified aquifer in which the equivalent transmissivity has again been found to be the arithmetic mean (Terzaghi and Peck, 1967).

7. Conclusions

In summary, the transmissivity measured in a well test is not only dependent upon local conditions in the area of the well, but it is equal to a weighted average of transmissivities in the whole flow domain. The weighting function decreases with the distance to the well. We present a second-order expansion for the flow to a well in a two-dimensional heterogeneous medium assuming constant head boundary conditions at the wellbore and an external radius. To first-order the final expression agrees with that of Desbarats, but the second-order terms differ. The perturbed solution is similar to a generalized form of the Craft and Hawkins (1959) solution but contains a term dependent upon the Green's function of the domain. This term may be very important under preferential flow conditions. The solution is verified by comparison with the exact solution for a radial case and with numerical simulations using a periodic Y function. Ranges in which the zero, second, and higher-order terms of the perturbed solution are necessary for accurate calculation of the numerical solution were observed in this example. A simple heuristic method of obtaining postulated higher-order terms of the perturbed solution that are not dependent upon the Green's function is proposed. Finally, we note that the formulation is general and not dependent upon any uni- or multivariate distribution of Y. In a future paper we will use the expression derived in this paper to examine the statistical properties of Q under some simplifications such as multinormality in Y.

Acknowledgements

This work was funded by ENRESA (Spanish Nuclear Waste Management Company).

References

Butler, J. J., Jr., Pumping tests in nonuniform aquifers - The radially symmetric case, Journal of Hydrology, 101, 15-30, 1988.

Butler, J. J., Jr., A stochastic analysis of pumping tests in laterally nonuniform media, Water Resour. Res., 27(9), 2401-2414, 1991.

Cardwell, W. T., and R. L. Parsons, Average permeabilities of heterogeneous oil sands, Trans Am Inst Min Metall Pet Eng, 160, 34-42, 1945.

Craft, B. C., and M. F. Hawkins, Applied Petroleum Reservoir Engineering, 437 pp., Prentice-Hall, Englewood Cliffs, N. J., 1959.

Dagan, G., Flow and Transport in Porous Formations, 465 pp., Springer-Verlag, New York, 1989.

Desbarats, A. J., Spatial averaging of transmissivity in heterogeneous fields with flow toward a well, Water Resour. Res., 28(3), 757-767, 1992.

Matheron, G., Elements Pour une Theorie des Milieux Poreux, Masson et Cie, Paris, 166 pp., 1967.

Schvidler, M. I., Filtration Flows in Heterogeneous Media (a Statistical Approach), Consultants Bureau, New York, 1964.

Terzaghi, K., and R. B. Peck, Soil Mechanics in Engineering Practice, 2nd ed., John Wiley & Sons, New York, 1967.

Trederman, C. R., P. A. Hsieh, and S. B. Christian, Characterization of a high-transmissivity zone by well test analysis: Steady state case, Water Resour. Res., 31(1), 27-37, 1995.

Weinberger, H. F., Partial Differential Equations, Blaisdell Publishing Company, New York, 1965.

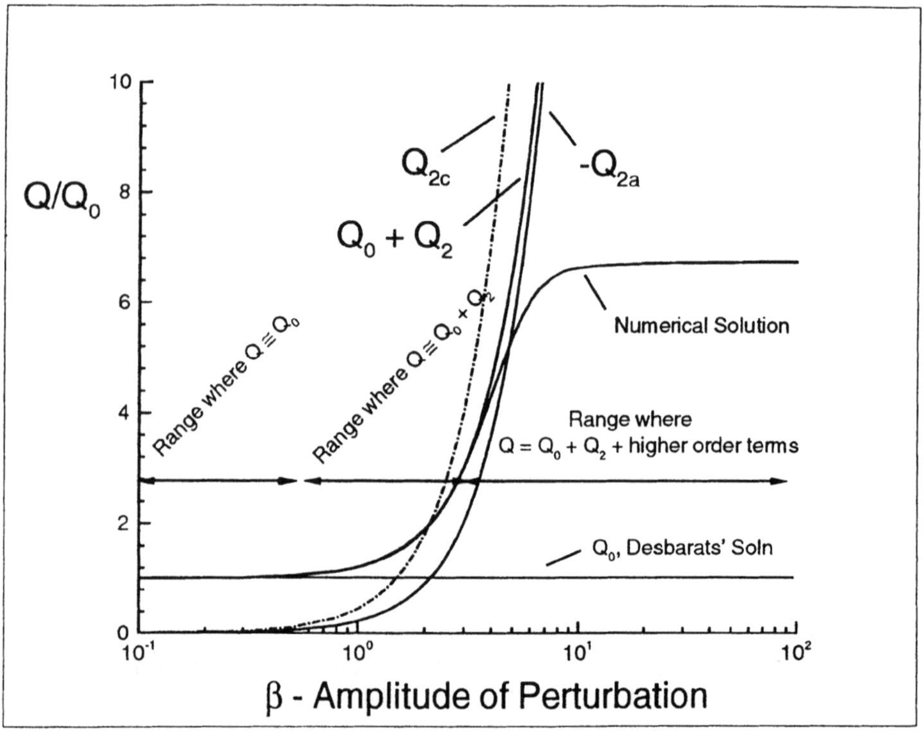

Figure 1: Normalized Q vs. perturbation amplitude in the analytical and numerical solutions for the example.

INVERSE MODELLING OF GROUNDWATER FLOW INCORPORATING PARAMETER, BOUNDARY AND INITIAL CONDITIONS AND MODEL ERROR UNCERTAINTY

JOHAN VALSTAR
Delft University of Technology, Faculty of Civil Engineering
DENNIS MCLAUGHLIN
Mass. Institute of Technology, Faculty of Civil and Environmental Engineering, Parsons Laboratory
AND
CHRIS TE STROET
Delft University of Technology, Faculty of Civil Engineering

Abstract. A new inverse method for groundwater flow is presented which can optimize large scale inverse problems efficiently. All parameters, initial and boundary conditions and driving forces are allowed to vary in space, while boundary conditions and driving forces can vary in time as well. The objective function, which is minimized, is a weighted least squares function with prior terms. The method minimizes the objective function by directly solving the Euler-Lagrange equations in an iterative way by using a basis function expansion with representers. The method was applied successfully in a 2 dimensional transient problem where it had to optimize 9225 unknowns.

1. Introduction

In recent years many attempts have been made to characterize soil parameters by inverse modelling. Most of these studies have focussed on characterizing spatially variable hydraulic conductivity using head and conductivity measurements (Gómez-Hernández et al., 1995 [6]) and, more recently, solute concentration measurements (Reid, 1996 [7]). When there are sources of uncertainty other than hydraulic conductivity variability, algorithms which focus on conductivity estimation must compensate by artificially adjusting

A. Soares et al. (eds.), geoENV I–Geostatistics for Environmental Applications, 13–24.

conductivity values. Moreover the posterior covariances of the hydraulic conductivity and predicted state variables are too optimistic, because the model neglects the other sources of uncertainty in covariance update equations.

Some investigators have included more sources of uncertainty in groundwater characterization studies, but only in examples where a limited number of parameters are estimated (Carrera and Neuman, 1986 [2], Sun, 1994 [9]). One of the main reasons for this limitation is the enormous size of the covariance of the parameters, when discretized on the pixel scale, which has to be inverted and stored in a computer algorithm. These operations become computationally demanding when large numbers of parameters are estimated. The zonation methods typically used to reduce the number of estimated parameters may not be appropriate in some applications, such as the design of groundwater remediation strategies.

Te Stroet, 1995 [10] has classified methods which ignore model errors as deterministic. Schweppe, 1973 [8] has shown that such methods produce biased estimates when model errors are actually present. Stochastic methods explicitly account for model error by adding a new source of uncertainty to the underlying model. One would expect such methods to provide more accurate and more robust estimates, since they do not need to compensate for model error by artificially-adjusting hydraulic conductivity estimates.

We have derived a stochastic method which allows us to update many different sources of uncertainty, such as hydraulic conductivity, specific storage, initial and boundary conditions, and model error (which can be regarded physically as time dependent uncertainty in the driving forces, such as recharge and pumping rates). The method minimizes a functional weighted least square objective, where the weights are determined by the inverses of the prior covariances of all uncertain parameter errors. In order to solve the minimization problem we use the Euler-Lagrange equations and introduce a convenient parameterization which relies on a basis function expansion. This parameterization reduces the dimensionality of the discretized optimization problem to the number of measurements. Our method is an extended version of a data assimilation technique used in oceanography (Bennett, 1992 [1]).

2. Model definition

The groundwater flow is described by the following model:

$$S\frac{\partial h}{\partial t} - \nabla(e^f \nabla h) = Q + q_h \qquad x \epsilon D, 0 < t \leq T \tag{1}$$

$$h = h_{i_0} + \Delta h_i \qquad x \epsilon D, t = 0 \tag{2}$$

$$h = h_{b_0} + \Delta h_b \qquad x \epsilon \partial D_1, 0 < t \leq T \tag{3}$$

$$e^f \frac{\partial h}{\partial n} = q_{b_0} + \Delta q_b \qquad x \epsilon \partial D_2, 0 < t \leq T \tag{4}$$

$$S = S_0 + \Delta S \qquad x \epsilon D \tag{5}$$

$$f = f_0 + \Delta f \qquad x \epsilon D \tag{6}$$

where S is the specific storage; h is the real hydraulic head; f is log- conductivity; Q is recharge; q_h is the model error and q_b is boundary flux. The subscript 0 indicates the mean of the prior distribution and the prefix Δ indicates the difference between the real value and the prior mean.

The measurement equations describe heads measurement only:

$$z = F(h) + \xi_z \tag{7}$$

where z is a measurement vector; F is the linear measurement operator and ξ_z denotes measurement error. Other types of measurements like log-conductivity or specific storage measurements could also be included, but are ignored in order to make the derivation clearer.

The variables q_h, Δh_i, Δh_b, Δq_b, ΔS, Δf and ξ_z are zero mean normally distributed random functions with known covariance functions $C_{q_h q_h}$, $C_{h_i h_i}$, $C_{h_b h_b}$, $C_{q_b q_b}$, C_{SS}, C_{ff} and $C_{\xi_z \xi_z}$ respectively. The variance of the specific storage is assumed to be small enough to keep the values of the real specific storage between 0 and 1.

The objective functional J, which has to be minimized is:

$$\begin{aligned} J = \frac{1}{2}[& \int_0^T \int_D \int_0^T \int_D q_h(x,t) C_{q_h q_h}^{-1}(x,t,x_2,t_2) q_h(x_2,t_2) dx dt dx_2 dt_2 + \\ & \int_D \int_D \Delta h_i(x) C_{h_i h_i}^{-1}(x,x_2) \Delta h_i(x_2) dx dx_2 + \\ & \int_0^T \int_{\partial D_1} \int_0^T \int_{\partial D_1} \Delta h_b(x,t) C_{h_b h_b}^{-1}(x,t,x_2,t_2) \Delta h_b(x_2,t_2) dx dt dx_2 dt_2 + \\ & \int_0^T \int_{\partial D_2} \int_0^T \int_{\partial D_2} \Delta q_b(x,t) C_{q_b q_b}^{-1}(x,t,x_2,t_2) \Delta q_b(x_2,t_2) dx dt dx_2 dt_2 + \\ & \int_D \int_D \Delta S(x) C_{SS}^{-1}(x,x_2) \Delta S(x_2) dx dx_2 + \\ & \int_D \int_D \Delta f(x) C_{ff}^{-1}(x,x_2) \Delta f(x_2) dx dx_2 + \xi_z^T C_{\xi_z \xi_z}^{-1} \xi_z] \end{aligned} \tag{8}$$

3. Euler-Lagrange equations

The minimum of the objective functional of (8) is the solution to the Euler-Lagrange equations. For a more detailed description of the derivation of these equations we refer to Sun, 1994 [9] or Bennett, 1992 [1].

In order to minimize the objective functional we adjoin the state equations (1) to (7) and set the first variation of the resulting expression equal to zero.

This yields the following set of coupled nonlinear integro-differential equations (The Euler-Lagrange equations):
The head variation optimality condition gives the adjoint equation:

$$-\check{S}(x)\frac{\partial\lambda(x,t)}{\partial t} - \nabla(e^{\check{f}(x)}\nabla\lambda(x,t)) - (z - F(\check{h})^T C^{-1}_{\xi_z\xi_z}$$

$$F(I(x,t)) = 0 \quad x\epsilon D \quad 0 \le t < T \tag{9}$$

$$\lambda(x,T) = 0 \quad x\epsilon D \tag{10}$$

$$\lambda(x,t) = 0 \quad x\epsilon\partial D_1 \quad 0 \le t < T \tag{11}$$

$$\frac{\partial\lambda(x,t)}{\partial n} = 0 \quad x\epsilon\partial D_2 \quad 0 \le t < T \tag{12}$$

where
λ is the adjoint variable; check notation $(\check{f})$ indicates the optimal estimate and the indicator function $I(x,t)$ is defined to be:

$$I(x,t) = 1 \quad x\epsilon D, \quad 0 \le t < T \tag{13}$$

The adjoint variation optimality condition gives the forward equation:

$$\check{S}(x)\frac{\partial\check{h}(x,t)}{\partial t} - \nabla(e^{\check{f}(x)}\nabla\check{h}(x,t)) - Q(x,t) -$$

$$\int_0^T\int_D C_{q_hq_h}(x,t,x_2,t_2)\lambda(x_2,t_2)dx_2dt_2 = 0 \quad x\epsilon D, 0 < t \le T \tag{14}$$

$$\check{h}(x,0) = h_i(x) + \int_D C_{h_ih_i}(x,x_2)\lambda(x_2,0)\check{S}(x_2)dx_2 \quad x\epsilon D \tag{15}$$

$$\check{h}(x,t) = h_{b_0}(x,t) - \int_0^T\int_{\partial D_1} C_{h_bh_b}(x,t,x_2,t_2)$$

$$e^{\check{f}(x_2)}\frac{\partial\lambda(x_2,t_2)}{\partial n}dx_2dt_2 \quad x\epsilon\partial D_1, 0 < t \le T \tag{16}$$

$$e^{\check{f}(x)}\frac{\partial\check{h}(x,t)}{\partial n} = q_{b_0}(x,t) - \int_0^T\int_{\partial D_2} C_{q_bq_b}(x,t,x_2,t_2)$$

$$\lambda(x_2,t_2)dx_2dt_2 \quad x\epsilon\partial D_2, 0 < t \le T \tag{17}$$

The specific storage variation optimality condition yields:

$$\check{S}(x) - S_0(x) + \int_D \int_0^T C_{SS}(x, x_2)\lambda(x_2, t_2)\frac{\partial \check{h}(x_2, t_2)}{\partial t_2} dx_2 dt_2 = 0 \quad x\epsilon D \tag{18}$$

The log-conductivity variation optimality condition yields:

$$\check{f}(x) - f_0(x) + \int_D C_{ff}(x, x_2) \int_0^T e^{\check{f}(x_2)} \nabla \check{h}(x_2, t_2).\nabla\lambda(x_2, t_2) dt_2 dx_2 = 0 \quad x\epsilon D \tag{19}$$

These Euler-Lagrange equations have been used extensively to calculate the gradients of the objective functional with respect to the unknown parameters (Carrera and Neuman, 1986 [3], Sun, 1994 [9], among others). These gradients are typically used to guide iterative searches for the solution to the Euler-Lagrange equations.
In contrast to these methods we have equated all gradients to be zero to obtain the direct optimization Equations (18) and (19). Unfortunately these systems of equations (9) - (19) are all coupled and therefore difficult to solve.

4. Parameterization

In order to solve the infinite dimensional variational optimization problem posed above a kind of parameterization is needed. Many different kinds of parameterization have been used in literature, such as zonation (Carrera and Neuman, 1986 [2]) and pilot points (de Marsily et al., 1984 [4]). These parameterizations are easy to implement, but they do not optimize our objective function. By reducing the degrees of freedom they force the parameter fields to have a predefined shape, which is unlikely to be optimal.

In oceanography basis function expansions with representer functions for each measurement have been used, see Bennett, 1992 [1], Eknes and Evensen, 1995 [5]. However they only use basis functions expansions for their dependent variable, the stream function, and its adjoint. This is sufficient as long as the uncertainty originating from the parameters is negligible compared to the uncertainty originating from the model error. In oceanography this assumption is often valid. However, for groundwater this assumption does not apply in many cases.

Basis function expansions with representers in groundwater flow and transport have been introduced by Reid, 1996 [7] who used representer functions on the log-conductivity only. She proved that this parameterization is optimal for linear optimization problems when choosing the head log-conductivity cross-covariance as representers. For non-linear optimization problems the cross-covariance will change during the iterative search. Reid simplified the problem by using an analytical solution for the representer function and kept it constant during all iterations.

We derive the optimal parameterization by solving the Euler-Lagrange equations with basis function expansions for all random variables and parameters. By using an iterative search technique we can decouple the Euler-Lagrange equations.

The basis function expansions are:

$$\check{h} = h_F + b^T r \tag{20}$$

$$\lambda = \lambda_F + b^T s \tag{21}$$

$$\check{f} = f_0 + b^T v \tag{22}$$

$$\check{S} = S_0 + b^T w \tag{23}$$

where h_F and λ_F denote the prior estimates of the head and its adjoint respectively. h_F can be calculated by solving equations 14-17 by using the prior estimates for the parameters initial and boundary conditions and model errors. λ_F is always 0. b is a vector of coefficients with the dimension of the number of measurements. The functions r, s, v, and w are called representers and have the dimension of the number of measurements times the number of grid cells. The representers s, v and w indicate the optimal way to change the adjoint variable and parameters in order to obtain a unitary change of each individual measurement estimate. The representer r indicates the influence on the entire head field for each unitary change of the head measurement estimate. The vector of coefficients b gives weights to these representer functions in order to minimize the objective.

In Appendix A the decoupled system for every representer function is derived as well as an expression for the vector of coefficients b. It has to be mentioned that the equations are only exact for the optimal solution and have to be solved by iteration starting with any other initial estimate. When the solution converges, the final estimate satisfies the Euler-Lagrange equations and therefore is the optimal solution of the minimization problem, at least if the objective is unimodal.

The equations of the representer function of the adjoint are:

$$-\hat{S}(x)\frac{\partial s(x,t)}{\partial t} - \nabla(e^{\hat{f}(x)}\nabla s(x,t)) = F(I(x,t)) \tag{24}$$

$$s(x,T) = 0 \quad x \epsilon D \tag{25}$$

$$s(x,t) = 0 \quad x\epsilon\partial D_1 \quad 0 < t < T \tag{26}$$

$$\frac{\partial s(x,t)}{\partial n} = 0 \quad x\epsilon\partial D_2 \quad 0 < t < T \tag{27}$$

The representers for heads are obtained from:

$$\hat{S}(x)\frac{\partial r(x,t)}{\partial t} - \nabla(e^{\hat{f}(x)}\nabla r(x,t)) = -w(x)\frac{\partial h_F(x,t)}{\partial t} +$$
$$\nabla(e^{f_0(x)}v(x)(\sum_{j=0}^{\infty}\frac{(\hat{f}(x)-f_0(x))^j}{(j+1)!})\nabla h_F(x,t)) +$$
$$\int_0^T\int_D C_{q_h q_h}(x,t,x_2,t_2)s(x_2,t_2)dx_2dt_2 \quad x\epsilon D \quad 0 < t \leq T \tag{28}$$

$$r(x,0) = \int_D C_{h_i h_i}(x,x_2)\hat{S}(x_2)s(x_2)dx_2 \quad x\epsilon D \tag{29}$$

$$r(x,t) = \int_0^T\int_{\partial D_1} C_{h_b h_b}(x,t,x_2,t_2)e^{\hat{f}(x_2)}\frac{\partial s(x_2,t_2)}{\partial n}dx_2dt_2$$
$$x\epsilon\partial D_1 \quad 0 < t \leq T \tag{30}$$

$$e^{\hat{f}(x)}\frac{\partial r(x,t)}{\partial n} = \int_0^T\int_{\partial D_2} -C_{q_b q_b}(x,t,x_2,t_2)s(x_2,t_2)dx_2dt_2 -$$
$$e^{f_0(x)}v(x)(\sum_{j=0}^{\infty}\frac{(\hat{f}(x)-f_0(x))^j}{(j+1)!})\frac{\partial h_F(x)}{\partial n} x\epsilon\partial D_2 \quad 0 < t \leq T \tag{31}$$

For practical problems the infinite sums can be truncated after a limited number of terms. This limited number will depend on the degree of non-linearity of the exponent of the log-conductivity update.

The representer functions for the parameters are:

$$w(x) = -\int_0^T\int_D C_{SS}(x,x_2)s(x_2,t_2)\frac{\partial\hat{h}(x_2,t_2)}{\partial t_2}dx_2dt_2 \tag{32}$$

$$v(x) = -\int_D C_{ff}(x,x_2)\int_0^T e^{\hat{f}(x_2)}\nabla\hat{h}(x_2,t_2)\nabla s(x_2,t_2)dt_2dx_2 \tag{33}$$

where the hat notation means the estimate obtained from the last iteration or the prior mean during the first iteration.

The vector of coefficients b can be calculated using:

$$(C_{\xi_z\xi_z} + F(r^T))b = (z - F(h_F)) \tag{34}$$

In summary our method consists of the following steps:

1. Solve the forward model: Equations 1-6 with the best estimates for the parameters, model error and perturbations of initial and boundary conditions. The best estimates are the prior mean during the first iteration and the estimates of the last iteration afterwards.
2. For all measurements
   - Solve representer s: Equations 24 - 27
   - Solve the representers w and v: Equations 32 and 33
   - Solve representer r for heads: Equation 28 - 31
3. Calculate vector b: Equation 34
4. Calculate adjoint: Equations 9 - 12
5. Update parameters: Equations 22 and 23
   go back to 1 until convergence
6. Calculate posterior heads: Equations 14 - 17

5. Example

The algorithm has been implemented in a computer code, which includes a modified version of MODFLOW. It was tested on a 2 dimensional transient problem with random log-transmissivity and model error. The grid size was 15 by 15, while the simulation lasted 40 time steps. The heads at the boundary were constant at 10 meters during the simulation. The initial heads were 10 meters in the entire domain. The grid cell was 100 times 100 m, while the timesteps was 5 days. The field was forced by a pumping well in the center of the domain during the first 10 timesteps with a discharge of 10000 m^3/day.

Synthetic log-transmissivity and model error fields were generated using sequential Gaussian simulation. The log-transmissivity field is normally distributed with mean $ln(100) m^2/day$ and an exponential covariance with variance 1.0 and correlation length of 10 block lengths. The model error field is zero mean normally distributed and has an exponential covariance with variance 10^{-6} m^4/day^2 and correlation length of 15 block lengths. The model error is uncorrelated in time. The simulated log-transmissivity field is shown in figure 1. Using the random fields a simulation was run.

During timesteps 5, 10, 20, 30 and 40, noisy head measurements were taken at 9 measurement locations. The measurement locations are at the intersection points of the lines through rows and columns 5, 8 and 11, see the diamonds in figures 3 to 6. The measurement errors are zero mean with variance 10^{-2} m^2 and are uncorrelated.

Using these 45 measurements, an optimization run was made to update the log-transmissivity and model error fields. The initial estimates were the prior mean of the random fields. The procedure stopped after 12 iterations when the maximum absolute change of all transmissivities was less than $5x10^{-4}$ times their last estimates.

The posterior log-transmissivity field is shown in figure 2. Differences between prior and posterior head distributions with respect to the real head distribution at timesteps 10 and 40 are shown in figures 3 to 6.

The posterior estimates of the log-transmissivity field show the large scale trend of the real log-transmissivity distribution: A zone with low log-transmissivities at the lower right part of the domain and a zone with high log-transmissivities at the upper left part of the domain. However, the estimate does not reproduce the local irregular differences in log-transmissivities. This is due to the fact that smooth fields of log-transmissivity and model error can also approximately reproduce the measurement values and are less penalized by the objective function than the true random fields. The posterior head distributions, however, were dramatically improved compared to the initial head estimates. The algorithm gives even better estimates at the measurement locations than the measurements themselves, which is due to the fact that the algorithm uses additional information along with the measurements.

6. Conclusions

Inverse modelling using the basis function expansion with representers can incorporate all kinds of sources of uncertainty. It avoids large dimensionality problems by its efficient parameterization. However, the algorithm is still to be tested on a real large inverse problem. Moreover a more detailed analysis of the results will give better insight of the possibilities of this inverse technique.

7. Future work

Future research will consist of testing the algorithm on a large transient problem and analyzing the posterior covariances. The posterior covariance can be approximated by equations as mentioned in Chapters 5 and 10 of Bennett, 1992 [1]. Moreover the algorithm will be applied to groundwater mass transport, where a detailed description of heterogeneity would help to make better decisions for practical problems, such as remediation of contaminated groundwater.

A. Derivation of the representer equation

By putting the basis function expansions (20) - (23) in the Euler-Lagrange equation (9) - (19) the representer equations can be derived. We will give the derivations for the representers of the adjoint and flow equations here. The same derivation can be used to derive the other representer equations.

The representers for the adjoint of flow:

Insert the basis functions expansions (23) in (9) - (12) and use $\lambda_F = 0$. The result is:

$$-(S_0(x) + b^T w(x))\frac{\partial b^T s(x,t)}{\partial t} - \nabla(e^{f_0(x)+b^T v(x)}\nabla b^T s(x,t)) = (z - F(h_F + b^T r))^T C_{\xi_z\xi_z}^{-1} F(I(x,t)) \quad x\epsilon D \quad 0 < t < T \tag{35}$$

Approximate the parameters $S + b^T w$ and $f + b^T v$ in the left-hand side of the equation by their best available estimates $\hat{S}$ and $\hat{f}$ and define the identity:

$$b^T = (z - F(h_F + b^T r))^T C_{\xi_z\xi_z}^{-1} \tag{36}$$

This gives:

$$-\hat{S}(x)\frac{\partial s(x,t)}{\partial t} - \nabla(e^{\hat{f}(x)}\nabla s(x,t)) = F(I(x,t)) \quad x\epsilon D \quad 0 < t < T \tag{37}$$

The representers for flow:

Insert the basis function expansions (20) - (23) in (14):

$$(S_0(x) + b^T w(x))\frac{\partial(h_F(x,t) + b^T r(x,t))}{\partial t} - \nabla(e^{(f_0(x)+b^T v(x))}\nabla(h_F(x,t) + b^T r(x,t)) = Q(x,t) + \int_0^T \int_D C_{q_h q_h}(x,t,x_2,t_2) b^T s(x_2,t_2) dx_2 dt_2 \tag{38}$$

Use a Taylor series expansion for $e^{(b^T v(x,t))}$ in order to subtract (1) and rearrange:

$$b^T \Big[(S_0(x) + b^T w(x))\frac{\partial r(x,t)}{\partial t} - \nabla(e^{(f_0(x)+b^T v(x))}\nabla r(x,t)) + w(x)\frac{\partial h_F(x,t)}{\partial t} - \nabla(e^{f_0(x)}(\sum_{j=1}^{\infty} \frac{b^{T^{(j-1)}} v(x)^j}{j!})\nabla h_F(x,t)) - \int_0^T \int_D C_{q_h q_h}(x,t,x_2,t_2) s(x_2,t_2) dx_2 dt_2 \Big] = 0 \tag{39}$$

Requiring this equation to be fulfilled for nonzero b if the head representers satisfy the following equation and use (22) and (23) to approximate the remaining $b^T v$ and $b^T w$ by their last estimate $\hat{f} - f_0$ and $\hat{S} - S_0$ it gives:

$$\hat{S}(x)\frac{\partial r(x,t)}{\partial t} - \nabla(e^{\hat{f}(x)}\nabla r(x,t)) = -w(x)\frac{\partial h_F(x,t)}{\partial t} + \nabla(e^{f_0(x)}v(x)(\sum_{j=0}^{\infty}\frac{(\hat{f}(x) - f_0(x))^j}{(j+1)!})\nabla h_F(x,t)) + \int_0^T \int_D C_{q_h q_h}(x,t,x_2,t_2)s(x_2,t_2)dx_2 dt_2 \quad (40)$$

Acknowledgments

This research was supported by TNO Institute of Applied Geoscience, Delft The Netherlands.

References

1. Bennett, A.F. *Inverse Methods in Physical Oceanography.* Cambridge University Press, 1992.
2. Carrera, J. and Neuman, S.P. Estimation of aquifer parameters under transient and steady state conditions, 3, application to synthetic and field data. *Water Resour. Res.*, 22:228–242, 1986.
3. Carrera, J. and Neuman, S.P. Estimation of aquifer parameters under transient and steady state conditions, 2, uniqueness, stability, and solution algorithms. *Water Resour. Res.*, 22:211–227, 1986.
4. DeMarsily, G., Lavedan, G., Boucher, M., and Fasanino, G. Interpretation of interference tests in a well field using geostatistical techniques to fit the permeability distribution in a reservoir model. In Verley, G et al., editor, *Geostatistics for Natural Resources Characterization, Part 2*, pages 831–849, D. Reidel Publishing Company, 1984.
5. Eknes, M and Evensen, G. Parameter Estimation Solving a Weak Constraint Variational Problem. In *Second International Symposium on Assimiliation of Observations in Meteorology and Oceonography*, pages 93–97, Tokyo, Japan, March 1995. World Meteorological Organization.
6. Gómez-Hernández, J.J., Sahuquillo, A., and Capilla, J.E. Stochastic simulation of transmissivity fields conditional to both transmissivity and piezometric data. *Water Resour. Res., Submitted*, 1995.
7. Reid, L.B. *A Functional Invers Approach for Three-Dimensional Characterization of Subsurface Contamination.* PhD thesis, Massachusetts Institute of Technology, 1996.
8. Schweppe, F.C. *Uncertain dynamic systems.* Prentice Hall. 1973.
9. Sun, N.-Z. *Inverse Problems in Groundwater Modeling.* Kluwer, Dordrecht, Netherlands, 1994.
10. Te Stroet, C.B.M. *Calibration of stochastic groundwater flow models. Estimation of system noise statistics and model parameters.* PhD thesis. Univ. of Techn. Delft, Delft, 1995.

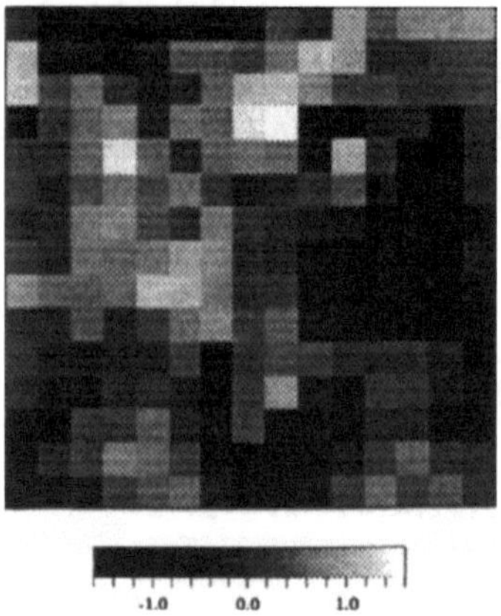

Figure 1. Synthetic mean removed log-transmissivity field

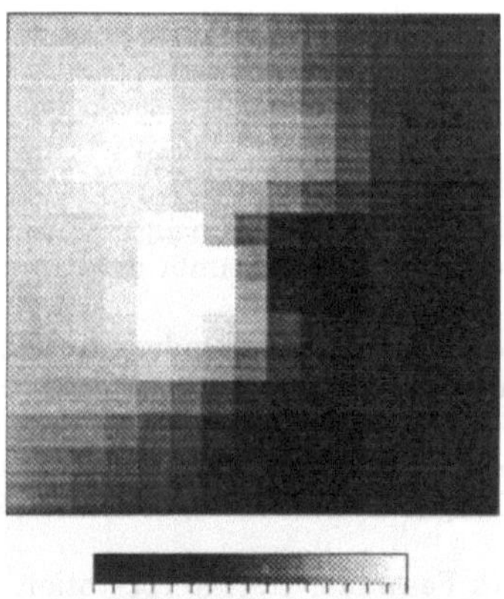

Figure 2. Difference between real and updated log-transmissivity

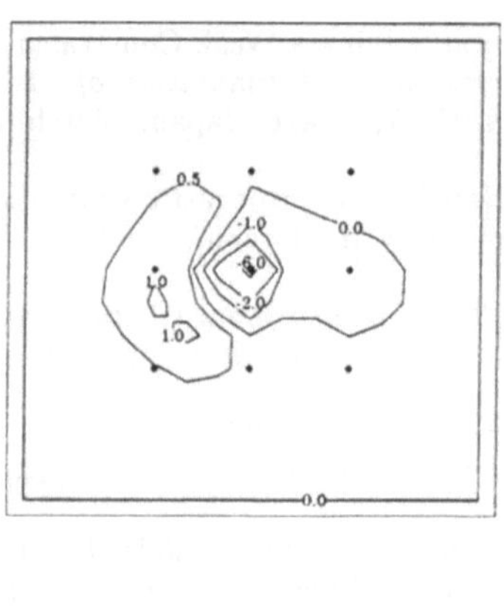

Figure 3. Difference between prior and real heads after 10 timesteps

Figure 4. Difference between prior and real heads after 40 timesteps

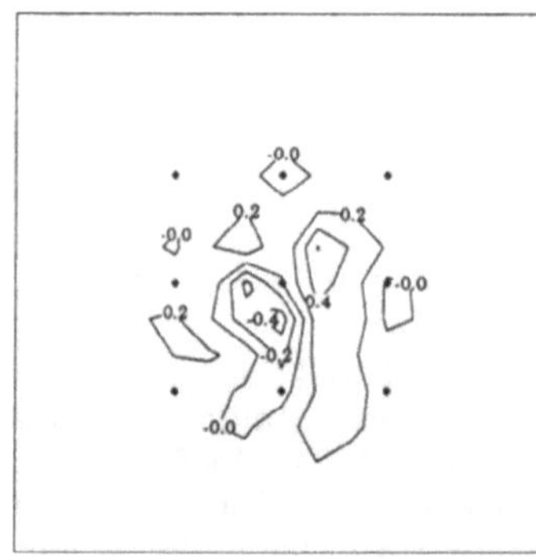

Figure 5. Difference between posterior and real heads after 10 timesteps

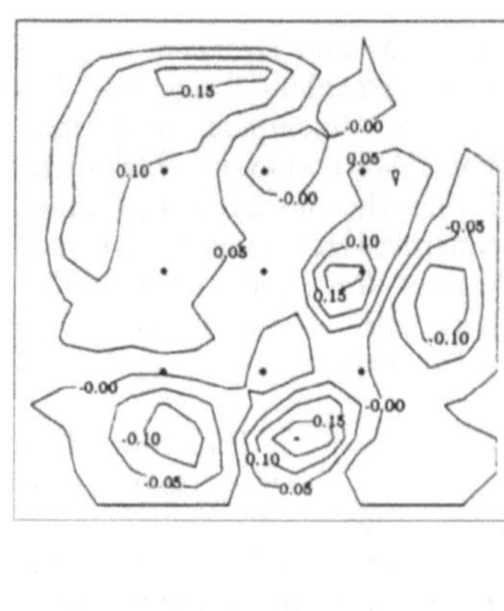

Figure 6. Difference between posterior and real heads after 40 timesteps

GEOSTATISTICAL ANALYSIS OF FRACTURED ROCK CHARACTERISTICS

R.J. LUNN & R. MACKAY
Centre for Land Use and Water Resources Research
University of Newcastle, Porter Building, Newcastle upon Tyne, UK.

Abstract

A programme of research is being carried out on behalf of Her Majesty's Inspectorate of Pollution, UK to investigate the hydraulic and transport characteristics of the rocks beneath Sellafield, Cumbria. The aims of the research are twofold: First, to establish the data requirements to identify potential hazard from the long term disposal of radioactive waste in the deep fractured rock formations of the Borrowdale Volcanic Group (BVG). The Sellafield site is currently being investigated by UK Nirex Ltd. as a possible location for a deep radioactive waste repository in the UK. Second, to develop a model of the site hydrogeology, consistent with the observed information.

This paper concentrates on investigating the vertical correlation structure of the flowing features in the BVG. This involves not only analysis of the permeability data, but also a geostatistical investigation of the Nirex sampling procedure. Numerical experiments are performed to investigate the effects of regular interval sampling on an assessment of correlated fault permeabilities. This enables an informed interpretation of the Nirex measurement data. Using statistical inference techniques, these results will form part of the basis for generation of hydraulic property realisations for use in flow and transport calculations. Accurate characterisation of the geometry and hydraulic properties of the faults within the BVG will have a significant effect on the reliability of future radionuclide transport predictions in the Sellafield area.

1. Introduction

This work has arisen out of recent developments within the radioactive waste research programme managed by Her Majesty's Inspectorate of Pollution* , UK (HMIP), to develop an integrated site model to test the impact of providing increasing levels of data on the assessment of risk from deep disposal of waste. Within this programme, a synthetic site model was developed for the regional geological system centred on a

* HMIP has now merged with the National Rivers Authority and the Waste Regulatory Authorities in England and Wales to form the new Environment Agency.

A. Soares et al. (eds.), geoENV I–Geostatistics for Environmental Applications, 25–37.

hypothetical radioactive waste repository at Harwell, Oxfordshire, UK (Mackay, 1993a). The synthetic site model is based on the integration of structural, lithological and hydrogeological models of the site, derived from the available measurement data. The application of this synthetic site model as a surrogate for a real site has shown that it is able to provide important information about the behaviour and variability of the groundwater regime in the vicinity of the site applicable to a site hydrogeological assessment (Mackay, 1993b). The same modelling approach is being applied here for the Sellafield site which has been targeted by Nirex UK Ltd. as a possible location for a deep radioactive waste repository in the UK.

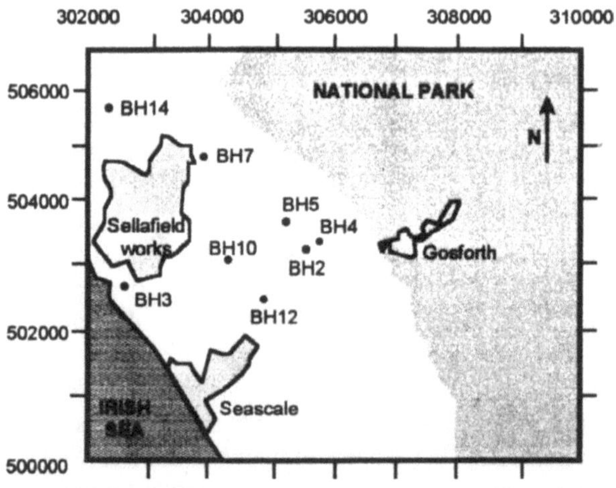

Figure 1 Plan of the Sellafield site area showing location of Nirex deep boreholes

The risk assessment programme for Sellafield requires flow and transport modelling of both the regional and the detailed near-repository environment. This modelling is to be implemented within a stochastic framework to provide a methodology for examining the implications of our uncertainty about the variability in radionuclide transport pathways from the site. One of the largest sources of model uncertainty is in the characterisation of the hydrogeological properties of the underlying strata, in particular, the Borrowdale Volcanic Group (BVG) within which the repository is to be sited. Flow within the BVG predominantly occurs through a network of conducting faults and fractures. In order to construct a flow model of the BVG, it is necessary to characterise the underlying vertical and horizontal permeability correlation structures, and to assess the connectivity of flowing features. This paper concentrates on investigating the vertical correlation structure of the flowing features identified within the Nirex data. This involves not only analysis of the available borehole permeability data, but also a geostatistical investigation of the Nirex sampling procedure. Numerical experiments are performed to investigate the effects of regular interval sampling on an assessment of correlated fault permeabilities. This will enable an informed interpretation of the available measurement data. Using statistical inference techniques, the results of this interpretation will form part of the basis for generation of hydraulic property realisations for use in flow and transport calculations. Accurate characterisation of the geometry and hydraulic properties of the faults within the BVG will have a significant effect on the outcome and reliability of future radionuclide transport predictions in the Sellafield area.

2. Summary of the Nirex Deep Borehole Data

The Sellafield site area with the locations of the Nirex deep boreholes is shown in figure 1. There are ten boreholes, (full permeability data sets are not available for any other boreholes) five of which (2, 4, 5, RCF2 and RCF3) lie within the potential repository zone. The data available for each borehole (Nirex Report Number 524, 1993 and Nirex Report Number S/94/008, 1994) includes lithology; fault locations, orientation and dip; fracture density; Environmental Pressure Measurement (EPM) single packer tests, which provide permeability measurements over 50m intervals; and locations of flowing features.

The general lithological sequence apparent from the borehole logs is drift; Omskirk sandstone; Calder sandstone, St. Bees sandstone; St. Bees shale; St. Bees evaporites; Brockham; carboniferous limestone and the Borrowdale Volcanic Group (BVG). The proposed repository is to be located within the BVG at a depth below the ground surface >650m and the work presented here concentrates on characterisation of this formation. The location of the top of the BVG in the Nirex boreholes dips from the North-East down to the South-West ranging from approximately 100m to 1600m below sea level. It consists of metamorphosed volcanic rocks and, as such, is fractured at all scales. Most of the fractures are healed or infilled (sealed) but a subset are at least partially open and provide pathways for groundwater flow. The field hydraulic permeability measurements range over eight orders of magnitude, from $8x10^{-14}ms^{-1}$ to $1x10^{-6}ms^{-1}$.

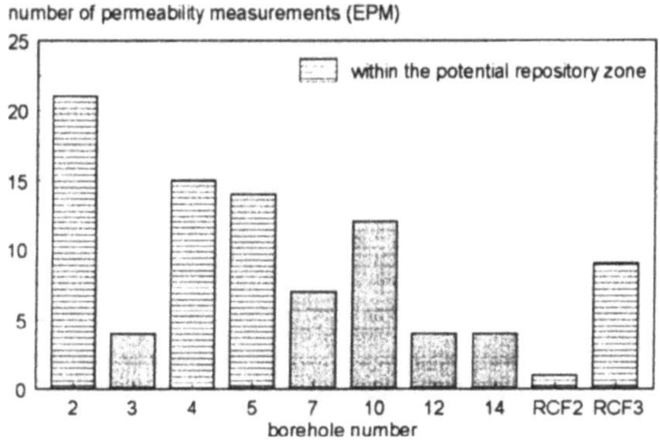

Figure 2 Number of EPM measurements taken within each Borehole

2.1 SUMMARY OF THE BOREHOLE DATA FOR THE BORROWDALE VOLCANIC GROUP

2.1.1 The Permeability Data

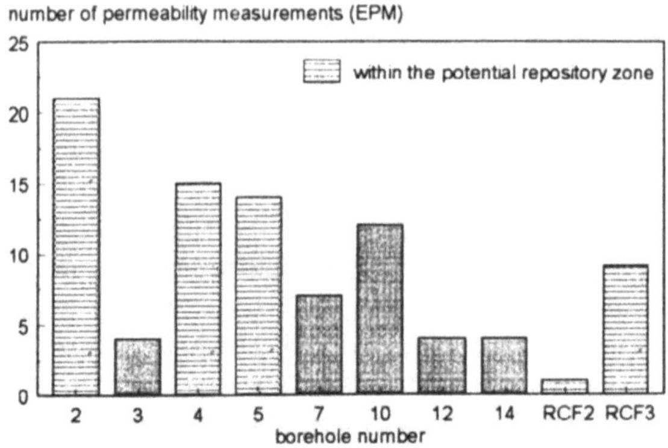

Figure 3 Number of EPM measurements taken within each Borehole

The most extensive permeability data set is that of the Environmental Pressure Measurement (EPM) tests. These are generally taken over consecutive 50m intervals down each of the boreholes (a few intervals are larger). A summary of the EPM data available for each borehole is provided in figure 2. The number of measurements within each of the other boreholes reflects the depth of penetration into the BVG. There are significantly more data from those boreholes within the region of the potential repository zone than from the regional boreholes. RCF2 and RCF3 penetrate the BVG to a similar depth to that of borehole 2, but have small data sets because unlike all the other boreholes, EPM tests were conducted only over a small section of their total depth.

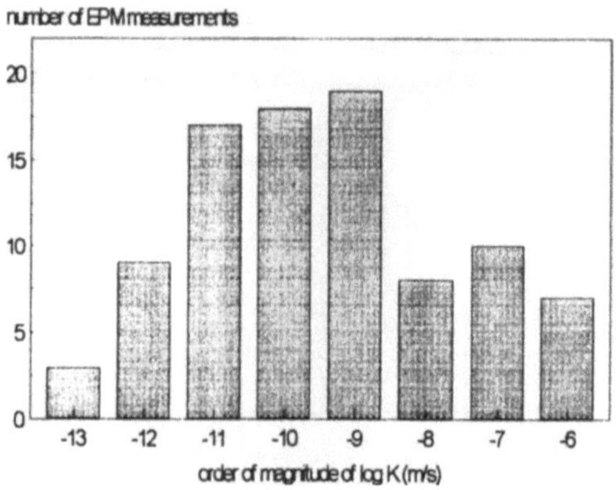

Figure 4 Frequency of EPM Observations within each Order of Magnitude for Hydraulic Conductivity

The distribution of the order of magnitude of the EPM data over all the boreholes is shown in figure 3. Whilst the peak has order of magnitude 10^{-9}, the distribution is not sharp, and significant numbers of observations are higher than this value.

2.1.2 The Flowing Feature Data and their Relationship to Permeability

The location of flowing features with each borehole was determined by Nirex using discrete extraction tests. The test interval is isolated with straddle packers with a downhole valve and gauges present to control and monitor the test. Flow is initiated by nitrogen injection through coiled tubing deployed at depth inside the testing string. The relationship between flowing features and permeability over all boreholes is demonstrated by figure 4. This shows the proportions of flowing features within each EPM interval for the range of magnitudes found within the permeability data. The likelihood of encountering at least one flowing feature within an EPM interval increases with an approximately linear relationship to the magnitude of the permeability data. Similarly, the likelihood of finding 2, 3, or 4 features is even more strongly related to the high permeability intervals.

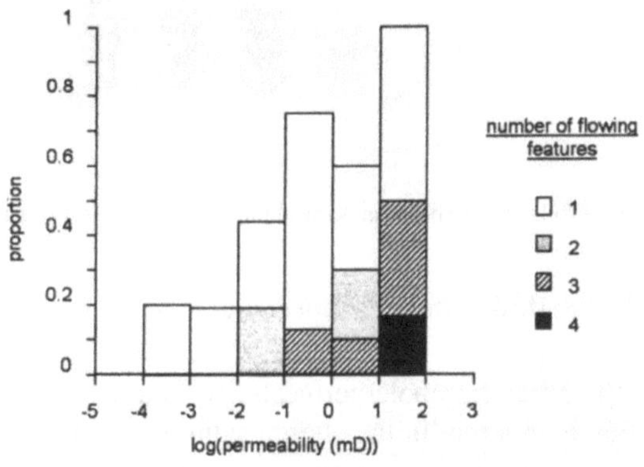

Figure 5 Probability of an EPM interval containing a given number of flowing features

Although a strong link between flowing features and high permeability is apparent, it is not one-to-one and therefore a significant proportion of the flow within the BVG must also be travelling through the rock matrix or a fine network of micro fractures within it. This dual flow regime is discussed in more detail within Lunn and Mackay (1996).

2.1.3 The Fault Data

The locations of faults have been interpreted from wireline logging data for each borehole. Information includes orientation of the downward dipping fault plane (measurements increasing clockwise with 0° aligned to the East) and dip measured as the angle below the horizontal. A strong relationship between the presence and orientation of faults and the permeability data is identified within Lunn *et al.* (1996).

It has not proved possible to match individual faults to the location of flowing features from the Nirex data available. The coincidence of faults and flowing features within each order of magnitude of the EPM intervals is shown within figure 5. Whilst they are obviously correlated, intervals remain that contain either only faults or only flowing features. Given the relationship between permeability and both faults and features, this suggests the possibility of some error in the identification of flowing features down the borehole. This possibility must be accounted for in any statistical investigation of the borehole permeability data.

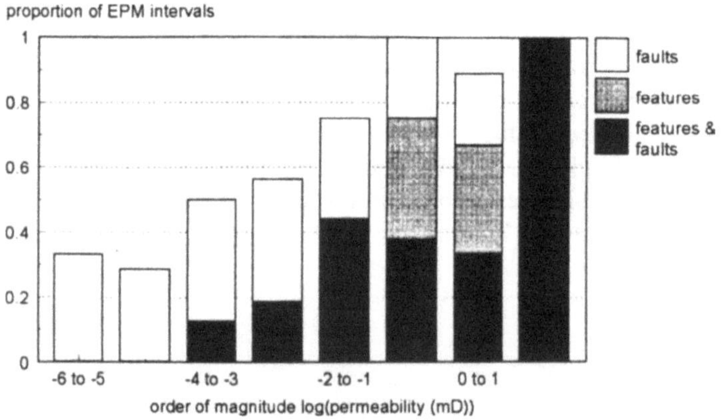

Figure 6 Relationship between permeability, faults and flowing features

3. Simulation of Permeability Down a Borehole

In order to analyse the Nirex borehole permeability data, it is necessary to investigate any statistical bias that is inherent in the chosen sampling method. To assess this bias, numerical experiments have been performed by simulating multiple realisations of permeability down a borehole and resampling at regular 50m intervals. Permeability is simulated down a vertical borehole assuming the dual permeability regime apparent in figure 5 consisting of flow within faults or features and flow within the underlying rock matrix. In the first instance, it was decided to investigate the effects of sampling a field with correlated feature permeabilities and uncorrelated underlying matrix permeability.

It is important to state here that the aim of this research is not to attempt to simulate the Nirex data itself, but to investigate all the possible interpretations of the derived experimental correlation statistics. It is expected that this numerical and theoretical investigation of the sampling method will show that several conceptual models could produce these statistics, and that for each model the effect of the sampling method on the experimental statistics must be investigated. The aim therefore is to assess the full range of possible interpretations of the data.

3.1 GENERATION ON THE CORRELATED FAULT PERMEABILITIES

The distance between consecutive flowing features is shown in the form of a cumulative distribution in figure 6. This has been calculated using data from all the boreholes. An exponential distribution is also shown on the figure and describes the data very accurately. Using the parameters derived from this exponential model, the locations of fractures are generated down the borehole as a Poisson process.

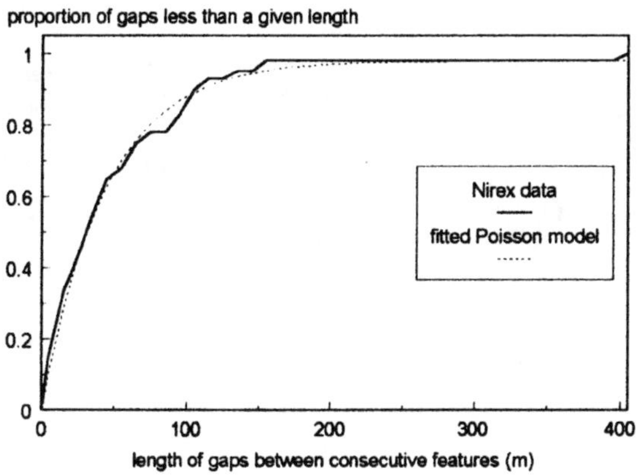

Figure 7 Fitting a Poisson process to the Nirex data describing gaps between consecutive flowing features

The correlated permeability field is generated using the spectral density method described within Tompson *et al.* (1989). Here, a fast Fourier transform technique is used to construct a realisation of the line process at each point by integration of a series of random spectral components over the whole frequency domain. This produces a one-dimensional correlated permeability field evaluated at regular intervals on the line. The correlation structure used was an exponential variogram that takes the form,

$$C(h) = \sigma^2 e^{-\lambda h} \tag{1}$$

with the corresponding spectral density function,

$$S(k) = \frac{\sigma^2 \lambda}{2\pi(\lambda^2 + k^2)} \tag{2}$$

where h is the lag interval, σ^2 is the variance and $1/\lambda$ is the correlation length. The permeability values are assigned to each fault by sampling the correlated permeability field at the Poisson generated location of each fault. The simulations are not conditional since as discussed in section 3, the aim is not to reproduce the Nirex data itself but to investigate all the possible interpretations of the experimental statistics.

3.2 SAMPLING THE PERMEABILITY (EPM) DATA

An example of a realisation of matrix and fault permeabilities for a 1000m section of a single borehole is shown in figure 7. Here, the single packer EPM data has been simulated by assuming that the total interval transmissivity is equal to the sum of the underlying rock matrix transmissivity and the fault transmissivity. In order to calculate this the fault aperture was estimated from an average of those measurements provided by Nirex. The mean and variance of the underlying matrix permeability was estimated using the Nirex EPM intervals without flowing features. The mean and variance for the correlated fault permeability field was estimated using the Nirex EPM intervals containing flowing features with the same assumption of superposition of transmissivities.

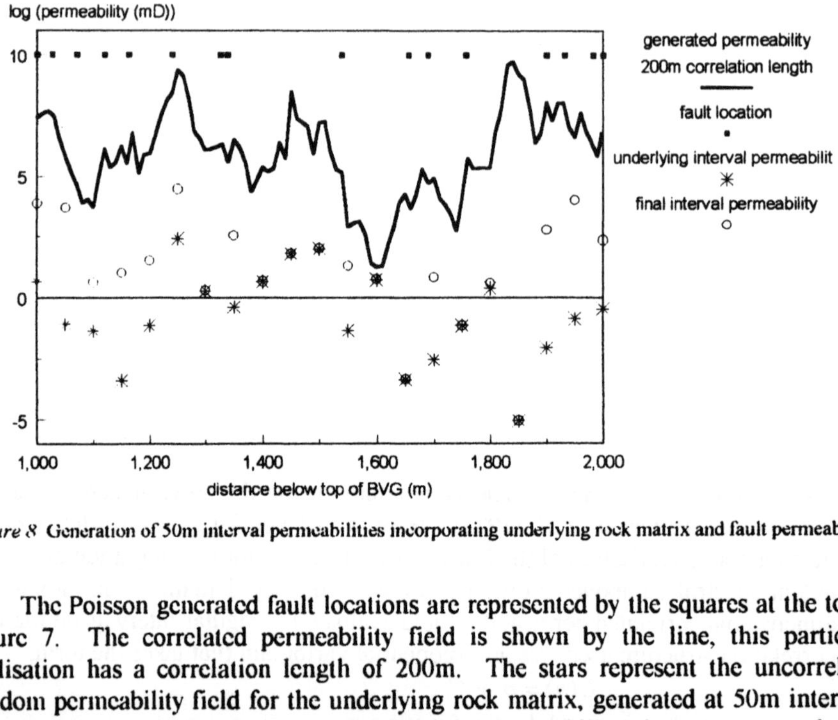

Figure 8 Generation of 50m interval permeabilities incorporating underlying rock matrix and fault permeabilities

The Poisson generated fault locations are represented by the squares at the top of figure 7. The correlated permeability field is shown by the line, this particular realisation has a correlation length of 200m. The stars represent the uncorrelated random permeability field for the underlying rock matrix, generated at 50m intervals. In order to calculate the circles (the final interval permeability) the presence or absence of any faults within the interval is determined. In the case of no faults, the stars are coincident with the circles and the interval permeability is that of the underlying matrix. If there is a fault within the interval, the final permeability is calculated by superposition of the fault and matrix transmissivities.

TABLE 1. Model parameters for borehole permeability simulations

Model Parameter	Estimated Value
Variance of matrix log(permeability (mD))	1.91
Mean matrix log(permeability (mD))	-0.72
Poisson distribution parameter (m)	43.8
Variance of flowing feature log(permeability (mD))	2.67
Mean flowing feature log(permeability (mD))	5.03
Fracture aperture (m)	0.005

3.3 RESULTS OF MONTE CARLO EXPERIMENTS OF DOWN-HOLE PERMEABILITY CALCULATIONS

Monte Carlo simulations have been completed using the parameter values estimated from the Nirex borehole data given in Table 1. For each set of simulations, one hundred runs were executed (a larger number was not found to significantly alter the results). Permeability measurements were simulated over 38500m for different correlation lengths of the fault distribution. The average 50m permeabilities were then calculated and sampled to produce an overall variogram.

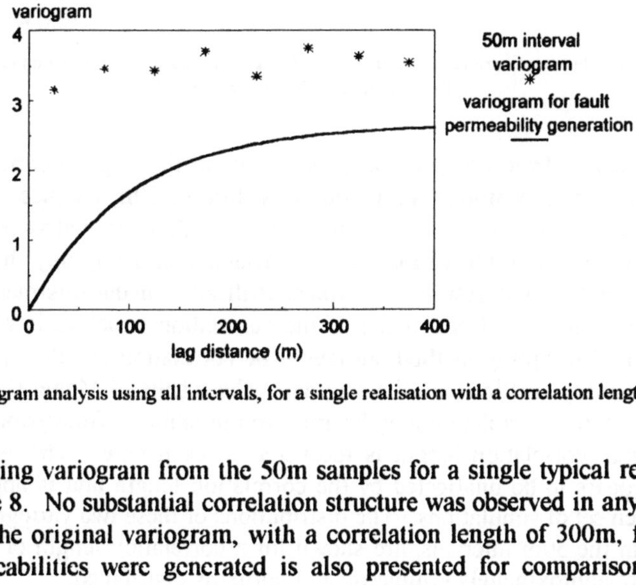

Figure 9 Variogram analysis using all intervals, for a single realisation with a correlation length of 300m

The resulting variogram from the 50m samples for a single typical realisation is shown in figure 8. No substantial correlation structure was observed in any of the 100 realisations. The original variogram, with a correlation length of 300m, from which the fault permeabilities were generated is also presented for comparison. Similar results were obtained for correlation lengths of 100, 200, 400 and 500 metres.

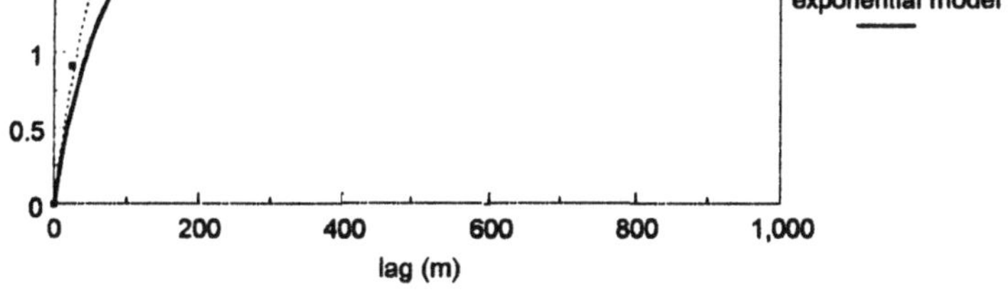

Figure 10 Variogram analysis using only intervals containing faults, for a single realisation with a correlation length of 300m

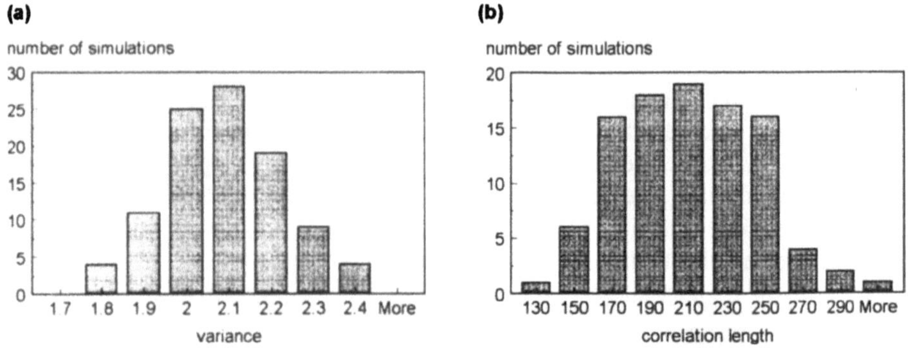

Figure 11 Distribution of variogram parameters for 100 simulations with a correlation length of 300m for (a) the variance and (b) the correlation length

It is apparent from figure 8 that almost none of the original correlation structure within the fault permeabilities is reproduced within the 50m samples. The correlation structure is being obscured by the variation in the underlying matrix permeability. To overcome this problem, the variogram was recalculated using only the 50m intervals that contain faults. The results of a typical realisation using this method of analysis are shown in figure 9. Here, much of the correlation structure is recovered by the regular interval sampling method, although the correlation length tends to be shorter and the variability smaller than the original. The results of Monte Carlo simulations using four different correlation lengths are given in table 2. Approximately two thirds of the original correlation length is recovered in each case. The estimated overall variance appears to be unaffected by the correlation length and is approximately the same for each set of simulations. The distributions of these two variogram parameters, derived from the 50m intervals, are shown for a correlation length of 300m in figure 10. They are approximately symmetric and normally distributed.

TABLE 2 Results of 100 Monte Carlo simulations for differing correlation lengths, using only intervals containing faults for calculation of the variogram parameters

Correlation length (m)	Recovered correlation length		Recovered variance		Ratio of correlation lengths
	mean	variance	mean	variance	
200	140.9	33.4	2.02	0.13	0.7
300	200.8	35.2	2.04	0.13	0.67
400	268.0	63.4	2.06	0.16	0.67
500	331.7	62.8	2.07	0.13	0.66

4. Variogram Analysis of the Nirex Data

Experimental variograms were calculated from the Nirex data using all the available boreholes by placing one above another as if they comprised a single borehole (correlations were not considered between consecutive boreholes). This approach assumes the data come from a single population, as there are insufficient data to apply any other method. Other analyses within Lunn *et al.* 1996 have indicated that this approach is valid, however this is almost certainly because the majority of the borehole data derives from the region of the potential repository zone. Any results that are derived from these analyses, therefore are assumed to characterise this region rather than represent the whole regional aquifer.

Results of the variogram analysis are presented in figure 11. Figure 11(a) describes the experimental variogram using all the 50m interval data. Correlation is apparent for lags of less than 200m and a variance at larger lags of approximately 3.5. In figure 11(b), the same analysis has been performed using only those intervals containing a fault and/or a flowing feature. A comparison of the two variograms shows that the variance is reduced to approximately 2.5 and the range to 150m. In addition, the correlation at lag intervals less than 150m is much stronger for the variogram using only those intervals that contain faults.

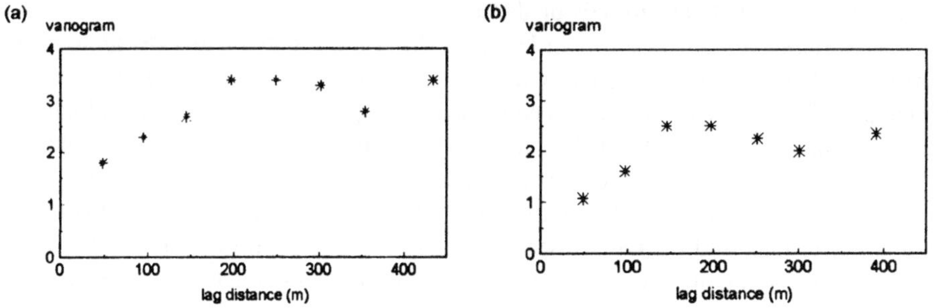

Figure 12 Experimental variograms from the Nirex borehole data using (a) all intervals and (b) only intervals containing faults and/or flowing features

4.1 INTERPRETATION OF THE DATA ANALYSIS IN LIGHT OF THE NUMERICAL INVESTIGATION OF THE SAMPING TECHNIQUE

Analysis of the Nirex sampling technique using Monte Carlo simulations demonstrated a number of numerical features that are inherent in the original sampling method and must be accounted for in the interpretation of the borehole data. First, analysis of the variogram calculated using all intervals, demonstrates that no underlying fault correlation can be expected to be recovered by this method. In addition, higher variability than either the initial fault or matrix variability is calculated from the interval variogram analysis. Second, calculation of the variogram using only those intervals which contain faults recovers a variability less than that within the underlying fault distribution. Third, the correlation observed using only the fault intervals in the analysis, is less than the true fault permeability correlation length.

A comparison of the analysis of the Nirex borehole data with the numerical investigation of the chosen sampling method highlights strong similarities between results. The Nirex data show a substantial reduction in variability from the variogram using all intervals (figure 11(a)) to that calculated from only intervals containing faults (figure 11(b)). This was also found to be the case in the numerical simulations. In addition, the borehole data demonstrate considerably stronger correlation when only the intervals containing faults are included in the variogram analysis. This effect is also observed with the numerical simulations.

There is only one significant difference between the analysis of the Nirex data and that of the numerical simulations; that some correlation is observed in the variogram analysis of the Nirex data for all the intervals. There are two possible reasons for this: the underlying matrix permeability distribution in the BVG could be correlated; or there could be significant errors in the Nirex assessment of which intervals contain flowing features. The higher the real proportion of intervals containing features, the more correlation is observed using all the intervals, since there are very few uncorrelated intervals remaining to confuse the analysis. The Poisson parameter describing average distance between consecutive features was calculated using only those features identified by Nirex as flowing. It is possible that there has been some error in the Nirex classification of such features and that they are more frequent than currently predicted.

Three important conclusions can be drawn from the results of this research. First, there are substantial similarities between the analysis of the borehole data and the numerical simulations. This confirms the derived conceptual hydrogeological model of both fault flow and flow within the underlying rock matrix (or an equivalent porous medium containing fine micro-fractures). Second, vertical correlation is apparent from analysis of the Nirex borehole data. Third, to parameterise this correlation in terms of the underlying distributions of fault and matrix permeability, it is necessary to incorporate the numerical effects of the sampling method on the variability and correlation length recovered from the variogram analysis.

5. Conclusions and Future Work

A numerical investigation of the regular interval sampling used by Nirex to determine borehole permeability has shown that this sampling method has a significant effect on the values of any parameters derived from variogram analyses of these data. A comparison of the results from these investigations with a geostatistical analysis of the borehole data have confirmed the derivation of a conceptual hydrogeological model of the BVG comprising flow within faults and the underlying rock matrix. Vertical correlation of fault permeabilities has been identified within the Nirex data, and the investigation of the sampling method has demonstrated that a standard approach cannot be applied to derive the statistical parameters of the underlying fault and matrix permeability distributions.

This research has proved essential to interpretation of the Nirex borehole data and further work is now being carried out to parameterise the BVG. More numerical simulations and analytical work are being performed to derive a generalised relationship between observed variogram parameters and those of the underlying distributions for the broarder case of correlated permeability fields in both the matrix and the features. In addition, the effect of incorporating different levels of error in the identification of flowing features is being analysed. These further investigations should allow determination of the range of possible statistical models for the BVG that are consistent with the observed permeabilities values, taking full account of the sampling strategy used to derive them.

6. Acknowledgements

The authors thank the Environment Agency for permission to publish this paper. The work may be used in the formulation of government policy but does not at this stage constitute UK government policy.

7. References

Lunn, R.J., Lunn, A.D. and Mackay, R. (1996) Development of a Hydrogeological Model of the Borrowdale Volcanics At Sellafield, *Groundwater*, (submitted).

Lunn, R.J. and Mackay, R. (1996) Development of a Probabilistic Model of the Three-Dimensional Groundwater Flow Regime in the Vicinity of a Proposed Waste Repository, in P.C. Cacciabue and I.A. Papazoglou (eds.) *Probabilistic Safety Assessment and Management*, ESREAL'96 - PSAM III, Springer-Verlag, London, pp. 801-806.

Mackay, R. (1993a) A Study of the Effect of the Extent of Site Investigation on the Estimation of Radiological Performance, Vol 1, Development of the Synthetic Site Model. HMIP report no. DoE/HMIP/RR/93.052.

Mackay, R. (1993b) A Study of the Effect of the Extent of Site Investigation on the Estimation of Radiological Performance, Vol 2, Analysis of sampling strategies. HMIP report no. DoE/HMIP/RR/93.075.

Nirex UK Ltd. (1994) Summary Geological Logs for the RCF and RCM Borehole, Nirex report no. S/94/008.

Nirex UK Ltd. (1993) The Geology and Hydrogeology of the Sellafield Area, Vol 4, Investigation Results, Nirex report no. 524.

Tompson, A.F.B., Ababou, R. and Gelhar, L.W. (1989) Implementation of the Three-Dimensional Turning Bands Random Field Generator, *Water Resources Research* **25**, 2227-2243.

ON THE APPLICATION OF STOCHASTIC INVERSE MODELLING TO THE FRACTURED SEMI-CONFINED AQUIFER OF BAGUEIXE, PORTUGAL

J.P. LOBO-FERREIRA
Dr.-Ing., Groundwater Research Group Head
Laboratório Nacional de Engenharia Civil, Lisbon, Portugal
Av. do Brasil, 101, P-1799 LISBOA Codex, Portugal.
Tel. (3511)8482131, Fax: (3511)8473845, E-mail: LOBOFERREIRA@LNEC.PT

1. Introduction

The aim of this paper is to present an application of stochastic inverse modelling for the computation of means and variances of logtransmissivities and piezometric heads of a semi-confined fractured aquifer located in the North-eastern region of Portugal, based on the methodology presented in Rubin et al. (1990), for the Rio Maior aquifer, in Portugal, and in the study published by Lobo-Ferreira and Leitão (1991) for the Bagueixe aquifer.

The selection of the fractured aquifer of Bagueixe, among the fractured aquifers of the North-eastern region of Portugal was due to the availability of hydrogeological data gathered (1) in eight pumping wells (Hidroprojecto et al., 1987) and (2) during field works aiming geological and hydrogeological assessments (Ribeiro, 1974).

The application of stochastic inverse modelling to the fractured aquifer of Bagueixe showed to be particularly effective in the identification of piezometric levels and of logtransmissivities of the case-study area, revealing itself as a powerful tool for aquifer parameter identification, complementing and surpassing the uncertainty caused by a limited number of available field measurements.

2. Hydrogeological Assessment Of Bagueixe Aquifer

2.1. BRIEF CHARACTERIZATION OF THE REGION OF BAGUEIXE

The region of Bagueixe is located in the North-eastern region of Portugal. The area selected for this hydrogeological case-study is 25 km² large (Figs. 1 and 2). The study area is located 15 km East of Macedo de Cavaleiros. Geomorphologically this area is characterized by moderate hills, not steep, having elevations ranging from 540 to 680 m.

The mean annual precipitation in the Bagueixe region is 750 mm. The mean annual potential evapotranspiration is approx. 750 mm, while the estimated real evapotranspiration is 460 mm per year. The mean annual temperature is 12.5°C.

2.2. GENERAL CHARACTERIZATION OF THE BAGUEIXE AQUIFER AIMING ITS MATHEMATICAL MODELLING

In Lobo-Ferreira and Leitão (1991) a detailed hydrogeologic characterization of the fractured Bagueixe aquifer is presented. The information gathered will be hereinafter

A. Soares et al. (eds.), geoENV I–Geostatistics for Environmental Applications, 39–50.

synthesized. The information was complemented with extra information required by the flow and transport modelling study developed for the Bagueixe region, and published in Lobo-Ferreira and Leitão (1991) and Lobo-Ferreira (1995).

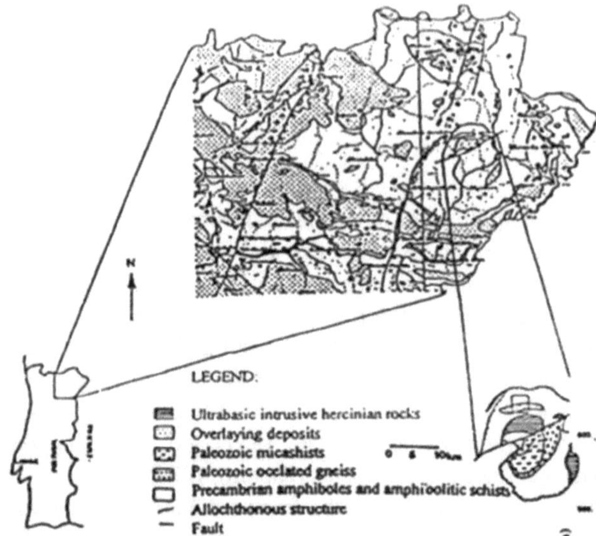

Figure 1. Geological map of Maciço de Morais, in the North-eastern region of Portugal

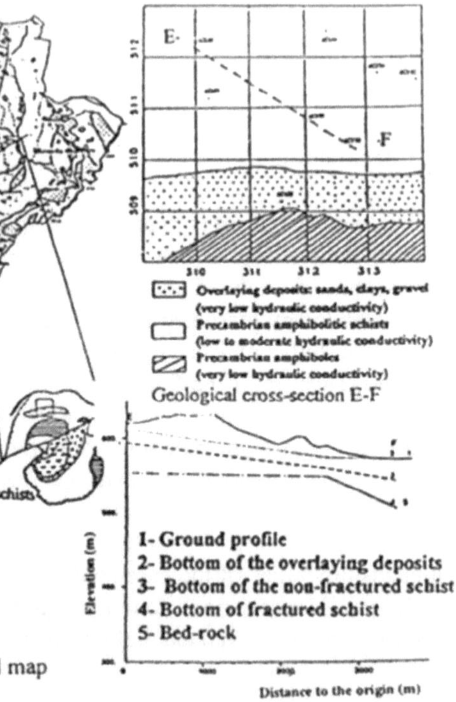

Figure 2. Hydrogeological map of Bagueixe aquifer

The drilling of the pumping wells and the development of pumping discharge experiments were performed by ACavaco firm during the 80's.

The aquifer is supposed to be located at the level of amphibolitic schist having a saturated depth of about 20 m.

We computed for each pumping well the values of transmissivity, of storativity and of hydraulic conductivity, using the Jacob method. The results are shown in Table 1:

Table 1. Evaluation of Bagueixe aquifer parameters

Pumping well	AC126	AC127	AC129	AC132	AC134	AC135	AC136	AC138
Meridian	309.92	310.20	311.55	313.90	313.18	312.08	312.34	312.80
Parallel	512.15	511.15	509.10	511.55	511.70	510.70	512.50	510.18
Elevation (1)	618	595	545	608	625	588	662	570
Piez. level (2)	0.3	0.3	0.3	0.3	0.3	0.3	0.3	0.3
Water table (3)	+ 0.3	+0.3	- 1.59	- 0.91	- 4.32	- 5.53	- 5.92	- 3.70
Piez. elevation (1)+(2)+(3)	618.60	595.60	543.71	607.39	620.98	582.77	656.38	566.60
Discharge (m^3/s)	0.00138	0.0014	0.00143	0.00145	0.00291	0.0033	0.00122	0.0055
Transmissivity (m^2/d)	1.67	7.10	37.68	17.36	19.20	7.90	3.57	38.65
Storativity	-	1.03×10^{-4}	1.5×10^{-4}	6.46×10^{-3}	0.084	-	6.38×10^{-3}	-
Hydr. conduct. (m/d)	0.08	0.36	1.88	0.87	0.96	0.40	0.18	1.9

From those results we have concluded and defined the following:

- The values of transmissivity (*T*) and subsequently those of hydraulic conductivity (*K*) due vary strongly across the case-study area, showing a clear increase towards South-east.
- The value for storativity (*S*) was defined as being the mean of the values obtained in the wells, i.e. equal to 0.1.

3. Estimation Of The Hydraulic Parameters Of Bagueixe Aquifer By Using Stochastic Inverse Modelling

3.1. GENERAL CONSIDERATIONS

In the framework of a scientifically co-operation on mathematical inverse modelling between the Tel-Aviv University (TAU), Israel, and the National Laboratory for Civil Engineering of Portugal (LNEC), partially sponsored by AID, the U.S. Agency for International Development, theoretical background on stochastic inverse modelling and an application to the Rio Maior aquifer, in Central Portugal, was developed and published by Rubin *et al.* (1990). These knowledge and experience was latter applied by Lobo-Ferreira and Leitão (1991) to the Bagueixe aquifer, aiming the mapping of means and variances of estimation of regional transmissivities and piezometric heads, in the framework of the NATO PO-RIVERS Project "Methodologies for Water Resources Development and Management in Portugal. Application to the Northern Region", Activity K: "Groundwater Pollution Modelling and Control". It should be mentioned that this was the first time that the method developed in Tel-Aviv University was applied to a fractured aquifer. The analysis was developed during a stay of the author and of Mrs. Teresa Eira Leitão in 1989 and 1990 in TAU, co-ordinated by Prof. Gedeon Dagan, in the framework of the above mentioned AID-TAU-LNEC project.

3.2. DEVELOPMENT OF THE THEORETICAL INVERSE MODELLING APPROACH

A brief review of the theoretical approch on the inverse modelling, as developed by Rubin and Dagan (1987a, b) and Dagan and Rubin (1988), is hereinafter presented, extracted from Rubin *et al.* (1990).

The aim of this section is to give a general, though brief, review of the major theoretical considerations taken towards the formulation of the approach. For a more comprehensive and detailed review, the reader is referred to the studies of Rubin and Dagan (1987a, b) and Dagan and Rubin (1988). In this study we are concerned with groundwater saturated flow at the regional scale, i.e. through a medium with horizontal extent much larger than its thickness. This flow is generally described by the model equation satisfied by the time dependent piezometric head $H_t\,(\mathbf{x}, t)$ as follows:

$$\frac{\partial}{\partial x}\left(T\,\frac{\partial H_t}{\partial x}\right) + \frac{\partial}{\partial y}\left(T\,\frac{\partial H_t}{\partial y}\right) = -R + S\,\frac{\partial H_t}{\partial t} = -R_{ef} \qquad (1)$$

where $\mathbf{x}\,(x, y)$ are Cartesian co-ordinates in the flow domain D, t is the time, $T\,(x, y)$ is the transmissivity, S is the storativity, and R is a distributed source. Eq. (1) pertains to untapped aquifers (this may be considered the case of Bagueixe aquifer), whereas the presence of pumping wells requires introduction of additional term to (1).

Eq. (1) is generally supplemented by boundary and initial conditions for H_t, e.g. H_t or its derivatives are given on the boundaries of D and H_t is given at $t = 0$. The solution of (1) for H_t, for given T, S and R as well as given boundary and initial conditions, is known as the direct problem and is well documented in the literature.

Dealing with the inverse problem, the given information is contained in: (i) values of measured transmissivities $T_j = T(x_j)$, at $j = 1,..., M$ points within D; (ii) records of the head $H_{t,lk} = H_t(x_l, t_k)$ at $l = M+1,..., N$ points and at various $k = 1,..., P$ times; (iii) boundary and initial conditions for H_t. Thus we arrive at the definition of the inverse problem, which, stated simply, is the determination of the spatial distributions of $T(x)$, R and S based on the above information, and in accordance with (1). A general review of existing solutions of the inverse problem is beyond the scope of this study, and the reader is referred to Yeh (1986).

The derivation of the model inverse solution here is based on some basic assumptions, given herein:

(i) T is regarded as a random space function of lognormal distribution, i.e. the values of $Y_j = ln\ T_j$ at an arbitrary set of points x_j, form a multivariate normal vector. The stationary unconditional joint p.d.f. is characterized completely by the expected value $<Y> = m_Y = const$ and the autocovariance $C_Y(x_m, x_n) = <Y'(x_m)\ Y'(x_n)> = C_Y(r'_{mn})$, where $r' = |r'|$ is the distance between two points and $Y'(x) = Y(x) - <Y>$. In particular we shall adopt here the isotropic exponential covariance:

$$C_Y(r') = w\,[1 - H(r')] + \sigma_Y^2 \quad exp(-r'/I) \tag{2}$$

where w represents an uncorrelated component of Y (a nugget), $H(r')$ is here the Heaviside step function ($H(0) = 0$; $H(r') = 1$ for $r' > 0$), σ_Y^2 is the variance of spatially correlated Y values and I is the correlation scale. Hence, Y statistical structure is represented in terms of four parameters: $<Y>$, w, σ_Y^2 and I.

(ii) A first-order approximation is adopted for the flow Eq. (1). With $h_t(x, t) = H_t - <H_t>$ and $Y' = Y - <Y>$, the head and logtransmissivity residuals, respectively, a first-order approximation is adopted for H_t, resulting in the following two equations which replace (1):

$$\nabla^2 <H_t> = -\frac{R_{ef}}{T_G} \tag{3}$$

$$\nabla^2 h_t = -\nabla.(Y'\nabla <H_t>) \tag{4}$$

where $T_G = exp(m_Y)$ is the geometric mean. The first-order approximation is employed since it simplifies considerably the flow equation with variable, random coefficients (1), leading to the simpler, linear forms (3) and (4). The expansion in Y' is valid in theory for small σ_Y^2 and Dagan (1985) showed indeed the first-order approximation to be valid for σ_Y^2 smaller than unit. However, in the study by Rubin and Dagan (1989a,b), the first order approximation was compared with results of the

Monte-Carlo simulations of Smith and Freeze (1979), showing a good agreement even for σ_Y^2 larger than unity.

(iii) We regard R_{ef} as a function of time only. This amounts to assuming that the piezometric surfaces at different times $H_t(\boldsymbol{x})$ and $H_{t'}(\boldsymbol{x})$ where $t \neq t'$, are essentially parallel. This is a limiting assumption, since by no means it is the general situation observed in nature. This assumption could be relaxed by incorporating an additional term in (4) which originates from the spatial variability of R. Such a generalization would introduce additional parameters in the parameter vector θ, related to the structure of R, that may be identified based on the model and the measurements of Y and H by employing a maximum likelihood procedure. Not having available periodically taken piezometric measurements in the Bagueixe case-study, contrary to the case-study of Rio Maior published by Rubin *et al.* (1990), the evaluation of the Bagueixe aquifer's recharge was not developed in this study.

(iv) The covariances of Y and H at any two points are evaluated under the assumption of an unbounded domain. There are some justifications for assumption. First, the joint Y, H correlation structure is dictated by the integral scale I, which is assumed to be much smaller than the areal extent of the flow domain. Second, in many cases boundary locations and boundary conditions are poorly defined, as the assumption of unboundedness is as good as any. Most important, however, is that a rigorous study of the effect of boundaries on the Y, H correlation structure (Rubin and Dagan, 1988) have shown it to be limited to a zone of a finite extent, of the order of an integral scale normal to the boundary, and even in this zone it is rather small.

(v) The effects of the presence of the boundaries of the flow domain on the flow regime is accounted for in an indirect approximate manner. $\langle H \rangle$ is represented by a quadratic polynomial trend, which is a function of the space coordinates, in the form:

$$\langle H(\boldsymbol{x}, t)\rangle = H_0(t) - \boldsymbol{J}_0(t).\boldsymbol{x} - \frac{1}{2}\sum_{i=1}^{2}\sum_{j=1}^{2} c_{ij}x_i x_j \tag{5}$$

where H_0 is the head H at the origin, $\boldsymbol{J}_0$ is its gradient there, c_{ij} are three independent coefficients (c_{11}, $c_{12} = c_{21}$, c_{22}). Eq. (5) does not satisfy exactly Eqs. (1) or (3) unless R, $\boldsymbol{J}_0$ and c_{ij} are time independent and H_0 is a linear function of time. This leads to the inherent assumptions that the head surfaces at different times are essentially parallel. Substitution of (5) into (3) results in the simple relationship:

$$c_{11}(t) + c_{22}(t) = \frac{R(t) - S\,(dH_o/dt)}{T_G} \tag{6}$$

generalizing the similar one for steady flow (Rubin and Dagan, 1987a, Eq.14). Thus, $c_{11} + c_{22}$ account for the effects of any spatially constant source. The additional freedom offered by the two remaining independent parameters of the quadratic trend may be relevant, taking into account the non-uniformity of average flow, due to boundary conditions.

3.3. A SCHEME FOR PARAMETER IDENTIFICATION

The aim of the present section is to outline the procedure of identifying θ, the vector of parameters describing the spatial variability of Y and of the mean head surface.

Following the geostatistical approach, the aquifer is regarded as a realization of an ensemble whose statistical structure is characterized by $<Y>$ and C_Y (Eq. 2). H and Y are thus regarded as space random functions, which are defined completely by their statistical moments. According to the findings of some field studies (e.g. Hoeksema and Kitanidis, 1985), Y is assumed normal, and the same is true in the first order approximation for the residual h, which satisfies the linear equation in Y' (4). Hence, the entire statistical structure of the bivariate normal variables $Y(\boldsymbol{x})$ and $H(\boldsymbol{x}, t)$ is expressed by the expected values $<Y>$, $<H>$ and by the unconditional covariances $C_Y(\boldsymbol{x}, \boldsymbol{x}') = <Y'(\boldsymbol{x})\, Y'(\boldsymbol{x}')>$, $C_{YH}(\boldsymbol{x}, \boldsymbol{x}', t) = <Y'(\boldsymbol{x})\, h(\boldsymbol{x}', t)>$ and $\Gamma_H(\boldsymbol{x}, \boldsymbol{x}', t, t') = (1/2) <[h(\boldsymbol{x}, t) - h(\boldsymbol{x}', t')]^2>$, where $\boldsymbol{x}$ and $\boldsymbol{x}'$ are the vector coordinates of two arbitrary points. By unconditional moments we refer to the ensemble of formations in which the constraints imposed by measurements at the measurement points are not taken into account. Furthermore, all the above moments are for given values of S and $R(t)$, which are regarded as independent of the space coordinates.

Under the assumptions of Section 3.2, we are able to write the above moments in terms of a series of parameters, by following closely the procedure of Rubin and Dagan (1987b). Thus, $<Y>$ and the given $C_Y(r')$ (2) are expressed in terms of the four parameters:

$$\theta_1 = <Y>\ ;\ \theta_2 = \sigma_Y^2\ ;\ \theta_3 = I\ ;\ \theta_4 = w \tag{7}$$

It is emphasized that other forms of covariances rather than (2), e.g. semispherical, could be selected, with a somewhat different parametrization.

The expected value $<H>$ is given in (5) in terms of the additional parameters:

$$\theta_5 = J_{0,x}(t);\ \theta_6 = J_{0,y}(t);\ \theta_7 = c_{11}(t);\ \theta_8 = c_{12}(t);\ \theta_9 = c_{22}(t);\ \theta_{10} = H_0(t) \tag{8}$$

The remaining two moments, the logtransmissivity-head covariance C_{YH} and the head variogram Γ_H, are derived by using the equation of flow (4) for h. The closed form analytical solutions are given in Eqs. (16)-(21) of Rubin and Dagan (1987a) and are not reproduced here. These covariances are expressed in terms of the parameter vector θ (Eq. 7 and 8).

In the stage of parameter identification, the objective is to identify the matrix of parameters θ_{jk} (Eq. 8) where k stands for the time $t = t_k$. This is achieved by the procedure suggested by Kitanidis and Vomvoris (1983) and Hoeksema and Kitanidis (1985) and followed by Rubin and Dagan (1987a, b), namely by a maximum likelihood procedure (MLP) applied simultaneously to the multivariate normal vector Y_j and H_{jk}, whose dimension is $M + (N - M)\, .\, P$. This results in large matrices to be handled in the ML process. Instead, the MLP can be applied sequentially, i.e., identifying the vector $\theta_k (k = 1,..., P)$ at each t_k with the aid of the head measurements at the same time only. Besides simplicity, this procedure is supported by the assumption that the flow is slowly varying in time, and the sequential application of a quasi-steady analysis is applicable. Furthermore, when the time intervals $t_k - t_{k-1}$, are large enough and considering measurement errors, the head data at each t_k can be considered as a different and independent realization of the same stochastic process. Carrera and Neuman (1986) suggested an autoregressive autocorrelation model to account for the time dependence, leading to the same conclusion.

The estimates $\hat{\theta}$ are obtained from the minimization of the likelihood function applied to the measurement vector, leading also to the error covariance matrix Σ of $\hat{\theta}$. The matrix Σ, the inverse of the Fisher information matrix, is sometimes referred to as the Cramer-Rao lower bound. Although it constitutes the lower bound of the parameter covariances, it was found in many cases (cf., Schweppe, 1973) to be asymptotically (in N) close to the true covariances.

With given Σ, the confidence limits of the parameters can be evaluated. Small variances and intervals of confidence of the parameters are partially indicative of a correct model parametrization, and hence of the suitability of the suggested approach to the problem at hand (an in-depth discussion about establishing optimal model parametrization is given in Carrera and Neuman, 1986).

At the completion of the stage of model validation, the recharge can be calculated, by employing Eq. (6), after replacing dH_0/dt by its finite difference approximation $[H_{0,k+1} - H_{0,k}]/\Delta t$, where $\Delta t = t_{k+1} - t_k$, and by using prior knowledge of S, if pumping wells of known discharges are absent. By the procedure outlined in Rubin and Dagan (1987a), the impact of the estimation errors of θ are accounted in evaluating the estimation variance of R.

The time sequence of $\hat{R}_k$ $(k=1,..., P)$ can supply useful information. For example, multi-seasonal water balances offer the possibility to estimate the relation between precipitation and natural recharge. If one is interested in the estimates of R, of the log-transmissivity and of the head statistical moments only, the computation may be terminated at this stage.

3.4. THE CONDITIONING STAGE

The unconditional moments derived previously define stochastically the ensemble of heterogeneous formations. Only in a subset of realizations do Y and H assume the measured values at the measurement points. The conditional p.d.f. of Y and H is defined as the one based on this subset, the conditioned Y and H being multivariate normal as well.

The method for computing the conditional moments of $Y(\boldsymbol{x})$, i.e. $Y^C(\boldsymbol{x})$ and $\sigma_Y^{2,C}(\boldsymbol{x})$, which relies on Gaussian conditional probability, is given in the study of Rubin and Dagan (Section 3, 1987a). It is similar to cokriging and it consists of expressing Y as a linear function of the head and transmissivity residuals (with respect to the unconditional mean) at the measurement points $\boldsymbol{x}_i$ $(i = 1,..., N)$. The coefficient for each residual is a function of its relative position with regard to $\boldsymbol{x}_i$. The coefficients are computed by solving a linear system of equations, which require the knowledge of the unconditional second moments. In general, conditioning must be carried out simultaneously for $\mathbf{Z}$, the vector of all measurements at all times. However, consistent with our separation of the identification process sequentially in time, conditioning is carried out on $\mathbf{Z}_k$ measurements at time t_k, rather then $\mathbf{Z}$. The conditional mean of Y at different times t_k should be ideally the same, since Y is supposed to be time independent. Since they might differ only slightly if the model is correct, their value can be averaged over time.

Finally, the computation of the conditional variance of Y is carried out according to the method outlined in the previous study of Rubin and Dagan (1987a), and it includes the effect of parameters variances of estimation.

The computation of the conditional moments of H is very similar to that of Y. The general relationships between the conditional moments are given in Dagan (1985) and further developed to account for parameters uncertainty in Rubin and Dagan (1987a). They can be applied directly by using the expressions derived here for the unconditional moments.

It is worthwhile to recall here the main features of conditioning or kriging: at the measurement points the expected value of Y is equal to the measured one and the variance is zero, the conditional variance is smaller or equal than the unconditional one and the conditioning by a measurement affects an area of radius of the order I around the point. Measurement errors are usually accounted for by inclusion of nuggets in C_Y and in the head variogram. In this case, the conditional variances at the measurement points will not be equal to zero but to the nugget. In the present application a nugget term was included in C_Y but not in the head variogram.

4. Results Obtained For The Bagueixe Aquifer

In order to obtain the regional mapping of mean transmissivities for Bagueixe aquifer several runs of the stochastic mathematical model were executed, associated with the characterization of the regional distribution of the logtransmissity field. The parameters related with the *drift* and with the *nugget* were researched and computed. The results from the analysis of several regional variances fields of the logtransmissivity, showed to be more plausible if the consideration of a linear *drift* was made simultaneously with a zero *nugget*.

The results computed for the Bagueixe aquifer, using the inverse modelling approach, are shown in Fig. 3a,b for piezometric heads, and in Fig. 4a,b for transmissivity values. From those figures the following conclusions may be drawn:

- The values of piezometric heads range between 543 m and 657 m, presenting a clear linear decreasing pattern towards SSE (Fig.3a,b).
- The computed values of logtransmissivity show an increasing linear drift towards SE ranging from 0.5 m^2/d, i.e T= 1.67 m^2/d, to 5.5 m^2/d, i.e. T= 148.4 m^2/d (Fig. 4a,b).
- Aiming a first analysis of the groundwater flow field, based on the stochastic inverse modelling outputs, values of Darcy's velocity (V_D) were evaluated using the formula $V_D = K.i$, where i is the hydraulic gradient, obtained from the piezometric lines of Fig. 3a, selecting for K the mean value of the hydraulic conductivity. The computed mean Darcy velocity was equal to 0.022 m/d.
- Aiming a first analysis of mass transport, based on the stochastic inverse modelling outputs, the interstitial velocity of a potential pollutant is larger than the one of Darcy's, as the flow and mass transport are not flowing across all the vertical cross-section of the aquifer but only through the area of voids located between the solid parts of the aquifer's, i.e. grain, faults, etc. So the real velocity of a conservative pollutant (i.e. its intrinsic or interstitial velocity V_r) was considered equal to $V_r = V_D/S_y$, being the value of the specific yield S_y equal to 10%. The computed value of the real

velocity was equal to 0.22 m/d.

The computed values of transmissivity were afterwards considered as input data for a finite element groundwater flow and mass transport mathematical model (TRAFRAP_WT) aiming the evaluation of flow and mass transport breakthrough curves for different scenarios. From the results presented in Lobo-Ferreira (1995) the evaluated piezometric surface is presented in Fig. 5. Comparing Figs. 3a and 5 one may observe a good agreement between both results.

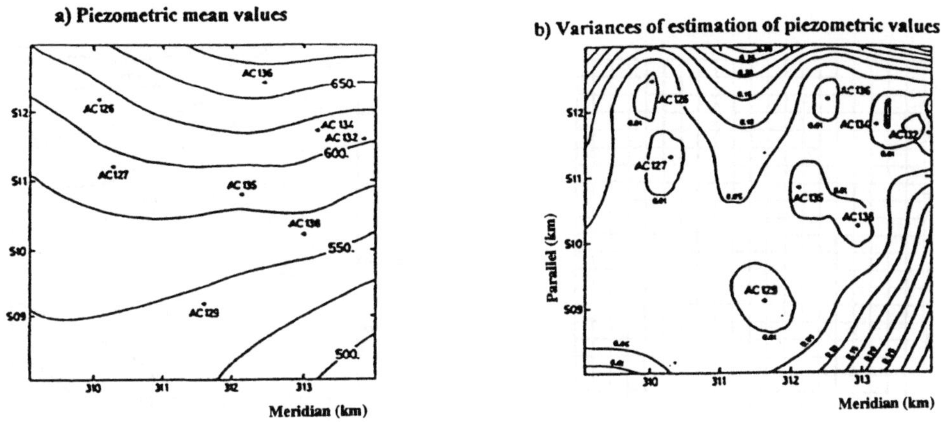

Figure 3. (a) Piezometric mean values and (b) variances of estimation computed for the Bagueixe aquifer with the inverse modelling approach

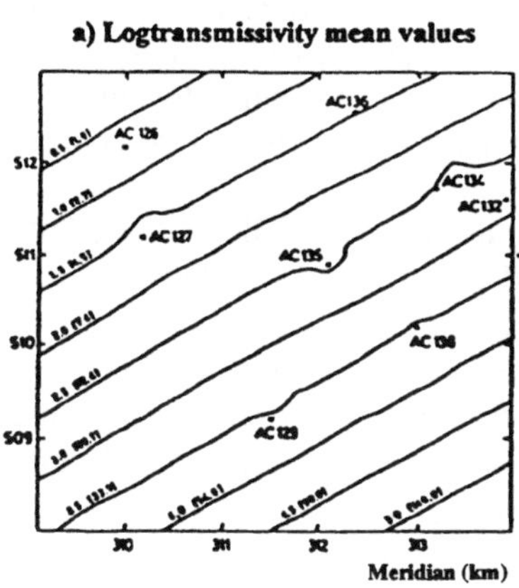

WELL	T (m^2/d)
AC 126	1.67
AC 127	7.1
AC 129	37.1
AC 132	17.36
AC 134	19.2
AC 135	7.9
AC 136	3.57
AC 138	38.65

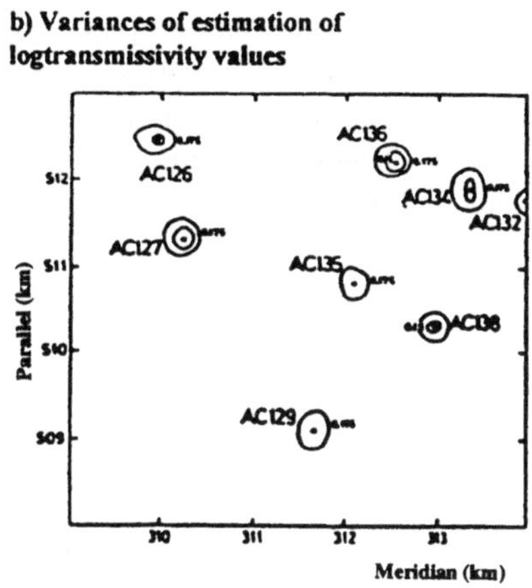

Figure 4. (a) Logtransmissivity mean values and (b) variances of estimation computed for the Bagueixe aquifer with the inverse modelling approach

5. Geological And Structural Hypothesis That May Justify The Results Computed Using The Stochastic Inverse Approach Developed For The Bagueixe Aquifer

Based on the data gathered in the pumping wells, three transverse profiles were plotted intersecting the pumping wells, allowing the drawing of possible piezometric lines for each profile. We could see that in some locations the piezometric lines have elevations above the ground surface, in places where there are no lakes or other sources of surface water. We have therefore considered that the Bagueixe aquifer is not phreatic.

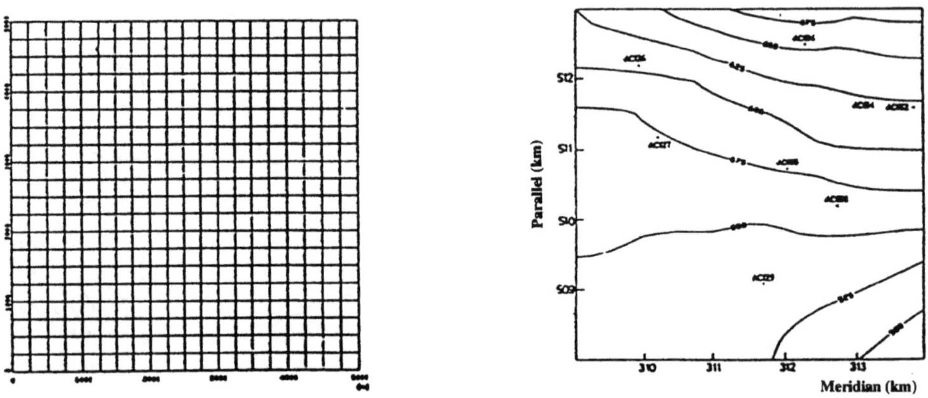

Figure 5. **Finite element grid selected and piezometric surface computed for the Bagueixe aquifer with the groundwater flow TRAFRAP_WT model**

The lithological characterization of the aquifer, obtained from analysis of lithological columns in the wells, served as the supporting background for classifying different lithological levels, based on their degrees of fracturing and weathering. We divided the profiles in four different levels (cf. Fig. 2): (1) covering deposits, with a thickness not exceeding ten meters; (2) weathered amphibolitic schists with only small fractures: (3) very weathered and fractured amphibolitic schist with quartz intrusions and (4) compact amphibolitic schist.

The projection of these elements in the above mentioned profiles, allowed the correlation between the different lithological levels and the inference of an hypothesis for the general pattern representing the Bagueixe aquifer, constituted by the following levels: a bed-rock base composed by amphibolitic schist, overlaid by a level of weathered and fractured schist with quartz intrusions, followed by a level of schist showing only small fractures. Some times over this layer we may observe the occurrence of covering deposits which depth varies, but is thinner than ten meters (cf. Fig. 2).

We could also observe, based on the data gathered in the wells, that the level where the water is being pumped corresponds to the strongly weathered and fractured level, confirming the possible existence of a fractured semi-confined aquifer at this level.

The flow and infiltration of water proceed, as is well known, preferentially through media that presents smaller hydraulic resistance. In fractured geological formations these paths are materialized by discontinuities as faults, joints, cleavage and fractures.

In the case of the fractured aquifer of Bagueixe the orientation that the discontinuities show, present attitudes located in NW-SE quarter, bent towards South. This observation allows the perception of flow favourable conditions, developing preferentially along these discontinuities and directions, as one may observe analyzing the piezometric surface of Bagueixe aquifer (Fig. 3a), as computed by stochastic inverse modelling.

The analysis of the mean values computed for the transmissivity, shown in Fig.4a, allows the observation that the linear variation of the logtransmissivity field develops according to the NW-SE direction.

The increasing pattern of the logtransmissivity field towards SE can eventually be explained considering that the opening of the fractures and/or the intensification of the number of discontinuities develops in that direction. This supposition could be supported considering the fact that in Bagueixe the aquifer's folds are more developed towards SE, and that the folds normally are associated and come together with an increase in cleavage density and also with an increase of fracturation.

The presence of a larger number of discontinuities could have as consequence an increase in the value of the hydraulic conductivity, and also, what is not considered the major reason in the case of Bagueixe aquifer, an increase in the value of the saturated depth, both having as effect an increase in the values of transmissivity.

6. Conclusions

The introduction of the stochastic inverse modelling approach for fractured domains showed to be particularly effective in the identification of piezometric levels and of logtransmissivities of the case-study area, revealing itself as a powerful tool for aquifer parameter identification complementing and surpassing, in a certain way, the uncertainty caused by a limited number of available field measurements.

Acknowledgements

We acknowledge the financial support received from NATO and from the US AID for this study. Mrs. Teresa Eira Leitão, geologist and Ph. D. candidate at the Groundwater Research Group of LNEC, developed most of the studies related to the geologic, structural and tectonic characterization of the Bagueixe aquifer, having also contributed to the development of most TRAFRAP_WT computations presented in Lobo-Ferreira and Leitão (1991) and Lobo-Ferreira (1995).

We acknowledge the co-operation of Dr. José Delgado Rodrigues, Principal Research Officer at LNEC, for his help during the hydrogeologic interpretation of the results obtained using the stochastic inverse modelling approach.

References

Carrera, J. and Neuman, S.P. (1986) Estimation of Aquifers Parameters under Transient and Steady-State Conditions: 1. Maximum Likelihood Method Incorporating Prior Information. *Water Resour. Res.*, 22 (2), 199-210.

Dagan, G. (1985) Analysis of flow through homogeneous random aquifers 2. Unsteady flow in confined formations. *Water Resour. Res.*, 18 (5), 1571-1585.

Dagan, G, and Rubin, Y. (1988) Stochastic Identification of Recharge, Transmissivity and Storativity in Aquifer Transient Flow: a Quasi-Steady Approach. *Water Resour. Res.*, 24 (10), 1968-1710.

Hidroprojecto, ACavaco and Tahal (1987) *Estudo de Viabilidade da Rega do Vale de Chaves e seus Vales Secundários e Prospecção de Águas Subterrâneas em algumas Zonas de Trás-os-Montes; C - Prospecção de Águas Subterrâneas em algumas Zonas de Trás-os- Montes. Direcção Regional de Agricultura de Trás-os-Montes.* Lisboa.

Hoeksema, R.J. and Kitanidis, P.K. (1985) Analysis of the Spatial Structure of Properties of Selected Aquifers. *Water Resour. Res.*, 21 (4), 563-572.

Kitanidis, P.K. and Vomvoris, E.G. (1983) A Geostatistical Approach to the Inverse Problem in Groundwater Modeling (Steady-State) and One-Dimensional Simulations. *Water Resour. Res.*, 19 (3), 677-690.

Lobo-Ferreira, J.P., and Leitão,T.E. (1991) *Aplicação da Modelação Estocástica Inversa e de um Modelo de Escoamento e de Transporte de Poluentes em Meios Fracturados à Região de Bagueixe.* Lisboa, Laboratório Nacional de Engenharia Civil, Memória n. 768.

Lobo-Ferreira, J.P. (1995) *On the Application of Stochastic Inverse Modelling, Aquifer Vulnerability Assessment and Groundwater Flow and Mass Transport Modelling in the Fractured Semi-Confined Aquifer of Bagueixe, in Portugal.* Key-Lecture presented to the "NATO Advanced Research Workshop on Regional Approaches to Water Pollution in the Environment", held in Liblice, Czech Republic, Sept. 95.

Ribeiro, A. (1974) *Contribuition à L'Étude Tectonique de Trás-os-Montes Oriental.* Tese de Doutoramento. Lisboa., Serviços Geológicos de Portugal, Memória n. 24 (Série Nova).

Rubin, Y. and Dagan, G. (1987a) Stochastic identification of transmissivity and effective recharge in steady flow: 1. Theory. *Water Resour. Res.*, 23 (7), 1855-1192.

Rubin, Y. and Dagan, G. (1987b) Stochastic identification of transmissivity and effective recharge in steady flow: 2. Case study. *Water Resour. Res.*, 23 (7), 1193-1200.

Rubin, Y. and Dagan, G. (1989a) Stochastic analysis of boundary effects on head spatial variability in heterogeneous aquifers: 1. Constant head boundary. *Water Resour. Res.*, 24 (10), 1689-1697.

Rubin, Y. and Dagan, G., (1989b) Stochastic analysis of boundary effects on head spatial variability in heterogeneous aquifers: 2. Impervious boundary. *Water Resour. Res.*, 25 (4), 707-712.

Rubin Y., Lobo-Ferreira J.P., Rodrigues J.D., and Dagan G. (1990) Estimation of the Hydraulic Parameters of the Rio Maior Aquifer in Portugal by Using Stochastic Inverse Modeling. *Journal of Hydrology*, 118 (1990), 257-279.

Schweppe,F.C. (1973) *Uncertainty Dynamic System.* Prectice-Hall, Englewood Cliffs,NJ.

SGP (1972) *Carta Geológica de Portugal à Escala 1/500 000.* Lisboa, Serviços Geológicos de Portugal.

Smith, L. and Freeze, R.A. (1979) Stochastic analysis of steady state groundwater flow in a bounded domain: 2. Two-dimensional simulations. *Water Resour. Res.*, 15 (6), 1543-1559.

Yeh, W.W.G. (1986) Review of parameters identification procedures in groundwater hydrology: The inverse problem. *Water Resour. Res.* 22 (2), 95-108.

SOLUTE TRANSPORT IN SINGLE AND MULTIPLE SCALE HETEROGENEOUS FORMATIONS: NUMERICAL EXPERIMENTS

A.M.M. ELFEKI
G.J.M. UFFINK
F.B.J. BARENDS
Delft University of Technology, Civil Engineering, Geotechnical Laboratory, P.O.Box 5048, 2600 GA Delft, The Netherlands.

Abstract. The heterogeneity of natural formations has a significant impact on the spreading characteristics of groundwater contaminants. This study focuses on the influence of different geological settings and characterization methods on transport characteristics in heterogeneous porous formations. Several patterns of geological structures have been synthesized by different techniques and then the two dimensional form of flow and solute transport equations are solved numerically in these structures. The generated fields considered in these experiments are stationary Gaussian fields, Markovian fields developed by [Elfeki, Uffink and Barends, 1995] and combiend random fields (Gaussian-Markovian fields).
From the experiments, one may distinguish two transport regimes: Fickian regime and non-Fickian (Anomalous) regime. The non-Fickian regimes can be divided into superdiffusive dispersion, in which, the growth of longitudinal variance of particles cloud with time is faster than linearity, or subdiffusive dispersion, in which, the growth of longitudinal variance of particles cloud is slower than linearity.

1. Introduction

Most methods for generation of artificial heterogeneous media are based on classical Gaussian stochastic theories. Recently the concept of fractals is applied [e.g. Cushman *et al.*, 1993; Sahimi, 1995]. In this paper a technique is used, based on the theory of Markov chains. Markov chains were used to generate sequences of geological layers by Krumbein [1967]. The method used in this paper is a two-dimensional extension. With this method the structure may be generated of an aquifer that consists of several geological units. To each geological unit a single value for the conductivity may be assigned, resulting in a Markovian random field. In the present paper we go one step further and use a Gaussian generator to assign a spatially varying conductivity field within each unit, however using different statistics for different units. This hybrid technique will be referred to as the Markovian-Gaussian model. The Markovian method was applied earlier in studying effective properties for groundwater flow in heterogeneous media [Elfeki et al., 1995; Elfeki, 1996]. In this paper the attention is focused on transport behaviour.

A. Soares et al. (eds.), geoENV I–Geostatistics for Environmental Applications, 51–63.

2. Experimental Set-up

Several types of heterogeneity are examined:

1. Homogeneous background with a single inclusion.
2. Gaussian heterogeneity (stationary random field).
3. Markovian heterogeneity (generation of the geometric structure of several geological units)
4. Combined Markovian-Gaussian heterogeneity.

3. Generation of Aquifer Heterogeneity

The Coupled Markov Chain Model (CMCM) developed by Elfeki, *et al.* [1995] is used to generate what is called here Markovian heterogeneity. A field generated by this method displays structures, that may be identified as geological units. Key-parameters in the CMCM-method are the transition probabilities between the units. The CMCM generates aquifers with a discrete type of heterogeneity. The turning bands method (TBM) [Tompson *et al.*, 1989] is used to generate a log-normally stationary two-dimensional hydraulic conductivity fiex with a prescribed mean, variance and exponential autocovariance function. This type of heterogeneity is called Gaussian heterogeneity. A third type of heterogeneity is obtained by a combination of CMCM and TBM. These aquifers consist of several distinct units, but inside each unit the conductivity varies continuously in a Gaussian way. This type of heterogeneity is referred to as Markovian-Gaussian.

4. Flow and Transport Simulation

In the generated aquifers the flow equation is solved using an implicit finite difference scheme. The computed potential head is differentiated to derive the velocity field. Simulation of contaminant transport is based on the classical advection-dispersion equation. The transport equation is solved using the random walk method. The particle tracking algorithm used here is similar to the one described by Uffink [1990]. The concept of dispersion is assumed to be valid at the microscopic scale. The analysis of the simulation results will be focused on the question whether the concept of Fickian dispersion is valid on a higher scale. The flow and transport model have been tested for some academic cases for which analytical solutions exist and it gives good results.

5. Evaluation

A convenient way to study the spreading behaviour from a particle model is from the development in time of the spatial moments of a particle cloud. The first and second particle spatial moments are calculated which correspond to the displacement of the plume centroid and the spreading around that centroid respectively. The macro-dispersion coefficient is calculated as half of the rate of change of the second moment. Spatial concentration fields are determined by superimposing a rectangular grid of dimensions Δx, and Δy in x and y direction respectively and calculating the distribution of the particles. The grid volume-averaged concentration is estimated.

6. Experimental Results

6.1 SINGLE INCLUSION

In this experiment, an inclusion is considered to get an insight in the mixing mechanisms. Two types of inclusions are investigated: a high permeable inclusion (with permeability of the inclusion $K_i=10$ m/day and permeability of the surrounding medium $K_m=1$m/day) and a low permeable one ($K_i=1$ m/day and $K_m=10$ m/day). Figure 1 shows the flow field and plume evolusion. The longitudinal and transverse spatial moments and dispersion coefficients are presented in Figure 2 for both cases. These figures show that due to the heterogeneity transport may advance (in case of converging flow lines) or retard (in case of diverging flow lines) along streamlines. As a consequence the transport behaviour may change from normal to anomalous diffusion or vice versa. In Figure 2 all kinds of mixing regimes (diffusive, subdiffusive and superdiffusive [Sahimi, 1995]) can be seen from this example. In real aquifers, however, more than one inclusion exist and their influences interfere in a complex fashion. The following experiments show complicated cases.

6.2 MARKOVIAN FIELD WITH LOW AND HIGH CONTRAST IN CONDUCTIVITY

In these experiments, the influence of a contrast in conductivity is investigated. A two-dimensional Markovian field has been generated with four geological materials. The generated field is characterized by the transition probabilities given in table 1. Two cases are considered: case 1, corresponds to low contrast in conductivity and case 2, corresponds to high contrast in conductivity. The conductivities in case of low contrast are 80, 50, 20 and 10 m/day for geological units 1, 2, 3 and 4 respectively.

For high contrast the conductivities are 100, 10, 1, 0.1 m/day for geological units 1, 2, 3 and 4 respectively. The results of both cases in terms of snapshots of the plume, longitudinal spreading and maco-dispersion are displayed in Figure 3 and Figure 4, respectively. The simulations (Figure 3 and Figure 4) show that the plume is moving slower in the case of high contrast and the spreading is covering a larger area.

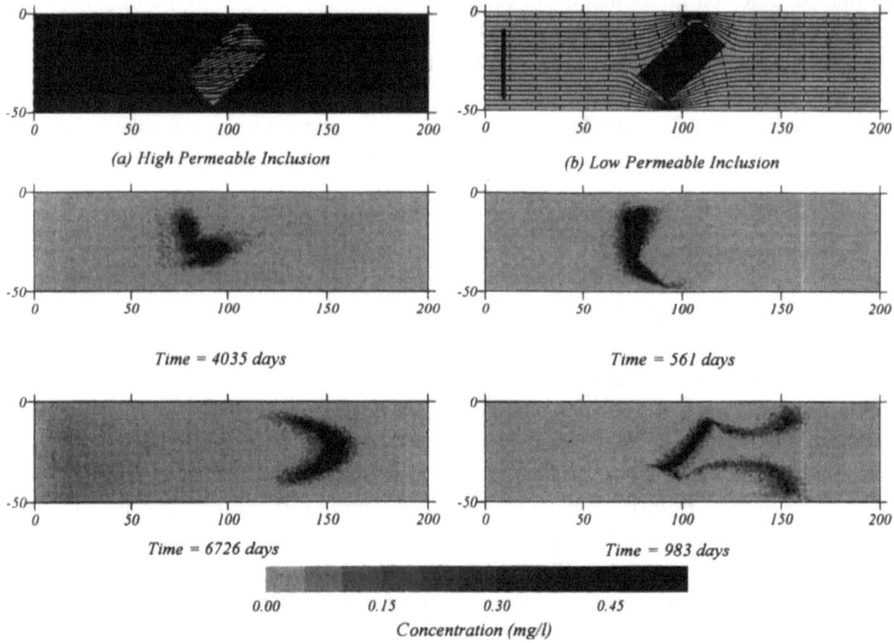

Figure 1 Plume evolution in single inclusion.

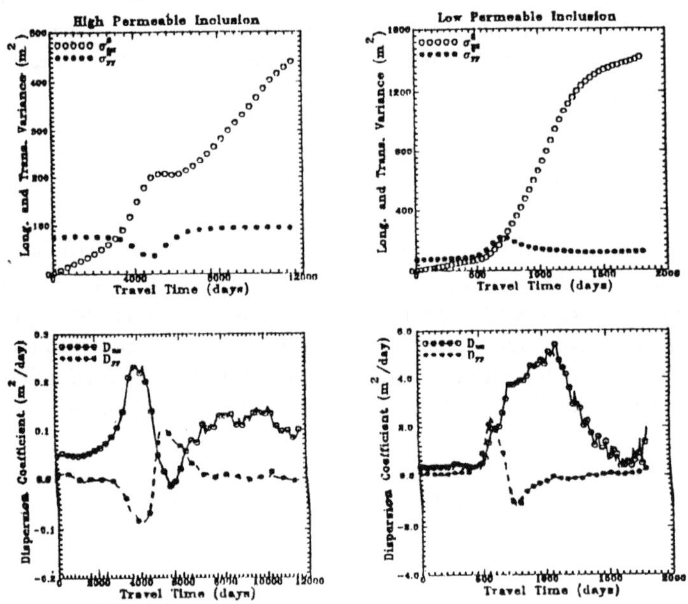

Figure 2 Spatial moments of the plume in case of single inclusion.

This is due to low permeable zones where the particles may be caught up causing elongated tails, while particles in the high permeable zones are moving relatively faster. The macro-dispersion, Figure 4, shows a higher value of the macro-dispersion coefficient in case of high contrast. It seems also that the plume has reached asymptotic macro-dispersion within 400 days in case of low contrast. However, the asymptotic macro-dispersion has not been reached in the case of high contrast in the same period. Thus the degree of contrast in conductivity has a remarkable influence on the transport characteristics.

TABLE 1. Transition probabilities of figure 3.

Sampling interval in X-axis (m)= 3.
Sampling interval in Y-axis (m)= 2.

	Input Statistics				Calculated Statistics			
Horizontal Transition Probability Matrix								
Unit	*1*	2	3	4	1	2	3	4
1	0.980	0.050	0.050	0.100	0.851	0.044	0.037	0.067
2	0.100	0.980	0.050	0.050	0.114	0.808	0.051	0.027
3	0.100	0.050	0.980	0.050	0.103	0.030	0.817	0.050
4	0.100	0.050	0.050	0.980	0.099	0.044	0.036	0.821
Vertical Transition Probability Matrix								
Unit	*1*	2	3	4	1	2	3	4
1	0.400	0.200	0.200	0.200	0.540	0.121	0.145	0.194
2	0.200	0.400	0.200	0.200	0.330	0.358	0.126	0.186
3	0.200	0.200	0.400	0.200	0.189	0.145	0.388	0.278
4	0.200	0.200	0.200	0.400	0.331	0.170	0.189	0.310

6.3 EFFECT OF VARIOUS CHARACTERIZATION METHODOLOGIES (MARKOVIAN VERSUS GAUSSIAN)

The influence of various types of characterization methodologies on flow and transport behaviour are considered. The stationary Gaussian fields and the Markovian fields are investigated. The statistical equivalence between the two fields is established by the following procedure. The spatial mean, variance and correlation lengths of the conductivity fields generated by CMCM in the previous section are computed. The parameters obtained are used to generate stationary Gaussian random field with the turning bands method. Then the flow and the transport have been calculated in both fields with given identical boundary and initial conditions. The results of this test in terms of flow nets and snapshot of the plumes are shown in Figure 5. This test is may be comparable with its statistically equivalent Markovian field for low and high contrast displayed in Figure 3. The second spatial moments of the plume in both low and high contrast for Gaussian and Markovian fields are shown in Figure 6 and Figure 7.

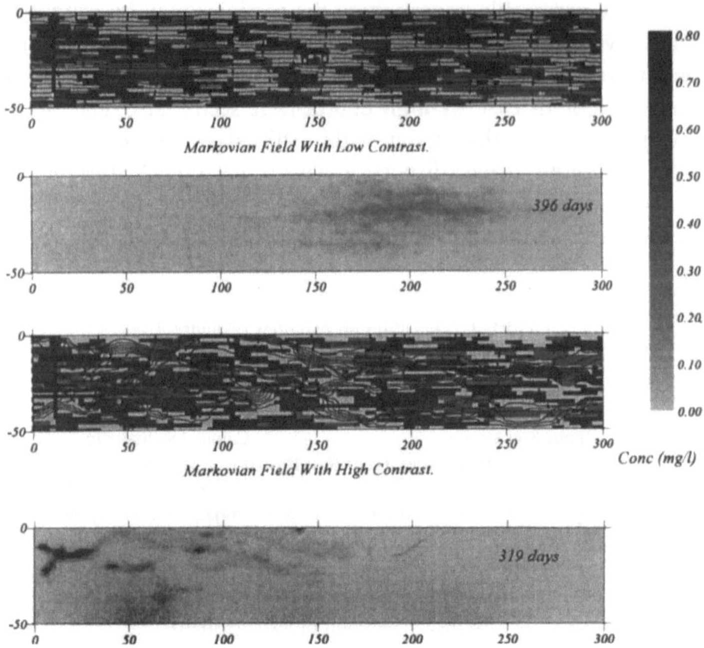

Figure 3 Snapshot of the plume in Markovian field with low and high contrast.

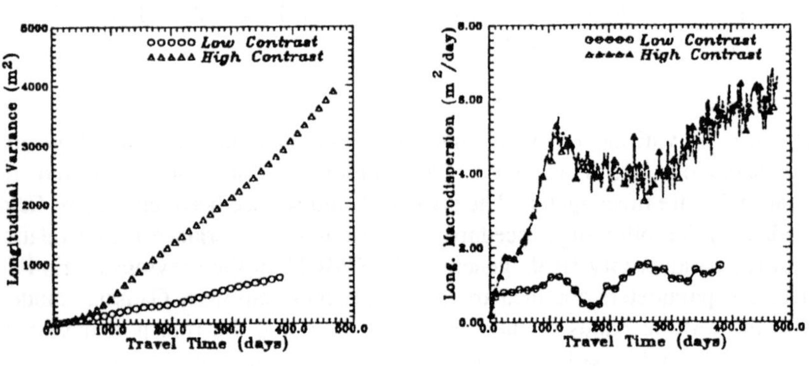

Figure 4 Comparison of second spatial moment and macro-dispersion.

Figure 6 shows the longitudinal spreading of the plume which is larger in Markovian field than in Gaussian field especially in case of high contrast. The reason is the discrete nature of Markovian heterogeneity. The plume faces sharp contrasts in conductivity between adjacent cells that causes high degree of spreading.

However, in Gaussian field the contrast between the adjacent cells are smooth due to the influence of continuous autocorrelation. Figure 6 shows the macro-dispersion coefficient in case of low contrast. The curves show an oscillation in the magnitude of the dispersion with some upward trend indicating an increase of the dispersion coefficient due to the heterogeneity. Both curves, globally, indicate approximately the same behaviour, whereas, in case of high contrast, Figure 7 shows that there is quite a distinction both in the magnitude and the global trend. The Gaussian field seems to reach an asymptotic average value of about 2 m^2/day while the Markovian field does not seem to reach an asymptotic value within the time of the experiment. However, if an average value could be estimated, it would be around 6 m^2/day which is higher than the corresponding Gaussian field.

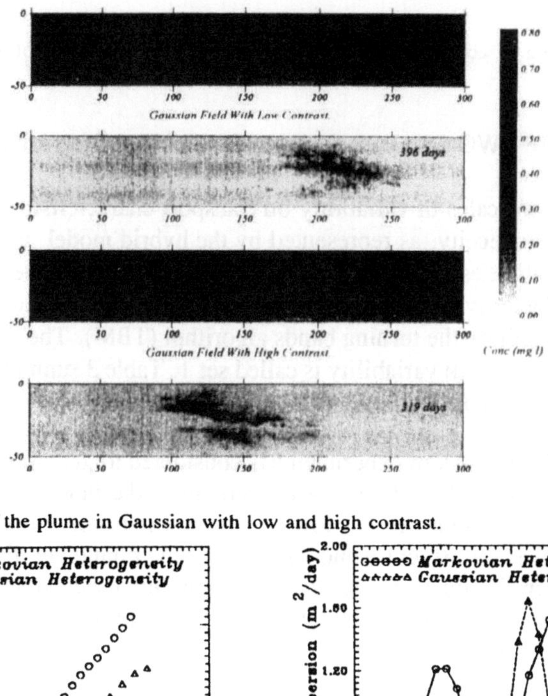

Figure 5 Snapshot of the plume in Gaussian with low and high contrast.

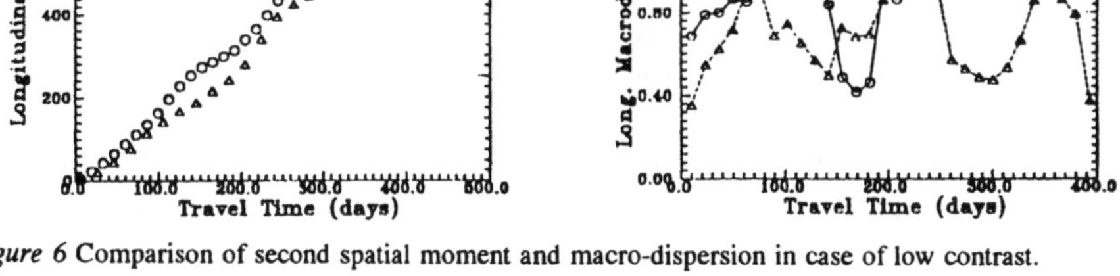

Figure 6 Comparison of second spatial moment and macro-dispersion in case of low contrast.

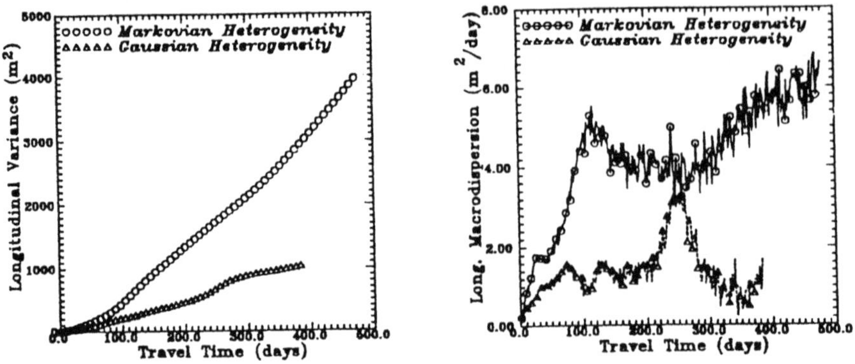

Figure 7 Comparison of second spatial moment and macro-dispersion in case of high contrast.

6.4 EFFECT OF TWO SCALES OF VARIABILITY (COMBINED FIELDS)

The effects of two scales of variability on transport characteristics are investigated. The aquifer heterogeneity, as represented by the hybrid model, is used for this test. Transition probabilities given in table 2 are used to generate the structures. Auto-correlated conductivity values drawn from log-normal distributions are assigned to each geological unit by the turning bands algorithm (TBM). The generated structure with and without internal variability is called set 1. Table 3 summarizes the statistics of the conductivity fields.

Another set of this sort of experiments is considered to get more insight about the transport behaviour in these fields. An inversion of the fields, mentioned in set 1, is considered to represent set 2. Flow and transport results in terms of flow nets superimposed over the structure and snapshot of the plume are displayed in Figure 8 for set 1 and in Figure 10 for set 2. The longitudinal second spatial moment and macro-dispersion of the plume in set 1 and set 2 are displayed in Figure 9 and Figure 11 respectively.

In order to interpret the results of these experiments, one may decompose transport into two parts, a convective part which takes place through the preferential flow paths and a dispersive part which takes place due to local scale heterogeneity. The superdiffusion regime is due to convection. In this regime the tracer finds permeable zones as it travels through the systems. Most of the particles enter the high conductivity zone near the source (preferential flow path) leading to a convection dominant transport.

TABLE 2. Transition probabilities of the structure in figure 8.

Sampling interval in X-axis (m)= 2.
Sampling interval in Y-axis (m)= 1.

	Input Statistics				Calculated Statistics			
	Horizontal Transition Probability Matrix							
Unit	*1*	*2*	*3*	*4*	*1*	*2*	*3*	*4*
1	0.960	0.010	0.010	0.020	0.975	0.007	0.002	0.016
2	0.010	0.970	0.010	0.010	0.008	0.977	0.005	0.010
3	0.010	0.020	0.960	0.010	0.024	0.023	0.948	0.005
4	0.010	0.010	0.010	0.970	0.022	0.007	0.007	0.965
	Vertical Transition Probability Matrix							
Unit	*1*	*2*	*3*	*4*	*1*	*2*	*3*	*4*
1	0.930	0.010	0.030	0.030	0.922	0.012	0.018	0.048
2	0.100	0.800	0.050	0.050	0.072	0.880	0.027	0.021
3	0.050	0.050	0.800	0.100	0.024	0.125	0.732	0.119
4	0.050	0.050	0.030	0.870	0.040	0.041	0.006	0.913

TABLE 3. Conductivity statistics in set 1.

Parameter Unit	w_i	⟨K⟩ m/day	σ_K m/day	⟨Y⟩	σ_Y	λ_x (m)	λ_y (m)
1	0.34	100.	100.	4.3	0.83	20.	5.
2	0.26	10.	10.	1.96	0.83	20.	5.
3	0.10	1.0	1.0	-0.35	0.83	20.	5.
4	0.30	0.1	0.1	-2.65	0.83	20.	5.

The parameters in table 3: w_i is the marginal probability of unit i, ⟨K⟩ is the arithmetic mean the conductivity in unit i, σ_K is the standard deviation of the conductivity in unit i, ⟨Y⟩, σ_Y are the logarithmic transformation of $<K>$ and σ_K respectively, and λ_x and λ_y are the correlation lengths in x and y direction.

This behaviour depends upon the existence of these preferred paths in the generated field and whether or not the tracer particles can migrate into that zone. Other particles may move relatively slowly towards the low permeable zones or may be caught up in these zones as shown in Figure 8. A considerable time is needed for the particles to cross these low permeable zone and to get released into the high permeable zones. Figure 8 shows that the plume is slower in case of introducing internal variability. This behaviour reflects the influence of log-normal distribution used to generate the internal variability. The values of the conductivity less than the arithmatic mean is more likely to occur than values higher than the arithmatic mean. This means that the aquifer units are dominated by conductivities less than the arithmatic mean. Due to that fact, the transport is slower. Thus chosen another probability distribution of the conductivity will influence the results.

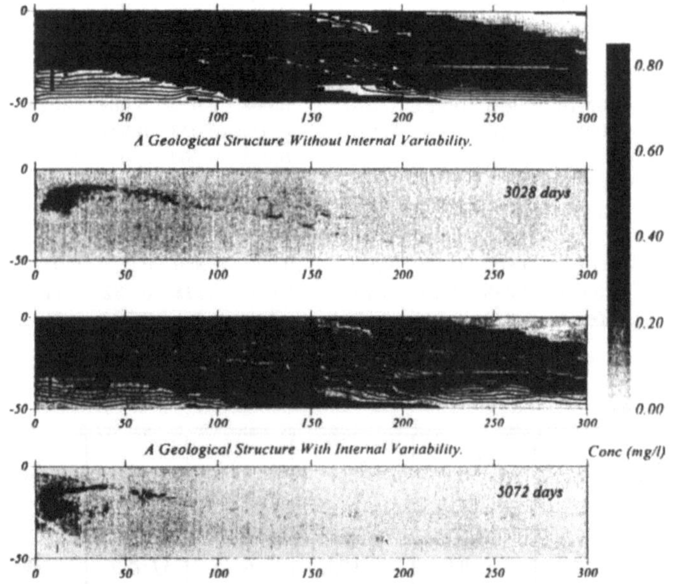

Figure 8 Snapshot of the plume in combined field set 1.

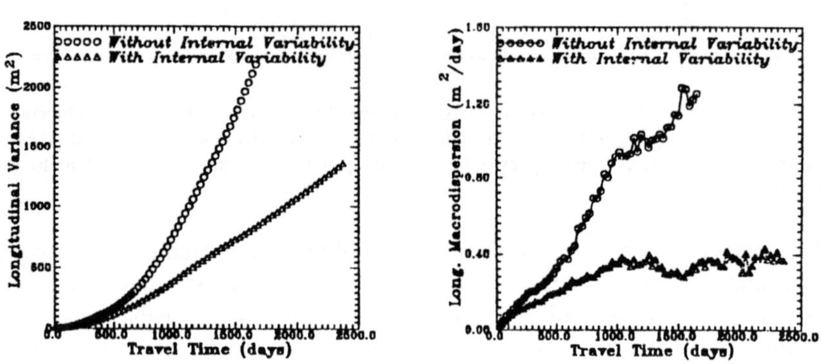

Figure 9 Comparison of second spatial moment and macro-dispersion in case set 1.

Plume statistics of set 1, displayed in Figure 9, show that in absence of internal variability, the macro-dispersion coefficient grows without bounds indicating superdiffusive regime within the time limit of the experiment. If the internal variability is present, the macro-dispersion coefficient reaches an asymptotic value within approximately the same time period. This behaviour can be accounted for by the internal variability which enhances the mixing process and leads to an asymptotic dispersion coefficient.

On the other hand, plume statistics of set 2 presented in Figure 11 shows that in case of absence of the internal variability, the macro-dispersion coefficient grows until it reaches more or less an asymptotic value within the time limit of the experiment whereas, with presence of the internal variability the macro-dispersion fluctuates introducing different transport regimes within the same time period. One may distinguish different stages. In the first stage, an asymptotic value of the macro-dispersion coefficient has been reached after about 200 *days* from the time of injection indicating a diffusive regime. A second stage appears, where the plume goes faster in the high permeable zone (superdiffusive regime) till it reaches another asymptotic dispersive regime after 800 days. Later on, the macro-dispersion coefficient decreases due to divergence of the flow lines at the low permeable zone indicating a subdiffusive regime. The general behaviour can be explained as follows. In case of the absence of internal variability the only heterogeneity encountered by the plume are those near the source. As long as the plume propagates in the system it gets only in touch with the heterogeneity near the source. Major parts of the plume are advancing with relatively high velocity in the homogeneous highly permeable zones, while other parts of the plume are moving very slowly in the low permeable zones. Since the plume meets only this heterogeneity, it gets into a dispersive regime. However, in case of introducing internal variability the plume gets in touch with the heterogeneity in the system in the earlier stage. Later on the plume meets the variability inside the high permeable zone which is moderately high. This variability introduces more tortuosity of the flowlines. As a consequence, it causes the transport regime to change according to the divergence and convergence of the flowlines. As long as the flow pattern does not repeat itself in a relatively regular manner, the transport process will not become Fickian.

7. Conclusions

In the proposed numerical experiments single realization hydraulic conductivity fields are generated by several methods and the associated flow and transport problems are solved numerically under specified initial and boundary conditions. The first two spatial moments of the particle cloud which describe the displacement of the plume centroid and the spread around the centroid and macro-dispersion coefficients are estimated from the numerical experiments. The various transport regimes are observed in these experiments. From the results the following conclusive remarks are made:

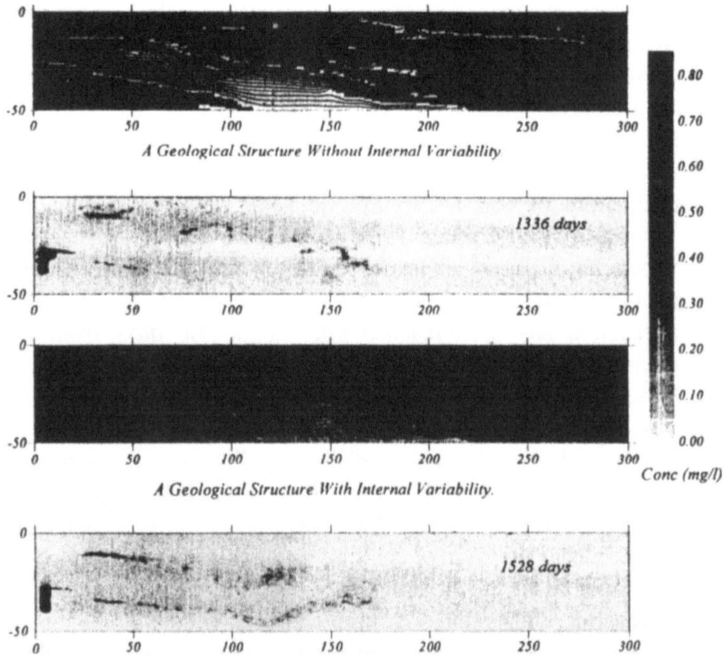

Figure 10 Snapshot of the plume in combined field set 2.

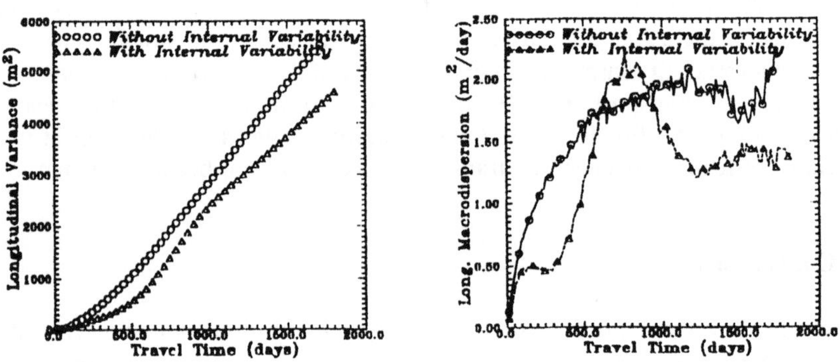

Figure 11 Comparison of second spatial moment and macro-dispersion in case set 2.

1. The numerical experiments give a conceptual understanding of the effects of various characterization methods of conductivity fields on the transport behaviour. For many geological patterns the concentration fields are no longer Gaussian, particularly when strong heterogeneity and preferential flow paths are present in the

system.

2. The experiments emphasize that the macro-dispersion coefficients are not equivalent to those used at the microscopic level. The higher the contrast of conductivity of the individual units, the larger the value of the macro-dispersion coefficient.

3. The experiments confirm that dispersion, in general, in many geological structures is non-Fickian at least for short to intermediate travel times investigated in these experiments. The existence of high conductivity channels in geological systems is the main cause of superdiffusive regime, while a subdiffusive dispersion regime can be observed in some patterns.

4. The assumption of log-normal *PDF* of conductivity inside the geological units is reflected in the results. This reflection is manifested as retardation in the plume evolution and consequently in its spatial moments. The reason is that zones of permeability less than the arithmetic mean of conductivity are more likely to appear in the field than highly permeable zones.

5. In combined field experiments the internal variability contributes mostly in the mixing processes by introducing tortuosity of the flowlines. This mimics the pore scale dispersion process. In some geological setting the internal variability changes the transport regimes from superdiffusive to diffusive.

6. It appears that reliable predictions of solute transport in aquifers require sufficiently detailed knowledge of the major geological features, specially preferential flow paths, within the aquifers. It has been shown that the characterization method either Markovian, Gaussian, combined or otherwise has a strong influence on transport behaviour. The choice of the appropriate method for field applications should be tested carefully. The Markovian and the hybrid methodologies used in this article may help in that respect.

8. References

Cushman, J. H. & Ginn, T. R. (1993) Nonlocal dispersion in media with continuously evolving scales of heterogeneity. *Trans. Porous Media* **13**, 123-138.

Elfeki, A. M. & Uffink, G. J. M. & Barends, F. J. B. (1995) Stochastic simulation of heterogeneous geological formations using soft information, with an application to groundwater. In *Groundwater Quality: Remediation and Protection, QG'95*. Edited by Kovar, K. and Krasny, IAHS Publication No. 225.

Elfeki, A. M. (1996) *Stochastic characterization of geological heterogeneity and its impacts on groundwater contaminant transport*. Ph.D. Thesis. TU Delft. The Netherlands.

Tompson, A. F. B. & Ababou, R. & Gelhar, L. W. (1989) Implementation of the three-dimensional turning bands random field generator. *Water Resour. Res.* **25**(10): 2227-2243.

Krumbein, W. C. (1967) Fortran computer program for Markov Chain experiments in geology. *Computer Contribution 13,* Kansas Geologic Survey, Lawrence, Kansas.

Sahimi, M. (1995) *Flow and Transport in Porous Media and Fractured Rocks: From Classical Methods to Modern Approaches*. VCH Verlagsgesellschaft. mbH, D-69451 Weinbeim.

Uffink, G. J. M. (1990) *Analysis of Dispersion by The Random Walk Method.* Ph.D. Thesis. Delft University of Technology. The Netherlands.

NUMERICAL ASPECTS OF THE UNIVERSAL KRIGING METHOD FOR HYDROLOGICAL APPLICATIONS

CARLOS LÓPEZ
Centro de Cálculo
Facultad de Ingeniería, CC 30, Montevideo, Uruguay
JAVIER SAMPER
E.T.S. de Ingenieros de Caminos, Canales y Puertos.
Campus de Elviña, 15192 La Coruña, Spain

Abstract. Many hydrological variables usually show the presence of spatial drifts. Most often they are accounted for by either universal or residual kriging and usually assuming low order polynomials. The Universal Kriging matrix ($\mathbf{M}$), which includes in it the values of the polynomials at data locations (matrix $\mathbf{F}$), in some cases may have a too large condition number and can even be nearly singular due to the fact that some columns are close to be linearly dependent. These problems are usually caused by a combination of pathological data locations and an inadequate choice of the coordinate system. As suggested by others, we have found that an appropriate scaling can alleviate the problem by significantly reducing the condition number of $\mathbf{M}$ (cond($\mathbf{M}$)). This scaling, however, does not affect the linear independence of this matrix. We show that a QR factorization of matrix $\mathbf{F}$ leads to a significant improvement on cond($\mathbf{M}$). An alternative to drift polynomials consists on using a set of functions derived from the eigenvectors of the variogram matrix (Γ). Although their potential usefulness as interpolating functions remains to be ascertained, they are optimal from the point of view of optimizing cond($\mathbf{M}$). In fact we are able to provide a rigorous proof for an upper bound of cond($\mathbf{M}$). Applications of the theoretical developments to hydraulic head data from an alluvial aquifer are also presented.

1. Introduction

The sensitivity of the kriging method in general, and the universal kriging in particular, either to input data or to variogram models have been analyzed by Diamond and Armstrong (1984), Posa (1989) and O'Dowd (1991). Diamond and Armstrong (1984) reported results on the relationship between the relative change in the kriging weights and the condition number of the ordinary kriging matrix. We remind the reader that the

A. Soares et al. (eds.), geoENV I–Geostatistics for Environmental Applications, 65–76.

condition number of the matrix of a system of equations measures the sensitivity of the solution to modifications on the input data (Atkinson, 1989). If the condition number is too large, the significance of the results (in this case, the universal kriging weights) can be completely lost. Posa (1989) studied the condition number of the ordinary kriging method for various well-known covariance models. He noticed that unstable situations occur when Gaussian models without a nugget are used. O'Dowd (1991) showed that for ordinary kriging the condition number of the matrix $\mathbf{M}$ has a lower bound, which depends on the covariance model and data location. He also showed some evidence that a suitable way to reduce cond($\mathbf{M}$) consists on scaling the sill (when defined) or conversely substituting the column and row of 1's to a different constant value. His findings were based upon theoretical results for a pure nugget variogram. Here we analyze different alternatives to lower the condition number of the universal kriging (UK) matrix. We first summarize briefly the UK theory and discuss some of the problems and inconsistencies of the standard UK matrix. Alternative approaches to overcome such problems include: (1) the factorization of matrix $\mathbf{F}$, and (2) the use of the eigenvectors of the variogram matrix as drift functions. The performance of these approaches is evaluated in terms of both the condition number and the estimation capabilities. To that purpose we use hydraulic head data from the Andújar alluvial aquifer in Southern Spain.

2. Universal kriging theory

Let Z(x) be a non-stationary random function. Its mathematical expectation $E[Z(x)] = m(x)$ exists and can be modeled as a linear combination of basis functions $f_l(x), l = 1,...,P$. The difference $\varepsilon(x) = Z(x) - m(x)$ is assumed to be a zero mean second order stationary random function having a variogram γ_ε (Samper and Carrera, 1990):

$$Z(x) = m(x) + \varepsilon(x);\ E(\varepsilon(x)) = 0 \text{ and } m(x) = \sum_{l}^{N} a_l f_l(x) \tag{1}$$

where a_l are the drift coefficients. Eq. (1) is assumed to hold locally (otherwise, residual kriging might be applied). It is also assumed that the variogram is known. The UK estimator is a linear combination of the data values,

$$Z^*(x) = \sum_{I}^{N} \lambda_i Z(x_i) = \sum_{I}^{N} \lambda_i Z_i$$

In order to be unbiased, this estimator must satisfy:

$$E(Z^*) = E(\sum_{I}^{N} \lambda_i Z_i) = E(Z) = E[m(x) + \varepsilon(x)] = E[m(x)] = m(x)$$

This means that the kriging coefficients λ_i must satisfy the following equations

$$\sum_{1}^{N} \lambda_i m(x_i) = m(x) \tag{2}$$

Assuming that m(x) can be appropriately described by a function subspace named f, Eq. (2) can be rewritten as (Samper and Carrera, 1990):

$$\sum_{i=1}^{N} \lambda_i \left(\sum_{l=1}^{P} a_l f_l(x_i) \right) = \sum_{l=1}^{P} a_l \left(\sum_{i=1}^{N} \lambda_i f_l(x_i) \right) = \sum_{l=1}^{P} a_l f_l(x)$$

which holds if

$$\sum_{i=1}^{N} \lambda_i f_l(x_i) = f_l(x) \qquad l = 1..P$$

Kriging equations are obtained after minimizing the estimation error

$$\sum_{j=1}^{N} \lambda_j \gamma_\varepsilon(x_i - x_j) + \sum_{l=1}^{P} \mu_l f_l(x_i) = \gamma_\varepsilon(x_i - x) \qquad i = 1,\dots,N(x)$$

where μ_l is the l-th Lagrange multiplier and N(x) is the size of the kriging neighborhood which is not necessarily equal to N because typically only the closest points are considered. Our calculations assume that N(x)=N, therefore our estimator is global. The system to be solved can be written in matrix form as

$$\mathbf{M}\begin{bmatrix} \underline{\lambda} \\ \underline{\mu} \end{bmatrix} = \begin{bmatrix} \Gamma & \mathbf{F} \\ \mathbf{F}^T & 0 \end{bmatrix}\begin{bmatrix} \underline{\lambda} \\ \underline{\mu} \end{bmatrix} = \begin{bmatrix} \underline{\gamma} \\ \underline{f} \end{bmatrix} \tag{3}$$

where $\underline{\lambda}$ is the vector of kriging coefficients, $\underline{\mu}$ is the vector of Lagrange multipliers, $\mathbf{F}$ is a rectangular matrix with entries $f_l(x_i)$, $\underline{\gamma}$ is a vector of variogram values $\gamma_j = \gamma_\varepsilon(x - x_j)$ and $\underline{f}$ is the vector of values of $f_l(x)$ at data locations.

3. Some inconsistencies of the standard approach

This standard approach has some problems which are discussed below.

3.1. LACK OF HOMOGENEITY ON THE ENTRIES OF KRIGING MATRIX

The units of matrix Γ are the same as those of the variogram. In the standard Universal Kriging method, $\mathbf{F}$ is formed by evaluating monomials at data points. Therefore, the entries of matrix $\mathbf{F}$ are derived from powers of distance, i.e., their units are different than those of Γ. Moreover, the entries of matrix $\mathbf{F}$ have very different absolute values. For this reason, numerical instabilities may arise.

3.2. STRONG DEPENDENCY ON THE ORIGIN OF COORDINATES

For a two-dimensional variable, the entries of matrix $\mathbf{F}$ are $f_{il} = f_l(x_i) = (x_i)^m(y_i)^{l-m}$. Their values depend strongly on the origin of coordinates. In fact, the position of the origin affects strongly the condition number of the system which in turn affects the precision of the estimator Z^* (Diamond and Armstrong, 1984). In some exploratory calculations using synthetic data, we performed a standard change of variables by mapping every (x,y) location to the range (-S,S),(-S,S) being S the sill, and used such standardized coordinates for creating $\mathbf{F}$. Such simple mapping leads to a decrease of the condition number of the full matrix from 10^8 to 10^4.

3.3. LINEAR INDEPENDENCE OF THE COLUMNS OF $\mathbf{F}$

The columns of matrix $\mathbf{F}$ are built from the values of known functions (typically monomials) at data locations. They should be linearly independent in order to uniquely determine the drift coefficients a_l. Such property is usually not checked because most programs use monomials, which are supposed to be independent by themselves. However, when considering arbitrary (x,y) data coordinates, such property does not necessarily hold. It should be noticed that there are *at most* N(x) independent column vectors (monomials) in the N(x) dimension space, so P is strictly bounded.

The problems discussed in Sections 3.1 and 3.2 lead to a large cond($\mathbf{M}$), while that of Section 3.3 may preclude to get a meaningful solution.

4. Suggested approaches

The problems discussed in the previous section can be overcome in several ways. Here we have considered three possibilities:

1. A proper transformation of the coordinate system in order to balance the columns of $\mathbf{F}$. This can be done by transforming the coordinate system (translation and scaling) so that any every point (x,y) is mapped into a square whose x and y limits are within the range (-S,S). Actual results illustrating the effect of this mapping are presented in Section 5.
2. Choosing the monomials in such a way that the orthogonality of the columns of $\mathbf{F}$ is ensured. Since there is no unique option for such a choice, we suggest to build the columns of $\mathbf{F}$ from a set of linearly independent and orthogonal vectors which are dimensionally consistent with matrix Γ. Such a set can be derived from the singular value decomposition (eigenvectors and eigenvalues) of matrix Γ. With this set of functions, the problems discussed in Sections 3.1, 3.2 and 3.3 are overcome. However, the resulting drift does not have a simple structure as that described in Eq. (1). With this method it is possible to optimize cond($\mathbf{M}$). In fact, in Section 4.1 we derive theoretical bounds for cond($\mathbf{M}$).

3. A QR factorization of matrix $\mathbf{F}$. With this approach it is possible to solve directly the problems mentioned in Sections 3.2 and 3.3. The lack of homogeneity of the entries of matrix $\mathbf{M}$ (Section 3.1) is managed in an indirect manner.

The first procedure is fairly simple and needs no additional explanation. We will describe with some detail the other two approaches. The performance of all three approaches using hydraulic head data is evaluated in Section 5.

4.1. SINGULAR VALUE DECOMPOSITION OF MATRIX Γ

Let $\mathbf{E}$ and Λ be the matrices containing the eigenvectors and eigenvalues of matrix Γ, respectively. These matrices are related through $\Gamma = \mathbf{E}\Lambda\mathbf{E}^{-1} = \mathbf{E}\Lambda\mathbf{E}^{T}$. Notice that $\mathbf{E}$ is orthogonal because Γ is symmetric. Matrix Λ is a diagonal matrix containing the eigenvalues. Let $\mathbf{D}_{PxP}$ be a square diagonal matrix holding P eigenvalues and let $\mathbf{G}_{NxP}$ be a matrix formed using the corresponding P columns of $\mathbf{E}$. We suggest to build matrix $\mathbf{F}$ such that

$$\mathbf{F} = \mathbf{G}.\mathbf{D} \tag{4}$$

Since the eigenvalues of Γ share its units, so does matrix $\mathbf{F}$. The columns of $\mathbf{G}$ are orthogonal among themselves, $\mathbf{G}^T.\mathbf{G} = \mathbf{I}_P$ ($\mathbf{I}_P$ is the identity matrix of size P) and orthogonal to all the columns of $\mathbf{E}$, that is, $\mathbf{G}^T.\mathbf{E} = [\mathbf{I}_P \quad \mathbf{0}]$ (here $\mathbf{0}$ denotes a matrix having all its entries equal to zero). Another relevant property is

$$\mathbf{D}[\mathbf{I}_P \quad \mathbf{0}]\Lambda^{-1}[\mathbf{I}_P \quad \mathbf{0}]^T = \mathbf{I}_P$$

From Eq. (3) it follows that

$$\begin{cases} \Gamma\underline{\lambda} + \mathbf{F}\underline{\mu} & = \underline{\gamma} \\ \mathbf{F}^T\underline{\lambda} & = \underline{f} \end{cases} \Rightarrow \begin{cases} \underline{\lambda} + \Gamma^{-1}\mathbf{F}\underline{\mu} & = \Gamma^{-1}\underline{\gamma} \\ \mathbf{F}^T\underline{\lambda} & = \underline{f} \end{cases}$$

Premultiplying the first row by $\mathbf{F}^T$, and using the second, one obtains

$$\mathbf{F}^T\Gamma^{-1}\mathbf{F}\underline{\mu} = \mathbf{F}^T\Gamma^{-1}\underline{\gamma} - \underline{f} \tag{5}$$

This equation holds for any matrix $\mathbf{F}$. By using Eq. (4), the left-hand side reduces to

$$\mathbf{F}^T\Gamma^{-1}\mathbf{F} = \mathbf{D}^T\mathbf{G}^T\mathbf{E}\Lambda^{-1}\mathbf{E}^T\mathbf{G}\mathbf{D} = \mathbf{D}^T[\mathbf{I}_P \quad 0]\Lambda^{-1}\begin{bmatrix}\mathbf{I}_P \\ 0\end{bmatrix}\mathbf{D} = \mathbf{D}^T\mathbf{D}^{-1}\mathbf{D} = \mathbf{D}$$

The first term on the right-hand-side of Eq. (5) can be rewritten as

$$\mathbf{F}^T\Gamma^{-1}\underline{\gamma} = \mathbf{D}^T\mathbf{G}^T\mathbf{E}\Lambda^{-1}\mathbf{E}^T\underline{\gamma} = \mathbf{D}^T\mathbf{G}^T\mathbf{E}\Lambda^{-1}\mathbf{E}^T\mathbf{G}\,\mathbf{G}^T\underline{\gamma} = \mathbf{G}^T\underline{\gamma}$$

Consequently, $\underline{\mu}$ can be derived easily from the following diagonal system

$$\mathbf{D}\underline{\mu} = \mathbf{G}^T\underline{\gamma} - f$$

The eigenvalues and eigenvectors of matrix $\mathbf{M}$ can be expressed in terms of those of Γ in a closed form. Appendix 1 contains the details of such expressions. This appendix contains also a proof for the upper bound of the condition number of the kriging matrix:

$$cond(\mathbf{M}) \leq 2.6\, cond(\Gamma)$$

O'Dowd (1991) stated that cond($\mathbf{M}$) cannot be less than cond(Γ). Here we were able to prove that cond($\mathbf{M}$) has a well defined upper bound for the approach based on the singular value decomposition of Γ.

4.2. QR ORTHOGONALIZATION OF MATRIX F

The QR factorization of matrix $\mathbf{F}$ renders a unique set of matrices $\mathbf{Q}$ and $\mathbf{R}$ which satisfy

$$\mathbf{F} = \mathbf{QR}$$

$$\mathbf{Q}^T\mathbf{Q} = \mathbf{I}$$

where R is an upper triangular matrix. It is straightforward to rewrite the system of equations (3) in a slightly different manner so that the overall matrix has a lower condition number. The new form of the system of equations reads as

$$\begin{bmatrix} \Gamma & \mathbf{Q} \\ \mathbf{Q}^T & 0 \end{bmatrix} \begin{bmatrix} \underline{\lambda} \\ \underline{\mu}^* \end{bmatrix} = \begin{bmatrix} \underline{\gamma} \\ \underline{f}^* \end{bmatrix} \qquad (6)$$

where $\underline{\mu}^*$ and $\underline{f}^*$ are defined in the following manner

$$\underline{f}^* = \left(\mathbf{R}^{-1}\right)^T \underline{f}\ ; \qquad \underline{\mu}^* = \mathbf{R}\ \underline{\mu} \qquad (7)$$

This means that instead of solving Eq. (3) one can solve Eq. (6) and obtain the expressions of $\underline{\lambda}$ and $\underline{\mu}^*$. From Eq. (7) one can derive $\underline{\mu}$ from $\underline{\mu} = \mathbf{R}^{-1}\underline{\mu}^*$. Notice that the orthogonality property of matrix $\mathbf{Q}$ has not been used yet. In fact, it is possible to re-scale its columns by a factor s in order to balance the entries of the system matrix. Moreover, the matrix can be factorized using the fact that $\Gamma = \mathbf{E}\Lambda\mathbf{E}^T$. Therefore, after introducing the scale factor s the system matrix can be rewritten as

$$\begin{bmatrix} \Gamma & s\mathbf{Q} \\ s\mathbf{Q}^T & \mathbf{0} \end{bmatrix} = \begin{bmatrix} \mathbf{E} & \mathbf{0} \\ \mathbf{0} & \mathbf{I} \end{bmatrix} \begin{bmatrix} \Lambda & s\mathbf{E}^T\mathbf{Q} \\ s\mathbf{Q}^T\mathbf{E} & \mathbf{0} \end{bmatrix} \begin{bmatrix} \mathbf{E}^T & \mathbf{0} \\ \mathbf{0} & \mathbf{I} \end{bmatrix} = \mathbf{BLB}^T \qquad (8)$$

Matrix $\mathbf{L}$ has the same eigenvalues as matrix $\mathbf{M}$ because $\mathbf{B}$ is orthogonal (Atkinson, 1989). Notice that $\mathbf{L}$ has the eigenvalues λ_i of Γ in the diagonal and terms containing the scale factors in the rest of the first N rows. This fact can be exploited by using the Gershgorin theorem [see Dahlquist *et al.* (1980)] which states that matrix $\mathbf{L}$ has N eigenvalues $e_i(s)$ which verify:

$$e_i(s) = \lambda_i + O(s) \qquad i = 1,\ldots,N \tag{9}$$

where $O(s)$ is an infinitesimal of first order on s (the limit of $O(s)/s$ as s goes to zero is a constant). The eigenvalues corresponding to the P remaining rows (those with zeros in the diagonal) behave like

$$e_i(s) = O(s) \qquad i = N+1,\ldots,N+P \tag{10}$$

Dahlquist *et al.* (1980) also give a proof that the eigenvalues of a perturbed matrix are continuous in terms of the perturbation factor. From Equations (9) and (10) it is possible to relate the condition numbers of matrices Γ and $\mathbf{L}$. For small s, the largest eigenvalues of $\mathbf{L}$ will coincide with those of Γ, while the smallest eigenvalues will be on the order of s. Therefore, for small s the condition number of $\mathbf{L}$ will behave like O($1/s$). For large enough s, the condition number grows as $\sqrt{s}$. For intermediate values of s, we noticed that the evolution of the minimum eigenvalue in terms of s has an absolute minimum s^* for some moderate s. We tried to estimate such a value in terms of some norm of matrix Γ, and for some limited analyses we concluded that a good guess is simply $s^* = \|\Gamma\|_1 / N$, where the term in the numerator is the norm of matrix Γ (the maximum row sum of the absolute values). The optimum value s^* is bounded by the sill (when it is defined).

5. Results for the Andújar case

Previous theoretical analyses have been applied to hydraulic head data from the Andújar alluvial aquifer in Southern Spain where a uranium mill tailings has caused a uranium plume in the aquifer (Samper and Carrera, 1995). This case study is a small-scale field problem in which spatial variability plays a major role. Geostatistical analyses of hydraulic heads in this aquifer have been carried out by ENRESA (1992) for the purpose of optimizing the head monitoring network and by Juanes and Samper (1996) who used residual kriging for estimating the hydraulic head field. The study area covers about 12 km^2 and is bounded by the Guadalquivir river on the North and a low permeability mountain on the south. The actual data set used to test the three procedures is that of Juanes and Samper (1996) which consists of 82 head data measured in April 1991. In their geostatistical analysis they identified a global linear drift. Residuals fitted an exponential semivariogram with a sill of 0.23 m^2 and an effective range of about 1 Km. Using the same data, we have performed universal kriging analyses using four different methods: (1) linear polynomials, (2) scaled linear polynomials, (3) the method of eigenvalues , and (4) linear polynomials with a QR factorization. The upper-left plot in Figure 1 shows a contour map of the residuals (raw data minus spatial drift). This plot was prepared using generalized least squares with first order polynomials.

Figure 1 shows also the plots of the spatial distribution of three eigenvectors (factors) of matrix Γ. These surfaces show some general patterns. In general, the eigenvectors associated to the largest eigenvalues show the largest spatial variation with clear global drifts (see the upper right plot in Figure 1). Eigenvectors corresponding to the smaller eigenvalues tend to show erratic spatial patterns. The interpretation of the spatial distribution of other eigenvectors is not straightforward because they do not reflect any spatial feature of the random function, but rather are determined by the semivariogram and the data locations. Their interpretation deserves additional analyses which have not yet been addressed and will be presented elsewhere.

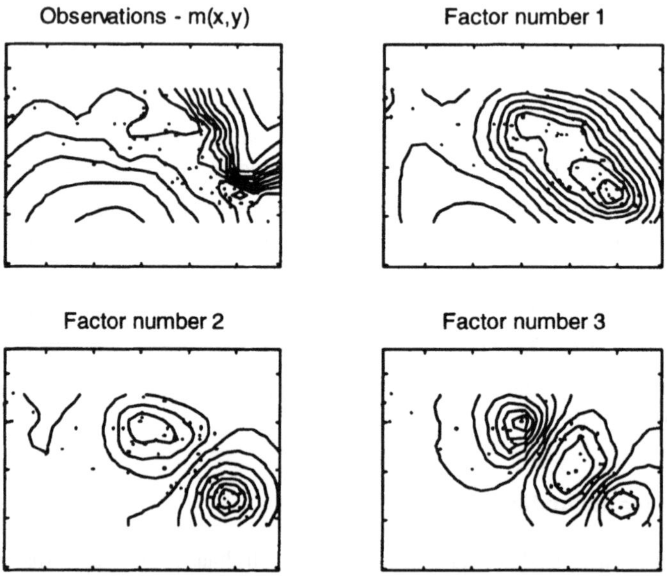

Figure 1. Contour map of Andújar raw hydraulic head data (upper left plot) and contour maps of three eigenvectors (factors) used to represent the spatial drift in the eigenvector method (remaining plots). Dots indicate data locations.

In applying the QR factorization method, we investigated the optimum value of the scale factor "s". Figure 2 shows a plot of the minimum eigenvalue of $\mathbf{M}$ as a function of s/s* (s* is our suggested estimate for the optimum). In this figure the star indicates the true optimum value. One can see that the true optimum is slightly greater than s*. The condition number of $\mathbf{M}$ as a function of s is shown in Figure 3.

Table 1 summarizes the results obtained with the four approaches. The fourth column in this table indicates the average value of the ratio between cond($\mathbf{M}$) and cond(Γ). Unscaled monomials lead to much greater condition numbers (eight orders of magnitude in this case!). The other methods give similar results. The results for scaled monomials are slightly better that those of the QR factorization, which in turn are very close to those obtained with eigenvectors.

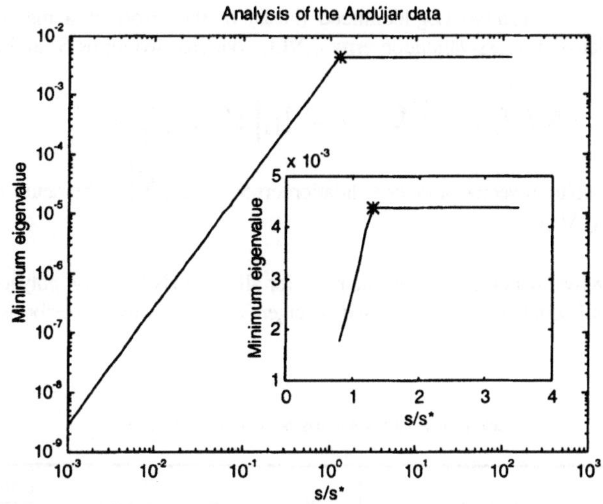

Figure 2 Evolution of the minimum eigenvalue for different choices of the s/s* ratio for the Andújar dataset. s stands for the scale factor used, while s* is the value suggested.The star in the figure indicates the optimum value.

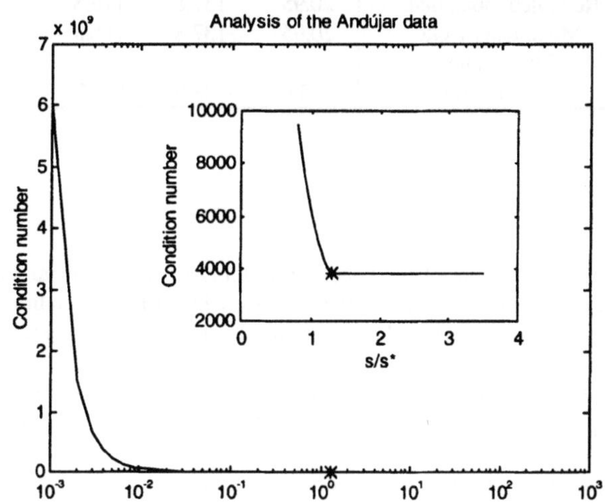

Figure 3 Evolution of the condition number for different choices of the s/s* ratio for the Andújar dataset. s stands for the scale factor used, while s* is the value suggested.The star in the figure indicates the optimum value.

For testing and comparing the estimation capabilities of the different approaches we have resorted to cross-validation. Cross-validation provides the estimation errors e_i and their corresponding standard deviations σ_i at all selected locations. Following Samper and Neuman

(1989), we use the Negative Log-likelihood (NLL) of the errors as a measure of model goodness. For Gaussian cross-validation errors, NLL contains two terms S_1 and S_2 which are defined as:

$$NLL = e^T \mathbf{V}^{-1} e + \ln|\mathbf{V}| = S_1 + S_2$$

where e is the column vector of cross-validation errors, and $|\mathbf{V}|$ is the determinant of their covariance matrix $\mathbf{V}$.

As expected, cross-validation results are identical for all methods based on polynomials. Their NLL values are all equal. The method based on eigenvectors, however, performs much worse than the others.

Table 1 Cross validation results the Andújar head data set.

Method	S_1	S_2	*NLL*	$\frac{cond(\mathbf{M})}{cond(\Gamma)}$
Monomials	20.95	-137.8	-116.8	5.598e+008
Re-scaled monomials	20.95	-137.8	-116.8	1.016
Monomials + QR factorization	20.95	-137.8	-116.8	1.672
Eigenvectors	253.6	-138.9	114.7	1.618

6. Conclusions

We have analyzed the problem of the condition number of the universal kriging method. We have compared the performance of four methods in terms of condition number and estimation capabilities by analyzing the likelihood of cross-validation errors. Our analyses indicate that:

1. very important improvements can be attained by re-scaling the coordinates of the data to the range (-S,S),(-S,S) being S the sill when defined, or the maximum element on matrix Γ in other case.
2. a general theoretical result for the condition number of kriging matrix has been obtained by using the eigenvectors of the variogram matrix. This approach performed worse than the others, but there is room still for some improvements. The shape of the fields derived from the eigenvectors have to be studied further because they could exhibit some interesting properties yet unexplored.
3. QR orthogonalization of matrix $\mathbf{F}$ has proved to be a nearly optimum method. While the orthogonal property of the columns of $\mathbf{F}$ after the QR factorization is independent of the data set, the same may not always be the case for the re-scaling method.

7. References

Atkinson, K. 1989 An introduction to numerical analysis, John Wiley and Sons 692 pp. New Jersey, 573 p.

Dahlquist, G. and A. Bjorck 1974 Numerical Methods. Prentice-Hall, Inc. Englewood Cliffs, , n.5, p. 721-739

Diamond P. and Armstrong, M. 1984 Robustness of Variograms and Conditioning of Kriging Matrices. Math. Geol., (16): 809-822.

ENRESA, 1992 Estudio Geoestadístico de la Piezometría del Acuífero de la Fábrica de Uranio de Andújar (Internal Report), 93 pp.

Juanes, R. and J. Samper 1996 Estudio geoestadístico de los niveles piezométricos en el acuífero de la F.U.A. (Internal Report)

O'Dowd, R.J., 1991 Conditioning of coefficient matrices of ordinary kriging, Math. Geol., 23(5):721-739.

Posa, D., 1989 Conditioning of the stationary kriging matrices for some well-known covariance models. Math. Geol., 21(7):755-765

Samper, J. and S. P. Neuman, 1989 Estimation of Spatial Covariance Structures by Adjoint State Maximum Likelihood Cross Validation. 1. Theory. Water Resour. Res. 25(3):351-362

Samper, J. and J. Carrera, 1990 Geostadística: aplicaciones a la Hidrología Subterránea. Ed. CIMNE. Barcelona. Spain, 484 pp.

Samper, J. and J. Carrera, 1995 Numerical Modeling of Remedial Actions for an Uranium Mill tailings in Spain: from model construction to prediction uncertainty. In: Groundwater Quality: Remediation and Protection. Praga, IAHS Pub. N. 225, pp. 299-310.

Appendix 1: upper bound for cond (M)

Let $\underline{x}$ be an eigenvector of Γ and λ its corresponding eigenvalue. Let $\underline{z}$ be a vector defined as:

$$\underline{z} = [\underline{x} \quad 0 \quad \cdots \quad \alpha \quad 0 \quad \cdots]^T.$$

were α is a scalar to be determined. We will find the appropriate value of α by imposing that $\underline{z}$ to be an eigenvector of $\mathbf{M}$. Notice that $|\underline{z}|^2 = \underline{z}^T \underline{z} = \underline{x}^T \underline{x} + \alpha^2 = 1 + \alpha^2$. If we denote by N the size of Γ and N+P the size of $\mathbf{M}$, one has at most P different choices for $\underline{x}$ to be proportional to a column of $\mathbf{F}$, and N-P choices for $\underline{x}$ to be orthogonal to a column of $\mathbf{F}$.

In the first case, $\mathbf{F}^T \underline{x} = [0 \quad \cdots \quad \lambda \quad 0 \quad \cdots]^T$, therefore for $\underline{z}$ to be an eigenvector of $\mathbf{M}$ it should verify

$$\mathbf{M}\underline{z} = [(\lambda + \alpha\lambda)\underline{x} \quad 0 \quad \cdots \quad \lambda \quad 0 \quad \cdots]^T = \lambda^* \underline{z} = \lambda^* [\underline{x} \quad 0 \quad \cdots \quad \alpha \quad 0 \quad \cdots]^T$$

which holds if

$$\begin{cases} \lambda(1+\alpha) = \lambda^* \\ \lambda = \lambda^* \alpha \end{cases} \Leftrightarrow \begin{cases} 1+\alpha = \lambda^* / \lambda \\ 1/\alpha = \lambda^* / \lambda \end{cases} \Leftrightarrow 1 + \alpha = 1/\alpha \Leftrightarrow \alpha_{1,2} = \frac{-1 \pm \sqrt{5}}{2}$$

Therefore for each λ there are two valid eigenvalues λ_1^* and λ_2^*. This renders in turn 2P valid eigenvectors $\underline{z}_1$ and $\underline{z}_2$, which are mutually orthogonal since $\underline{z}_1^T \underline{z}_2 = \underline{x}^T \underline{x} + \alpha_1 \alpha_2 = 1 - 1 = 0$.

For the second case, $\mathbf{F}^T \underline{x} = 0$ therefore $\mathbf{M}\underline{z} = \lambda \underline{z}$ and α=0. This implies that for such cases, $\lambda = \lambda^*$ and the eigenvalues are the same. There are N-P new eigenvectors which are also orthogonal to those already found.

Let λ_{max} and λ_{min} be the largest and the smallest eigenvalues of Γ, respectively. Depending on whether the corresponding eigenvectors are or are not columns of $\mathbf{F}$, the largest eigenvalue of $\mathbf{M}$ will be bounded by $\lambda_{max}/\alpha_1 \left(\alpha_1 < 1\right)$ while the smallest will be bounded by λ_{min}/α_2. Therefore the condition number of $\mathbf{M}$ is also bounded by

$$cond(\mathbf{M}) \leq \frac{\lambda_{\max}/\alpha_1}{\lambda_{\min}/\alpha_2} = \frac{\alpha_2}{\alpha_1} cond(\Gamma) = \frac{\alpha_2^2}{\alpha_1\alpha_2} cond(\Gamma) = \frac{\alpha_2^2}{-1} cond(\Gamma) =$$

$$= \left(\frac{-1-\sqrt{5}}{2}\right)^2 cond(\Gamma) \approx 2.6 cond(\Gamma)$$

which is valid for any choice of eigenvectors included in $\mathbf{F}$.

Acknowledgments

We want to thank the Uruguay National University, the University of La Coruña and the Galician Foundation for Civil Engineering (La Coruña, Spain) for their support and funding of the travel expenses of the first author to Spain and of the second author to Uruguay.

THE USE OF CONSERVATIVE MASS TRANSPORT ANALYSIS AS A QUICK LOOK TOOL TO RANK GEOSTATISTICAL REALIZATIONS OF HETEROGENEOUS STRUCTURES

M. VERT [(1)], P. BIVER [(2)], T. PTAK [(3)]
[(1)] *CSNS Elf, Institute of Geology, University of Tübingen, Germany*
[(2)] *Elf Aquitaine Production, Pau, France*
[(3)] *Institute of Geology, University of Tübingen, Germany*

Abstract :

A numerical transport simulator, combining a steady state flow model and a conservative particle tracking scheme, is used as a quick look tool for ranking stochastic images. Two conservative mass transport algorithms are considered, one based on a semi analytical technique (Pollock, 1988) used in the standard MODPATH (Pollock, 1994) software and the other one based on a node to node tracking algorithm (Desbarats, 1990). The tool is tested in a 3D case, based on a channelized oil reservoir. Results show that, due to artificial dispersion of the node to node algorithm, the rank given by the two particle tracking schemes are different and seems to be more definite for the Pollock scheme. But, when both are compared to the results given by a multiphase flow model, the node to node looks more effective to rank stochastic realizations than the Pollock method.

1. Introduction

One of the goals of geostatistics is to represent explicitly reservoir heterogeneities which may influence drastically the dynamic behaviour in terms of miscible and immiscible transport. In field case studies, pollutant transport may be affected by complex mechanisms, such as non linear absorption, chemical reactions and multiphase flow. Hence, to represent such physics, complex and time consuming numerical models are necessary. In case of a stochastic analysis, where an ensemble of geostatistical realizations of the heterogeneous structure, usually built with a large number of blocks, has to be considered, such complex models cannot be applied to all images.

In such a case, as suggested by Journel (1989) and Teutsch (1990), one could use simplified transport models to identify the geostatistical realizations with e.g. the shortest, the mean and the longest transport time to reach a certain concentration level at a given point within the simulation domain. Then, the complex model would be applied

A. Soares et al. (eds.), geoENV I–Geostatistics for Environmental Applications, 77–87.

only to the identified realizations to obtain an 'envelope' of the possible results, thus saving a lot of computing time.

In order to rank the geostatistical realizations, we therefore need to find a faster and simpler transport modeling tool. As far as ranking is concerned, we compare here the effectiveness of two conservative transport models applied in a steady state flow. These models are based either on the particle tracking technique used in MODPATH (Pollock, 1994) or on the node to node routine one presented in Desbarats (1988). As a preliminary work, these particle tracking schemes are compared in a 2D synthetic case, using spatial moments and particle distribution maps. Then, both methods are used to rank 3D geostatistical images, based on a real channelized oil reservoir.

As the ranking tool is not built to investigate the physics of solute transport, we will not get into details of transport processes. We are mainly interested in the ranking abilities of each particle tracking algorithm.

2. The conservative mass transport simulator

The simulator is composed of two main parts : The first one calculates the pressure head distribution within the stochastic image in a single phase steady state flow, the second one uses the head distribution in order to simulate conservative solute transport. This simulator is based on MODFLOW (McDonald and Harbaugh, 1988) and MODPATH as well as on the node to node algorithm (Desbarats, 1990) and is adapted to deal with the unit system used in hydrogeology and oil industry.

2.1. THE TWO CONSERVATIVE TRANSPORT SIMULATION METHODS

2.1.1. *Pollock algorithm*

Used in MODPATH, this semi analytical technique by Pollock (1988) assumes a linear velocity interpolation within a cell and computes then a pathline followed by the particle (Fig. 1a) and the corresponding time. Purely advective, this enhances that particles starting at the same point follow the identical pathline and have then the same travel time within the simulation domain.

2.1.2. *Node to Node tracking routine (hereinafter NNR)*

This particle tracking scheme (Desbarats, 1990) is based on routing particles from a cell to an adjacent cell, which is selected randomly. The probability of the particle going to this cell is the ratio of the water flow in that direction over the outflow budget of the cell (Fig. 1b) and the travel time is calculated with a velocity corresponding to the total outflow of the cell. This method assumes that a total mixing occurs within the cell and particles starting at the same point will not automatically follow the same path.

This method has been commented by Goode and Shapiro (1991) and their remarks were strongly focused on the numerical dispersion caused by the randomness of the technique, which is commented by Desbarats (1991) as the uncertainty caused by the ignorance of the particle positions within a cell.

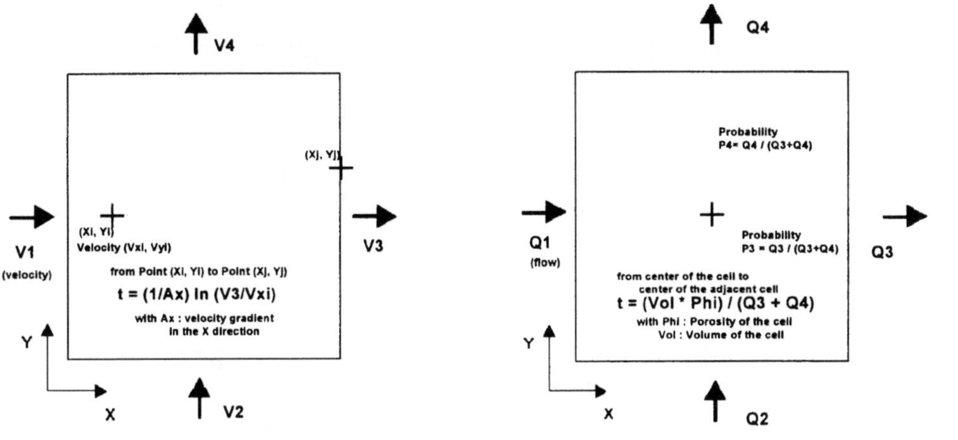

Figure 1. Schematic figure showing the path followed by a particle when using Pollock algorithm (a) and the Node to Node algorithm (b)

2.1.3. *Injecting and stopping particles.*

If we would like to compare the results given by both techniques, we have to look at the injection and stopping conditions for particles, since they are different.

The criterion to stop particles has the same definition, which is when the particle arrives in a predefined 'arrival cell' or when the travel time exceeds a limit defined by the user, however the particles do not stop at the same point when looking at both algorithms. In the Pollock algorithm, they finish their travel at the border of the cell, while in the node to node tracking scheme, they go to the center. The travel times are, according to that, different but, as it is part of the algorithms, we decided not to add any correction factor.

As for injection, in the node to node tracking scheme the particle starts at the center of the cell. In the Pollock scheme, the coordinates within the starting cell are also requested. Therefore, as a complete mixing is assumed in the cell, in the Pollock algorithm, we select random starting points within the cell.

2.2 COMPARISON IN A TWO-DIMENSIONAL SYNTHETIC CASE

Before applying both particle tracking methods in a more complex 3D model, a two dimensional synthetic case was investigated in order to point out the main differences, with respect to transport.

First, tests were performed in a homogeneous situation with a uniform flow and a continuous injection of particles. The direction of the flow had an angle of 45 degrees with the mesh. The test parameters are given in Table 1. The particles were injected in a single cell upstream (upper left zone of the image) and were asked to stop when encountering downstream a control plane perpendicular to the direction of flow (lower right zone). Figure 2 shows the evolution of the second order spatial moment (e.g. Kreft and Zuber (1978)) for both particle tracking methods.

TABLE 1. Parameters used for the homegeneous test

Parameter	Value
Hydraulic conductivity	0.001 m/s
Porosity	0.2
Model size (number of cells)	50 * 50
Size of the cells	10*10 (meters)
Pressure head gradient	0.1 m_H_2O / m

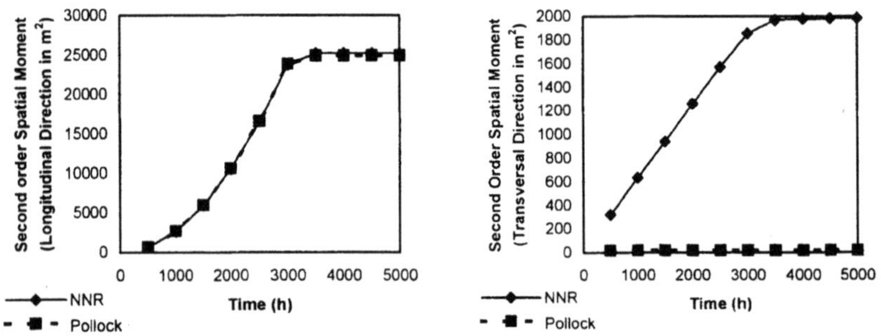

Figure 2. Second order spatial moment curves for the transversal and longitudinal direction of the flow after a continuous injection of particles

As expected, in the Pollock algorithm, the particles do not diffuse in the domain. In contrary, using the NNR where in this case the induced dispersion is significant, particles spread in the domain. This difference can be evaluated by calculating the second order transversal and longitudinal spatial moments (longitudinal direction is the direction of flow). The longitudinal second order moment curves are the same but, for the transversal second order spatial moment, the difference is more pronounced, as it is nearly zero for the Pollock routine and about three orders of magnitudes higher for the NNR scheme.

Moreover, as it is well described by Goode and Shapiro (1991), this artificial diffusion is directly induced by the grid orientation according to the flow direction, being equal to zero when the flow direction is parallel to one of the mesh axes.

In Desbarats (1990), it is advised to use the NNR scheme in a highly heterogeneous medium. We then tested both particle tracking schemes in a 2D stochastic image described in Table 2. For the steady state flow, we imposed the same boundary head conditions as in the homogeneous case and the same injection and arrival (control plane) cells. The results are shown in Figure 3.

TABLE 2. Parameters used for the heterogeneous test

Parameter	Value
Mean hydraulic conductivity (K)	0.001 m/s
Mean porosity (Φ)	0.2
Φ-K law	Φ = 0.1*Log(K) + 0.5
Model size (number of cells)	50 * 50
Size of the cells	10*10 (meters)
Pressure head gradient (boundary conditions)	0.1 m_H_2O / m

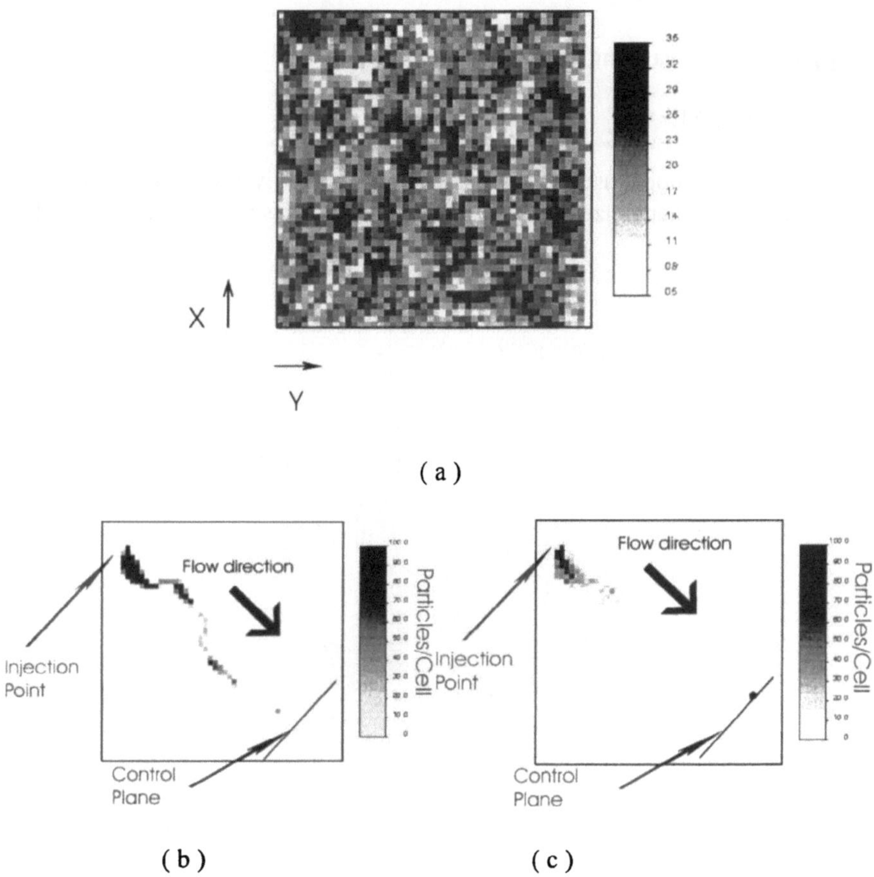

Figure 3. Particle distribution maps at 2500 hours after first particle injection in a heterogeneous 2D model (porosity map is given in (a)) when using Pollock algorithm (b) and node to node particle tracking scheme (c)

In the Pollock scheme, the particles follow a 'streamtube' and do not leave it whereas in the NNR scheme, they diffuse really quickly, though they tend to take the same streamtube near the injection point. This difference can be seen in the control plane where all particles using the Pollock scheme arrive at the same area while using the NNR algorithm they arrive at various points of the control plane.

The artificial dispersion which characterizes the NNR method is the main difference between the two particle tracking scheme and changes drastically the paths followed by the particles. It occurs also in heterogeneous images which may cause a big difference in ranking stochastic images, using the particle tracking methods. The best way to show this is to apply both particle tracking algorithms in a case based on a real study.

3. Application for ranking realizations

3.1. THE 3D CASE

The geostatistical model used to test the ranking quality given by both tracking schemes is based on a real channelized oil reservoir. The model was conditioned by data from ten wells.

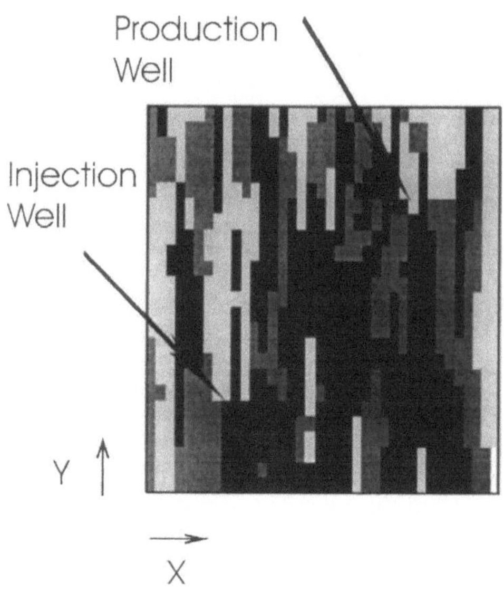

Figure 4. Layer 5 of the 3D facies model with channels (light gray), levees (medium gray) and flood plains (dark)

TABLE 3. Parameters used for the 3D application

Parameter	Value Facies 1 (channel)	Value Facies 2 (levees)	Value Facies 3 (Flood Plains)	Number of Realizations
Mean permeability	1 Darcy	10 mD	1 mD	
Mean porosity	0.2	0.15	0.1	
A priori proportions Set 1	25 %	20 %	55 %	10
A priori proportions Set 2	20 %	19 %	61 %	3
A priori proportions Set 3	30 %	21 %	45 %	3

First, an indicator model was built using 3 facies, one for the channels, one for the levees and one for the flood plains (see Fig. 4). Three different proportions of facies were studied. Then, according to them, a multigaussian simulation was performed yielding a porosity distribution. Subsequently, using relationships of porosity to permeability, a permeability distribution was obtained. A total number of 16 realizations were generated, divided into three sets of respectively 10, 3 and 3 images, corresponding to the three different proportions studied. All parameters are given in Table 3.

The geostatistical model was built in order to be run by the multiphase oil simulator ECLIPSE (Intera, 1993). Big sized cells (60*100*1 meters) and a low number of cells (38*25*16) were considered.

3.2. RESULTS OF THE PARTICLE TRACKING SIMULATIONS

To test the ranking quality of the two particle tracking algorithms, we used the 16 different stochastic realizations of the geostatistical model, where we placed an injection and pumping well at two opposite quarters of the domain (Fig. 4). Considering a bounded and confined reservoir and full of water, we simulated a single phase steady state flow by injecting and pumping at the same rate (500 m^3/day). The wells were fully penetrating, and the input and output flow budget of the well cells were proportional to cell transmissivities (hydraulic conductivities*height).

A continuous input of particles was applied at the injection well, proportional to the input budget of each cell, and the times when the particles arrive at the pumping well were calculated for both particle tracking schemes. The computing time for simulating the travel of 100 000 particles was about 3-4 minutes per image and per tracking scheme.

We finally obtained two sets of breakthrough curves which are shown in Figure 5. First it can be recognized, that at a given time after the first particle injection, the number of particles arrived at the pumping well when using each algorithm is quite different. The ratio of Pollock's particles arrived at the pumping well above NNR's ones is about 0.4 and depends on the image. As the first arrival times are equivalent, this seems to mean that many more particles take the fast ways in the NNR scheme than in the Pollock scheme.

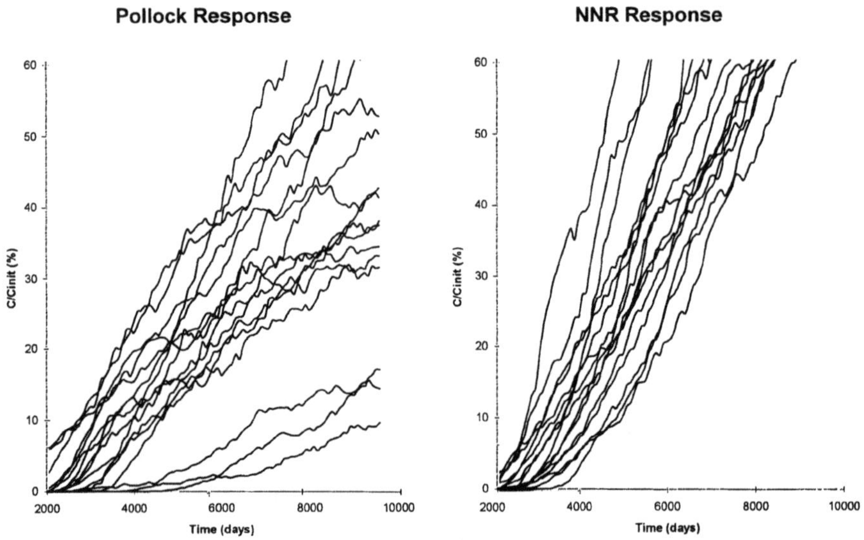

Figure 5. Breakthrough curves of the particle transport simulation when using the two studied particle tracking schemes

Secondly, when we look at the breakthrough curves, the NNR curves have about the same shape. They are smoother and more close together. This is not the case for the Pollock scheme where the curves look more different.

The differences between the two sets of breakthrough curves for the two tracking schemes show that the ranking of the geostatistical images depends strongly on the tracking method. Table 4 shows the ranking given by both methods by considering a cutoff time which is related to the injection concentration Co : At what time do we see at the pumping well a concentration of particles equal to one twentieth of Co? For the two schemes we see some images ranked in the first quarter for one method and ranked in the last quarter for the other method. This means that the ranking from the two methods is different.

Ranking is more difficult when based on the NNR scheme, since the NNR breakthrough curves are closer together compared to the Pollock method. The range of the cutoff times is indeed much smaller when using the NNR scheme (Tab. 4).

On the other hand, using the Pollock algorithm, the curves differ much more, which makes ranking easier.

The best way to check which method is more effective in ranking images is to simulate multiphase flow in the different images, see the ranking by using water production curves (water pumped at the oil pumping well) and compare this ranking with the particle tracking results.

TABLE 4. Comparison of the cutoff time and the rank given by the two studied particle tracking schemes (the extreme cutoff times are given in bold)

Set of Realizations	Realization Number	Rank given by Pollock algorithm		Rank given by NNR algorithm		Rank Discrepancy
		Cut-off time (days)	rank	Cut-off time (days)	rank	
1	1	3650	8	3100	2	6
	2	4600	12	4050	11	1
	3	3500	7	3150	3	4
	4	4850	13	4550	15	2
	5	**2250**	1	3250	4	3
	6	3150	4	4350	13	9
	7	3200	5	3350	6	1
	8	4500	11	**4650**	16	5
	9	3000	3	3750	7	4
	10	3400	6	4000	10	4
2	1	6500	14	3800	8	6
	2	4150	10	3900	9	1
	3	2650	2	**3000**	1	1
3	1	**8850**	16	4400	14	2
	2	4100	9	4150	12	3
	3	7650	15	3300	5	10

3.3. COMPARISON WITH MULTIPHASE SIMULATION RESULTS

In this case, we considered the reservoir bounded, confined and full of oil. In each image studied a transient multiphase flow simulation was performed by injecting water and pumping oil. The injection rate was 500 m^3/day and the production rate (gas, water and oil) was calculated in order to have a stable reservoir pressure. The water production curves at the well are given in Figure 6. The simulations took 3-4 hours per image, that is two orders of magnitude higher than the time needed by a particle tracking scheme.

This set of curves was the reference to study the ranking ability of the two tracking schemes. We took as cutoff time for ranking the time when 5 % of water is produced and evaluated the effectiveness of both particle tracking methods by calculating for each image the difference between the rank given by the particle tracking algorithm (Pollock or NNR) and the rank given by the ECLIPSE simulator (Tab. 5).

The mean discrepancy is smaller for the NNR algorithm. In order to avoid using the ECLIPSE simulator to study a whole set of geostatistical images, which is the aim of the tool, the NNR particle tracking scheme seems to be more effective to rank images than the Pollock scheme, in order to obtain an 'envelope' of the distribution of some predicted values, e.g. some predefined proportions of water produced at an oil production well. But neither the Pollock scheme nor the NNR algorithm state that the three images with low proportion of channels (set 2) come first in the ranking, as given by the Eclipse simulator.

Though there is some discrepancy in ranking, it has to be considered that there is an extreme saving of computing time when the ranking of stochastic images based on particle tracking is performed prior to the Eclipse simulations.

As well, we have to take into account that the characteristics of the studied 3D model may hinder the ranking of stochastic images by particle tracking. Indeed, the big size of the cells, the high permeability contrast due to prior facies modeling and the low number of stochastic images may certainly have an influence on the ranking results of transport simulations.

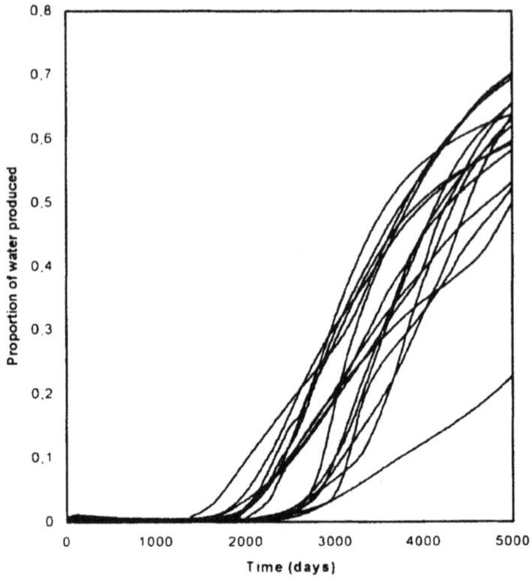

Figure 6. Set of water cut curves given by the ECLIPSE simulator (water cut : Ratio of water production over water and oil production)

TABLE 5. Rank given by ECLISPE and discrepancy with the rank given by the studied particle tracking algorithms (Mean discrepancy per set and per method is also given)

Set of Realizations	Realization Number	Rank given by Eclipse		Rank discrepancy with			
		Cut-off time (days)	rank	Pollock	Mean	NNR	Mean
1	1	2210	7	1		5	
	2	2230	8	4		3	
	3	3170	16	9		13	
	4	2850	14	1		1	
	5	2160	4	3		0	
	6	2770	11	7		2	
	7	2330	9	4		3	
	8	3030	15	4		1	
	9	2740	10	7	*4,6*	3	*3,3*
	10	2820	12	6		2	
2	1	2150	3	11		5	
	2	2020	2	8	*6*	7	*4*
	3	1770	1	1		0	
3	1	2840	13	3		1	
	2	2190	6	3	*5,3*	6	*2,3*
	3	2180	5	10		0	

4. Conclusion

Conservative mass transport simulations were used to rank geostatistical realizations in order to be compared with the ranks given by a multiphase flow simulator. Two particle tracking methods, one based on Pollock (1988) and the other based on a node to node routine algorithm (Desbarats, 1990) were studied and compared by testing them in a 3D model, based on a real channelized reservoir.

First, due to artificial dispersion of the node to node ranking algorithm, the ranks given by the two particle tracking schemes are different. The breakthrough curves using the node to node method are more similar and the ranking is more difficult compared to the Pollock scheme. Though, when both are compared to the results given by a multiphase simulation, the node to node ranking scheme looks more effective to rank stochastic realizations than the Pollock method.

Moreover, as a large number of realizations has to be considered to predict low probability events (Teutsch, 1990), computing more stochastic images for the particle tracking schemes could improve the ranking efficiency.

References.

Desbarats, A. J. (1990) Macrodispersion in sand-shale sequences. *Water Resources Research*, **26 (1)** 153-163.

Desbarats, A. J. (1991) Reply, *Water Resources Research*, **27 (1)**, 141-143.

Goode, D. J., Shapiro, A. M. (1991), Comment on „Macrodispersion in sand-shale sequences" by A.J.Desbarats, *Water Resources Research*, **27 (1)**. 135-139.

INTERA (1993) *ECLIPSE 100 Reference Manual*, Intera Information Technologies Limited, Henley-on-Thames

Journel, A. (1989) Fundamentals of Geostatistics in Five Lessons, *Short Course in Geology*, **8**, American Geophysical Union, Washington DC, 40 p.

Kreft, A., Zuber, A. (1978) On the physical meaning of the dispersion equation and its solution for different initial and boundary conditions, *Chem. Eng. Sci.*, **33**, 1471-1480.

McDonald, M. G., Harbaugh, A. W. (1988) *A modular three-dimensional finite-difference ground-waterflow model*, U.S. Geological Survey, Reston.

Pollock, D. W. (1988) Semianalytical computation of path lines for finite difference models, *Ground Water*, **26 (6)**, 743-750.

Pollock, D. W. (1994) *User's Guide for MODPATH/MODPATH-PLOT, Version3*, U.S. Geological Survey, Reston.

Teutsch, G. (1990) Stochastic groundwater transport simulation using replacement transfer functions (RTFs), Proc. Int. Work. Geostatistical Methods: Recent Developments and Applications in Surface and Subsurface Hydrology, Karlsruhe, 32-39

SIMPLIFIED RENORMALIZATION: A NEW QUICK UPSCALING TECHNIQUE

P. RENARD[1], G. LE LOC'H[2] AND E. LEDOUX[1]
École des Mines de Paris
[1] *Centre d'Informatique Géologique*
[2] *Centre de Géostatistique*
35 Rue Saint Honoré, 77305 Fontainebleau, France
Email: renard@cig.ensmp.fr, leloch@cg.ensmp.fr, ledoux@cig.ensmp.fr

AND

G. DE MARSILY
Université Paris VI, URA 1367 of CNRS,
4 place Jussieu, 75252 Paris Cedex 05, France
Email: gdm@ccr.jussieu.fr

Abstract. A new technique for the quick upscaling of absolute permeabilities is presented. It is shown that this method and the tensorial renormalization are more reliable than the standard renormalization, the Kruel-Romeu formula and the algebraic means, for the two high anisotropy ratios, for the two different block sizes and for the two geostatistical models we have tested. The comparison of the execution time for the different techniques leads us to recommend the use of the new method in the case where the non-diagonal terms of the permeability tensors are not necessary. If the non-diagonal terms are necessary we recommend the use of the tensorial renormalization.

1. Introduction

To simulate flow through heterogeneous porous media, a very important task is the estimation of grid-block permeabilities. One possible approach is to use the geostatistical tools to build a fine geological description of the aquifer. Because the permeability behaviour is non additive, geostatistical studies have to be conducted at the scale of the measurement. That implies that the grid describing the aquifer rapidly contains several billions of nodes. Flow simulators of today cannot handle such large grids and it becomes necessary to scale-up from the fine geological grid to a coarser flow simulator grid. A fairly large review of this problem and the

A. Soares et al. (eds.), geoENV I–Geostatistics for Environmental Applications, 89–100.

different techniques used to its resolution is presented in Renard and de Marsily (1997).

At the local scale the simplest flow equations are Darcy's law and the conservation equation (written here for steady and incompressible flow).

$$\vec{u} = -\overline{\overline{k}} \cdot \mathbf{grad}\,(h) \quad \text{and} \quad \mathrm{div}(\vec{u}) = 0 \tag{1}$$

$\vec{u}$ represents the filtration velocity, $\overline{\overline{k}}$ the hydraulic conductivity tensor (latter called *permeability*) and h the hydraulic head. Several authors (Matheron, 1967; Quintard and Whitaker, 1987; Sáez et al., 1989) using different mathematical formulations have shown that, under the assumptions of statistical homogeneity and uniform flow, a large scale Darcy's law emerges.

$$\vec{U} = -\overline{\overline{K}} \cdot \mathbf{grad}\,(H) \quad \text{and} \quad \mathrm{div}(\vec{U}) = 0 \tag{2}$$

Under the above mentioned assumptions, the upscaled permeability $\overline{\overline{K}}$ is an intrinsic property of the porous medium. It is independent of the boundary conditions and of the shape of the sample. If the above assumptions are not verified then, the upscaled permeability depends both on the boundary conditions used to calculate it and on the criterion used to define it (equality of the fluxes, of the mean velocities or of the energies dissipated by the viscous forces).

In this paper, we present a new technique, based on a real space renormalization procedure and a heuristic formula, to calculate the upscaled permeability $\overline{\overline{K}}$. The renormalization procedure is very different from the standard one (King, 1989), which solves iteratively a flow problem on an elementary group of cells (and then need to fix boundary conditions at the local scale). The new procedure is based on iteratively applying the theoretical Cardwell and Parsons bounds (Cardwell and Parsons, 1945) on the elementary group of cells. We get two values, less distant than the Cardwell and Parsons bounds calculated directly on the global mesh, and we propose a heuristic formula to estimate a unique upscaled permeability from this two values. This technique is compared to seven common techniques including the standard algebraic means, Kruel-Romeu's formula, and the standard and tensorial renormalization. The numerical resolution of the flow equations provides the reference permeability values. To evaluate the reliability of the upscaling techniques, tests are conducted at two different scales and on two types of geostatistical models: the truncated gaussian random functions and the boolean model.

The paper is divided into four sections. Section 1 presents the new simplified renormalization method. Section 2 describes the methodology used to compare the new method with the other upscaling techniques which are listed in section 3. Finally, section 4 summarizes and discuss our results.

2. Simplified renormalization

The new technique is based on a simple renormalization procedure and on a heuristic formula adapted from the Kruel-Romeu's formula (see section 3.2).

The renormalization procedure has been proposed by Le Loc'h (1987). The cells are grouped two by two, and the permeability of the cell pairs is calculated by applying the theoretical results in a one dimensional space: the harmonic mean if the cells are in series, or the arithmetic mean if the cells are in parallel (see figure 1).

The upscaled permeability is calculated by grouping alternatively the cells in parallel and in series (see figure 2). If the algorithm starts with a grouping in parallel, the final value obtained at the end of the iterative procedure is noted c_{max}. This is strictly equivalent to compute the Cardwell and Parsons's maximum bound on an elementary group of 4 (in 2D) or 8 (in 3D) cells, and to iterate this calculation.

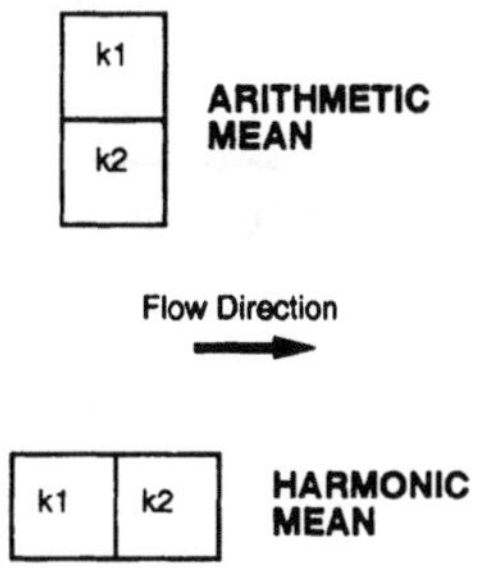

Figure 1. Local formula to calculate the upscaled permeability for a group of two cells.

Let us recall that Cardwell and Parsons (1945) used an electrical analogy and Le Loc'h (1987) used a variational method to demonstrate that the upscaled permeability in a given direction is bounded by (figure 3): the arithmetic mean of the harmonic means of the local permeabilities, calculated on each cells line parallel to the given direction (lower bound K_1); the harmonic mean of the arithmetic means of the local permeabilities calculated over each slice of cells perpendicular to the given direction (upper bound K_2).

c_{max} is always higher than the final value c_{min} obtained when the algorithm starts with a grouping in series. The calculation of c_{min} is equivalent to calculate the Cardwell and Parsons's minimum bounds iteratively on the elementary cells.

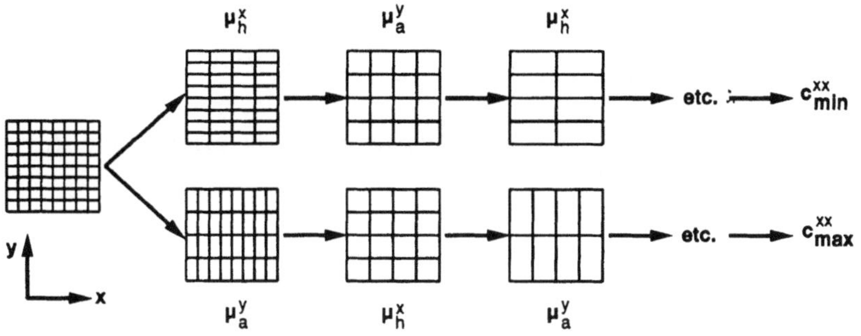

Figure 2. Simplified renormalization procedure.

Our numerical experiences (Galli et al., 1996) have shown that, depending on the anisotropy, the upscaled permeability can be close to c_{min} when the flow is perpendicular to flat cells or close to c_{max} when the flow is parallel to flat cells. Furthermore, these experiences have shown a smaller difference between c_{min} and c_{max} than between the Cardwell and Parsons bounds. Then, the use of c_{min} and c_{max} in a heuristic formula based on those proposed by Kruel-Romeu (1994) (see

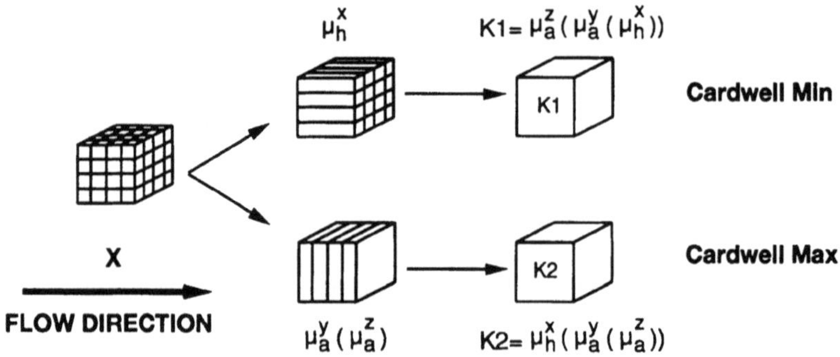

Figure 3. Calculation of the Cardwell and Parsons bounds.

equation 6), instead of the Cardwell and Parsons bounds, was expected to give a better estimation of the upscaled permeability.

Assuming the local permeability tensors and the upscaled permeability tensor diagonals, we propose to estimate the upscaled permeability tensor component along the i direction by the following formula:

$$c^{ii} = \left(c^{ii}_{\max}\right)^{\alpha} \cdot \left(c^{ii}_{\min}\right)^{1-\alpha} \tag{3}$$

with α, expressed as a function of the anisotropy ratios a^i_j and a^i_k.

$$\alpha(a^i_j, a^i_k) = \frac{u(a^i_j) + u(a^i_k) - 2u(a^i_j)u(a^i_k)}{1 - u(a^i_j)u(a^i_k)}, \qquad i \neq j \neq k \in \{x, y, z\}$$

$$a^i_j = \left(\frac{d_i}{d_j}\right)^2, \qquad u(t) = \frac{\arctan\sqrt{t}}{\pi/2}$$

a^i_j is the anisotropy factor due to the flattening of the grid: d_i and d_j represent the size of the cells in direction i and j.

3. Method to compare the upscaling techniques

To test an upscaling technique we have to compare its results with the true values of the upscaled permeability. Then we need 1) a given number of test cases, 2) a reference value for each test case, and 3) some criteria to measure the distance between the references and the results of the tested technique.

3.1. SYNTHETIC PERMEABILITY FIELDS

A geological reservoir is always imperfectly known due to its underground nature, it is then impossible to calculate the upscaled permeabilities for a real permeability field. Furthermore, even if it would be possible on a few cases, it will not have any statistical meaning. Thus, we built synthetic permeability fields as test cases. More precisely two types of model, have been considered: the truncated gaussian random functions (Galli et al., 1994) and the boolean.

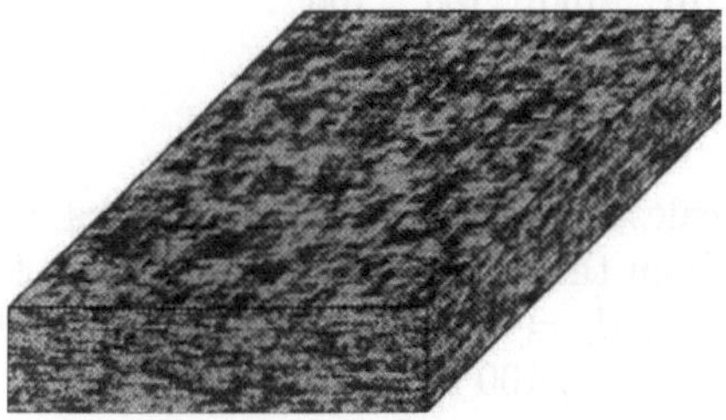

(a) Heresim, short correlation length, constant proportion along the depth

(b) Heresim, long correlation length, varying proportion along the depth

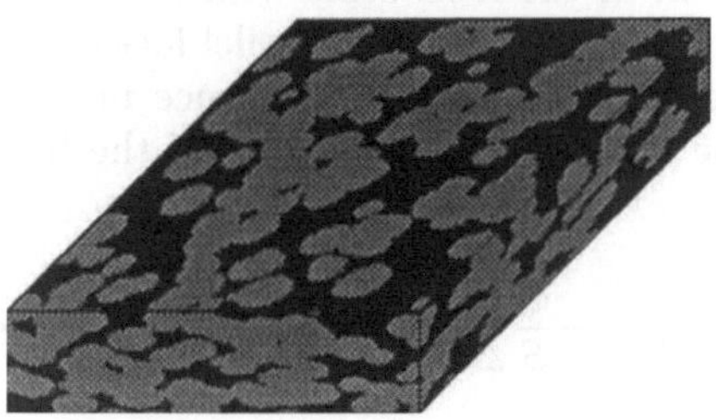

(c) Boolean, ellipsoidal grains

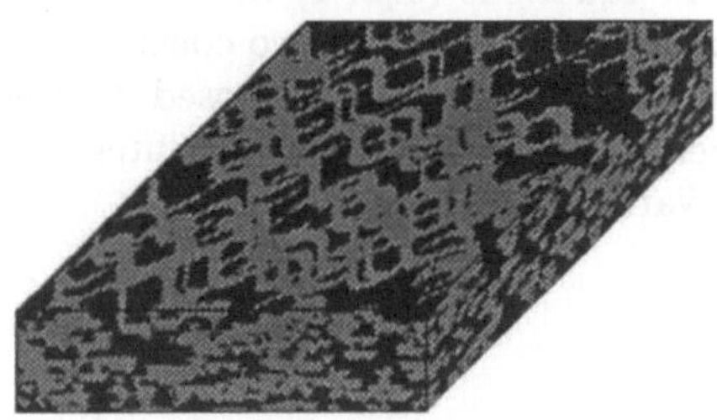

(d) Boolean, sinusoidal grains

Figure 4. Examples of the different types of synthetic media (large blocks)

The truncated gaussian permeability fields have been generated using the *HERESIM* software package[1]. The statistical parameters, for this type of model, are the ranges of the covariance model and the proportion of each lithotype as a function of the depth. Figures 4 (a) and (b) show two realizations generated using two different sets of parameters. All the truncated gaussian permeability fields present four lithotypes, whose permeabilities have been set to 1000, 500, 10 and 0.1.

For the boolean model, the parameters are the inclusion shape, the density of the inclusion and its size statistics. Figures 4 (c) and (d) show two realizations with the two shapes used for the inclusions (ellipsoidal and sinusoidal). Six permeability couples – one value for the matrix, one for the inclusions – have been used for each realization:

Finally, about two thousand permeability fields containing $128\times128\times64 \simeq 1$ million cells have been created with 21 different sets of "realistic" parameters. To

[1]Institut Français du Pétrole/Centre de Géostatistique de l'Ecole des Mines de Paris

TABLE 1. Permeability couples used for the boolean media.

K matrix	100	100	100	0.1	1	10
K inclusion	0.1	1	10	100	100	100

test the upscaling rules at two different scales, about 67 000 small blocks with 16×16×8=2 048 cells have been extracted from the preceding "large blocks". For all these synthetic media, the size of the cells is $d_x = 10$, $d_y = 10$ and $d_z = 1$. The anisotropy ratios a_z^x and a_z^y are, therefore, equal to 100 and $a_x^z = a_y^z = 0.001$.

3.2. REFERENCE UPSCALED PERMEABILITY

We have calculated the reference upscaled permeability by solving numerically the flow equations (eqn. 1) through each medium at the local scale with the no-flow boundary conditions: two constant heads are imposed on two parallel faces of the block and no flow is imposed on the other boundaries. The reference upscaled permeability is obtained by integrating the fluxes at the boundary of the block (Warren and Price, 1961):

$$Q = \int_S \vec{u} \cdot \vec{n} ds, \quad K_{ref} = -\frac{Q}{S}\frac{L}{\Delta H} \tag{4}$$

where L is the distance between the two sides of the block, ΔH is the head difference and S is the cross-section traversed by the total flow Q.

This approach poses the problem of the error due to the numerical technique (finite difference, finite element, etc.). The calculated permeability is not the *true* one (Kruel-Romeu and Nœtinger, 1995) and the estimation of the error is not simple. As expected, the error depends on the numerical technique, the scale of discretization and the type of heterogeneous medium (log-normal, chess-board, etc.). In order to evaluate this effect, two numerical techniques have been used (finite element and finite difference). The classification of the upscaling techniques has been shown to be the same using both finite element and finite difference[2] as reference values (Galli et al., 1996). Then we present only in this paper the comparison with reference values calculated with the multigrid finite element code *TRIMULEC* (Goblet, 1992).

3.3. COMPARISON CRITERIA

Two criteria have been used to classify the quick upscaling techniques: the relative bias e and the linear correlation coefficient ρ. They are calculated on the logarithm

[2]The finite difference simulations were conducted by D. Griffin (University of Newcastle)

of the permeabilities $Y = \log_{10} K$ because the order of magnitude is more important than the value by itself. A technique is accurate when the bias e is around 0 and when the linear correlation coefficient ρ is close to 1.

$$e = \frac{\left|\sum_i \left(Y^i_{ref} - Y^i_{quick}\right)\right|}{\sum_i Y^i_{ref}} \qquad \rho = \frac{\sum_i \left(Y^i_{ref} - \bar{Y}_{ref}\right)\left(Y^i_{quick} - \bar{Y}_{quick}\right)}{\sqrt{\sum_i \left(Y^i_{ref} - \bar{Y}_{ref}\right)^2}\sqrt{\sum_i \left(Y^i_{quick} - \bar{Y}_{quick}\right)^2}} \tag{5}$$

i is an index over all the synthetic permeability fields, $Y^i_{ref} = \log_{10} K^i_{ref}$ is the reference value for the ith synthetic medium and Y^i_{quick} the estimation of the upscaled permeability with a given quick technique. $\bar{Y}$ represents the mean over all the media $\bar{Y} = \frac{1}{n}\sum_{i=1}^{n} Y^i$.

Scatter diagrams of K_{ref} as a function of the different quick upscaling results (in log scaling) are also examined in details (figure 6).

4. Quick-upscaling rules

4.1. ALGEBRAIC MEANS

The algebraic means are the most simple techniques to determine upscaled permeability. They assume the local permeabilities to be isotropic, and they do not take into account their spatial distribution.

Four algebraic means have been tested. The *arithmetic mean* and the *harmonic mean* are respectively the upper bound and the lower bound for upscaled permeability (Cardwell and Parsons, 1945; Matheron, 1967), and the exact value in the case of a stratified medium with flow respectively parallel or perpendicular to the strata. The *geometric mean* is the exact upscaled value in two dimensions, if the normalized local permeabilities $\frac{k}{E(k)}$ and their inverses $\frac{k^{-1}}{E(k^{-1})}$ have the same probability density function (pdf) and if this pdf is invariant by rotation of 90°, (Matheron, 1967). This is true in particular in the case of a chess board and of a log-normal medium. The *power mean with an exponent of* $\frac{1}{3}$ is an approximate value in the case of a three dimensional log-normal isotropic medium (Desbarats, 1992; Nœtinger, 1994).

4.2. KRUEL-ROMEU'S FORMULA

Kruel-Romeu (Duquerroix et al., 1993; Kruel-Romeu, 1994) proposed a heuristic formula to compose the Cardwell and Parsons bounds with an exponent function of the anisotropy ratio.

$$k_r^{xx} = K_1^{(\theta_{y2}\theta_{z3}+\theta_{z2}\theta_{y3})} K_2^{(1-\theta_{y3}-\theta_{z3})} K_3^{(1-\theta_{z2})\theta_{y3}} K_4^{(1-\theta_{y2})\theta_{z3}} \tag{6}$$

with K_1 and K_2 are the Cardwell and Parsons bounds, K_3 and K_4 are two intermediate values.

$$K_1 = \mu_h^x(\mu_a^y(\mu_a^z)) = \mu_h^x(\mu_a^z(\mu_a^y)) \qquad K_2 = \mu_a^y(\mu_a^z(\mu_h^x)) = \mu_a^z(\mu_a^y(\mu_h^x))$$

$$K_3 = \mu_a^y(\mu_h^x(\mu_a^z)) \qquad K_4 = \mu_a^z(\mu_h^x(\mu_a^y))$$

The exponents are derived from a small perturbation calculation of the equivalent permeability (Kruel-Romeu and Nœtinger, 1995) in the case of an anisotropy due to flat cells and/or due to a constant anisotropy of directional permeability in the domain.

$$\theta_{y2} = \frac{\arctan\sqrt{a_y}}{\pi/2} \qquad \theta_{z2} = \frac{\arctan\sqrt{a_z}}{\pi/2} \qquad \theta_{y3} = \frac{\theta_{y2}(1-\theta_{z2})}{1-\theta_{y2}\theta_{z2}} \qquad \theta_{z3} = \frac{\theta_{z2}(1-\theta_{y2})}{1-\theta_{y2}\theta_{z2}}$$

$a_y = \frac{k^{yy}}{k^{xx}}\left(\frac{dx}{dy}\right)^2$ and $a_z = \frac{k^{zz}}{k^{xx}}\left(\frac{dx}{dz}\right)^2$ represent the anisotropy factors. The ratios $\frac{k^{yy}}{k^{xx}}$ and $\frac{k^{zz}}{k^{xx}}$ are assumed constant over the wole domain.

4.3. STANDARD RENORMALIZATION

The standard real space renormalization r_s, applied to the upscaling of permeability, is due to King. The general idea is to use a recursive algorithm. The upscaled permeability is calculated by a series of successive aggregations. The basic calculation is to compute the permeability of an elementary cell of eight meshes (in 3D) with a finite difference approximation and with standard no-flow boundary conditions.

4.4. TENSORIAL RENORMALIZATION

The tensorial renormalization r_t is a recent improvement of the renormalization algorithm (Gautier and Nœtinger, 1994). The modifications are : a new finite difference "direct" scheme and periodic boundary conditions allowing the calculation of a complete tensor of permeability.

5. Results - Discussion

5.1. EFFICIENCY OF THE TECHNIQUES

Table 2 shows the mean cpu time for each upscaling technique and the time for full solution using the finite element method.

The standard renormalization is slower than the tensorial renormalization technique because it has to solve a linear system three times and the tensorial renormalization has to solve only one linear system. The simplified renormalization does not involve the solution of a linear system of equations and is therefore faster. Obviously, the algebraic means are the fastest.

The comparison of the execution times for the quick techniques and for the finite element shows a gain between twenty and several hundred.

5.2. RELIABILITY OF THE TECHNIQUES

The four diagrams in figure 5 summarize our results. In these diagrams, each point represents a technique in the accuracy criteria space. The coordinate of a point are the relative bias and the linear correlation coefficient. Hence, the closer the point is to the upper-left corner ($e = 0$, $\rho = 1$), the more accurate is the technique.

TABLE 2. Mean cpu time in seconds on a Sparc 10 workstation (taking into account the input and output)

Name of the method	small blocks (2 048 cells)	large blocks (1 048 576 cells)
Finite Element	461	10 286
Algebraic means	0.04	27
Kruel-Romeu's estimator	0.04	42
Simplified Renormalization	0.07	49
Tensorial Renormalization	0.36	215
Standard Renormalization	1.04	445

In figure 5, a group of methods (the renormalization techniques: r_s, r_t, c) are almost always the closest to the upper-left corner and a second group corresponding to the algebraic means (μ_a, μ_h, μ_g et $\mu_{1/3}$) are more distant to the upper-left corner. Kruel-Romeu's formula k_r falls in between.

The algebraic techniques can present a small bias (for example μ_g for K_z, see figure 5 (b) and (d)). But, in general, they present a large dispersion ($\rho << 1$) of the estimated permeability (figure 6). The explanation is fairly straightforward. Because the algebraic means do not depend on the spatial distribution of the local permeabilities but depend only on the values of the local permeabilities, two media with the same value of the local permeabilities but with different spatial distributions will have the same algebraic means, but not the same upscaled permeabilities. The renormalization techniques takes into account the spatial distribution and then reduces dispersion. Kruel-Romeu's formula is intermediate because it takes into account approximatively the spatial distribution (by using the Cardwell and Parsons bounds).

The harmonic and the arithmetic means are the less reliable techniques. They can have very important bias and poor linear correlation coefficient (figure 5). But, the fundamental inequality $\mu_h < K_{ref} < \mu_a$ is verified whatever the size of the block and the direction of flow (figure 6). We have also checked that the Cardwell and Parsons bounds are always verified.

In conclusion, it appears from these results that the three best techniques are the standard renormalization, the tensorial renormalization and the simplified renormalization.

6. Conclusion

A new upscaling technique, the simplified renormalization, has been developed.

Eight techniques were tested under uniform flow conditions. Among them, the four algebraic means show an important dispersion in almost all the experiments. Then, we cannot recommend the use of such a formula.

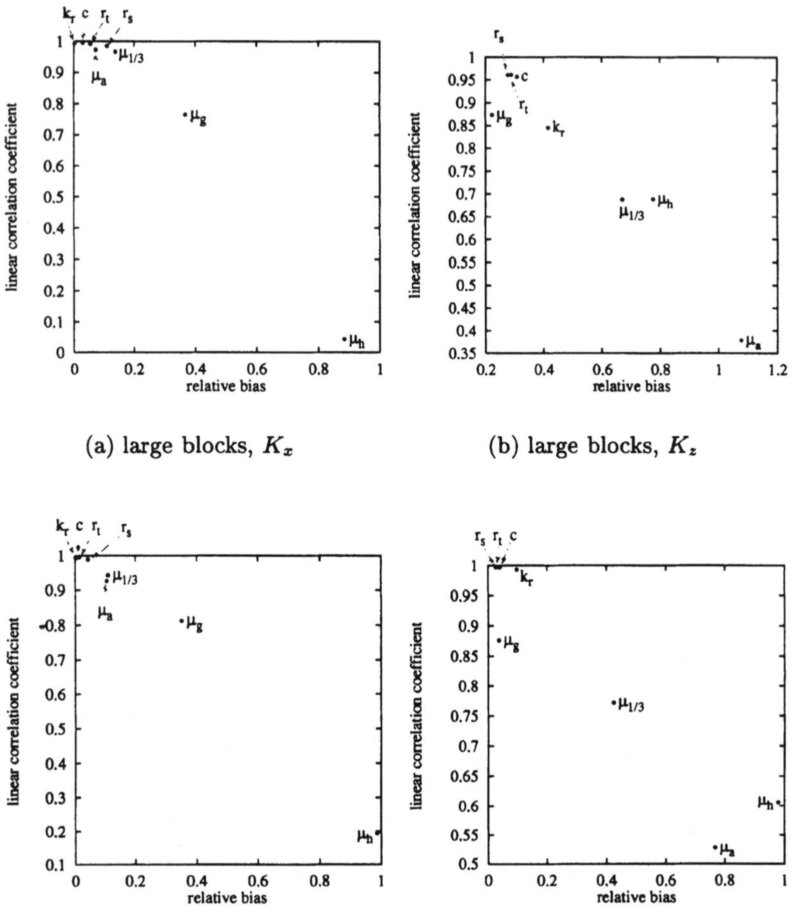

(a) large blocks, K_x (b) large blocks, K_z

(c) small blocks, K_x (d) small blocks, K_z

Figure 5. Reliability of the techniques.

Focusing on the four remaining techniques, we eliminate Kruel-Romeu's formula because it is the worst technique for K_z.

The comparison of the execution time for the three remaining techniques leads us to recommend the use of the new simplified renormalization in the case where the non-diagonal terms of the permeability tensors are not necessary. If the non-diagonal terms are necessary we recommend the use of the tensorial renormalization.

These two techniques are considered as the most accurate and the most reliable techniques we have tested for the two different anisotropy ratios and for the two different block sizes.

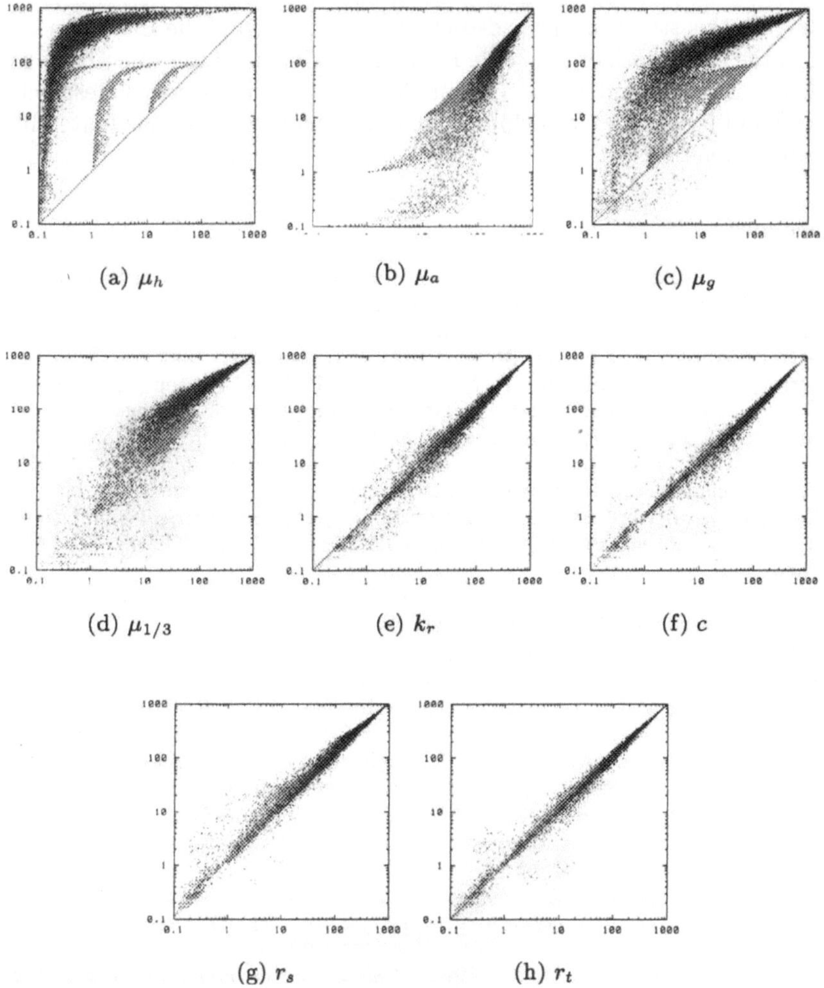

(a) μ_h (b) μ_a (c) μ_g

(d) $\mu_{1/3}$ (e) k_r (f) c

(g) r_s (h) r_t

Figure 6. Scatter diagrams of the **horizontal** reference permeabilities as a function of the quick upscaling methods.

In our point of view, these are the best of the tested techniques, but that does not involve they always produce accurate results. In particular, for the large blocks and for K_z, none technique is very accurate. Furthermore, no test has been conducted yet under non uniform flow conditions (radial for example).

Acknowledgements

Funding for this study was provided by the Commission of the European Communities in the framework of the Joule II Reservoir Engineering Project. The authors thank Rae Mackay and David Griffin (Water Resource Systems Research Unit, University of Newcastle) for helpfull discussions and providing the finite difference flow simulations.

References

Cardwell, W. T. and R. L. Parsons (1945). Average permeabilities of heterogeneous oil sands. *Trans. Amer. Inst. Min. Met. Pet. Eng. 160*, 34–42.

Desbarats, A. J. (1992). Spatial averaging of hydraulic conductivity in three-dimensional heterogeneous porous media. *Math. Geol. 24*(3), 249–267.

Duquerroix, J.-P. L., P. Lemouzy, B. Nœtinger, and R. Kruel-Romeu (1993). Influence of the permeability anisotropy ratio on large-scale properties of heterogeneous reservoirs. *68th Annual Tech. Conf. and Exhb. of the SPE, Houston, SPE 26648*, 29–40.

Galli, A., H. Beucher, G. Le Loc'h, B. Doligez, and Heresim Group (1994). The pros and cons of the truncated Gaussian method. In M. Armstrong and P. Dowd (Eds.), *Geostatistical simulation, Proceedings of the geostatistical simulations workshop, Fontainebleau, France, 27-28 May 1993*, pp. 217–233. Kluwer Academic Publishers.

Galli, A., P. Goblet, D. Griffin, E. Ledoux, G. L. Loc'h, R. Mackay, and P. Renard (1996). Upscaling of flow and transport related parameters. Technical Report LHM/96/RP/11, École des Mines de Paris - CIG, Fontainebleau.

Gautier, Y. and B. Nœtinger (1994). Preferential flow-paths detection for heterogeneous reservoirs using a new renormalization technique. In P. R. Thomassen (Ed.), *4th European Conference on the Mathematics of Oil Recovery*, Røros, Norway.

Goblet, P. (1992). Programme metis 7, Application d'une technique d'éléments finis multigrille à la modélisation d'un milieu fortement hétérogène. Technical Report LHM/RD/92/2, École des Mines de Paris - CIG, Fontainebleau.

King, P. R. (1989). The use of renormalization for calculating effective permeability. *Trans. Porous Media 4*, 37–58.

Kruel-Romeu, R. (1994). *Écoulement en milieu hétérogène: prise de moyenne de perméabilité en régimes permanent et transitoire.* Ph. D. thesis, Université Paris VI.

Kruel-Romeu, R. and B. Nœtinger (1995). Calculation of inter-nodal transmissibilities in finite-differences models of flow in heterogeneous media. *Water Res. Res. 31*(4), 943–959.

Le Loc'h, G. (1987). *Étude de la composition des perméabilités par des méthodes variationnelles.* Ph. D. thesis, Ecole Nationale Supérieure des Mines de Paris.

Matheron, G. (1967). *Eléments pour une théorie des milieux poreux.* Paris: Masson.

Nœtinger, B. (1994). The effective permeability of a heterogeneous porous medium. *Trans. Porous Media 15*, 99–127.

Quintard, M. and S. Whitaker (1987). Ecoulement monophasique en milieu poreux : effets des hétérogénéités locales. *Journal de Mécanique Théorique et Appliquée 6*(5), 691–726.

Renard, P. and G. de Marsily (1997). Calculating equivalent permeability : A review. *Adv. Water Res. To be published.*

Sáez, A. E., C. J. Otero, and I. Rusinek (1989). The effective homogeneous behaviour of heterogeneous porous media. *Trans. Porous Media 4*(3), p. 213–238.

Warren, J. and H. Price (1961). Flow in heterogeneous porous media. *SPE Journal 1*, 153–169.

Impact of Random Function Choice on Groundwater Mass Transport Modelling

H.J.W.M. Hendricks Franssen and J.J. Gómez-Hernández
Departamento de Ingeniería Hidráulica y Medio Ambiente
Universidad Politécnica de Valencia
C/ Camino de Vera s/n
Apdo. Correos 22012
46080 Valencia, Spain

Abstract

Stochastic analysis of 2-D groundwater flow requires adopting a random function (RF) model for transmissivity. The parameters of this random function are decided upon after analysing measured data and other relevant information. Decisions on whether the RF is multiGaussian or not, stationary or not, multimodal or not, must be taken. We analyse the impact that some of these decisions may have in modelling groundwater flow in the Culebra dolomite formation at the WIPP site. The data set, consisting of 36 measurements, displays an apparent spatial trend and an apparent bimodality. Four RF models are selected, all of them MultiGaussian; in two of the models, the normally distributed random variable is log-transmissivity, in the other two, the random variable is the normal-score transform of transmissivity; in two of the models, the RF is stationary, in the other two the RF is non-stationary in the mean, explicitly accounting for the apparent spatial trend. For all four RF models, 200 conditional realizations of transmissivity are generated, and groundwater travel times from the centre of the domain to the southern boundary are computed. In addition, transport was modelled with and without intra-cell dispersion. Regarding RF model choice, the results show little difference between the use of stationary and non-stationary RF models, the main reason being the control that the conditioning data exert on all realizations. On average, the arrival positions at a given control plane are very close; only the variance of arrival positions differs, being smaller for the non-stationary RF than for the stationary one. When normal scores are used, the bimodality observed in the sample data is imposed in all realizations. If not, the sample data are supposed to belong to a normal population with same mean and variance, departure of the sample data histogram from the normal distribution is considered sampling fluctuation. We conclude that data conditioning implicitly accounts for the apparent trend, however, whether to extend the conditioning so as to reproduce exactly the same sample data histogram is a decision that must be carefully evaluated. Regarding intra-cell dispersion, the resulting breakthrough curves are very different depending on whether it is accounted for or not. For this data set, this is the single most influential decision.

A. Soares et al. (eds.), geoENV I–Geostatistics for Environmental Applications, 101–110.

1. Introduction

Groundwater contamination has become an important environmental problem. In order to predict the pollutant concentration in space and time the coupled partial differential equations for groundwater flow and mass transport have to be solved. In addition, the controlling parameters in the equations (e.g. the hydraulic conductivity) vary in a complex manner in space. It is impossible to characterise this spatial variation deterministically. It is common to use stochastic approaches to attack the problem.

Geostatistical methods are used for modelling the spatial variability of hydraulic properties. In this study, only transmissivity (*T*) is considered to be variable in space. Hydraulic conductivity and transmissivity are the most consequential parameters for flow and transport (Smith and Schwartz, 1981a, 1981b). For illustration purposes we have used the WIPP (Waste Isolation Pilot Plant) data set. It consists of transmissivity data from a confined aquifer in a dolomite rock formation in New Mexico (USA) overlying a potential repository for transuranic waste. The data set displays an apparent E-W trend and an apparent bimodality on its histogram. The trend can be accounted for by using a RF model which explicitly incorporates it, such as the RF model used for universal kriging (Journel and Huijbregts, 1978), or else, the trend can be implicitly incorporated in the model through conditioning and a zonal anisotropic variogram. The bimodality of the sample can be dealt with either by adopting a single-mode Gaussian histogram with same mean and variance as the sample data and accept that all other deviations from the multiGaussian model are due to sampling fluctuations, or else, take a normal-score transform of the data, generate the realizations in the Gaussian space and backtransform the resulting realizations assuring almost perfect reproduction of the sample histogram in all realizations.

This study analyses the results obtained after processing the realizations through a complex non-linear transfer function (groundwater flow and mass transport equations) with the aim of evaluating the impact of RF model choice. Other studies have compared the performances of similar geostatistical models for linear estimation, namely the interpolation of an environmental variable. Journel and Rossi (1989) show that interpolation of a variable using both ordinary kriging and universal kriging yields nearly exact the same results. However, they use for both ordinary kriging and universal kriging the same variogram model, while in practice the variogram models will differ for the two cases. On the contrary, in some specific case studies it is found that universal kriging gives better results than ordinary kriging (e.g. Webster and Burgess, 1980).

Four different sets of 200 realizations of log*T* have been generated. The four sets are coded as follows depending on the RF model choice: OK-NSCORE (no trend, variate is normal score of data), OK-UNTRAN (no trend, variate is log*T*), UK-NSCORE (linear trend, variate is normal score of data), UK-UNTRAN (linear trend, variate is log*T*). In addition, the four different sets of log*T* realizations are compared in two mass transport situations (with and without intra-cell dispersion).

2. Theory

Flow and mass transport in a confined aquifer is considered. Flow is assumed to be at steady-state with constant density and to take place in 2-D. Transmissivity is spatially variable. The governing flow equation is:

$$\frac{\partial}{\partial x}\left(T(x,y)\frac{\partial H(x,y)}{\partial x}\right)+\frac{\partial}{\partial y}\left(T(x,y)\frac{\partial H(x,y)}{\partial y}\right)=0 \qquad (1)$$

where (x,y) are the spatial co-ordinates $[L]$, T is transmissivity $[L^2/T]$ and H is vertically-averaged hydraulic head $[L]$. Particles are transported in the aquifer by convection and dispersion. Linear adsorption has not been taken into account since this process will not influence the relative differences between the breakthrough curves in the different sets of simulated fields apart from a multiplicative factor. Mass transport is governed by the following equation:

$$\frac{\partial}{\partial x}\left(D_{xx}\frac{\partial c}{\partial x}+D_{xy}\frac{\partial c}{\partial y}\right)+\frac{\partial}{\partial y}\left(D_{yy}\frac{\partial c}{\partial y}+D_{yx}\frac{\partial c}{\partial x}\right)-\left[\frac{\partial}{\partial x}(v_x c)+\frac{\partial}{\partial y}(v_y c)\right]=\frac{\partial c}{\partial t} \qquad (2)$$

where D_{xx} , D_{yy} , D_{xy} and D_{yx} are the components of the dispersion tensor $[L^2/T]$, c is mass concentration $[M/L^3]$, v_x and v_y are the components of groundwater pore velocity $[L/T]$, and t is time $[T]$.

Transmissivity fields for the four different RF models are generated by sequential simulation. For a detailed description of the generation of multiGaussian fields by sequential simulation the reader is referred to Gómez-Hernández and Journel (1993) and Gómez-Hernández and Cassiraga (1994). The generation of realizations for the RF models with trend differs only slightly from the simulation of multiGaussian fields as described by Gómez-Hernández and Journel (1993). The local conditional probability distribution functions (cpdfs) are assumed to be Gaussian with mean and variance identified to the universal kriging estimate and the universal kriging variance.

Because of the clustered sampling it is not possible to test whether the logtransform of the data follows a Gaussian histogram or not. For this reason, normal-score transformed data and log-transformed data have been used. The differences between the four sets of simulated fields are as follows:

- OK-NSCORE. A variogram of the normal scores is computed. The cpdfs are obtained by solving the ordinary kriging equations, conditional to normal score data. The final realization is backtransformed to the original space.
- OK-UNTRAN. A variogram of the log transformed data is computed. The cpdfs are calculated by solving the ordinary kriging equations, conditional to the logtransformed data. The final realization is exponentiated to produce T realizations.
- UK-NSCORE. A variogram of the normal score residuals is determined (this is the variogram of the normal score transformed log$T$ data, after the linear spatial trend has been filtered out). The cpdfs are calculated by solving the universal kriging

equations, conditional to the normal score transformed data. A normal score backtransform yields the T realization.

- UK-UNTRAN. A variogram of the residuals of the logtransformed data is computed. The cpdfs are calculated by solving the universal kriging equations, conditional to the log$T$ data. Exponentiation results in the T field.

3. Software used

The transmissivity fields are generated with the SGSIM program from GSLIB (Deutsch and Journel, 1992). For the simulation of fields from RF´s with trend a modified version of the algorithm is used. This modified version of the program has been tested and it was found that over a large amount of realizations the ensemble local mean and the ensemble local variance are approximately equal to the UK-estimate and UK-variance, respectively. Four sets of 200 fields have been generated. For each field equation 1 has been solved using standard five-point block-centred finite differences as implemented in the program FLOW (Wen, 1995). Along the four edges of the field, prescribed head boundaries have been used, as given by LaVenue *et al* (1990). Equation 2 has been solved by means of the program TRANSP (Wen and Kung, 1995). This program implements both a particle tracking algorithm for dispersion-free transport and a random walk for transport with dispersion. Particles are released at a certain location in the field at the beginning of the simulation (instantaneous injection) and breakthrough curves at a given control plane are determined. To better account for the heterogeneity of the transmissivity field without the requirement of a too-large time discretization a constant displacement step instead of a constant time step has been used (Wen and Kung, 1995; Wen and Gómez-Hernández, 1996).

4. Results

4.1. THE DATA SET

From the original WIPP data set, as reported by LaVenue *et al* (1990), 36 measurements have been retained. They correspond to the data falling within the simulation domain, with the exception of the datum corresponding to well P-18 that has been removed because the value seems to be unreliable low for dolomite rock (Freeze and Cherry, 1979). The simulation domain corresponds to an area of about 20 km in the x-direction and 30 km in the y-direction, discretized into 43 by 61 square cells of 500 by 500 m. The drillings are not equally spaced over the region, but preferentially located in the centre part of the region overlying the panels that will host the waste if the site is licensed. The experimental histogram of all 36 data (plus the anomalous low datum) is displayed in Figure 1. The histogram displays an apparent bimodality with two populations segregated at a cut-off of -5.5 log(m^2/s).

4.2. VARIOGRAMS

The experimental variogram of log*T* for the *x*-direction shows an exponential rise (indicating a possible trend in this direction). In the *y*-direction, however, a sill value is reached after some distance. This behaviour is noticed both for the log-transformed values and for the normal-score transformed values.

For the two RF models in which the trend is accounted for implicitly through the variogram a spherical model with both geometric and zonal anisotropy is fitted to the experimental variograms. The model variograms for the log-transformed data are shown in Figure 1 (similar models are fitted to the normal-score data). For the RF models in which a linear trend is considered, the variogram of the residuals is taken to be isotropic and equal to the variogram of the data in the N-S direction, since this is the direction in which the trend is not apparent.

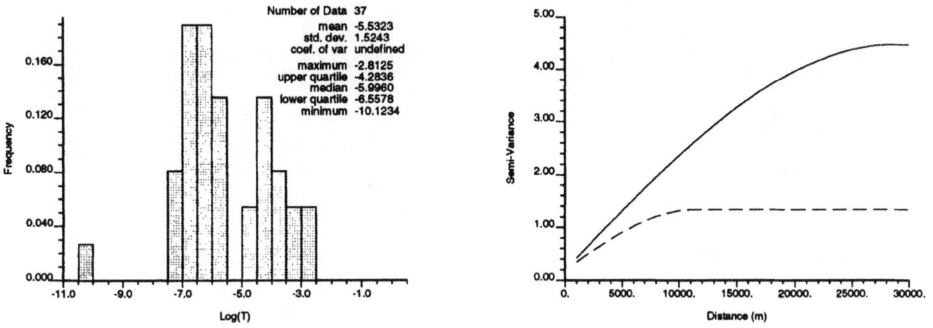

Figure 1. Histogram of log transmissivity data (left). Model variograms for log transmissivity in the North-South (dash) and East-West (solid) directions (right).

4.3. GENERATION OF THE TRANSMISSIVITY FIELDS

Four sets of 200 log*T* realizations each are generated by sequential simulation. Figure 2 shows the first field generated in each method. The impact of the random function model choice is already apparent. The RF models that use the normal-score transform of the data produce realizations with a much more abrupt variation than the models that use the logtransform of the data. Both models are able to reproduce the E-W trend: the NSCORE models with a quasi-step function, the UNTRAN models with a more gradual variation. The differences between the OK and the UK simulations are not so violent. This finding confirms that conditioning will impose the trend in the realizations even if the RF model does not incorporate it explicitly. The main difference between the OK and the UK-simulations is the slightly larger continuity in the N-S direction displayed by

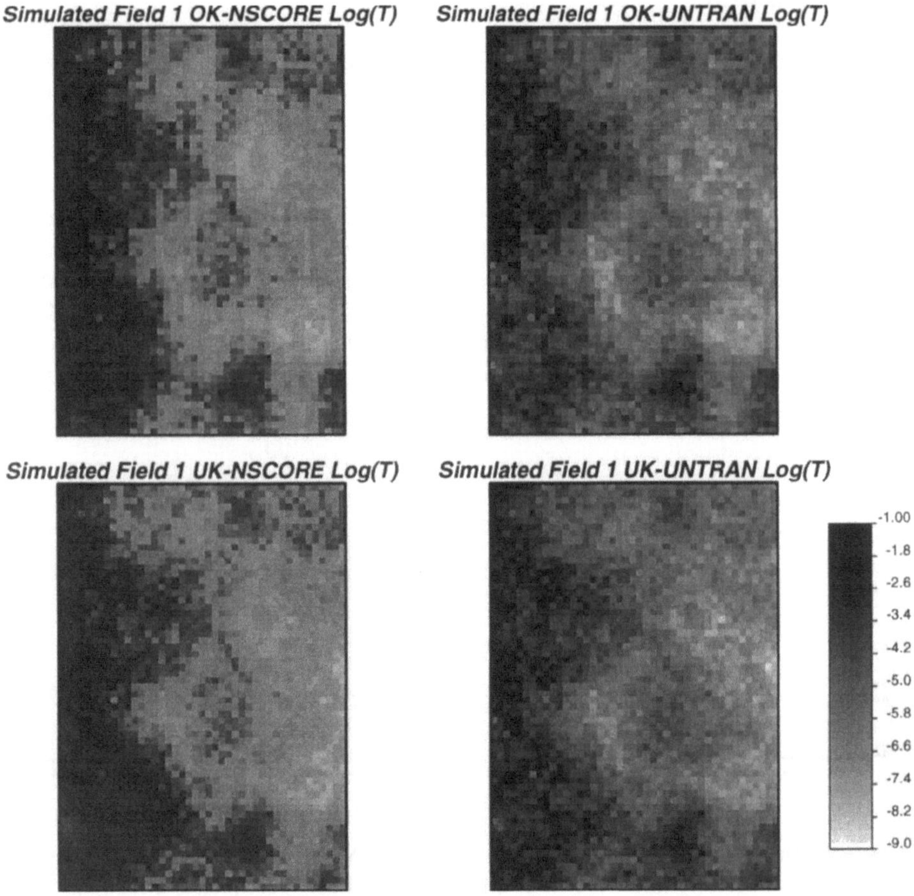

Figure 2. The first simulated log transmissivity field of each set

the OK-simulations. The reason for this enhanced N-S continuity lays on the zonal anisotropy used to model the variogram with different sills in the N-S and E-W directions.

4.4. FLOW AND TRANSPORT SIMULATION.

The four models are compared in two different transport situations. A release of particles in the geometric centre of the field is considered and calculations are made by particle tracking (no dispersion) and by random walk (with dispersion). This resembles the potential release from the nuclear waste repository at WIPP. The four RF models are compared in terms of the results of the transport simulations. Cumulative statistics of the arrival locations and arrival times of the released particles at a selected control plane are computed and compared for the different cases.

The groundwater flow equation has been solved for the 800 (4 x 200) log*T* fields. The prescribed head boundaries impose an average flow oriented on the northnortheast-southsouthwest direction (from the upper boundary to the lower boundary). Flow velocities in the *x*-direction are much lower than flow velocities in the *y*-direction. The highest flow velocities in both the *x*-direction and the *y*-direction are found near the left boundary, which is due to the high log*T* values in this part of the area. A highly permeable zone near the left boundary produces a channelling effect for the particles travelling near it. This high zone tends to be more extended near the lower boundary. Flow velocities are the lowest near the right boundary, especially in the right-lower corner.

For the mass transport simulations, 2000 particles are released in each of the 800 log*T* fields. The release is from the geometric centre of the field, and the cumulative breakthrough at the lower boundary is obtained. Two cases are analysed, one considering advective transport only, and another considering both advective and dispersive transport. For the latter case, the longitudinal and transversal dispersivities are set to 10 m and 1 m respectively (cell size is 500 by 500 m).

Table 1 summarises some basic statistics about the arrival locations of the particles to the control plane for the four RF models. In general, it can be seen that considering dispersion has little influence on the arrival locations of the particles. The value for the transversal dispersivity considered is too small to make a difference. For the UK-sets and the OK-sets the results show that accounting for the trend explicitly in the model (UK-sets) shifts the arrival locations to the right by an amount which is only of the order of a discretization cell. The differences in average arrival location of the particles from the -NSCORE and -UNTRAN sets are also on the order of the width of the discretization cell (~500 m) although slightly larger than the differences between the OK- and UK- sets.

More important differences can be noticed while comparing the spread of the arrived particles on the lower left boundary. This spread is an indication about the uncertainty in the transport simulations. The standard deviation of the particle arrival locations for the models explicitly accounting for the trend is lower than for the models without the trend. This effect is bigger for the models using the log-transform than the models using the normal-score transform. Furthermore, for the model without the trend, it seems to make little difference which transform is used.

TABLE 1. Statistics of particle arrival locations to the lower boundary, given in m from the lower left corner.

(in meters)	mean	std.dev.	max.	upper quartile	median	lower quartile	min.
OK-NSCORE	8885	1012	11016	9411	9012	8587	5231
UK-NSCORE	9302	673	11017	9693	9252	9027	6462
OK-UNTRAN	8284	1057	10528	9000	8446	7769	3879
UK-UNTRAN	8567	915	10513	9064	8775	8105	3554
OK-NSCORE **	8879	1018	11527	9436	9020	8585	4958
UK-NSCORE **	9301	677	11150	9701	9248	9033	5896
OK-UNTRAN **	8282	1062	10700	9000	8452	7760	3514
UK-UNTRAN **	8563	916	10600	9057	8767	8097	3343

**simulations including dispersion

The cumulative breakthrough curves for the different cases are given in Figure 3. The breakthrough curves for the OK-sets and UK-sets are very similar when intra-cell dispersion is considered and very close to having a lognormal distribution. Differences in log*T* values between the fields from the OK-sets and the fields from the UK-sets tend to have less impact on the breakthrough curves in case of intra-cell dispersion because transversal dispersion causes mass exchange between faster and slower flowlines, which results in more "averaged" travel times of the particles. This effect is similar when comparing the obtained breakthrough curves for the cases with intra-cell dispersion and the breakthrough curves obtained for the cases without intra-cell dispersion; intra-cell dispersion reduces the tails of the breakthrough curve due to the transversal mass exchange between fast and slow travel paths. The effect of including intra-cell dispersivity is considerable. The resulting breakthrough curves from the -NSCORE sets and the -UNTRAN sets are comparable, but differ more than the breakthrough curves for the UK- and OK-sets. The breakthrough curves from the -NSCORE sets are in general more abrupt. In terms of concentration this would mean that for the -UNTRAN sets the concentration starts to rise earlier, but that for the -NSCORE sets a higher peak value would be obtained.

5. Conclusions

(1) The differences in the transport simulations between the UK- and the OK- sets are in general small, particularly taking into account the distance travelled by the particles relative to the size of the simulation domain. One of the differences between the two sets is the smaller standard deviation in arrival locations for the UK- realizations than for the OK- ones in the transport situations considered. If we consider that the standard deviation is a measure of the uncertainty in the transport simulations, the UK- RF results in less uncertain simulations for this transport situation, partly due to the explicit incorporation of the trend in the RF model.

(2) Of the three effects analysed (trend, normal-score transform, and intra-cell dispersion), it seems that, for this particular data set, results are robust to the way the trend is modelled and also to the use of normal-scores. There are significant differences between results in which intra-cell dispersion is represented and those in which it is not. The breakthrough curves are more affected by the inclusion of intra-cell dispersion than by the choice of the random function model. If one takes into account that there are many more sources of uncertainty (e.g. uncertainty with respect to the prescribed head boundaries, probably nonsteady flow regime), which are not considered in this case study, one may conclude that the extra modelling effort and the difficult estimation of the variogram of the residuals are not worth the effort of accounting explicitly for the trend in the RF model. With respect to the use of the normal score transform in the simulation it can be stated that in spite of the clearly visible differences between the simulated fields, the impact has not been very big in these transport simulations. The most important impact was found on the breakthrough curves, especially in the tails of the breakthrough curves. However, the visible differences in simulated fields suggest that, for other transport situations, the normal score transform may have a larger impact. The decision on the use of the normal score transform is relevant in groundwater modelling where the normal distribution to model the sample data frequency distribution

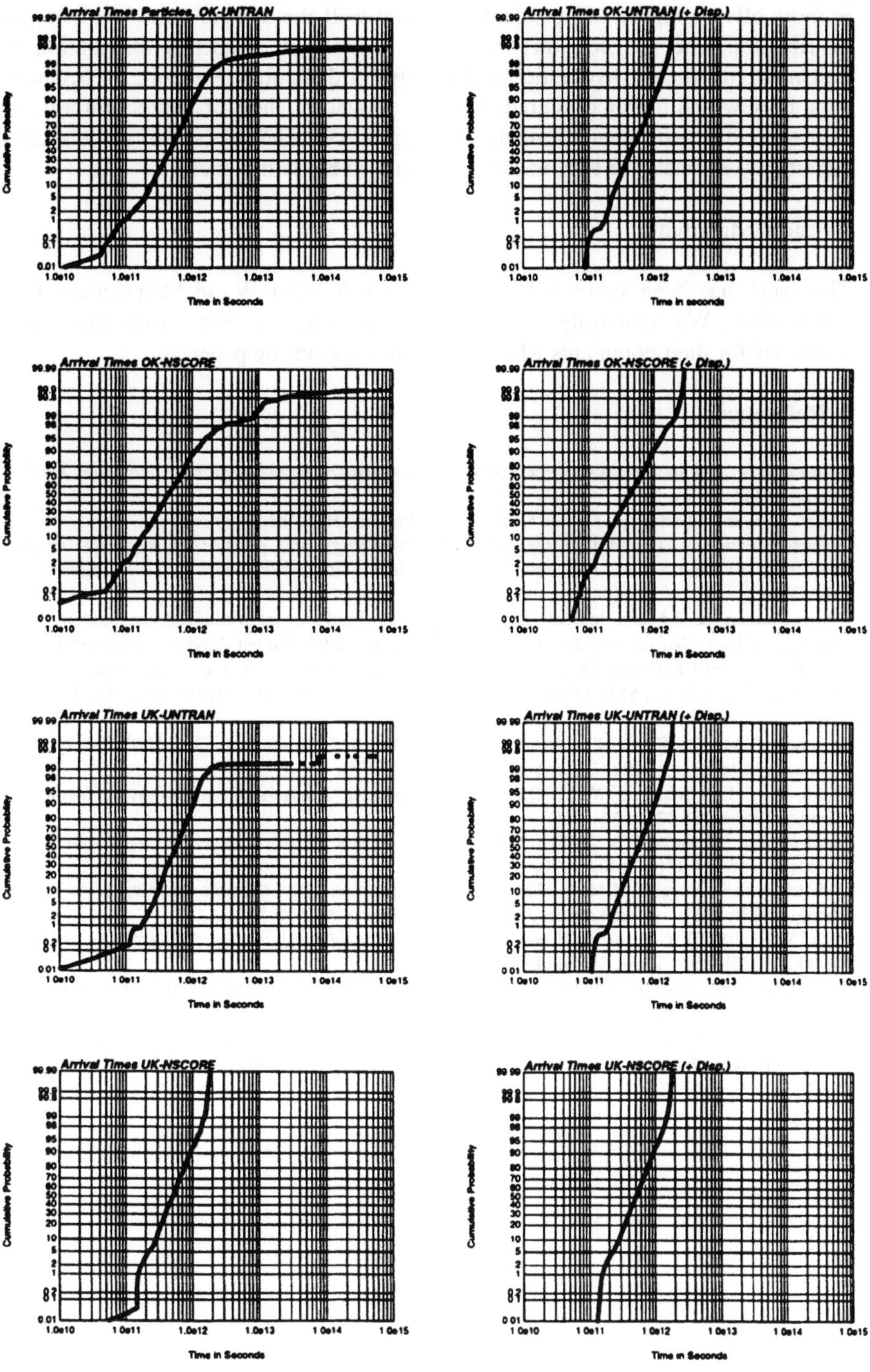

Figure 3. Breakthrough curves for the four different sets, both for non-dispersive mass transport and dispersive mass transport. The curves are obtained from 200 realizations.

is most often adopted independently of how well the sample data follow a Gaussian distribution. The decision on whether to use or not a normal distribution for log transmissivity (or for hydraulic conductivity) is difficult to support, particularly when data are few and highly clustered. In these occasions, short of indirect information that may support one or other RF model, we suggest considering the two models and use as uncertainty estimate the total uncertainty spanned by both models.

Acknowledgement

This work has been supported by the Spanish Nuclear Waste Management Company (ENRESA). We gratefully acknowledge this support. We thank the anonymous reviewers for their comments which helped to improve the paper.

6. References

Deutsch, C.V. and Journel, A.G. (1992). *GSLIB Geostatistical Software Library and Users Guide.* Oxford University Press, New York

Freeze, R.A. and Cherry, J.A. (1979). *Groundwater.* Prentice-Hall, Inc., New Jersey.

Gómez-Hernández, J.J. and Cassiraga, E.F. (1994). Theory and practice of sequential simulation. In Armstrong, M. and Dowd, P.A. (eds.), *Geostatistical Simulations*, pages 111-124. Kluwer Academic Publishers.

Gómez-Hernández, J.J. and Journel, A.G. (1993) Joint simulation of MultiGaussian random variables. In Soares, A., editor, *Geostatistics Tróia '92, volume 1*, pages 85-94. Kluwer Academic Publishers.

Journel, A.G. and Huijbregts, Ch. (1978). *Mining Geostatistics.* Academic Press, San Diego.

Journel, A.G. and Rossi, M.E. (1989) When do we need a trend model?, *Mathematical Geology* **21(7)**, 715-739.

LaVenue, A.M., Cauffman, T.L. and Pickens, J.F. (1990). *Groundwater Flow Modelling of the Culebra Dolomite: Volume I - Model Calibration.*

Smith, L. and Schwartz, F.W. (1981a) Mass transport, 2. Analysis of uncertainty in prediction, *Water Resources Research* **17(2)**, 351-369.

Smith, L. and Schwartz, F.W. (1981b) Mass transport, 3. Role of hydraulic conductivity in prediction, *Water resources research* **17(5)**, 1463-1479.

Webster, R. and Burgess, T.M. (1980) Optimal interpolation and isarithmic mapping of soil properties III Changing drift and universal kriging, *Journal of Soil Science* **31**, 505-524.

Wen, X.-H. (1995). *Geostatistical methods for prediction of mass transport in groundwater.* PhD thesis. Stockholm University Press, Stockholm.

Wen, X.-H. and Gómez-Hernández, J.J. (1996). The constant displacement scheme for tracking particles in heterogeneous aquifers, *Ground Water* **34(1)**, 135-142.

Wen, X.-H. and Kung, C.-S. (1995). A Q-BASIC program for modelling advective mass transport with retardation and radioactive decay by particle tracking, *Computers & Geosciences* **21(4)**, 463-480.

MAPPING SPATIAL AND TEMPORAL VARIATION OF GROUNDWATER QUALITY IN THE NETHERLANDS

EDZER J. PEBESMA[1], J.W. DE KWAADSTENIET
National Institute of Public Health and the Environment (RIVM)
P.O. Box 1, 3720 BA Bilthoven, The Netherlands

Abstract

Analysing groundwater quality monitoring network data, we are interested in the spatial pattern of groundwater quality and the steady temporal changes in this pattern. The spatial pattern is revealed by creating statistical maps of groundwater quality in the Netherlands, using measurements from available groundwater quality monitoring networks and maps of soil type and land use. Within soil-land use categories we adopted a model for spatial dependence between the measurements (stratified ordinary block kriging). To avoid misinterpretations of point estimates, 95% confidence intervals for 4 km × 4 km block median concentrations are presented on maps. Steady temporal changes in the spatial pattern of groundwater quality variables are revealed by interpolating short-term predictions, obtained by a linear regression model for each monitoring well screen. Yearly measurements from 1987-1993 are used to obtain predictions for 1980 and 2000. Statistical maps of the time-predicted (future and past) situations are created by modifying the within-category ordinary kriging procedure in order to account for location-specific time-prediction variances. Statistical indications of regional time-trends are obtained by showing differences between the confidence interval for 1980 and 2000 on maps.

1. Introduction

In the first half of the 1980s the national groundwater quality monitoring network (NGM) was installed for the monitoring of groundwater quality and systematic changes in time of groundwater quality in the Netherlands (Van Duijvenbooden *et al.*, 1985). For this purpose 370 monitoring sites, spread fairly evenly over the country, are sampled yearly at two depths (from well screens at about 10 and 25 metre below surface). At first samples were analysed for 19, and later 25, groundwater quality variables.

More recently (during the first half of the 1990s) the 12 provinces of the Netherlands started to install similar monitoring sites (the provincial groundwater quality monitoring

---

[1]present address: Landscape and Environmental Research Group, Faculty of Environmental Sciences, University of Amsterdam, Nieuwe Prinsengracht 130, 1018 VZ Amsterdam, The Netherlands

A. Soares et al. (eds.), geoENV I–Geostatistics for Environmental Applications, 111–122.

networks, the PGMs), in order to study groundwater quality at a denser spatial resolution. Currently, the PGMs have as many monitoring sites as the original NGM.

The current study aims at mapping groundwater quality variables, using point measurements obtained from the monitoring networks, in such a way that uncertainties in estimates are quantified statistically. In particular, the attention is focused on the mapping of groundwater quality at a measurement moment, as well as mapping changes in time of groundwater quality.

All variables measured in 1991 (Table A1, Appendix) were mapped, using the 1991 measurements from both monitoring networks. Maps of temporal changes in groundwater quality are made using the NGM measurements series from 1987 to 1993. Uncertainties in predicted values are quantified and presented on the maps. Full results are presented in separate reports (Pebesma and De Kwaadsteniet, 1994, 1995). In this paper we illustrate the methods used with results on one groundwater quality variable, namely potassium concentration.

From geohydrology we know that many factors influence groundwater quality, like soil type, land use, and geohydrological situation (e.g. depth of groundwater table, infiltration or seepage, marine influence, precipitation). Unlike other factors, soil type (Steur *et al.*, 1985) and land use (Thunnissen *et al.*, 1992) are available as digitized maps and so can be used to stratify the study area prior to the interpolation.

For each soil-land use category and for each groundwater quality variable, attention was paid to the level of, the variation in and the spatial dependence in the values measured (at a fixed moment) by using kriging interpolation (Journel and Huijbregts, 1978) within categories.

A groundwater quality monitoring network with a national or provincial extent by no means can provide accurate estimates of groundwater quality variables for specific unmeasured locations. Therefore, a coarser spatial resolution for the groundwater quality maps was adopted, and block mean values of half overlapping 4 km $\times$ 4 km square elements were estimated for a 2 km $\times$ 2 km grid covering the whole country using block kriging (Journel and Huijbregts, 1978). The size of the square elements was chosen with the spatial differentiation of soil type and land use in mind.

Because both starting level and time gradient are significant to the assessment of changes in groundwater quality, we chose to examine the subject of changes in time in terms of short-term predictions. Using a simple regression model, extrapolation from observed time series yields short-term predictions ('expected observations') for the monitoring screens, which can be used in the spatial interpolation. Spatially non-constant time-prediction uncertainties, expressed as estimation variances of 'expected observations' were incorporated in the spatial interpolation by using a slightly modified version of ordinary kriging (Delhomme, 1978).

The maps of (changes in) groundwater quality can be used for (i) assessment of diffuse groundwater contaminations (because the measurements do not represent strictly local contaminations) (ii) the validation of policy-supporting regional or national groundwater quality models.

Sections 2, 3, 4 and 5 describe the stratification, data selection, the changes in time at the well screens and the spatial interpolation. The results are discussed in Section 6 and we conclude with a discussion in Section 7.

2. Stratification of the study area

The available maps regarding soil type (Steur *et al.*, 1985) and land use (Thunnissen *et al.*, 1992) are so detailed that hundreds of combinations of classes can be made. For the purpose of this study such detail does not make sense: many categories would represent only a small fraction of the total area and would consequently contain too few (or no) measurements. Therefore, and analogous with considerations on the monitoring network design (Van Duijvenbooden *et al.*, 1985), we aggregated classes to the 13 categories—apart from built-up areas—shown in Table 1. Here, semi-natural vegetation comprises forests, heaths and nature reserves; the category 'dunes' is strictly spoken not a soil-land use category, but was obtained from a map with ecological districts (Klijn, 1988). Dominant soil type and land use categories for 2 km × 2 km cells are shown in Fig. 1. The soil-land use information for measurements was derived from this map.

Table 1. Basic information: (1) soil-land use category (2) area covered by the category, (3) number of selected measurements from the national groundwater quality monitoring network for each category and (4) number of selected measurements from the provincial groundwater quality monitoring network for each category. Fig. 1 shows the spatial information on (2), (3) and (4)

category	area		number	
	(%)	NGM	PGM	total
semi-natural vegetation on sand	10	36	26	62
grassland on sand	25	79	45	124
arable land on sand	7.1	29	15	44
dunes	2.8	11	4	15
semi-natural vegetation on fluvial clay	0.2	0	2	2
grassland on fluvial clay	8	25	26	51
arable land on fluvial clay	0.5	1	0	1
semi-natural vegetation on peat	0.3	1	4	5
grassland on peat	8.6	24	21	45
arable land on peat	0.2	2	1	3
semi-natural vegetation on marine clay	1	2	2	4
grassland on marine clay	9	16	10	26
arable land on marine clay	15.5	26	17	43
built up area	11.8			
total:	100	252	173	425

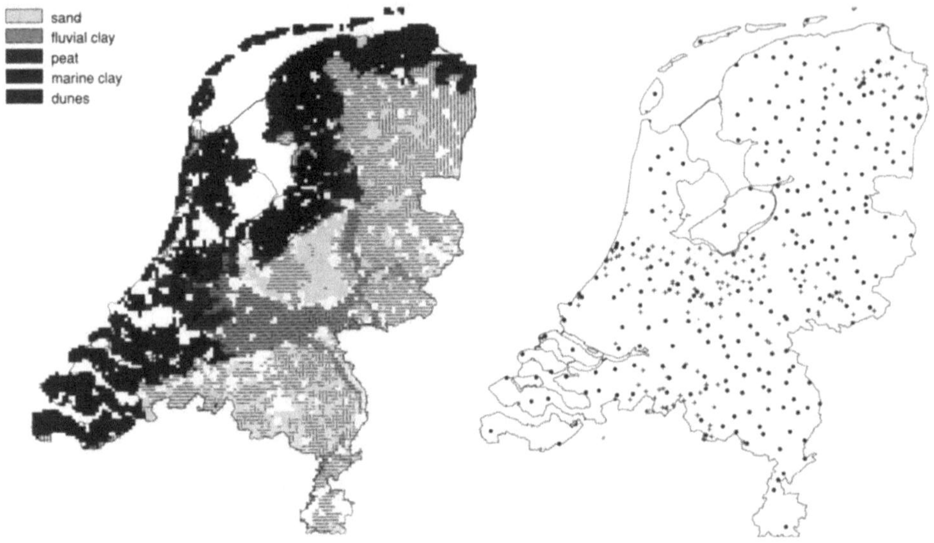

Figure 1. *Left:* Soil type and land use map (horizontal lines: grassland; vertical lines: arable land; no lines: semi-natural vegetation); *Right:* Monitoring network locations of the national groundwater quality monitoring network (•) and the provincial groundwater quality monitoring network (+) that were used in this study

3. Data selection

For the generation of groundwater quality maps a number of measurements were left out because they seemed unsuitable for this purpose. For instance, monitoring sites from the category 'built-up area' were left out because the soil map (Steur *et al.*, 1985) does not provide information about soil type in built-up areas. Interpolating the measurement information, taking soil type into account is therefore impossible in built-up areas with the information available, and no estimates for this class were made. Monitoring sites that could not be matched to a soil-land use category (Fig. 1) because they were influenced directly by water from rivers or the sea were left out too. Fig. 1 shows the locations of the 425 monitoring network sites that contributed to the groundwater quality maps.

The groundwater quality maps were based on measurements from monitoring screens completely between 5 and 17 metre below ground surface. For mapping groundwater quality in 1991, all 25 variables (table A1, Appendix) available in the NMG and PMGs of that year are used. For mapping changes in time of groundwater quality this was not possible because (i) many measurement sites have been installed only recently, and (ii) for the remaining sites the monitoring of several groundwater quality attributes was stopped or started only recently. Therefore, the mapping of changes in groundater quality concerns only 12 groundwater quality variables (variables marked with a * in Table A1, Appendix) and is based on a much smaller set of observations: the measurements of the provincial groundwater quality monitoring networks had to be left out (Table 1).

Detection limit values were assigned to measurements below the detection limit (using the highest value when different detection limits occurred). Full results are presented in two separate report (Pebesma and De Kwaadsteniet, 1994, 1995).

Some doubtful measurements were left out altogether, because they were considered as outliers. Outliers were assigned using both visual examination and simple distributional assumptions (Pebesma, 1996).

4. Changes in groundwater quality variables at the well screens

Because starting level and time gradient are both significant to the assessment of changes in groundwater quality, we chose to examine the subject of changes in groundwater quality in terms of short-term predictions. In particular, we are interested in the assessment of meaningful, steady (i.e. monotonous), long-term components in the change in time of groundwater quality variables.

For each groundwater quality variable and for each separate monitoring well screen, the variation of a concentration variable on the log-scale in time $y(t)$ is represented by a linear model with time, having independent, identically distributed errors

$$Y(t) = \alpha_0 + \alpha_1 t + e(t), \quad \mathrm{E}(e(t)) = 0, \quad \mathrm{Cov}(e(t)) = \sigma^2 I, \tag{1}$$

with I the identity matrix. Standard regression analysis (e.g., Draper and Smith, 1981) yields the ordinary least squares estimates of α_0 and α_1, and allows prediction of the mean value of $Y(t_{new})$, giving the short-term prediction at t_{new}, $\hat{y}(t_{new})$ and the corresponding prediction variance $s^2(t_{new})$.

Model (1) implies that on the log-scale we assume a constant increase or decrease of concentration in time: a first approach for a trend in which the deviation from a straight line is considered as 'white noise'. Using the log-transformation not only guarantees predictions that are strictly positive on the original scale, it also yields predictions at the monitoring network sites on the scale of the spatial interpolation (Section 5).

For creating maps of changes in groundwater quality the short-term predictions were taken as 'new observations' at the monitoring network sites. These 'new observations' were spatially interpolated. More specifically, for two moments (i.e. time periods) at equal times from the centre of the observation time series 1987-1993 (at t = 1980.5 and t = 2000.5), a set of 'new observations' was computed. For two 'new observations' at a specific monitoring site, this equitemporal choice of extrapolation times results—on the log-scale, and by absence of a trend on the original scale as well—in equal observation accuracy for both extrapolation moments. The choice of an extrapolation span of 1980-2000 aims to emphasize trends that are visible despite the observation inaccuracies.

Quantifying the observation accuracy for the *mean* value of $Y(t_{new})$ implies that the 'new observations' only relate to the situation *around* 1980 and *around* 2000. In this way, for 'new observations' the short-term components of temporal variation, the year-to-year deviations from the regression line, are filtered out.

5. Spatial interpolation

Maps of groundwater quality variables were obtained by means of ordinary block-kriging (Journel and Huijbregts, 1978, V.A.1) of log-transformed measurements within each of the soil-land use categories of Table 1, using the measurements available. For categories with only very few measurements a modified procedure was used, as described at the end of this section.

For the spatial interpolation of the measurements log-transformed measurements of a groundwater quality variable $z(x)$ were considered as a sample from a realization of the random field $Z(x)$ that is modelled as:

$$Z(x) = m + e(x), \tag{2}$$

with

$$m \text{ constant}, \quad \mathrm{E}(e(x)) = 0, \quad \tfrac{1}{2}\mathrm{E}[(Z(x_1) - Z(x_2))^2] = \gamma(\|x_1 - x_2\|), \quad \forall\, x, x_1, x_2 \in \mathbf{R}.$$

That is, within an area of size **R** inside a soil-land use category we used an additive model (on log-scale) with (i) a constant expected value and (ii) a spatial dependence between measurements that is a function of the distance h between the measurement locations only. Additionally we assumed that in a soil-land use category for a groundwater quality variable only one function $\gamma(h)$ is needed (independent of the locations of the area of size **R**).

Depending on the shape of the sample variogram the theoretical variogram $\gamma(h)$ was chosen from a nugget model, a spherical model, or a combination of these two models (Journel and Huijbregts, 1978). Fitting the variogram model to the sample variogram was done mostly by eye (only the sill was fitted using weighted least squares). Fig. 2 shows the variograms for potassium. The area size **R** (the kriging neighbourhood) was defined (in order of preference) as (i) the closest 20 observations when more than 20 observations were available within a distance of 100 km, (ii) 100 km when 10 to 20 observations were available within this distance, (iii) the 10 nearest observations.

For the mapping of changes in time of groundwater quality we additionally assumed that the variogram obtained for 1991 (Fig. 2) is valid for the 'new observations' as well. The interpolation procedure that was used for interpolating 'new observations' differs from standard block-kriging based on (2) in that location specific uncertainty for each 'new observation' was taken into account. In effect, this results in larger estimation errors and less weight assigned to 'new observations' with the largest prediction uncertainties (Delhomme 1978, Pebesma 1996).

The estimates given in the maps concern mean values on log-scale of 4 km × 4 km blocks. Estimates are presented as approximate 95% confidence intervals calculated by

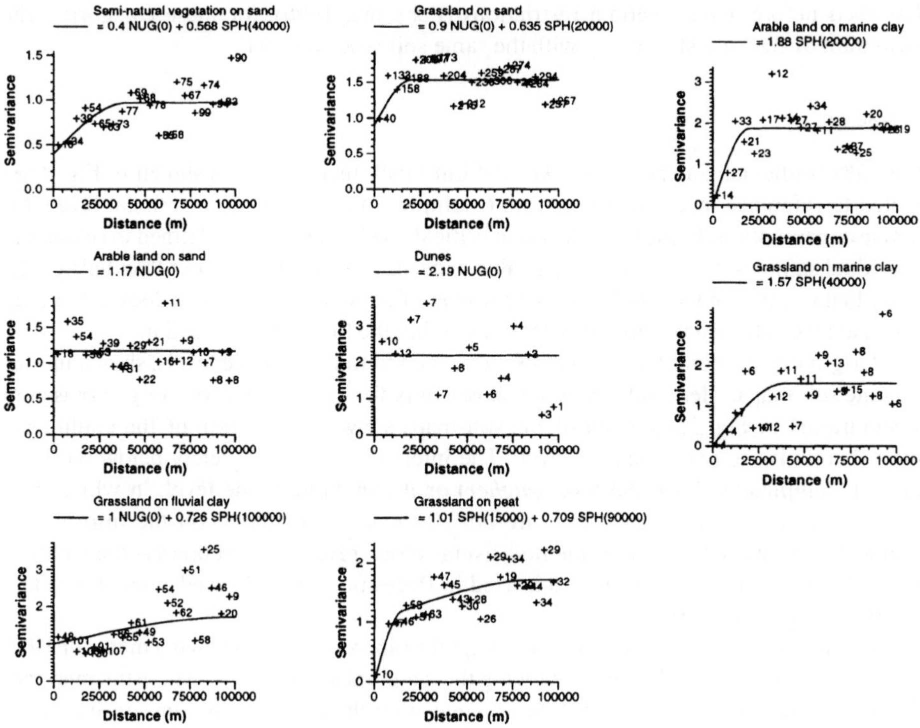

Figure 2. Variograms of $^{e}\log(K)$, stratified by soil type and land use. A number reflects the number of observation pairs used for an estimate (+) of a variogram point.

$$[\hat{z}(x_0) - 2\,\sigma_k(x_0)\,,\ \hat{z}(x_0) + 2\,\sigma_k(x_0)] \tag{3}$$

with $\hat{z}(x_0)$ the point estimate (kriging estimate) on log-scale, and $\sigma_k(x_0)$ the estimation standard deviation (kriging standard deviation) of the estimate (Journel and Huijbregts, 1978). Back transformation of this interval to the original scale (taking the exponent of both sides) yields an approximate 95% confidence interval for the block median value (see also a remark in section 7).

All 4 km × 4 km blocks considered are centred at the nodes of the 2 km × 2 km grid of Fig. 1 (assuming that soil-land use dominance of 2 km × 2 km cells still holds for the 4 km × 4 km blocks). The results have been presented in maps (e.g. see Figs. 3 and 4). Only the 2 km × 2 km centres of blocks for which estimates were made are shown on the maps.

For the small soil-land use categories with less than 10 observations on a variable the estimation procedure was modified, because too little information is available in these categories for modelling the variogram and for estimating block mean values in the way described before. Under certain restrictions (Pebesma, 1996), for these categories, the variogram of the largest category with the same soil type was used.

6. Results

The 95% confidence intervals for 4 km × 4 km block median values shown in Fig. 3 are related to reference levels obtained from (multiples of) target levels or critical levels for human consumption. These target levels or critical levels however, are defined in relation to individual samples. If a block median value exceeds a critical level, more than half of the individual samples in the block exceed this level. If a block median value does not exceed a critical level, still many individual samples within the block may exceed it.

Originally, the confidence intervals are related to four reference levels, shown in four separate sub-maps. Here, sub-maps are shown only for one or two levels (e.g. potassium concentration in Fig. 3, 4). Each of the sub-maps shows the position of the confidence intervals relative to one reference level. The interval can be completely below the level (*lower*), completely above the level (*higher*) or it can straddle the level, in which case estimated value and reference level are *not distinguishable*, based on the information available. (An alternative display method, using colour, reveals the accuracy—the width of the confidence intervals—better, and is used in Pebesma and De Kwaadstniet (1994) and presented in Pebesma (1996).)

For a variable, changes in groundwater quality are revealed by showing maps for 1980 and 2000 side by side, and by only showing their mutual differences—where the maps are identical, no values are shown. Fig. 4 shows a map with changes in potassium concentration. The map reflects the development of groundwater quality over a 20 year span, within the context of the current model structure. The differences shown arise directly from the slope of all or part of the regression lines, as fitted to the measured time series: if the slopes are zero the maps are identical. The indicated trend direction in a cell can be derived from the direction of shift in legend class from one extrapolation time to the other. In general, trend magnitude cannot be derived from this shift, for instance, in case of a single legend class shift. It can be stated that all trend indications shown are of vital importance in the sense that they concern a shift in estimates of 4 km × 4 km block median concentration that results in a different map for at least one of the reference levels.

7. Discussion

In this study, estimates of 4 km × 4 km block median values are presented as approximate 95% confidence intervals: a compact way of combining the presentation of expected values (point estimates) and uncertainty (estimation variances) that prevents misinterpretation of point estimates. If there is an area where confidence intervals are wide and where block median values cannot be distinguished from an important critical level, one can decide to place an additional monitoring site in or close to that area. On the other hand, where such a distinction can easily be made, one may decide that fewer measurements are sufficient.

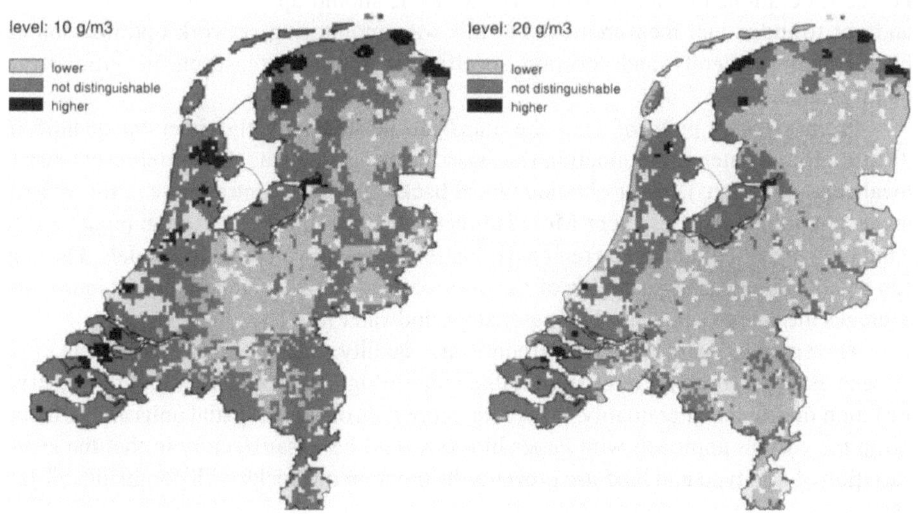

Figure 3. Map of potassium in the groundwater. 95% Confidence intervals for 4 km × 4 km block median values related to two concentration levels

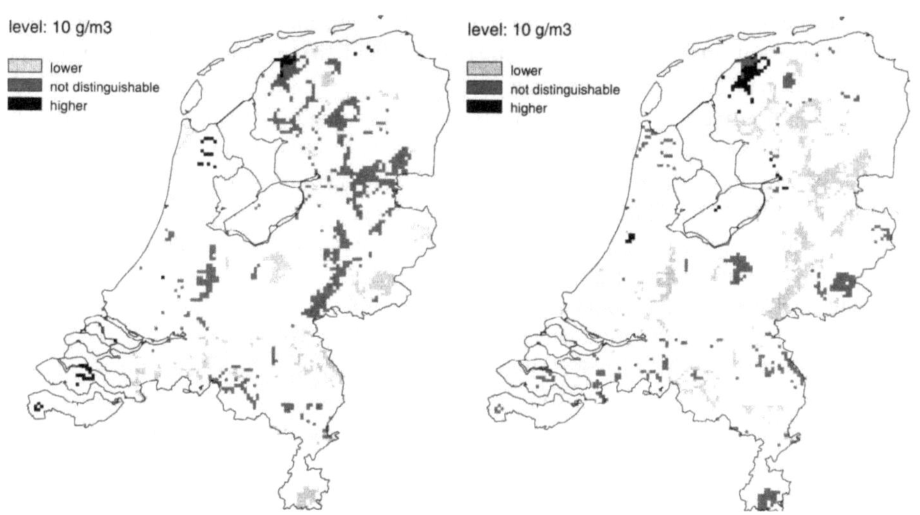

Figure 4. Map of potassium concentration in the groundwater around 1980 (left) and 2000 (right), only where they are mutually different. 95% Confidence intervals for 4 km × 4 km block median values, related to one concentration level

For considerations of this kind, the critical levels should apply to block median values and not to individual measurements. In this way, monitoring network optimization can be directed efficiently and location specifically to relevant aspects of groundwater quality maps.

From a policy point of view the maps can be used directly either for quantifying diffuse groundwater contamination (because the measurements do not represent strictly local contaminations), or for obtaining local background concentrations e.g. for assisting local contamination assessment. More indirectly the maps can be used as an input to or for validation of policy supporting regional or national groundwater quality models. The maps can be considered as a translation of the information the size of measurement into spatial averages the size of units, used in regional groundwater models.

The spatial resolution of the groundwater quality maps (4 km $\times$ 4 km blocks) can be seen as a maximum, given the available monitoring network density and the observed variation in groundwater quality. Where the properties of larger spatial units are sufficient, using the current approach with larger blocks would be unsatisfactory in that the spatial variation of soil type and land use prevents the creation of blocks with dominant soil-land use.

The stratification used for this study is not ideal. The soil type-land use map used for this study is a rough generalization, and information on geohydrology was lacking, which results in larger within-category variation (wider confidence intervals).

The changes in predicted groundwater quality as presented on the maps arise from (i) the measurement series from 250 monitoring sites and (ii) the model structure used. For the description of temporal change in a groundwater quality variable we used a first order linear regression model for measurements on the log-scale with time, assuming that the deviation from a regression line could be considered as 'white noise'.

Using the log-scale as the working scale yields (i) temporal predictions on the most obvious scale for spatial interpolation, (ii) location specific accuracy for the prediction at the monitoring sites, as obtained directly on the spatial interpolation scale, and (iii) strictly positive predictions on the original scale.

The adaption of the standard kriging procedure to handle observations that are subject to an (known, spatially and temporally uncorrelated) observation error was given in Delhomme (1978) and in Pebesma (1996). Compared to the situation where the observation error would have been ignored, this adaption represents a more careful approach that results in wider confidence intervals, and an in assigning less weight to observations with relatively large prediction errors.

For the variograms of 'new observations' we used the variograms of the 1991 measurements. The primary reason for this is that for studying spatial variation of a groundwater quality variable within a soil-land use category, in 1991 we had a much larger observation set: apart from the measurements from the national groundwater quality monitoring network we could use about 170 measurements from provincial monitoring networks. Distinguishing the 1991-situation and the 'around 1991-situation' with respect to the variogram specification was—with all other simplifications in mind—considered to be overdone.

It should be noticed that the maps of changes in time of groundwater quality do not pretend to show the actual changes that occur over the 20 year time span considered. The changes shown only concern extrapolations from observed changes in the measurement series 1987-1993, within the context of the model structure.

Mathematical aspects were kept simple, primarily because we aim for (a) an optimal accessibility of the various methodological items mentioned, and (b) the avoidance of more complex assumptions than necessary, based on the objective of this study and the information available. The simplicity refers to the model used (2), the choice and fit of variogram models, the choice of **R**, the way approximate confidence intervals for block median values were calculated, and the outlier procedure. Some general remarks about the mathematical aspects are:

(i) Kriging as applied in this study (using model (2) for each variable for each soil-land use category) can be considered as a first step beyond and an improvement of the classical statistical approach (for each variable for each category) in its four forms based on spatial dependence or independence, and one mean per category or local means of regions within a category, respectively. The kriging approach is an improvement on these classical statistical approaches because local information is used more effectively.

(ii) The (approximate) 95% confidence intervals presented in this study are consistent with interpolations on a log-scale: an obvious choice given the character of the measurements at hand. For a first quantification of uncertainties (the calculation of confidence intervals using (3)) the estimation error of block mean values on log-scale is assumed to be normally distributed, all second-order statistics (variograms, time-prediction uncertainties) are taken as deterministic quantities, and uncertainty of the (kriging) standard deviation is ignored. This implies that only a rough indication of confidence is obtained, and that the width of the confidence intervals should be considered as the lower bound to more carefully obtained confidence intervals.

(iii) For the interpretation of the intervals (3) as 95% confidence intervals for block median concentration it is necessary that on the log-scale the block mean and block median coincide. This is for instance true when for each block on the log-scale individual samples within the block are distributed symmetrically. When this assumption is too restrictive, an alternative and, with respect to within-block distribution non-parametric, interpretation of (3) is that of the block geometric mean.

(iv) In the current context (block kriging on a log-scale) the central measure that is estimated is the block median, not the block mean. When the block mean is needed (e.g. for comparing the maps with groundwater quality model results) first indications of block means can be obtained from the maps of block medians by using an approximate relation between block mean $m_{\square}$ and block median $M_{\square}$, given by

$$m_{\square} = M_{\square}\exp(s_{\square}^2/2), \tag{4}$$

where $s_{\square}^2$ is the short-distance variance, the short-distance value of the variogram for the relevant variable and category. Relation (4) is derived from the well known relation between mean and median of a lognormal distribution (Aitchison and Brown, 1957). First indications of 95% confidence intervals for 4 km × 4 km block mean values are obtained when the legend class boundaries of a map are multiplied with the factor of $M_{\square}$ from (4). For a groundwater quality variable, this will result in different legend class boundaries for different soil-land use categories.

Appendix

TABLE A1. The 25 Groundwater quality variables measured in 1991 (* 9 for 1987-1993)

Al, As, Ba, Ca*, Cd, Cl*, Cr, Cu, DOC, EC*, Fe, pH*, HCO_3*, K*, Mg*, Mn, Na*, NH_4-N*, Ni, NO_3-N*, Pb, P-tot*, SO_4*, Sr, Zn

References

Aitchison, J., J.A.C. Brown (1957) The Lognormal Distribution. Cambridge University Press, Cambridge.

Delhomme, J.P. (1978) Kriging in the Hydrosciences. Adv. Water Res., 1: 251-266

Draper, N, and H. Smith (1981) Applied Regression Analysis, Second Edition. John Wiley and Sons, New York.

Journel, A.G., Ch.J. Huijbregts (1978) Mining Geostatistics. Academic Press, London.

Klijn, F. (1988) Milieubeheergebieden. Deel A: Indeling van Nederland in Ecoregio's en Ecodistricten *(Environment management areas. Part A: subdivision of the Netherlands in ecological regions and ecological districts).* Rijksuniversiteit Leiden, mededelingen No. 37, 92 pp. (in Dutch).

Pebesma, E.J., J.W. de Kwaadsteniet (1994) Een landsdekkend beeld van de Nederlandse grondwaterkwaliteit op 5 tot 17 meter diepte in 1991 *(Maps of groundwater quality in the Netherlands at 5 – 17 metre depth, in 1991).* National Institute of Public Health and the Environment, Bilthoven. Report No. 714810014 (in Dutch).

Pebesma, E.J., J.W. de Kwaadsteniet (1995) Een landsdekkend beeld van veranderingen in de Nederlandse grondwaterkwaliteit op 5 tot 17 meter diepte *(Maps of temporal changes in groundwater quality in the Netherlands at 5 – 17 metre depth).* National Institute of Public Health and the Environment, Bilthoven. Report No. 714810015 (in Dutch).

Pebesma, E.J. (1996) Mapping Groundwater Quality in the Netherlands. PhD thesis, Utrecht University, Utrecht. (at *http://www.frw.uva.nl/~pebesma/*)

Steur, G.G.L., F. de Vries and C. van Wallenburg (1985) Bodemkaart van Nederland 1:250 000 *(Soil map of the Netherlands 1:250 000).* Stichting voor Bodemkartering, Wageningen (in Dutch).

Thunnissen, H., R. Olthof, P. Getz and L. Vels (1992) Grondgebruiksdatabank van Nederland vervaardigd met behulp van Landsat Thematic Mapper opnamen *(Land use database of the Netherlands constructed with the aid of Thematic Mapper images).* DLO-Staring Centrum, Wageningen. Report No. 168 (in Dutch).

Van Duijvenbooden, W., J. Taat, L.F.L. Gast (1985) Landelijk meetnet grondwaterkwaliteit: eindrapport van de inrichtingsfase *(The national groundwater quality monitoring network: final report of the installation).* National Institute of Public Health and the Environment, Bilthoven. Report No. 840382001 (in Dutch).

A MODEL OF AQUIFER HETEROGENEITY

G.L. MOLTYANER
Environmental Research Branch, Chalk River Laboratories, AECL
Chalk River, Ontario, Canada KOJ 1J0

1. Introduction

Groundwater modelling has become a reliable tool for solving groundwater budgeting and contamination problems. It has long been recognized, however, that the degree of reliability of this tool varies in proportion to the amount of information available on the heterogeneity of the flow system.

Conventionally, the information is given in terms of spatial variability in hydraulic conductivity. To solve groundwater contamination problems, it is more advantageous to describe heterogeneity in terms of spatial variability in groundwater velocity measured in a natural-gradient tracer test. The tracer method has a higher spatial resolution than other methods of heterogeneity sampling. Results of a number of large-scale tracer tests, performed at the Twin Lake research site [Killey and Moltyaner, 1988] to characterize the groundwater velocity field, support this argument.

In searching for alternative methods by which to characterize heterogeneity, a large number of sediment samples were extracted at the Twin Lake site, in order to: 1. match the tracer test in heterogeneity sampling, 2. estimate permeabilities from grain-size measurements, and 3. compare the grain-size-derived permeabilities with permeabilities calculated from the tracer test, and thereby demonstrate that the grain-size measurements are an alternative to the field tracer test.

Results of the preliminary sediment coring and sampling program reported earlier [Pfleiderer and Moltyaner, 1993] were encouraging. Therefore, a number of new boreholes were drilled at the Twin Lake site and an additional 487 sediment samples were analyzed to test the results previously reported. All grain-size data were combined to build a geometric model of aquifer heterogeneity. The model was compared with the geometric model constructed previously from the natural-gradient tracer test data [Moltyaner and Wills, 1993].

The good agreement between the two independent hydrostratigraphic models, and the simplicity of sediment grain analyses, motivated the development of a new groundwater flow model based

A. Soares et al. (eds.), geoENV I–Geostatistics for Environmental Applications, 123–137.

on the two-phase representation of aquifer properties. This paper describes how the new model was developed. Emphasis is placed on the measurement of, and relations between, individual phase properties, and the derivation of continuity and flow equations for the fluid phase.

2. Physical Properties of a Granular Geologic Medium

Physical properties describing the flow of fluids through a granular geologic medium fall into three classes: those describing the geologic medium, those describing the fluid, and those describing the flow. Geologic medium properties include the medium geometry, density and porosity. Fluid properties include the fluid density, viscosity, compressibility and pressure. Flow properties include the specific discharge and velocity.

Conventional methods of hydrogeology consider the geologic medium and fluid properties collectively using the hydraulic conductivity concept. No hypotheses are made about the texture and structure of a geologic medium. We go beyond the limitations of these conventional methods by building a two-phase model of aquifer properties and predicting the bulk properties of both the geologic medium and fluid from the grain-scale properties.

2.1. BULK PROPERTIES

2.1.1. *Homogeneity and Composition*

A medium is homogeneous if its properties and composition are uniform. A geologic medium is far from homogeneous and, at a scale larger than an individual grain, unconsolidated sediment is essentially a heterogeneous two-phase aggregate separable by mechanical means. Usually, it is assumed that the fluid phase is homogeneous but of a variable composition. The heterogeneity of the solid phase is accounted for through the heterogeneity in the hydraulic conductivity.

2.1.2. *Sediment and Fluid Bulk Density and Volume*

The amount of sediment in a sediment sample can be determined by the amount of space occupied by its bulk volume. The mass of grains in a sample and the sample's bulk volume are termed the sediment bulk density. Since the distribution of grain sizes may change from one sample to another, the sediment bulk density may change, and may be used to characterize the heterogeneity of the solid phase.

Because the distribution of sizes of sediment grains changes with the sample, so does the sample's pore volume available for groundwater. As a result, the groundwater mass changes from sample to sample, even though a sampler's bulk volume remains fixed. The two quantities are related in a derived quantity; the fluid bulk density may change with the sample's position. The fluid bulk density, therefore, may be used to characterize the heterogeneity of the fluid phase. The common bulk volume of sediment for which the fluid and sediment bulk densities can be defined is called the local-scale volume. The total mass density, ρ_t, of fluids and solids in the local-scale volume is given by the law of mechanical mixtures.

2.1.3. *Interactions*

The interaction between the solid and fluid phases affects the transport properties of groundwater, and can be studied using radioactive tracers. An in situ dector can be used to detect the quantity of gamma photons that are not absorbed by the medium. Since the non-absorbed radioactivity is related to the total mass density, changes in the total mass density are manifested in the measured tracer concentration. Then the groundwater velocity, estimated from the concentration-versus-time data, should be correlated with the total mass density (see Sec. 2.2.4).

2.2. SCALING RELATIONSHIPS

The statistical definition of bulk properties requires an assumption that sediment grain sizes form a continuous random variable. The assumption is satisfied, if in a local-scale neighbourhood of a sediment grain in the aquifer, there is a large number of grains. The requirement for a large number of grains in the local-scale volume is similar to that in fluid mechanics [Daily and Harleman, 1966], where a homogeneous fluid sample is composed of large quantities of molecules of one size. Sediment grains, however, unlike the molecules of a homogeneous fluid, have a distribution of sizes. At the Twin Lake site, for instance, a sediment sample contains approximately 10^6 grains with a normal distribution of sizes in the local-scale volume of 243 cm^3 [Pfleiderer and Moltyaner, 1993]. The mean size of these grains is 0.17 mm.

2.2.1. *Sediment Bulk Density*

In order to derive the local-scale properties from properties of individual sediment grains, we introduce grain number density, N_g/V, where N_g is the number of sediment grains in the local-scale volume, V. If m_g denotes the mass of the mean size grain, the sediment bulk density is given by $\rho_g=m_gN_g/V$.

2.2.2. *Fluid Bulk Density*

The pore number density can be defined as the number of pores, N_p, in the local-scale volume, N_p/V. If the distribution of pore sizes can be characterized by the mean pore size, then the fluid bulk density, ρ_f, is defined as $\rho_f=m_fN_p/V$. Here it is assumed that the medium is fully saturated and m_f is mass of fluid filling up a pore of the mean size.

2.2.3. *Porosity*

The pore number density is a discrete analog of porosity

$$n=V_p/V \tag{1a}$$

defined as the volume of pores, V_p, in the local-scale volume. For the spherical shape of a pore, the pore volume can be taken as

$$V_p=(1/6)\pi(d_{10})^3N_p \tag{1b}$$

where d_{10} is the tenth percentile of the grain size distribution. Then

$$n=(1/6)\ \pi\ (d_{10})^3N_p/V \tag{1c}$$

for the spherical sediment grain, the sediment volume can be estimated from the following relationship

$$V_g=(1/6)\ \pi\ (d_{50})^3N_g \tag{2a}$$

where d_{50} is the fiftieth percentile of the grain size distribution. The sediment proportion by volume is given as

$$V_g/V=1-n \tag{2b}$$

2.2.4. *Medium Mass Density*

The fluid and sediment bulk densities characterize the fluid and sediment proportions by mass. The total mass density, ρ_t, of fluids and solids in the local-scale volume is given by the law of mechanical mixtures

$$\rho_t=\rho_f+\rho_g \tag{3a}$$

The relationship between mass and volumetric properties is established through (1), (2) and (3)

$$\rho_t=n\rho+(1-n)\ \rho_s \tag{3b}$$

In (3b), ρ and ρ_s are the fluid mass density and sediment material mass density, respectively.

2.2.5. *Equivalent Heterogeneity Measures*

From (3) it follows that, under the assumptions of constant sediment material mass density and constant fluid mass density, volumetric porosity can be used to characterize the heterogeneity of a geologic medium. This is equivalent to the characterization based upon total mass density. Therefore, a geologic medium is homogeneous if its volumetric porosity does not vary in space.

According to (1), the effective grain diameter, d_{10}, considered as a continuous random variable at each point in space, can be used instead of porosity to characterize heterogeneity. Then, as discussed in the previous section, the groundwater velocity should be correlated with the effective grain diameter.

2.2.6. *Velocity*

A fluid particle is visualized as the volume of groundwater enclosed in the local-scale volume. It moves with velocity $\mathbf{u}(\mathbf{x},t)$, defined as the temporal rate of change of a fluid particle displacement vector along the particle trajectory, $\mathbf{x}(t)$

$$\mathbf{u}(\mathbf{x},t)=\partial\mathbf{x}/\partial t \tag{4}$$

The change in velocity of a local-scale particle is described by the concentration balance equation

$$(\partial c/\partial t)+(\mathbf{u}\cdot\mathrm{grad})c=0 \tag{5}$$

for different moments of time, t=T [Moltyaner, 1993]. In (5), $c(\mathbf{x},t)$ is the tracer concentration, grad is the gradient vector, and ($\mathbf{u}\cdot$grad) is the dot product. It is assumed that the diffusion and kinematic dispersion are negligibly small. The equation of concentration continuity (5) establishes a relationship between the groundwater velocity and medium density manifested in the measured tracer concentration. On the other hand, the equation of fluid continuity derived in the appendix establishes a relationship between the groundwater velocity and porosity.

2.2.7. *Flux and Specific Discharge*

A discrete analog of the fluid flux, the number flux vector is defined through the pore number density as $N_p\mathbf{u}/V$. The fluid flux vector is then a product of the bulk fluid density and local-scale velocity, $\rho_f\mathbf{u}$. The specific discharge vector, $\mathbf{q}(\mathbf{x},t)$, is defined as a product of the fractional fluid bulk density, equal to porosity, and the local-scale velocity

$$\mathbf{q}(\mathbf{x},t)=(\rho_f(\mathbf{x},t)/\rho)\mathbf{u}(\mathbf{x},t)=n\mathbf{u}(\mathbf{x},t). \tag{6a}$$

For the spherical type of pores,

$$\mathbf{q}(\mathbf{x},t)=(1/6)\,\pi\,(d_{10})^3N_p\mathbf{u}(\mathbf{x},t)/V \tag{6b}$$

It follows from (6) that the specific discharge vector changes with the change in the mean grain size.

2.2.8. *Piezometric Head*

Since the fluid phase properties are described at the local scale, the pore pressure is averaged over the local-scale volume and referred to as the local-scale pressure, p. The local-scale hydraulic head h(**x**,t) is defined according to the hydrostatic law

$$h(\mathbf{x},t)=z+p(\mathbf{x},t)/[\rho_f(\mathbf{x})g] \tag{7}$$

Where z is the elevation and g is the gravitational constant note the use of ρ_f in defining the hydraulic head.

2.2.9. *Darcy's Law*

Darcy's velocity or specific discharge vector, **q**(**x**,t), is calculated using Darcy's law [Bear, 1972]

$$\mathbf{q}(\mathbf{x},t)=-\underline{K}\mathrm{grad}\ h(\mathbf{x},t) \tag{8}$$

where $\underline{K}$ is the hydraulic conductivity tensor. The substitution of (6) into (8) leads to

$$\mathbf{u}(\mathbf{x},t)=\underline{K}\mathrm{grad}\ h(\mathbf{x},t)/n(\mathbf{x}) \tag{9}$$

3. Field Measurements of Physical Properties

The flow geometry in a geologic medium is controlled by lithology and stratigraphy. In this paper, lithology is quantified using the concept of an average grain size. The average grain size is determined by invoking the principle of geometrical similarity. According to this principle, particles with similar three-dimensional forms have similar sizes. The similarity in grain sizes is analyzed in a laboratory by mechanical sieving.

The same principle of similarity is employed in structuring the average grain-size data into stratigraphic units. Geostatistics is employed to establish similarity and determine the physical dimensions of a stratigraphic unit. A linear empirical relationship [Hazen,1893] is used to calculate the hydraulic conductivity from the effective grain-size data and relate stratigraphy to hydrostratigraphy. The high spatial resolution of grain-size data results in the high spatial resolution of hydraulic conductivity data, permitting detailed well-to-well correlations and analysis of hydrostratigraphy.

3.1. GRAIN-SIZE ANALYSIS

Figures 1d and 2d show the frequency distribution of two individual samples taken at random from the coarse and fine-to-medium sand units of the Twin Lake aquifer. There is a 20% difference between mean grain sizes. The vertical dimension of an individual stratigraphic unit is 0.21 m based on the preliminary grain-size analysis. Mean grain sizes from sediment samples collected within the individual layer are plotted as frequency distributions in Figures 1c and 2c. There is the same 20% difference between the means.

Figures 1b and 2b show frequency distributions constructed from 28 to 26 samples of individual layers collected at four borehole locations. The same 20% difference between means is observed. Figures 1a and 2a show frequency distributions from 227 and 362 samples collected from the coarse and fine-to-medium sands at four boreholes spaced at 1 m. A comparison of these figures illustrates that the distributions of grain sizes are similar in shape, and that the mean of effective grain diameters in the coarse sand unit is 20% larger than in the fine-to-medium sand unit. The variance within an individual unit is significantly less than the variance between units.

The vertical dimension of an individual sedimentary unit was estimated to be the integral correlation length scale of the effective grain diameter random function, $d_{10}(x,y,z)$. Figure 3 illustrates a sample correlogram of the random function fitted by the exponential functions at 19W 2S borehole. The average vertical dimension of a sedimentary unit is 0.39 m.

3.2. CORRELATIONS BETWEEN EFFECTIVE GRAIN SIZES AND FLOW VELOCITY

The correlation between flow geometry and stratigraphy was evaluated at the Twin Lake site using a natural-gradient dispersion test with radioiodine (^{131}I). The first step was to identify individual velocity layers. The dispersion of radioiodine was controlled by variability in the pore-scale velocity, which was correlated, at the local scale, with variability in the effective grain diameter. Individual velocity "layers" were mapped by analyzing vertical profiles of concentration distribution [Moltyaner and Wills, 1993]. For the tracer concentration, considered as a random function of space coordinates, the vertical dimension of the individual layer was 0.21 m, which is in excellent agreement with the vertical dimension of the individual layer estimated from a geostatistical analysis of grain sizes [Pfleiderer and Moltyaner, 1993].

Figure 4 compares the velocities of individual layers with the corresponding squared mean of the effective grain diameters of those layers. The regression line drawn through the data points illustrates the linear correlation between the mean of effective grain diameters and the mean velocity for each individual layer in the high-velocity zone. We can then adopt the straight capillary tube model [Bear, 1972] as the flow model in an individual layer. In a straight capillary tube of effective diameter d_{10}, the water flow is described by Poiseuille's law with the average flow velocity

$$u_p = -\rho g d_{10}^2/32\mu \text{ grad } h \tag{10a}$$

The linear, local-scale velocity, u_L in N_p such channels, as measured in a tracer test, is given by

$$u_L=L/(t_2-t_1) \tag{10b}$$

where L is the linear distance between two tracer concentration measurements and t_2-t_1 is the difference between the mean tracer concentration arrival time. We can then equate (10a) and (10b)

$$u_L=-A(\rho g d_{10}^2/\mu)\text{grad } h \tag{10c}$$

where A is introduced to account for the change in scale from a capillary tube to a layer. Owing to the correlation between the velocity and grain-size diameter, we take Hazen's formula [1893] in the form

$$K= Cd_{10}^2 =A(\rho g d_{10}^2/\mu)n \tag{11}$$

to arrive at Darcy's law.

4. Conclusions

The grain-size analysis method is a reliable alternative to the natural-gradient tracer test. The two-phase model of a fluid-saturated geologic medium developed in this paper provides a linkage between the alternative approaches. The main advantage of this model is that it explicitly establishes a relationship between flow velocity and stratigraphy, quantified in terms of the correlation structure of the effective grain diameter random field.

Confidence in the two-phase model was gained from the observed agreement between the hydraulic conductivities calculated from tracer tests and grain-size analysis data, and from the linear correlation established between the effective grain diameter and velocity.

The model was applied at the Twin Lake site. We were satisfied with the ability of this model to describe the groundwater flow and tracer migration at the site [Moltyaner et al., 1993]. Based on this experience, we encourage fellow hydrogeologists to use the two-phase model in computer simulations and field practice.

5. Acknowledgements

My thanks are due to C.A. Wills for graphs and calculations. The financial support for this work was provided by Atomic Energy of Canada Limited and the Japanese Atomic Energy Research Institute.

6. References

Bear, J. (1972) Dynamics of Fluids in Porous Media. *Elsevier Science*, New York.

Daily, J.W. and Harleman, D.R.F. (1966) Fluid Mechanics, Addison-Wesley, Reading, Mass.

Hazen, A. (1893) Some physical properties of sands and gravels with special reference to their use in filtration, Annu. Rep. Mass. Board Public Health, 541-546.

Killey, R.W.D. and Moltyaner, G.L. (1988) Twin Lake tracer tests: Setting, methodology and hydraulic conductivity distributions, *Water Resour. Res.*, 24(10), 1585-1613.

Moltyaner, G. L., (1993) Advection in geologic media, *Water Resour.* Res., 29(10), 3407-3415.

Moltyaner, G. L., and Wills, C.A. (1993) Characterization of aquifer heterogeneity by in situ sensing, *Water Resour.* Res., 29(10), 3417-3431.

Moltyaner, G. L., Klukas, M.H., Wills, C.A. and Killey, R.W.D. (1993) Numerical simulation of Twin Lake natural-gradient tracer tests: A comparison of methods, *Water Resour.* Res., 29(10), 3417-3431.

Pfleiderer, S. and Moltyaner, G.L. (1993) The use of velocity and conductivity data for the quantification of heterogeneity: a comparison, Water *Resour.* Res., 29(12), 4151-4156.

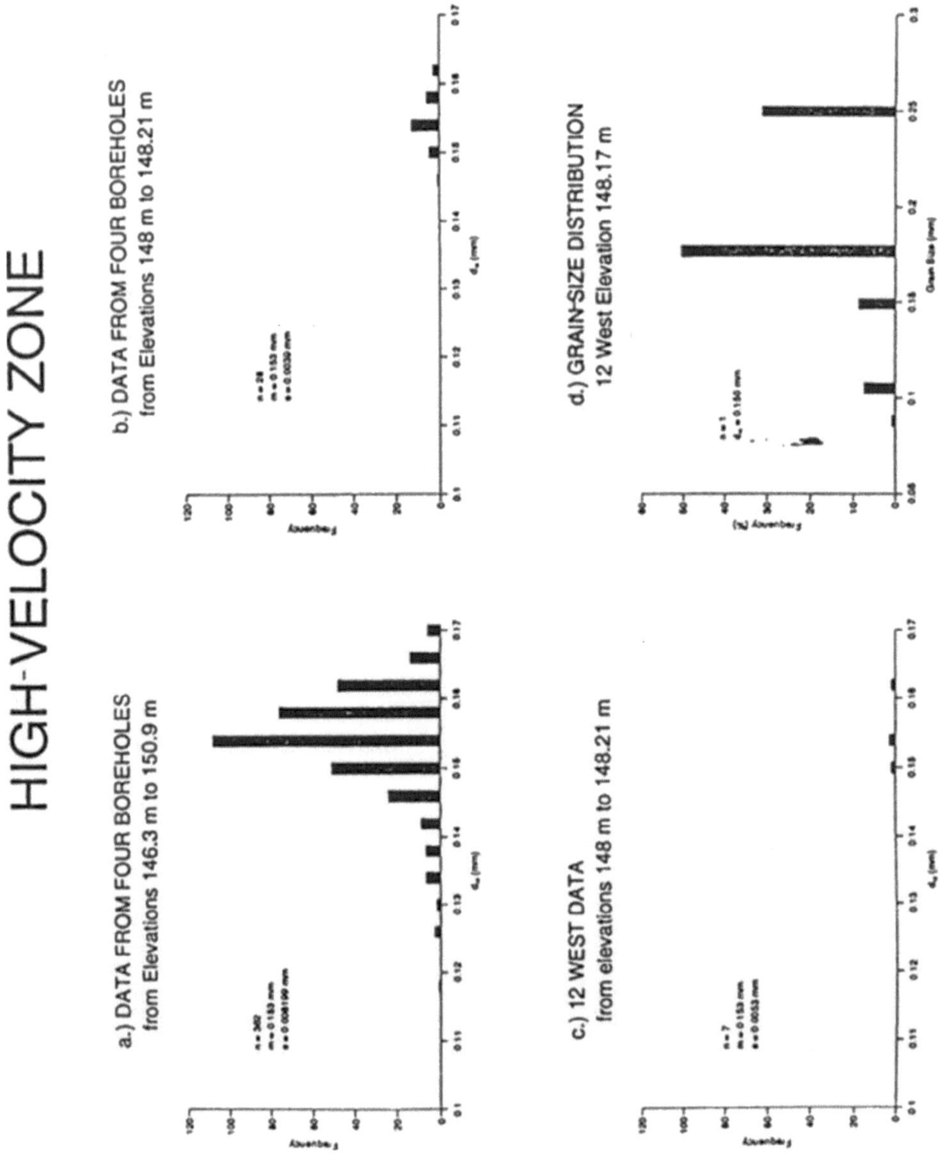

Figure 1. Histograms of Grain Size and Effective Grain Diameter Distributions for the High-Velocity Zone

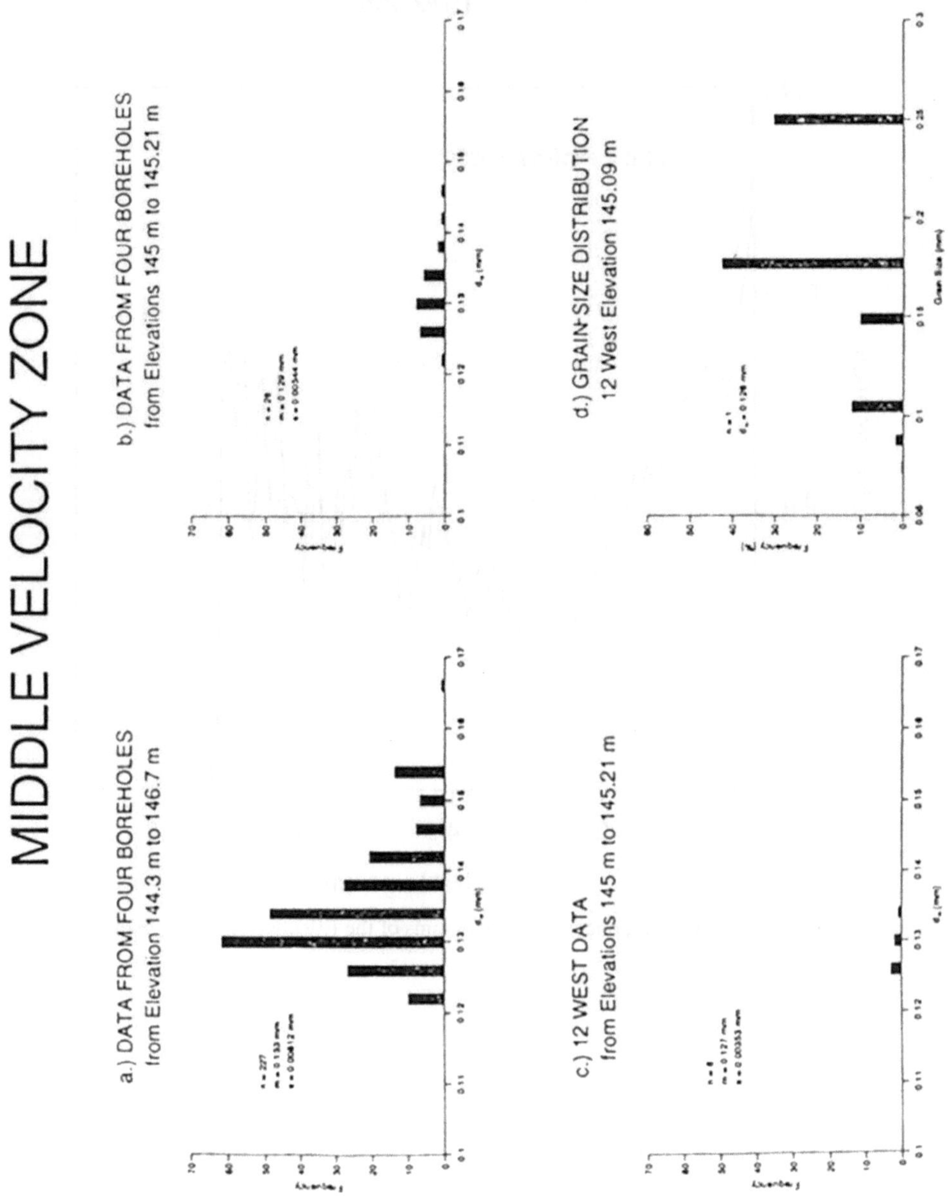

Figure 2. Histograms of Grain Size and Effective Grain Diameter Distributions for the Middle Velocity Zone

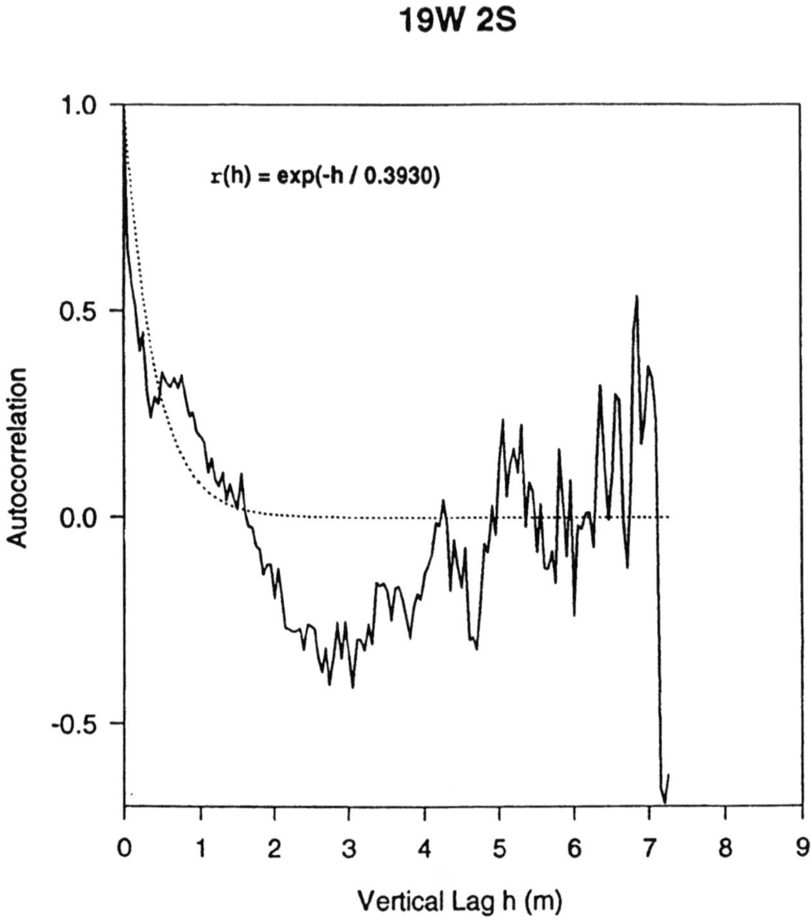

Figure 3. Sample and Fitted Exponential Correlograms of the Effective Grain Diameter Function

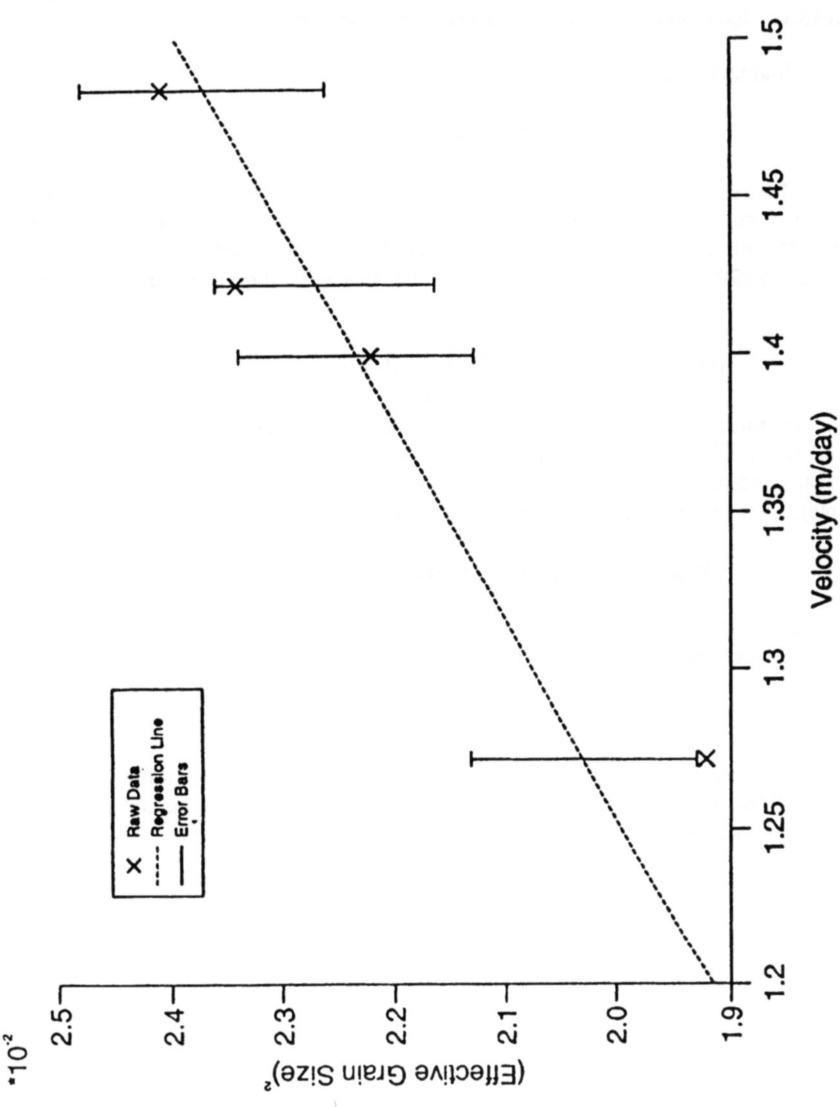

Figure 4. Correlation between Velocity and Effective Grain Size for Individual Layers

Appendix: Equation of Continuity and Groundwater Flow

Mass of fluid M_f in the local-scale volume is defined as

$$M_f=\iiint\rho_f(\mathbf{x})d\mathbf{x} \tag{1a),}$$

where $\mathbf{x}=\{x_1,x_2,x_3\}$ and the integration is over the local-scale volume. According to the principle of conservation of fluid mass, the groundwater flowing into the local-scale volume is accounted for by outflow from the volume, some change in bulk fluid density within the volume, or some combination of those processes. The time rate of change of fluid mass in the local-scale volume is

$$dM_f/dt=d(\iiint\rho_f(\mathbf{x},t)d\mathbf{x})/dt=\iiint(\partial\rho_f(\mathbf{x},t)/\partial t)d\mathbf{x} \tag{2a}$$

In (2a), the integral and differential commute, since the mass fluid content in the local-scale volume remains fixed in time and the partial derivative is used, instead of the total derivative, since M_f is independent of $\mathbf{x}$. The amount of fluid, $\rho_f\mathbf{u}ds$, that crosses surface S (of the local-scale volume), having element ds, is due to the local-scale velocity of flow. Then, by Gauss' theorem

$$\iint\rho_f\mathbf{u}ds=-\iiint \mathrm{div}(\rho_f\mathbf{u})dx=\iiint(\partial\rho_f(\mathbf{x},t)/\partial t)dx \tag{3a}$$

On the left-hand side of (3a) the integration is over surface S. Equating (2a) and (3a), we arrive at the continuity equation for the fluid phase

$$\partial\rho_f/\partial t+\mathrm{div}(\rho_f\mathbf{u})=0 \tag{4a}$$

Equation (4a) is mathematically identical to the continuity equation used in fluid mechanics. The only difference is that equation (4a) includes the fluid bulk density, instead of the mass density. For incompressible and homogeneous fluid, equation (4a), is reduced to

$$\mathrm{div}(n\mathbf{u})=0 \tag{5a}$$

The introduction of the specific discharge vector leads to the following equation

$$\mathrm{div}(\mathbf{q})=0 \tag{6a}$$

From (5a) and (6a) it follows that although the specific discharge vector is divergence-free in a heterogeneous geologic medium, this is not the case for the groundwater flow velocity, **u**. That is why a tracer test demonstrates velocity variability not observable with conventional hydrogeologic measurements. The groundwater velocity is divergence-free only in a homogeneous medium under a condition of constant porosity

$$\mathrm{div}(\mathbf{u})=0 \tag{7a}$$

The groundwater flow equation is obtained by the substitution of Darcy's law into (5a)

$$\mathrm{div}(K\ \mathrm{grad}\ h)=0 \tag{8a}$$

GEOSTATISTICAL STUDY OF NITRATE CONTENTS IN THE AQUIFERS OF *CAMPO DE DALIAS* (SE SPAIN)

JIMÉNEZ-ESPINOSA, R.[1]; MOLINA-SÁNCHEZ, L.[2]; PULIDO-BOSCH, A.[3]; NAVARRETE, F.[2]

(1) Dpto. Geología. Escuela Politécnica Superior. Univ. Jaén. Av. Madrid, 35, 23071, Jaén.
(2) Dpto. Hidrogeología y Química Analítica. Facultad de Ciencias Experimentales. Univ. Almería. Cañada de San Urbano, s/n. 04071, Almería.
(3) Dpto. Geodinámica. Facultad de Ciencias. Univ. Granada. Avda. Fuentenueva, s/n, 18071, Granada.

Abstract

The Campo de Dalías region is economically the most important agricultural area in the province of Almería (SE Spain). In this area, agricultural activities cover more than 200 km^2 and the water used in these farming activities is supplied from many wells drilled in the aquifers of the region. Seasonal nitrate contents in the aquifers of Campo de Dalías have been studied from 1988 to 1990. We have considered two aquifer systems, S1 and S2, in which we have calculated some exploratory statistics, experimental and theoretical variograms, maps of space-time evolution of nitrates using ordinary kriging. The deepest system S1 is not very polluted, but there is an evidence of a trend in rising pollution in 1990. In system S2 there appears a trend in reduction of concentrations through time. Pollution is mainly related to the use of fertilizers. Geostatistical techniques are very useful to estimate the behaviour of nitrates in Campo de Dalias waters and to locate areas with serious problems of farming pollution.

Key words: Campo de Dalías, agricultural activities, nitrates, variogram, ordinary kriging.

A. Soares et al. (eds.), geoENV I–Geostatistics for Environmental Applications, 139–151.

1. Introduction

The Campo de Dalías is in eastern Andalucia (south of Spain), and to the west of the Bay of Almeria. This area occupies nearly 330 km^2 The relief is smooth and the area lies between the Sierra de Gador, and an important carbonate mountain alignment, and the Mediterranean sea (Fig. 1). This region has a semi-arid Mediterranean climate. Dry, warm summers, and springs and autumns with local, violent rains, concentrated mainly in the Sierra de Gador mountains. This is the only way the aquifer layers in the area are refilled.

This zone has experienced an important economic development due to farming activities, being one of the most important agricultural areas in Andalucia, and even in Spain. The climate is favourable, with many hours of sunshine during the year. This allows the cultivation of different class of crops during all the year. Therefore, intensive greenhouse activities are highly spread out in this region. This increase in farming activities implies a parallel rise in the need of water and continuos damage in water quality. The agricultural water is supplied principally from many wells drilled in the aquifers of the zone. It is important to highlight that the averages yearly recharge for these aquifers are around 50 Hm^3, while approximately 100 Hm^3 are being pumped.

The main objective of this work is to establish the time and spatial evolution of the nitrate contents in the groundwater of the area. We have computed several basic statistics, experimental and theoretical variograms, and using these, we have estimated the maps of different periods using ordinary kriging.

2. Geological and hydrogeological framework

Geologically, the outcrops of the area can be grouped in two broad categories: pre-orogenic and post-orogenic materials (Fig. 1). In the former, called Alpujarride materials, two distinct groups of units appear: the Gador and the Felix. In the Gador units, from bottom to top, a member of phyllite and quartzite can be distinguished, upon which rests a carbonate series of about 1000 m in thickness. In the Felix units, there also are two members; the lower one comprising phyllites, schists and quartzites; and the upper carbonate member, composed mainly of dolostones. The age of these both formations is

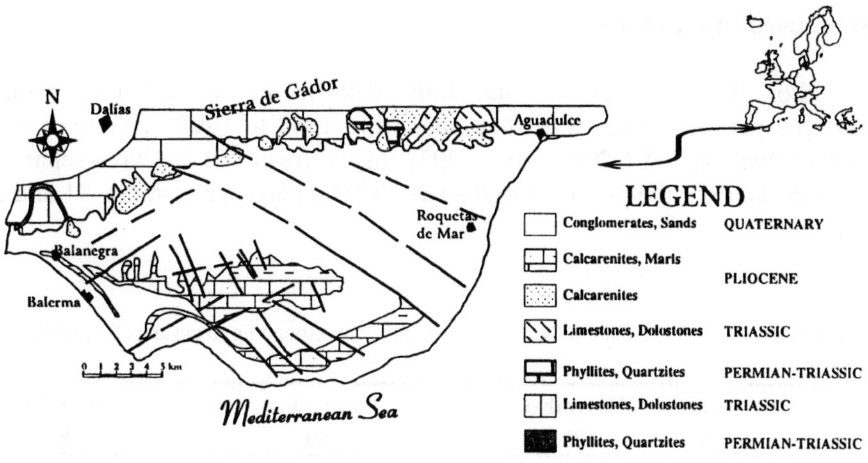

Figure1. Geographical location and geological map of the study area.

approximately Permian to Middle-Upper Triassic (Pulido et al., 1993). No deposits are known between Upper Triassic and Lower Miocene. The Upper Miocene materials outcrop in the northern margin of the Campo de Dalías, over limestones and dolostones and over phyllites and quartzites of both units. The Pliocene is represented by a lowermost conglomerate upon which a thick marl marine formation was deposited. Calcarenites, reaching over 100 m thick, are the top of this formation. The Quaternary materials are widely represented, with diverse facies: from alluvial fans to older marine Quaternary deposits (Fourniguet, 1975).

From a hydrogeological point of view, three different and complex aquifer units can be distinguished: Aguadulce, Balanegra and Balerma-Las Marinas (Pulido-Bosch, et al., 1989). However, the aquifer formations correspond mainly to: (i) carbonate materials from preorogenic sequences, (ii) Pliocene calcarenites and Quaternary deposits, each separated by impermeable formations of pelitic rocks. Therefore, in this study we have considered two aquifer systems: a) the system *S1*, made of Triassic limestones and dolostones; b) the system *S2* comprising calcarenites and detritic Quaternary materials.

3. Data used in the study

For the purposes of this study data from two different sample campaigns per year, May and November from 1988 to 1990, have been used (Molina-Sánchez, 1989; Navarrete, 1992). In the treatment we have separated data from each sample period in to the two different aquifer systems, S1 and S2 (Table 1).

TABLE 1. Number of data in the different months corresponding to systems S1 and S2.

	May 88	November 88	May 89	November 89	May 90	November 90
S1	45	51	47	41	44	46
S2	65	62	59	64	59	60

4. Basic exploratory statistics

Analysing the time evolution of the different basic statistics of S1 (Fig. 2), an analogous trend can be deduced for months from May'88 to May'90. Average values, first and second quartiles (Q1 and Q2) are close, with values less or equal to 20 mg/l, suggesting a general absence of high nitrate values in the study area for this period. On the other hand, maximum values for these months are quite far from third quartile (Q3), which can reveal that some wells have serious problems of pollution, with values clearly higher than 100 mg/l, mainly in the first years.

Data from Nov'90, shows a slight evidence of a different trend, with the basic statistics close. The nitrate average value (36 mg/l) is higher than the preceding ones, and its maximum is similar to Q3. These numerical features can show a generalized presence of high nitrate values in the area, showing a rising trend in the development of the pollutant.

About the aquifer system S2 (Fig. 3), data reveal a critical nitrate content, which give a classification of undrinkable to pumped water in this level. We find average values around 75 mg/l or higher. Values of Q3 are about

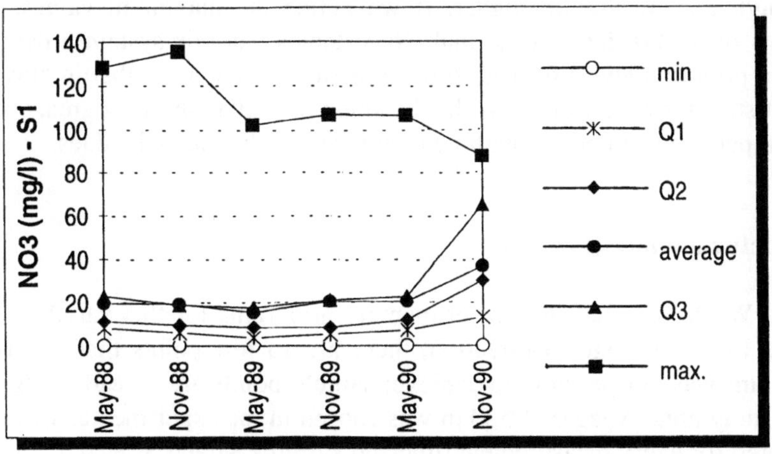

Figure 2. Time evolution of the basic statistics of system S1.

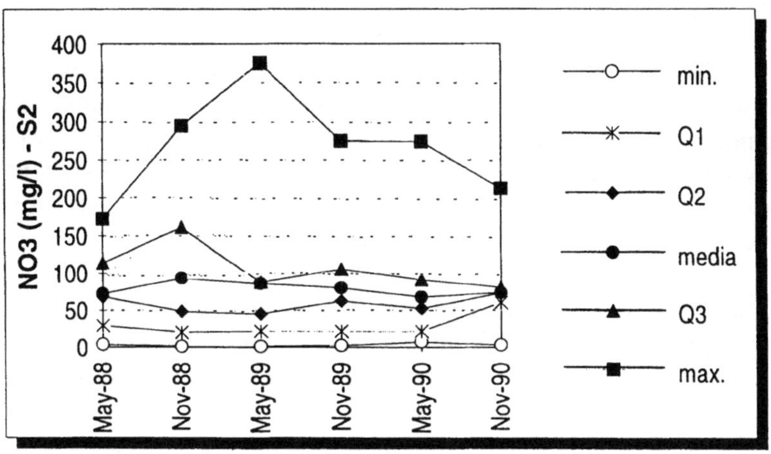

Figure 3. Time evolution of the basic statistics of system S2.

100 mg/l, and in Nov'88 higher. If we notice the maximum values, nitrate contents over 200 mg/l are normal. Nevertheless, in this system a decreasing trend in pollution levels through time pointed out. It appears that in Nov'90 all the statistics were very close, with all Campo de Dalías showing great values in the pumped wells, although with no anomalous high punctual values.

5. Structural analysis of NO_3

We have calculated experimental variograms for this variable only in N45°E. In the orthogonal direction, there are so few points that a consistent variogram was not possible to achieve; sample points are spread along N45°E in the study area. A lag of 1,500 m was chosen in the experimental variograms, computed for every measurement time and systems S1 and S2.

Parameters of the fitted models for these variograms are presented in Table 2 and the charts are shown in Figures 4 and 5, for S1 and S2, respectively. A nested model, composed of a nugget effect, a first spherical structure with a small range, a second spherical structure with a great range, is fitted for each variogram.

From a hydrogeological point of view, a distinction between the two aquifer systems must be made. System S1 presents more structured variograms with a nugget effect of around 25% of the experimental variance. Variograms from system S2 display values of a higher nugget effect, about 50% of the variance, showing an irregular behaviour of nitrates for this layer. Apart from C0, fitted parameters in S1 and S2 are quite similar:

a) A small structure corresponding to ranges about 1,500 m. That suggests a local variability for nitrates and represents two main high concentration areas, both in S1 and in S2, located near the Aguadulce and Balanegra villages, respectively.

b) A bigger structure with more fluctuating ranges (from 5,500 to 10,000 m). It indicates a variability for distances included between these two main polluted areas.

TABLE 2. Parameters of fitted variogram models in S1 and S2.

		C0	Model 1	C1	A1 (m)	Model 2	C2	A2 (m)
	May'88	223	sph.	20	1,500	sph.	250	7,500
	Nov'88	200	sph.	100	1,500	sph.	485	7,500
System	**May'89**	100	sph.	50	2,000	sph.	277	10,000
S1	**Nov'89**	185	sph.	50	1,500	sph.	500	10,000
	May'90	160	sph.	50	1,500	sph.	350	7,500
	Nov'90	157	sph.	75	1,500	sph.	500	7,500
	May'88	1,300	sph.	300	1,500	sph.	480	5,000
	Nov'88	5,000	sph.	1,500	1,500	sph.	2,035	10,000
System	**May'89**	4,000	sph.	750	1,500	sph.	3915	8,500
S2	**Nov'89**	3,000	sph.	750	1,500	sph.	1,800	10,000
	May'90	1,500	sph.	500	1,500	sph.	1,400	5,000
	Nov'90	200	sph.	100	1,500	sph.	840	10,000

6. Analysis of ordinary kriging maps

The final objective of this work is to calculate the shape of polluted zones based on the measurements in every time measurement. To infer the characteristics of nitrates we have assumed the variable in each period as second-order stationary. With this assumption, we have selected ordinary kriging to perform the different estimations of NO_3 through time.

The spatial analysis of system S1 in the different months (Fig. 6) shows the area of the aquifer of Aguadulce as that of greater concentrations in nitrates, concretely north of Roquetas de Mar. On the other hand, the highest values are located in west of the Campo de Dalías, around of the village of Balanegra (aquifer of Balanegra). In those areas an evolution from high contents in some specific wells to a generalized increasing trend over all the area are produced, appearing extensive zones with high values in 1990 November.

About the system S2 (Fig.7), it seems obvious that respect to the shallow hydrogeological system, the nitrate contents will be greater. In this system the evolution is different from S1, since it seems to have some very high concentrations in 1988 and 1989, with maxima very high, subsequently

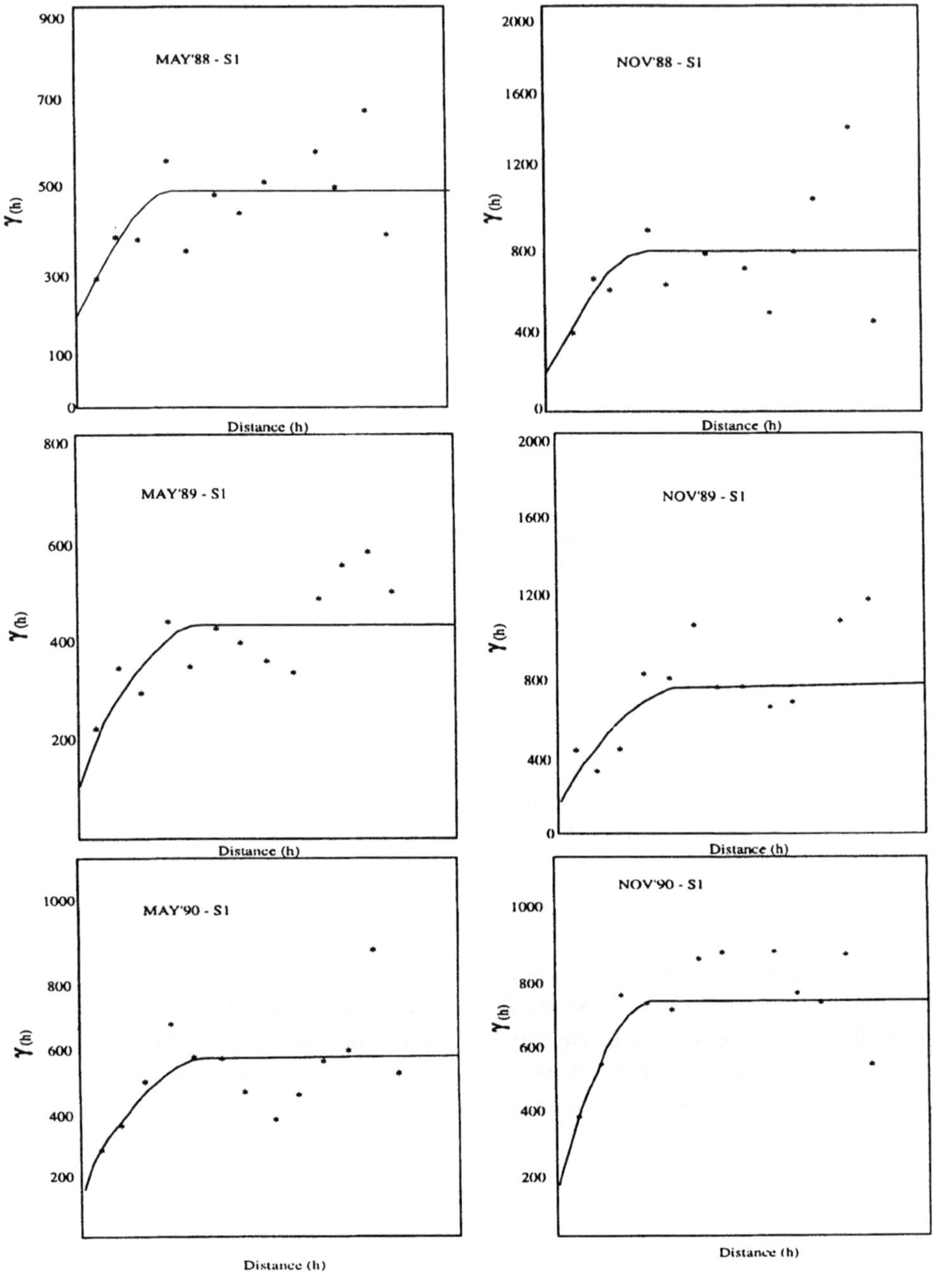

Figure 4. Variograms of system S1.

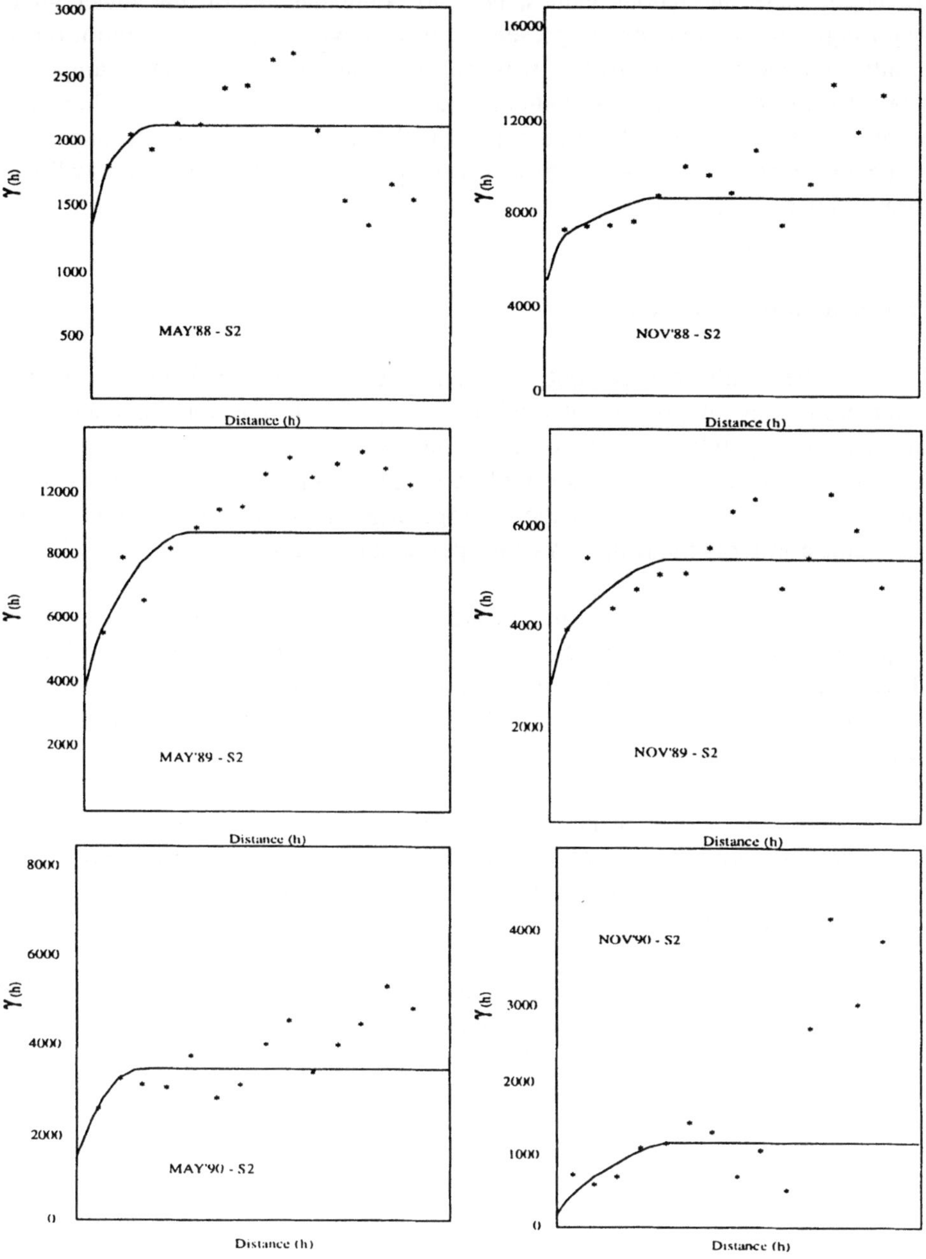

Figure 5. Variograms of system S2.

reducing these anomalous values, Nov'90 being the month that fewer areas with high nitrate contents are presented. In this system, great concentration of contours appears again toward western and eastern zones of the Campo de Dalías (area of Roquetas-Aguadulce and Balanegra, respectively). There are some wells that should correspond to the maxima, which produce the appearance of close, packed contours in maps of some months, specifically in May'90 is quite meaningful.

7. Discussion and conclusions

The study carried out has permitted the accomplishment of the characterization of groundwater of the Campo de Dalías for the content in nitrates during 1988 to 1990. We have considered two aquifer systems to unify data of all the Campo and to establish evolution maps of the pollutant. One must notice the complexity of the hydrogeological structure of the zone, since the same aquifer is made up of various permeable levels.

When we analysed the nitrate contents in the study area we can appreciate how the system S1 is not excessively influenced by pollution. However the trend is increasing, appearing already in 1990 areas with concentrations not excessively high, but generalized, and not just in some wells with local high values.

On the other hand, S2 presents a slight variation somewhat contradictory, with greater contents in the first years. Nevertheless, it presents a trend in reduction of the concentrations through time, allowing decreases in the nitrate values to appear.

The most important conclusion that is deduced from this study is that the temporary evolution of the contents in nitrates in these waters is growing in the deep aquifer levels. The origin of this pollution can be found in the use of fertilizers in the intense agriculture of the region.

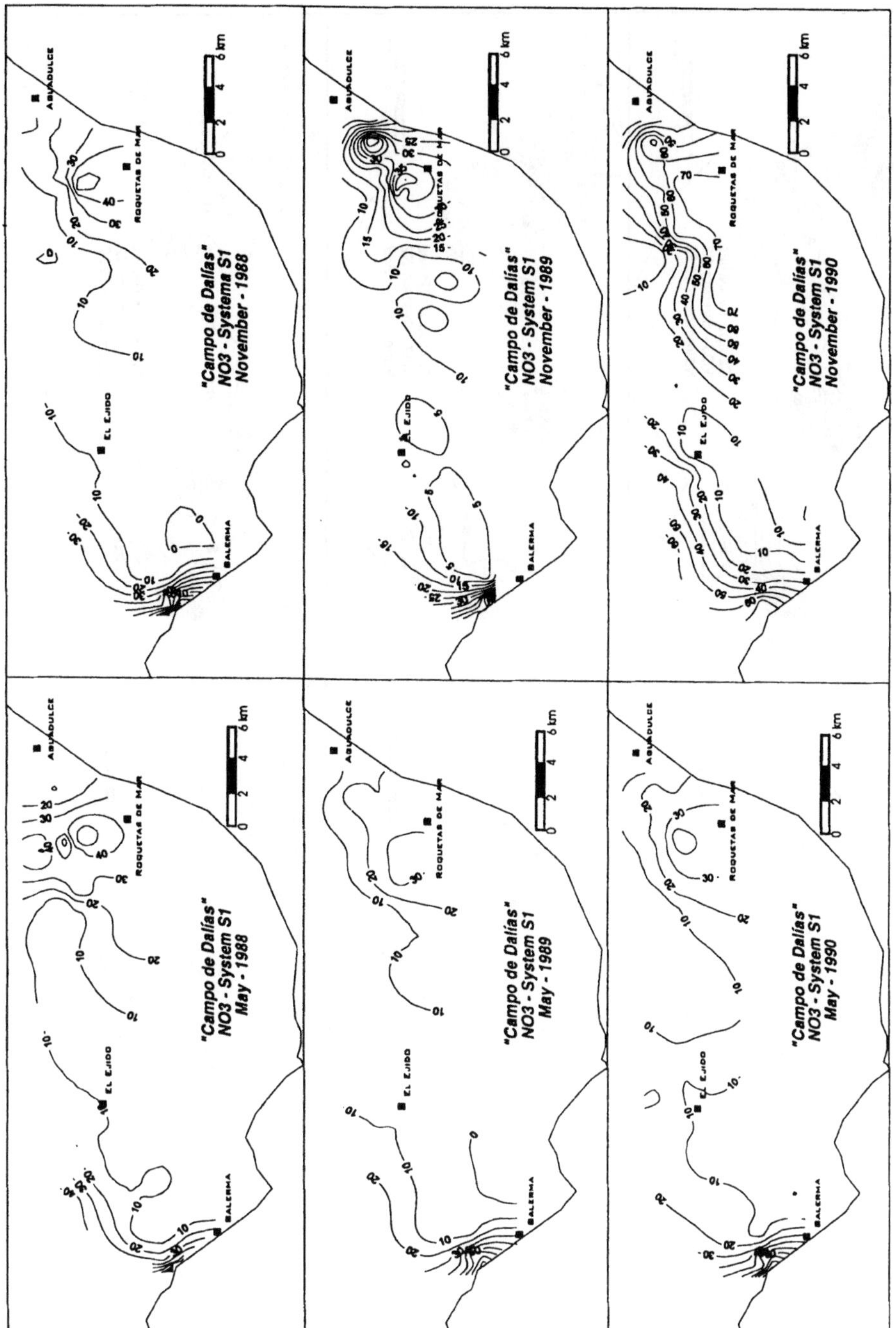

Figure 6. Ordinary kriging maps of system S1.

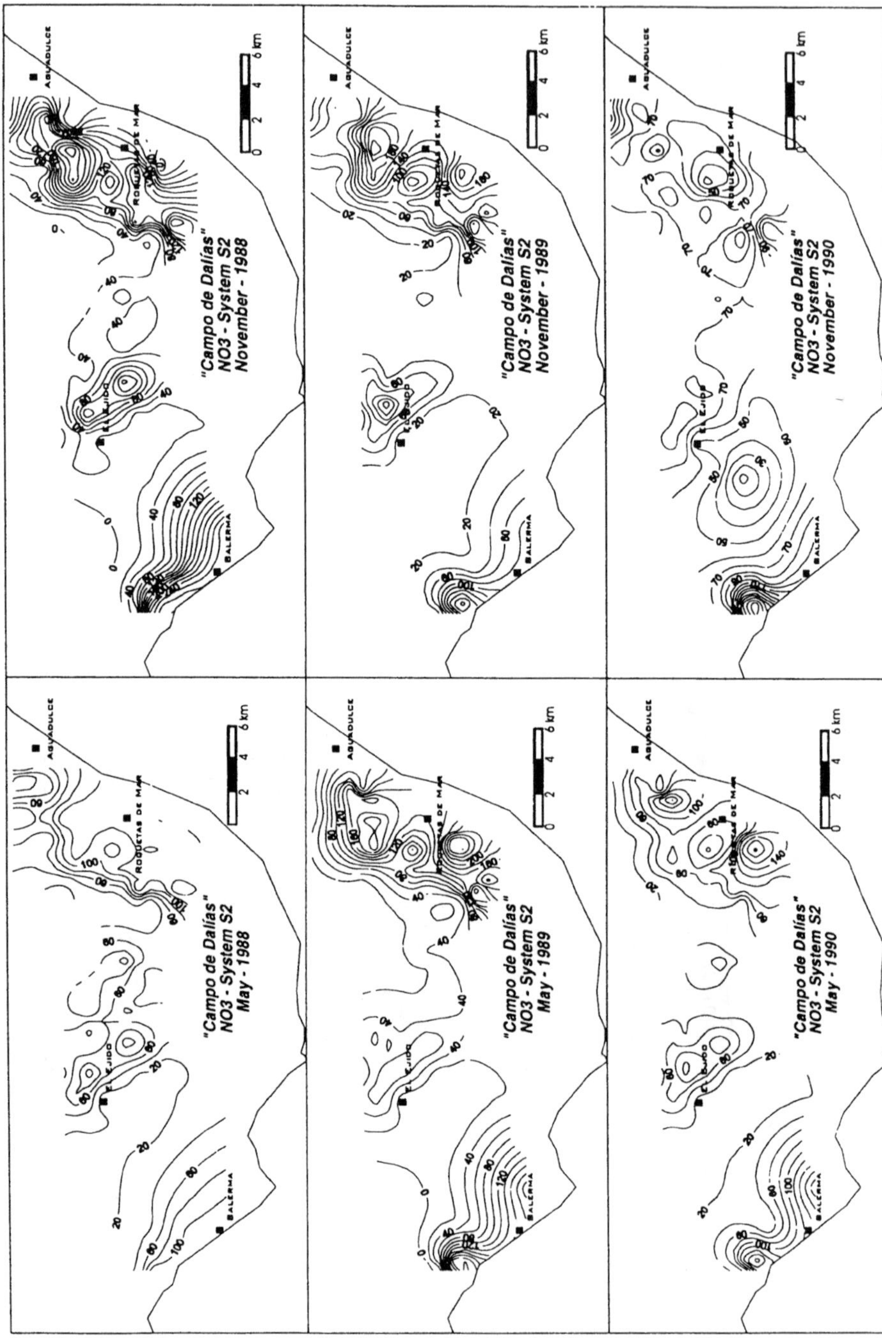

Figure 7. Ordinary kriging maps of system S2.

8. References

Aldaya, F.; Baena, J. and Ewert, K. (1983). *Memoria y Hoja Geológica de Adra (1:50.000). MAGNA (1057).* IG.M.E. Madrid.

Baena, J. and Ewert, K. (1983). *Memoria y Hoja Geológica de Roquetas de Mar (1:50.000). MAGNA (1058).* IG.M.E. Madrid.

Fourniguet, J. (1975). *Néotectonique et Quaternaire marin sur le litoral de la Sierra Nevada, Andalousie (Espagne).* Thèse Univ. Orléans.

Molina Sánchez, L. (1989). *Contribución al conocimiento hidrogeoquímico del sector oriental del Campo de Dalías (Almería).* Tesis de Licenciatura. Univ. Granada.

Navarrete López Cózar, F. (1992). *Contribución al conocimiento hidrogeoquímico del Campo de Dalías (Almería).* Tesis Doctoral. Univ. Granada.

Pulido Bosch, A.; Martínez Vidal, J.L.; Navarrete, F.; Alonso, C.; Macías, A.; Molina, L.; Padilla, A. y Benavente, J. (1989). *Caracterización hidrogeoquímica del Campo de Dalías (Almería).* I.A.R.A. Granada.

Pulido Bosch, A.; Navarrete, F.; Molina, L.; Vallejos, A.; Martín Rosales, W.; and Martinez Vidal, J.L. (1993). In: Pulido Bosch, A. (ed), *The Campo de Dalías aquifers. In: Some Spanish Karstic Aquifers*, pp. 143-158.

Acknowledgments

This work was supported by the CICYT Spanish Research Project AMB95-0493. We would like to thank to Prof. H.R. Brewer for the final revision of the English manuscript.

EVALUATION OF REACTIVE TRANSPORT PROCESSES IN A HETEROGENEOUS POROUS AQUIFER WITHIN A NON-PARAMETRIC NUMERICAL STOCHASTIC TRANSPORT MODELLING FRAMEWORK BASED ON SEQUENTIAL INDICATOR SIMULATION OF CATEGORICAL VARIABLES

THOMAS PTAK
University of Tübingen, Applied Geology, Sigwartstrasse 10,
D-72076 Tübingen, Germany
Tel.: +49/7071/29-76991; Fax: +49/7071/5059
E-mail: thomas.ptak@uni-tuebingen.de

Abstract A non-parametric Monte Carlo type stochastic transport model is applied to simulate the simultaneous transport of nonreactive and reactive (sorptive) tracers in groundwater as observed during a forced gradient tracer experiment at the 'Horkheimer Insel' test site. In the modelling approach, at first categorical variables are defined based on aquifer material grain size distribution curves. Then conditional sequential indicator simulation of categorical variables is used to generate grain size distribution curves. Subsequently, hydraulic conductivity values (K) are derived from the simulated grain size distribution curves, and spatially variable retardation factors (R) are computed based on grain size related sorption data from laboratory experiments. Multiple realizations of the K and R fields are generated to simulate the observed tracer transport stochastically. According to the results, the stochastic transport model seems suitable for highly heterogeneous aquifers such as at the 'Horkheimer Insel' test site.

1. Introduction

In order to investigate reactive solute transport processes within a highly heterogeneous porous aquifer and in order to obtain a detailed data base for the development of numerical transport models, forced gradient tracer tests at small to intermediate transport distances were conducted at the 'Horkheimer Insel' field site in South Germany. During one of the experiments, two tracers, one non-sorbing (Fluoresceine) and one sorbing (Rhodamine WT), were injected simultaneously into the same groundwater monitoring well, located 8.9 m from the pumping well. For the evaluation of the experiments, the computer model needed to simulate the near-source non-reactive (non-sorbing) and reactive (sorbing) solute transport in groundwater within the highly heterogeneous aquifer requires a detailed knowledge of the spatial distribution of aquifer parameters controlling the transport. At present, using available subsurface investigation techniques it is not possible to obtain a description of the subsurface structure and its properties in a resolution needed for a near-source deterministic reactive transport model

A. Soares et al. (eds.), geoENV I–Geostatistics for Environmental Applications, 153–164.

in case of a heterogeneous aquifer such as at the 'Horkheimer Insel' field site. Then modelling can be performed within a stochastic framework in order to account for the remaining parameter uncertainty.

From stochastic groundwater flow and transport theory, closed-form (first-order) analytical solutions or combined numerical-analytical methods are available for both nonreactive and reactive solutes (e.g. Dagan, 1989; Cvetkovic & Shapiro, 1990; Selroos and Cvetkovic, 1994; Bellin & Rinaldo, 1995 and Selroos, 1995). In order to obtain the closed-form solutions, a small variance of hydraulic conductivity ($\sigma^2_{\ln K}$) has to be assumed in general. Other limiting assumptions may be that the aquifer is infinite and the groundwater flow field is uniform and stationary, and the analytical solutions generally only provide ensemble mean results. Other authors employed numerical approaches for modelling of nonreactive and reactive solute transport. For example, Teutsch *et al.* (1990) and Ptak (1993) used a flexible Monte Carlo method for the analysis of nonreactive tracer transport. Tompson (1993) performed particle tracking simulations of reactive solute transport in physically and chemically heterogeneous porous media with a correlation of hydraulic conductivity and sorption properties. Burr *et al.* (1994) assumed that hydraulic conductivity and distribution coefficient fields are inversely correlated and that they have the same spatial correlation structure.

In this paper a different approach, accounting for parameter uncertainty and variability, is applied for the modelling of the simultaneous transport of the nonreactive tracer (Fluoresceine) and the reactive, sorptive tracer (Rhodamine WT), since no significant correlation of hydraulic conductivity and distribution coefficient values could be established from laboratory measurements. In this approach, grain size distribution curves are first generated stochastically using the sequential indicator simulation technique. The geostatistical parameters needed for the simulation are obtained from investigated aquifer material parameter sets using a categorical variable approach. Then hydraulic conductivity values are derived from the grain size curves using an empirical relationship, and spatially variable retardation factors are computed based on grain size related sorption data from laboratory batch experiments on aquifer material samples. All data are used in a Monte Carlo type non-parametric stochastic transport model to evaluate one of the tracer tests. According to the results, the transport model seems suitable for highly heterogeneous aquifers such as at the 'Horkheimer Insel' test site.

2. Field site

The 'Horkheimer Insel' field site is located in the Neckar valley, north of Stuttgart. Fig. 1 shows the site map and the location of the sampling and monitoring wells.

The aquifer is formed by a 2.5 - 4 m thick sequence of poorly sorted alluvial sand and gravel deposits of holocene age. From pumping tests, the geometric mean hydraulic conductivity was determined to 0.012 m s^{-1}. The underlying Triassic clay- and limestone bedrock formation is several orders of magnitude less permeable and can be considered hydraulically tight. During construction of the wells, 100 mm diameter core samples of the aquifer material were obtained. Based on point scale hydraulic conductivity values (K) from sieve analyses, the variance $\sigma^2_{\ln K}$ was estimated to 2.34 (Schad, 1993). This is characteristic for a heterogeneous aquifer.

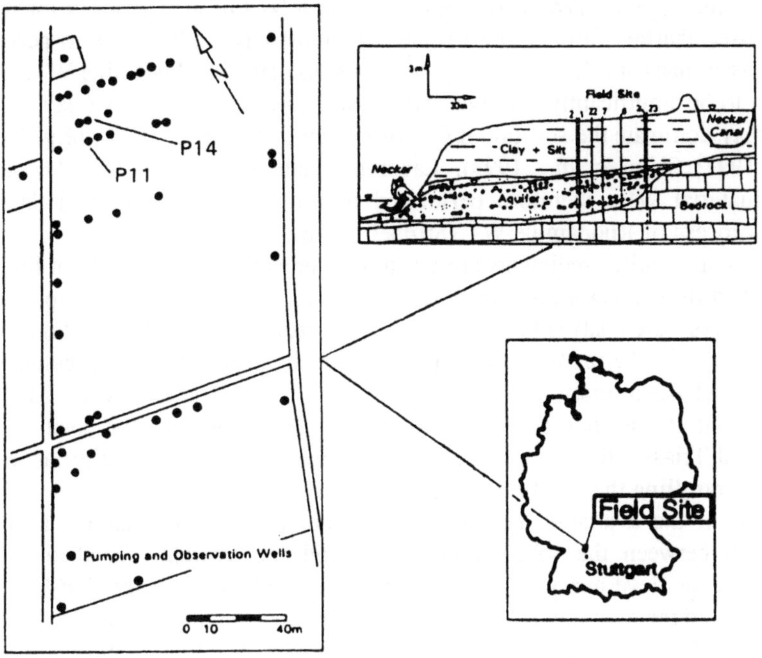

Figure 1. Site map and well locations.

3. Dual-tracer transport field experiment

A forced gradient tracer test method was applied for the short transport distance of 8.9 m investigated here. In the experiment (FGTT9), groundwater was pumped out of a monitoring well (P11, Fig. 1) at a constant rate of 3.02 l s^{-1}. After a quasi stationary radially convergent flow field was established, the two tracers (10 g Fluoresceine and 8 g Rhodamine WT) were injected instantaneously across the entire saturated aquifer thickness into a neighbouring monitoring well (P14, Fig. 1).

Seven multilevel breakthrough curves with a temporal resolution of 1 min, each representative of a vertical section of 0.3 m thickness, were measured within the pumping well for both tracers directly in-situ using fibre-optic fluorimeters. A detailed description of the experimental setup, the instrumentation, the results and interpretation is given in Ptak & Schmid (1996). The experimental results indicate the presence of physical and chemical aquifer heterogeneity, suggesting a three-dimensional transport modelling approach. Examples of the measurements will be presented together with results from the numerical simulation.

4. Laboratory batch experiments

Fluoresceine and Rhodamine WT are both polar and ionizable. From a practical point of view, if the organic carbon content (C_{org}) is low (as low as 0.2 mg/g for the Horkheimer Insel aquifer (Strobel, 1996)), it is in most cases admissible to accept Fluoresceine as a quasi ideal, non-sorbing tracer (Käss, 1992). Rhodamine WT, however, is known to be significantly sorptive. Adsorption on mineral surfaces (clay, carbonate, quartz) may dominate the sorption if C_{org} of the aquifer material is low (e.g. Sabatini & Austin, 1991; Shiau *et al.*, 1993). Laboratory studies using the Horkheimer Insel aquifer material were therefore performed in order to obtain detailed data on the sorption process of Rhodamine WT (Strobel, 1996).

In general, aquifer sediments are a heterogeneous mixture of different lithocomponents with different grain size fractions. Therefore aquifer material samples of about 5 cm thickness were analyzed petrographically in order to obtain the lithological composition. Sorption isotherms were then measured for the different lithocomponents as well as for the individual grain size fractions in batch sorption experiments. The analysis revealed, that it was not necessary to regard the lithological composition (quartz, Jurassic and Triassic limestones, Keuper and Bunter sandstones, calzite) as an aquifer property controlling the sorption process.

Based on the available data base it was not possible to establish a significant relationship between the sorption properties and the hydraulic conductivity values (estimated from sieve analyses with the empirical method by Beyer (1964)), which is probably due to the high nonuniformity of the heterogeneous aquifer material. However, a relationship between sorption capacity and grain size could be established based on a Freundlich isotherm model. The relationship is:

$$K_{Di} = \frac{1}{n_i} \cdot K_{Fri} \cdot c_{eq}^{\frac{1}{n_i}-1} = 0.85 \cdot 0.132 \cdot d_i^{-1.0} \cdot c_{eq}^{0.85-1} \qquad (1)$$

where K_{Di} is the distribution coefficient [l kg^{-1}] at sorption equilibrium, K_{Fri} is the Freundlich coefficient [l kg^{-1}], c_{eq} [μg l^{-1}] is the equilibrium concentration, n_i^{-1} [-] is the Freundlich exponent of a grain size fraction i, and d_i is the representative diameter [mm] of i.

Within the batch experiments, sorption equilibrium was achieved almost immediately, which is typical for surface sorption. Due to the resulting high Damkoehler number, local equilibrium sorption could be assumed within the transport simulations. Since equation (1), describing the sorption process, is based on grain size, it was necessary to provide information about the spatial distribution of grain size fractions as a model input, which will be described in the following.

5. Three-dimensional numerical stochastic transport simulation

The three-dimensional Monte Carlo type transport simulation technique comprises three main steps. First, a geostatistical analysis of grain size distribution curves obtained from the core samples is performed. Second, equiprobable three-dimensional realizations of the spatial distribution of grain size distribution curves are generated using an indicator based geostatistical simulation approach. From the simulated grain size distribution curves, obtained at each nodal position of the simulation domain, hydraulic conductivity values and local effective distribution coefficients are computed. Third, flow and transport simulations are performed for the two tracers within each generated hydraulic conductivity field, with initial and boundary conditions and geometry according to the tracer experiment. This yields an ensemble of realizations of the nonreactive and reactive transport processes, which can be compared with the field measurements.

5.1 GEOSTATISTICAL ANALYSIS OF GRAIN SIZE DISTRIBUTION CURVES

A structural analysis yielding parameters of the spatial structure of grain size distributions can be performed using categorical variables which are defined based on measured grain size distribution curves. This approach was used by Schad (1993). Categorical variables were considered as qualitative attributes of the individual aquifer material samples. Out of 243 measured grain size distribution curves (observations), curves with similar properties (shapes) were clustered into six groups (six clusters) with the multivariate statistical method of KMEANS clustering (McQueen, 1967). Each grain size distribution curve was discretized into seven classes, represented by class variables, which describe the mass fractional contributions of individual grain size fractions to the total sample mass (Fig. 2).

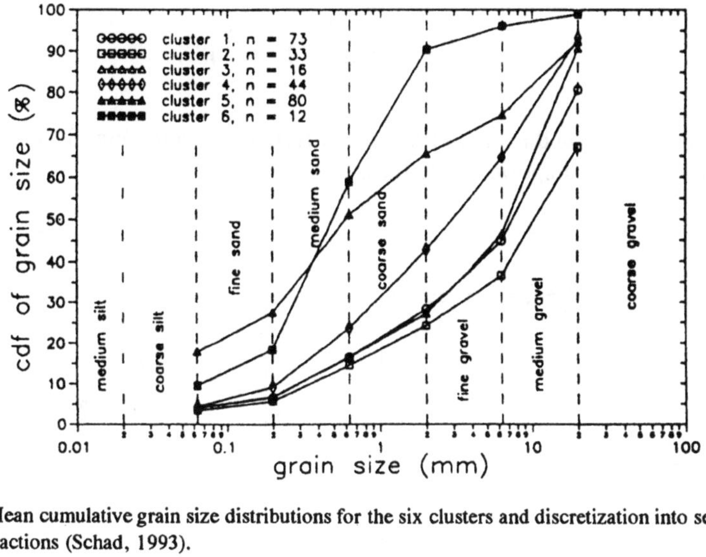

Figure 2. Mean cumulative grain size distributions for the six clusters and discretization into seven grain size fractions (Schad, 1993).

In the KMEANS clustering, the total sum of squares within a cluster is minimized:

$$\Phi = \sum_{i=1}^{k} \sum_{j=1}^{p} \sum_{m=1}^{n_i} (x_{\nu_{i,m},j} - \bar{x}_{i,j})^2 \qquad (2)$$

where $\nu_{i,m}$ is the row index of the m-th observation in the i-th cluster in a data matrix **X** (which is an a x b matrix with a observations and b class variables), $\bar{x}_{i,j}$ is the average of all values of a class variable j in a cluster i, n_i is the number of observations in a cluster i (number of rows in **X** assigned to cluster i), and where k is the number of clusters (here six) and p is the number of class variables (here seven).

After clustering, six binary categorical variables (cluster1 - cluster6) were defined for each observation. Since categorical variables are exclusive, one of these six variables was assigned the value 1 and the remaining five were assigned the value 0, thus yielding six binary data sets, one for each cluster. Then, the statistics (mean and variance) of the seven class variables were computed for each of the six clusters. Fig. 2 shows the resulting means for each cluster.

Subsequently, Schad (1993) investigated the spatial correlation of the six categorical variables by computing experimental variograms and fitting exponential variogram models in vertical and two horizontal directions.

The experimental variogram is defined by:

$$\gamma_i^*(\underline{h}) = \frac{1}{2N(\underline{h})}\sum_{i=1}^{N(\underline{h})}\left[z(\underline{x}_i+\underline{h}) - z(\underline{x}_i)\right]^2 \qquad (3)$$

$N(\underline{h})$ is the number of data pairs $z(\underline{x}_i+\underline{h})$, $z(\underline{x}_i)$ separated by the distance $\underline{h}$. The exponential variogram model is written:

$$\gamma(h) = c_o + \omega \cdot (1 - e^{-\frac{h}{\lambda}}) \qquad (4)$$

where c_o is the nugget, ω the sill and λ is the correlation length. Fig. 3 shows examples of variograms for the categorical variables.

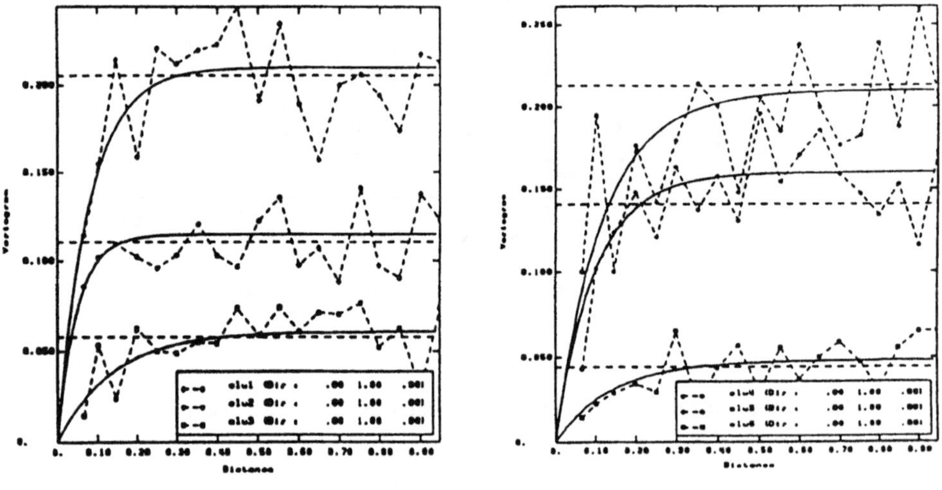

Figure 3. Example variograms for categorical variables, vertical direction (Schad, 1993).

Schad (1993) found for the clusters correlation lengths between 0.08 m and 0.12 m in vertical direction, and between 2 m and 10 m for the two principal horizontal directions, indicating a structural anisotropy of the heterogeneous aquifer.

5.2 GENERATION OF THREE-DIMENSIONAL HYDRAULIC CONDUCTIVITY AND DISTRIBUTION COEFFICIENT FIELDS

In the next step, the experimental histogram of the clusters (cluster1 - cluster6) and the cluster variogram models are used to generate conditioned equiprobable three-dimensional realizations of the categorical variable field. For the non-parametric approach used here, the three-dimensional conditional sequential indicator simulation method (SIS) for categorical variables (Deutsch & Journel, 1992) is applied. This method is based on an indicator kriging approach. It is able to honour extreme values and allows for consideration of more than one spatial structure of the investigated data.

Each simulated cluster from the categorical variable field represents a mean grain size distribution (Fig. 2), consisting of the means of the seven class variables. Subsequently, the means of these class variables are perturbed with a normally distributed random component following the individual class variable statistics estimated from the measurements. The resulting simulated grain size distribution curves are then evaluated in terms of local hydraulic conductivity values K and local effective distribution coefficients K_D. K is obtained using Beyer's (1964) empirical relationship between hydraulic conductivity and grain size distribution:

$$K = c(u) \cdot d_{10}^2 \tag{5}$$

where c(u) is an empirical constant, u is defined as d_{60}/d_{10} and d_{10} and d_{60} are the diameters of the grains where 90% respective 40% of the sample mass are retained in a sieve analysis. The local effective K_D value is obtained from:

$$K_D = \sum_{i=1}^{m} x_i \cdot K_{Di} \tag{6}$$

where m is the number of grain size fractions (here 7), x_i [-] is the mass fractional contribution of each grain size fraction and K_{Di} is the distribution coefficient following equation (1).

As an example, Fig. 4 shows a section from a layer of the generated hydraulic conductivity field. From Fig. 4 the heterogeneity and the structural anisotropy obtained from the SIS simulation of categorical variables can be recognized.

For the Monte Carlo analysis presented here, twenty realizations of the cluster field measuring 80 m x 80 m x 2.4 m were generated. Even though twenty realizations may be not enough for a full Monte Carlo analysis, it is possible to identify trends associated with parameter uncertainty and to assess the variability of predicted parameters.

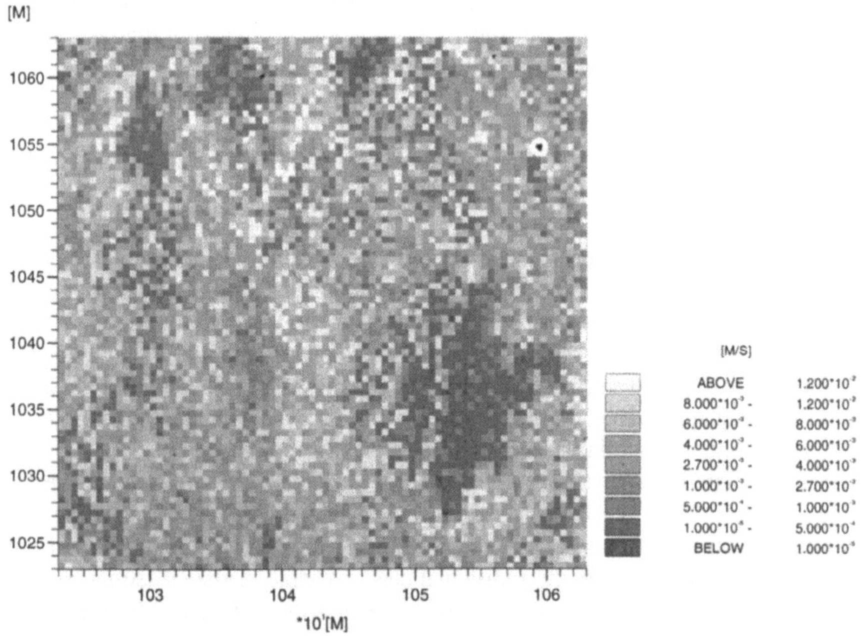

Figure 4. Section from a layer of the generated hydraulic conductivity field.

5.3 FLOW AND TRANSPORT SIMULATIONS

For the flow and transport modelling, the domain simulated with the SIS-method was extended up to a size of 2086 m x 2086 m x 2.4 m, using an numerically estimated uniform effective K-value. Then the three-dimensional flow field within the tracer test was computed for each stochastic aquifer realization first, using the finite-difference code MODFLOW (McDonald & Harbaugh, 1984). Subsequently, for each geostatistical realization breakthrough curves were simulated at the pumping well for Fluoresceine without sorption and with sorption for Rhodamine WT, using an extended version of the program MT3D (Zheng, 1991). Simulation routines were added to the original MT3D code to allow for the grain size based description of the sorption process, which is finally introduced by a retardation factor R [-]:

$$R = 1 + \frac{\rho_b}{\theta} \cdot K_D \tag{7}$$

where ρ_b [kg m^{-3}] is the bulk density of the aquifer material and θ [-] is the porosity. K_D is the concentration dependent distribution coefficient, computed with equation (6). A variable effective porosity, following a correlation function of porosity (estimated from permeameter measurements) with hydraulic conductivity, was used for the transport calculations. No model calibration was performed using the tracer experiment results, as it was an aim to test the model predictions based only on measured input parameters.

6. Comparison of simulated and measured results

Since concentrations were measured only within the pumping well, the comparison of simulated and measured tracer spread was performed using breakthrough curves. In this comparison, each measured and numerically simulated breakthrough curve is evaluated individually, and effective transport parameters (e.g. peak concentration, peak concentration arrival time, second temporal moment, mean transport velocity, etc.) derived from one measured breakthrough curve are compared with a numerically simulated stochastic ensemble. The variance of a transport parameter within the ensemble is a measure of prediction uncertainty. In the following, some examples of the comparison are given. Since for each realization the transport computations for Fluoresceine and Rhodamine WT were performed using the identical hydraulic conductivity field, the differences in transport behaviour can be attributed to the sorption process only.

In Fig. 5, measured and numerically simulated peak concentration arrival times (peak times) at the pumping well are compared. For Fluoresceine, it is seen from Fig. 5, that within the vertical profile most of the measured peak time values are close to the arithmetic ensemble mean from the stochastic simulation, and mostly within the range of the simulated values. No bias is recognized. For Rhodamine WT, later peak concentration arrival times are predicted by the numerical model on average. However, the results still seem comparable. It should be remembered, that the transport model was not calibrated with respect to the transport process.

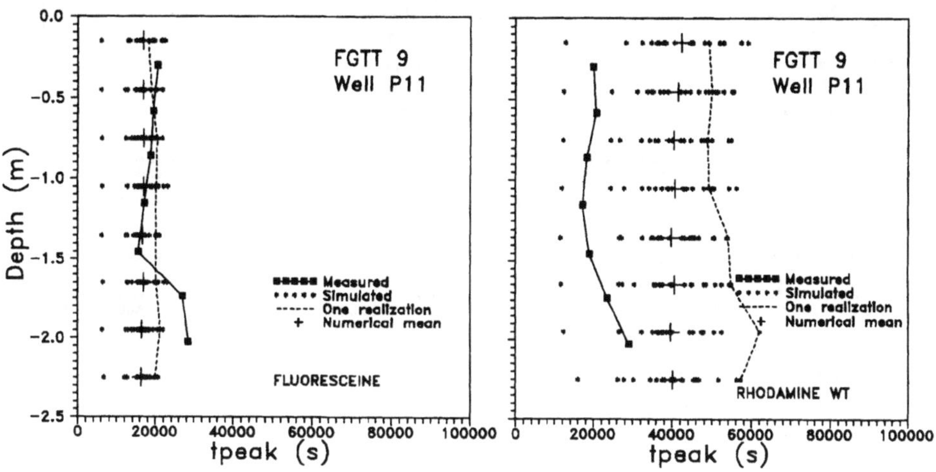

Figure 5. Comparison of measured and simulated tracer peak concentration arrival times at the pumping well (P11).

The variability, or prediction uncertainty, of the simulated peak time values is smaller for Fluoresceine compared to Rhodamine WT. In the case of Rhodamine WT, the variability of the peak times is increased by the variability of the stochastically generated effective retardation factors. Of course, a large variability cannot be expected, since the stochastic realizations are conditioned at the borehole locations and the transport distance of 8.9 m is shorter compared to the horizontal correlation lengths of some clusters, that can be following Schad (1993) up to 10 m in horizontal direction.

As an another example, Fig. 6 gives for both tracers the comparison of measured and simulated vertical profiles of peak concentration values at the pumping well. The measured peak concentration values are mostly close to the arithmetic ensemble means and mostly within the range of the numerically simulated concentration values. A deviation of measured transport parameters from their corresponding stochastic arithmetic ensemble means can be expected, because the measurements were performed in a near source, non-ergodic stage of tracer plume development.

The measured peak concentrations tend to be smaller compared to the simulated ensemble means. The deviations are larger for Fluoresceine. This might be caused by biodegradation of Fluoresceine, which was already observed in some former tracer experiments. The prediction uncertainty (variability of simulated values) is smaller for Rhodamine WT compared to Fluoresceine. A possible reason for this is the enhanced spread of the Rhodamine WT plumes due to sorption, which is recognized from the normalized second central temporal moments of the Rhodamine WT breakthrough curves, being about a factor 12 higher compared to Fluoresceine.

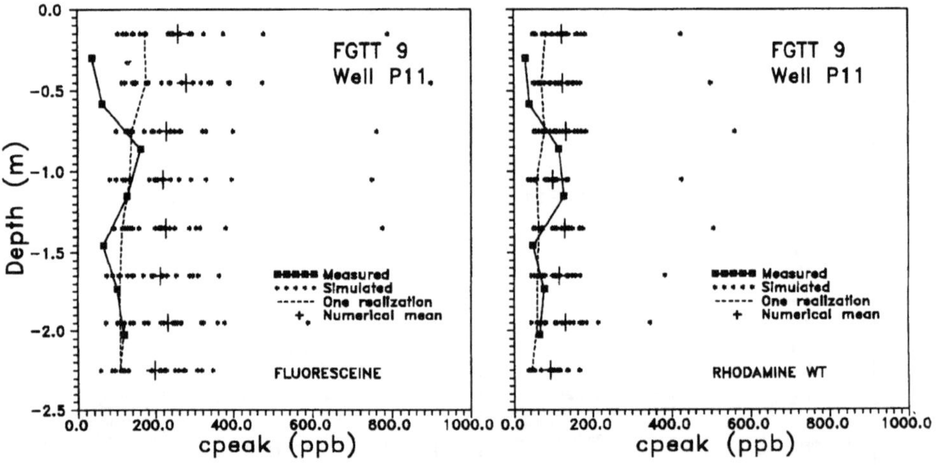

Figure 6. Comparison of measured and simulated breakthrough curve tracer peak concentrations at the pumping well (P11).

Due to the enhanced spread of the Rhodamine WT plume, its irregularities, caused by aquifer heterogeneity, are already more averaged out compared with the Fluoresceine plume. This (less non-ergodic situation) expresses in a reduced variability of the simulated peak concentration values.

Finally, Fig. 7 compares the measured and numerically simulated vertical profiles of effective retardation factors. The effective retardation factors were defined as quotients of the effective transport velocities of Fluoresceine and Rhodamine WT, estimated from the breakthrough curves. As expected from the comparison of the peak times, the numerical model yields at present mostly higher retardation factors compared to the measurements. However, the differences between measurements and numerical simulation still seem acceptable since the model was not calibrated with respect to transport.

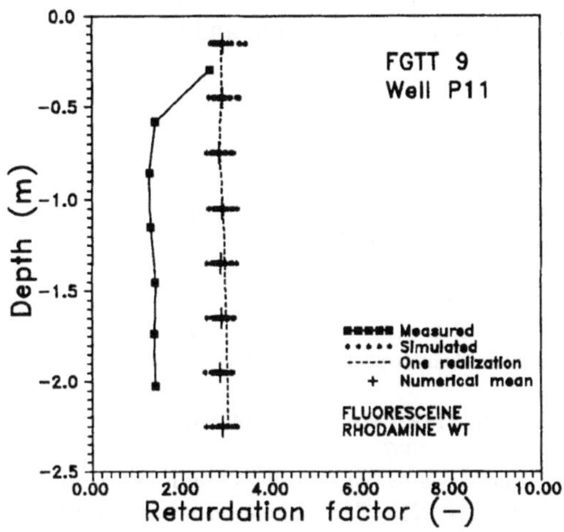

Figure 7. Comparison of measured and simulated effective retardation factors at the pumping well (P11).

A possible explanation for the deviation is that the numerical model is based on sorption related input parameters from batch experiments, which are known to yield systematically higher sorption capacities compared to a field situation. Performing column experiments might improve the results. Another reason might be attributed to the disturbance of the core samples during drilling activities, where the core is drained rapidly under gravity when the core barrel is lifted to the surface. The induced hydrodynamic forces most likely distribute fine grains within the core. On the other hand, within a heterogeneous aquifer such as at the 'Horkheimer Insel', near the source most of the tracer mass is transported in preferential zones, where almost no fine grains are present (open framework gravels). Since the sorption process described by equation (1) yields high K_D values for fine grains even if their mass fractional contribution is relatively small, the distribution of fine grains into these preferential transport zones during core sampling may express in higher simulated retardation factors based on laboratory measurements.

7. Conclusions

Following the first simulation results presented in this contribution, it is believed that the flexible numerical stochastic transport simulation technique based on categorical variables can be successfully applied in highly heterogeneous porous aquifers such as at the 'Horkheimer Insel' test site. The geostatistical approach using sequential indicator simulation of categorical variables and the grain size based formulation of the surface sorption process allows the generation of spatially variable hydraulic conductivity and retardation factor fields, without the need for an a-priori correlation function of hydraulic conductivities and distribution coefficients. The results from measurements and simulations are in a comparable range, even though the model was not calibrated with respect to transport. The stochastic simulation technique presented offers a broad field of applications, for example for the testing of upscaling of laboratory measurements to the field scale, or for the planning of remediation activities. Future work will comprise also further transport processes, e.g. diffusion and biodegradation.

8. References

Beyer, W. (1964) Zur Beschreibung der Wasserdurchlässigkeit von Kiesen und Sanden. *Zeitschr. f. Wasserwirtschaft-Wassertechnik*, **14**, 165-168.

Bellin, A. & Rinaldo, A. (1995) Analytical solutions for transport of linearly adsorbing solutes in heterogeneous formations. *Water Resour. Res.*, **31**(6), 1505-1511.

Burr, D.T., Sudicky, E.A. & Naff, R.L. (1994) Nonreactive and reactive solute transport in three-dimensional heterogeneous porous media: Mean displacement, plume spreading, and uncertainty. *Water Resour. Res.*, **30**(3), 791-815.

Cvetkovic, V.D. & Shapiro, A.M. (1990) Mass arrival of sorptive solute in heterogeneous porous media, *Water Resour. Res.*, **26**(9), 2057-2067.

Dagan, G. (1989) *Flow and transport in porous formations*. Springer, Berlin, F.R.G., 465 pp.

Deutsch, C.V. and Journel, A.G. (1992) *GSLIB Geostatistical software library and user's guide*. Oxford University Press, New York, 340 pp.

Käss, W. (1992) *Geohydrologische Markierungstechnik*. Lehrbuch der Hydrogeologie, Band 9, Verlag Gebrüder Borntraeger, Stuttgart, F.R.G.

McDonald, M.G. & Harbaugh, A.W. (1984) *A modular three-dimensional finite-difference groundwater flow model*. U.S.G.S. Open-File Report 83-875, National Center Reston, Virginia, USA.

McQueen, J. (1967) Some methods for classification and analysis of multivariate observations. *5th Berkeley Symposium on Mathematics, Statistics and Probability*, 1, 281-298.

Ptak, T. (1993) *Stofftransport in heterogenen Porenaquiferen: Felduntersuchungen und stochastische Modellierung*. Dissertation, Heft 80, Institut für Wasserbau, Universität Stuttgart, F.R.G.

Ptak, T. & Schmid, G. (1996) Dual-tracer transport experiments in a physically and chemically heterogeneous porous aquifer: Effective transport parameters and spatial variability. *J. Hydrol.*, **183**(1-2), 117-138.

Sabatini, D.A. & Austin, T.A. (1991) *Characteristics of Rhodamine WT and Fluorescein as adsorbing ground-water tracers*. Groundwater, **29**(3), 341-349.

Schad, H. (1993) Geostatistical analysis of hydraulic conductivity related data based on core samples from a heterogeneous fluvial aquifer. *International Workshop on Statistics of Spatial Processes*, Bari, Italy, 27.-30. Sept. 1993.

Selroos, J.-O. & Cvetkovic, V. (1994) Mass flux statistics of kinetically sorbing solute in heterogeneous aquifers: Analytical solution and comparison with simulations. *Water Resour. Res.*, **30**(1), 63-69.

Selroos, J.-O. (1995) Temporal moments for nonergodic solute transport in heterogeneous aquifers. *Water Resour. Res.*, **30**(7), 1705-1712.

Shiau, B.J., Sabatini, D.A. & Harwell, J.H. (1993) *Influence of Rhodamine WT properties on sorption and transport in subsurface media*. Groundwater, **31**(6), 913-920.

Strobel, H. (1996) *Sorption eines reaktiven Tracers (Rhodamin WT) in heterogenem Aquifermaterial vom Testfeld "Horkheimer Neckarinsel"*. Diplomarbeit, Geologisches Institut, Universität Tübingen, F.R.G.

Teutsch, G., Hofmann, B. & Ptak, T. (1991) Non-parametric stochastic simulation of groundwater transport processes in highly heterogeneous formations. *Proc. Int. Conference and Workshop on Transport and Mass Exchange Processes in Sand and Gravel Aquifers*, Oct. 1-4, 1990, Ottawa, Canada, AECL-10308, Vol. 1, 224-241.

Tompson, A.F.B. (1993) Numerical simulation of chemical migration in physically and chemically heterogeneous porous media. *Water Resour. Res.*, **29**(11), 3709-3726.

Zheng, Ch. (1991) *MT3D: A modular three-dimensional transport model*. S.S. Papadopulos and Associates, Inc., Rockville, Maryland, USA.

PARAMETER ESTIMATION FOR GROUNDWATER MODELS BY INDICATOR KRIGING

MARIA-TH. SCHAFMEISTER
Freie Universität Berlin
Malteserstr. 74-100, D-12249 Berlin, Germany

Abstract

Earlier studies have shown that indicator kriging provides reasonable estimates of the spatial distribution of hydraulic conductivity based on both quantitative 'hard' and qualitative 'soft' data. One crucial question is the optimal choice of threshold values which define the indicator variables. Another problem is to select a representative point estimator which extracts the most suitable K-value from the local cdf. Two methods to optimize this selection will be presented: a pure statistical cross check with measured data (i) and a verification of estimates by comparison with actual flow conditions (ii). The paper will illustrate these ideas by means of an actual data-set taken from pleistocene glacio-fluvial deposits in North Germany.

1. Introduction

The estimation of hydraulic parameters, e.g. hydraulic conductivity in space is one of the most crucial problems preceding the simulation of groundwater flow and contaminant transport. Variogram analyses can provide a suitable covariance structure and various kriging methods as well as stochastic simulation routines return spatial realizations of hydraulic parameters. All these methods depend on a sufficiently reliable database, concerning the number of data, their spatial distribution and quality.

In hydrogeological practice hydraulic conductivity is determined by pumping tests, which are time consuming and costly and which cannot reflect small scale variabilities which are of major importance concerning contaminant transport

A. Soares et al. (eds.), geoENV I–Geostatistics for Environmental Applications, 165–176.

problems (Schafmeister and de Marsily, 1994). K-values derived from grain-size analyses are cheaper and provide point information. However in general only few hard conductivity values derived from grain-size analyses are available. Therefore methods should be developed to include imprecise knowledge into spatial estimation processes.

Recently Piotrowski *et al.* (1994) used the Fuzzy approach, based on expert knowledge on regional hydrogeology and depositional processes as well as on nearby K-values to improve the spatial estimation of hydraulic conductivity for a glacio-fluvial aquifer in North Germany on a regional scale (tenth of kilometers). Considering the high small scale variability of hydraulic conductivity this approach might not be suitable for a more local study on possible contaminant flow paths.

In practice a limited number of quantitatively known hydraulic conductivity values but a bigger amount of qualitative descriptions of lithology of core samples is available. Using this information hydrogeologists assign more or less wide ranges of hydraulic conductivity to the sediments referring to tabulated values in common hydrogeological literature.

It was shown (Schafmeister and Burger, 1995) that the indicator approach is a promising tool for estimating hydraulic conductivity from quantitative (hard) and qualitative (soft) data simultaneously. A crucial problem is the selection of a point estimator of K from the estimated local cumulative probability density function (cdf) which is required in order to calculate flow conditions by means of a numerical model.

Initial modeling results showed that the mean value derived from the cdf does not represent the regional flow regime of the area. By using the median as point estimator more realistic flow results were produced. Repeated flow modeling based on different point estimators was carried out to find an optimal representative which is valid for the regional parameter field.

Additionally a statistical cross check comparing cell estimates and actual (hard) values was applied.

2. Methods

2.1. DEFINITION OF INDICATOR VARIABLES

In this approach hard hydraulic conductivity values $K(\boldsymbol{x})$ are coded as a set of indicator variables (Journel, 1989):

$$I_j := I(x;z_j) = \begin{cases} 1, & \text{if } Z(x) \leq z_j \\ 0, & \text{otherwise} \end{cases}, j = 1,..,m \tag{1}$$

with $Z(x) = K(x)$

Hydraulic as well as statistical considerations can help to choose threshold values z_c which optimally represent the variation spectrum of data. In the first case hydrogeologists can define intervals of K which reflect the hydraulic characteristics of porous media from practical experience, e.g. K-values less than $5*10^{-7}$ ms^{-1} representing almost impermeable materials whereas values above 10^{-3} ms^{-1} represent very high permeable sediments.

Pure statistical criteria which subdivide the variation spectrum of K-values into equally spaced probability classes are less biased but extremely dependent on the histogram as being a representative estimator of the true pdf. Declustering procedures should be applied prior to selection of threshold values if data are known to be spatially clustered.

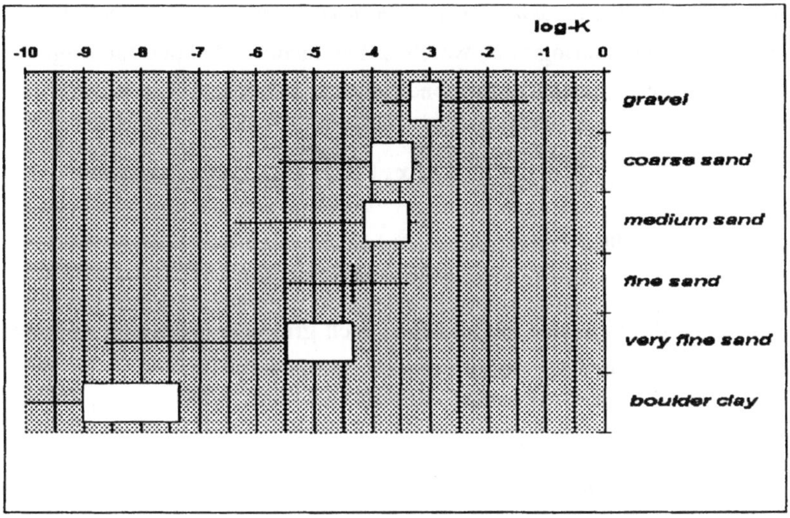

Figure 1. Variation spectra of K-values [ms^{-1}] for six lithotypes; boxes indicate the inner-quartile range.

Subsequently the qualitative descriptions of lithology must be coded into the same indicator variables. This is done as follows:

Hard and soft data are subdivided into few major groups of lithotypes, e.g. fine, medium or coarse sand according to their predominant grain-size compound. A range of possible K-values for a given lithotype can either be drawn from hydrogeological textbooks or directly derived from hard data in the actual area. In the latter case the variation spectra of K-values for each lithotype is calculated (Figure 1). Limits such as the inner-quartile range are chosen to define a representative range of K-values for each lithotype.

Indicator variables are then defined as follows:

$$I(\boldsymbol{x};z_j) = \begin{cases} 1, \text{ if } z_j \text{ is higher than specified K - range of sample in } \boldsymbol{x}, \\ 0, \text{ if } z_j \text{ is lower than specified K - range of sample in } \boldsymbol{x}, \\ \text{missing, if } z_j \text{ lies within the specified K - range of sample} \\ \qquad \text{in } \boldsymbol{x}, \text{ i.e } I(\boldsymbol{x},z_j) \text{ cannot exactly be defined.} \end{cases} \quad j = 1,m \qquad (2)$$

2.2. INDICATOR KRIGING AND CHOICE OF POINT ESTIMATOR FOR K

Indicator transforms of data can now be used to estimate the spatial autocovariance structure (indicator variograms) which is then incorporated into the kriging estimation. Kriging provides estimates of the conditional probability of K in $\boldsymbol{x}_0$:

$$\left[i(\boldsymbol{x}_0,z_j)\right]^* = \text{Prob}^*\left\{Z(\boldsymbol{x}_0) \le z_j\right\} = \text{E}\left\{I(\boldsymbol{x}_0,z_j)\right\} \quad j = 1,m \qquad (3)$$

with $Z(\boldsymbol{x}_0) = \text{K}(\boldsymbol{x}_0)$

Because the estimated cdf's of K at each grid node $\boldsymbol{x}_0$ cannot directly be used in flow simulation models an appropriate point estimator must be derived which represents the model cell. This can be the average (E-type estimator) or any appropriate quantile (Figure 2).

Multiple K-fields are produced running POSTIK (Deutsch and Journel, 1992) with different quantiles ranging between 0.4 and 0.6.

Two different methods can then be applied in order to select an optimal point estimator from the local cdf: hydraulic a-priori knowledge (i) and cross validation (ii).

Cross validation - better cross check - means here that the cell estimates for K are compared with composite values of the same cell calculated with 'hard' K-values from bore hole data and the mean square error is minimized. This method uses only the intrinsic statistical properties of the variable but not its physical meaning.

The first method requires multiple calculations of a 'forward' flow model and comparison of estimated and observed parameters until a sufficiently close fit to the field observations is reached.

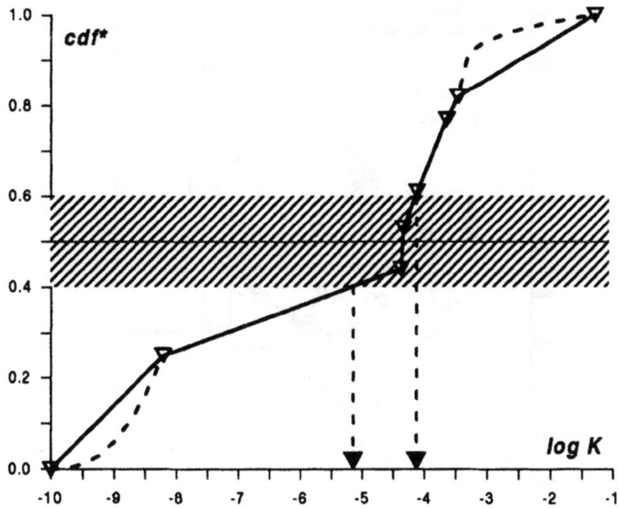

Figure 2. Example of point estimation from local cdf of K; dashed line indicates a power model interpolation for lower and upper tail.

3. Test Site

3.1. HYDROGEOLOGICAL SETTING

The test site is situated in the western outskirts of Berlin/Germany (Figure 3). The entire area covers about 24 km^2. The investigation presented here is limited to the uppermost aquifer which is primarily exposed to anthropogeneous damages. This aquifer consists of glacio-fluvial sands, gravels and boulder clays from weichselian and saalian glaciation periods.

The aquifer bottom is built by clayy and sapropelic sediments of the holstein-interglacial at a depth between -10 and 0 m a.s.l. The phreatic water table declines in southeastern direction towards the river Havel with a regional hydraulic gradient ranging between 0.001 and 0.0015. A lake at the western border of the area and the river Havel which serves as local gaining stream can be regarded as fixed hydraulic potentials. The average aquifer thickness is 30 m. At the banks of the Havel a waterwork produces groundwater mainly from deeper aquifers but also to a smaller extend from the uppermost aquifer and bank filtration (Sommer-von Jarmersted, 1992). In a distance of about 4 km upstream of the groundwater production wells three landfills are identified as potential contaminant sources (Kerndorff *et al.*, 1985).

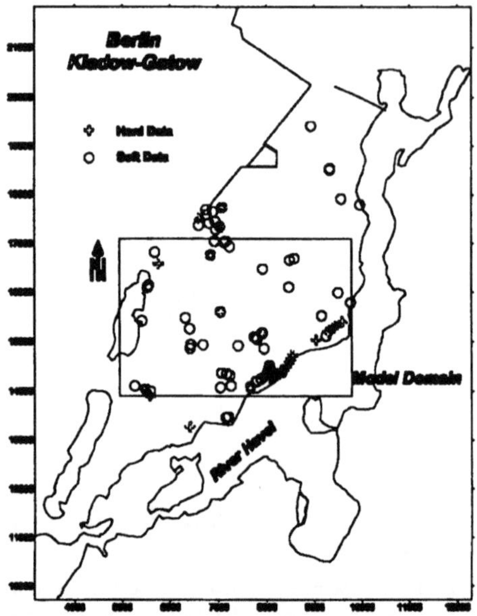

Figure 3. Test site and data locations; crosses: hard data, circles soft data.

3.2. STRUCTURE OF HYDRAULIC CONDUCTIVITY DATA

From detailed grain size analyses at production wells and at 31 scattered bore holes 466 hard K-values are obtained. K-values are derived by using the empirical formula of Beyer (1964) for unconsolidated porous media. Additional 2025 lithological descriptions are available at 153 additional bore holes. Since the drilling of bore holes and wells was performed on occasion of different successive investigation campaigns different geologists made their personal interpretation which might affect the quality of soft data.

Samples for both hard and soft information are taken according to the thickness of macroscopically distinguishable strata. Thus the vertical sampling distance varies in a range of few decimeters up to a few meters. Therefore the vertical resolution of information is high compared to the horizontal resolution with distances varying between 30 m and 1 km.

3.2.1. *Probability structure of K and choice of threshold values*

Hydraulic conductivity of hard data varies between 10^{-10} ms^{-1} for clay and $5*10^{-2}$ ms^{-1} for gravel. More than 70 percent of the samples are described as very fine, fine and medium sands with K-values in the order of 10^{-5} up to 10^{-3} ms^{-1}. About 10 percent of data represent boulder clays with hydraulic conductivity in the order of 10^{-7} ms^{-1}.

Statistical considerations lead to 7 conductivity intervals, i.e. 6 thresholds. In order to obtain reasonable as well as reliable indicator variograms, the cdf was divided into intervals covering 20 percent of data each. The marginal intervals, i.e. first 20 percent which represent nearly impermeable and very low permeable sediments, and the uppermost 20 percent, which represent highly permeable materials, are subdivided into 10 percent intervals. This way the following threshold values were chosen: $6.1*10^{-9}$, $4.2*10^{-5}$, $4.6*10^{-5}$, $7.6*10^{-5}$, $2.3*10^{-4}$ and $3.5*10^{-4}$ ms^{-1}. The fact that three thresholds fall within a close segment between 10^{-5} and 10^{-4} ms^{-1} reflects the very steep slope of the cdf of K (Figure 4).

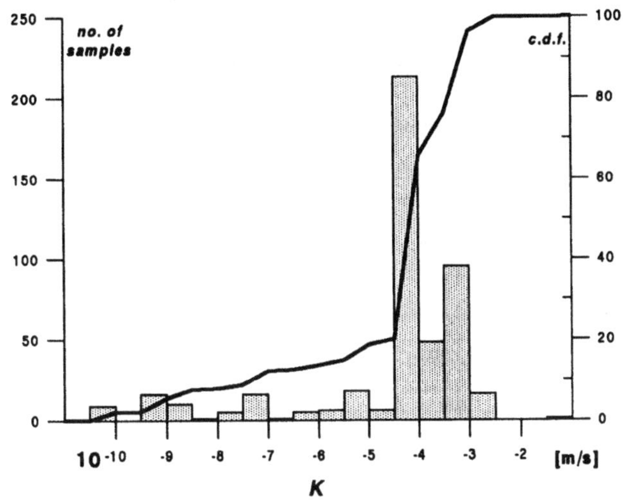

Figure 4. Histogram and cumulative distribution function of K; 466 hard values.

Although data - hard as well as soft data - are extremely clustered in space declustering doesn't result in significant changes of the pdf. Thus the raw pdf of K-values was used.

TABLE 1. Soft data: number and percentage of lithotypes.

lithotype	total number	percentage
clay and silt	293	14.5
very fine sand	60	3.0
fine sand	756	37.3
medium sand	624	30.8
coarse sand	159	7.9
gravel	133	6.6

The soft data are classified into five major lithotypes according to their predominant grain-size, i.e. clay and silt, very fine sand, fine sand, medium sand, coarse sand and gravel. Table 1 gives the number as well as percentage for each lithotype. The possible range of K-values for each group was derived from 466 hard data (Figure 1). Note that lithotypes 'very fine sand' and 'fine sand' were differentiated by the geologists but cover almost the same range of measured K-values, moreover, most samples (inner 50 percent) which were identified as 'fine sands' have a very narrow range of K-values, varying between 4.4 and $4.7*10^{-5}$ ms^{-1}.

TABLE 2. Variogram parameters of indicator variables.

cut-off z_c [ms^{-1}]	nugget variance as percent	vertical range [m] (exponential variogram scheme)
$6.1*10^{-9}$	10	1.7
$4.2*10^{-5}$	8	1.0
$4.6*10^{-5}$	15	2.0
$7.6*10^{-5}$	30	1.2
$2.3*10^{-4}$	15	1.4
$3.5*10^{-4}$	25	1.5

3.2.2. Spatial structure

The 6 indicator variograms calculated in the vertical direction exhibit nugget variances between 8 and 30 percent of total variance indicating a relatively high small scale variability of hydraulic conductivity which is also reported by Schafmeister (1990),

Schafmeister and de Marsily (1994) and Huggenberger *et al.* (1994). The exponential ranges vary between 1 and 2 m (Table 2).

The horizontal variograms of hard data cannot be modeled because the horizontal distances are too large to detect any spatial dependency. From previous studies vertical anisotropies between 1:15 and 1:80 are known. In this study an anisotropy of 1:67 was assumed in the indicator kriging procedure.

4. Results

Within the above described test site an area of 4.8 km by 3.2 km was selected for which the hydraulic condition should be simulated. The quasi 3-dimesional finite-difference numerical model MODFLOW (McDonald and Harbaugh, 1988) was used for modeling; preprocessing and postprocessing, including the calculation of flow paths was performed with PMWIN (Chiang and Kinzelbach, 1996).

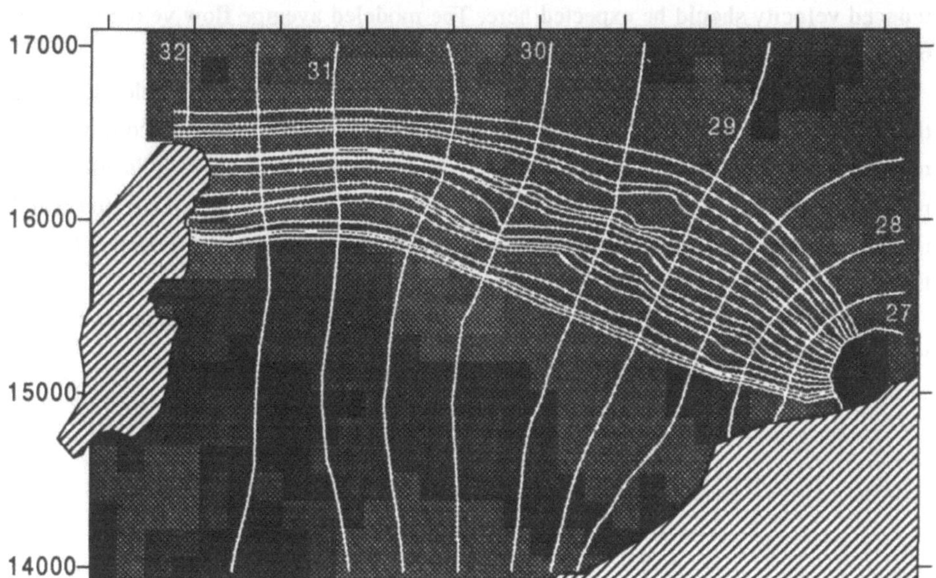

Figure 5. Equipotentials and horizontal projection of flow paths. Grey shadings indicate K-values (dark: less permeable, light: higher permeable); hatched areas: surface waters.

The model domain was discretisized into 30 by 20 cells with 160 m length both in x- and y-direction. Vertically the aquifer was modeled by three 10 m thick layers. Thus K-values at 1800 grid nodes had to be estimated. Based exclusively on hard data only about 50 percent of the model domain was covered with estimates of the local cdf of K due to wide areas where hard information is lacking. Retaining the search parameters more than 95 percent of the model domain was estimated using additional soft information.

From these local cdf's of K seven K-fields were built by selecting the 0.4, 0.42, 0.45, 0.47, median, 0.55 and 0.6 quantile as point estimator.

Based on these K-fields steady-state hydraulic conditions were calculated with the above described boundary conditions. 45 particles were positioned in the drawdown cone of the waterworks. Their paths were traced backwards towards the northwestern boundary (Figure 5). An average pore velocity was calculated for all 45 particles.

None of the different flow fields renders the expected average pore velocity of 0.05 md^{-1} which is calculated by means of DARCY's law based on given hydraulic conditions with regionally specific K-value of $1*10^{-4}$ ms^{-1} and spatially constant porosity of 0.2. Since about 70 percent of the data in the test site lie below this value a reduced velocity should be expected here. The modeled average flow velocities range between 0.033 and 0.048 md^{-1} for the different K-fields (Table 3).

Since in the investigated area no direct measurements of flow velocities, e.g. through pumping test or through tracer tests, were available, another control value must be found: when the mode ($4.6*10^{-5}$ ms^{-1}) of the cdf (Figure 4) is used as a representative of K, an average pore velocity of 0.025 md^{-1} would be expected which is much smaller than the modeled velocities. But there is an apparent tendency that K-fields estimated by lower quantiles approach the expected value.

Likewise the cross-check of point estimated K-values with hard data renders the best agreement for the 0.45 quantile. The results are listed in Table 3.

TABLE 3. Results of cross check and flow modeling for selected point estimators of K.

Point estimator	statistical cross check	modeled average pore velocity [md^{-1}]
0.40	1.061	0.029±0.008
0.42	1.051	0.031±0.008
0.45	1.042	0.033±0.008
0.47	1.048	0.035±0.009
median 0.50	1.051	0.036±0.009
0.55	1.066	0.041±0.011
0.60	1.113	0.048±0.012

Although the resulting values of correspondence aren't very satisfying which is partly due to the lack of field measurements in the investigated area, the best point estimator of K for the regional conditions seems to be slightly less than the median.

5. Discussion and Conclusion

It was shown that the indicator approach is a promising tool to combine hard data with soft qualitative descriptions of lithology in order to obtain reasonable spatial estimates of flow parameters, i.e. hydraulic conductivity K. When applying this method the frequent problem of scarce quantitative information on hydraulic conductivity can be overcome. In the case study presented here an increase of about 80 percent of the estimation domain was achieved.

However, some problems remain. First, the choice of suitable threshold values is crucial. Pure statistical criteria as well as hydraulic criteria must be considered. In any case the selection of threshold values is strongly affected by (i) the variation spectrum of K, which can spread over more than 8 orders of magnitude in glacio-fluvial deposits, (ii) the geometry of sampling locations, which in hydrogeological practice is often preferential dependent on either the actual hydrogeological problem (landfills, waterworks) or the grain size, since in general sampling in clays and silts is avoided.

Another problem is the indicator coding of soft data since their quality is highly dependent on the reproducibility of lithological descriptions. Therefore a reasonable ratio of soft information and hard data is important. In the example presented here soft information is 4 times larger than hard information, thus the quality of estiamtion results depends almost exclusively on the quality of soft information.

The selection of a suitable point estimator from the local cdf can be supported by statistical measures but it is strongly recommended that it is controlled by a-priori knowledge on local flow conditions, which can be derived from a pumping test or a tracer test within the investigated area.

The verification by means of a numerical flow model must be done carefully. Especially an appropriate choice of cell dimensions compared to sampling distances is necessary.

Although the method presented here can be improved in many aspects it was shown that it provides reasonable results. Based on this method more detailed studies, e.g. concerning contaminant transport can be performed applying indicator simulation of hydraulic parameters to take into account their inherent small scale variability.

6. Acknowledgments

The author likes to thank Heinz Burger for fruitful discussions and for critically reading the manuscript. Thanks also for the valuable remarks given by an anonymous reviewer.

7. References

Beyer, W. (1964): Zur Bestimmung der Wasserdurchlässigkeit von Kiesen und Sanden aus der Kornverteilungskurve.-Wasserwirtsch.-Wassertechn. 14, pp. 165-168, Berlin.

Chiang, W. and W. Kinzelbach (1996): Processing Modflow, A simulation system for modeling groundwater flow and pollution.- user's manual.

Deutsch, C.V. and A.G. Journel (1992): GSLIB, Geostatistical Software Library and User's Guide. Oxford University Press, New York.

Huggenberger, P., M. Rauber, F. Stauffer (1994): Integration of geophysical and sedimentological information in the stochastic description of inhomogeneities in fluvial gravel deposits.- in Dracos and Stauffer (eds.) Transport and Reactive Processes in Aquifers.- IAHR Proceedings 5, Balkema Rotterdam, Brookflield.

Journel, A.G. (1989): Fundamentals of Geostatistics in Five Lessons.- Short Course in Geology, Vol. 8, AGU, Washington D.C., pp. 40.

Kerndorff, H., V. Brill, R. Schleyer, P. Friesel, G. Milde (1985): Erfassung grundwassergefährdender Altablagerungen - Ergebnisse hydrogeochemischer Untersuchungen.- WaBoLu Hefte 5/1985, pp. 175, Berlin.

McDonald, M.C. and A.W. Harbaugh (1988): MODFLOW, A modular three-dimensional finite difference ground-water flow model.- U.S. Geological Survey, Open-file report, 83-875, Chapter A1.

Piotrowski, J.A., F. Bartels, A. Salski, G. Schmidt (1994): Fuzzy Kriging of imprecise hydrogeological data.- Abstract Volume for the Technical Programs at the International Meeting of IAMG, Mont Tremblant, Québec, Canada, October 3-5, 1994, 282-288.

Schafmeister, M.-Th. (1990): Geostatistische Simulationstechniken als Grundlage der Modellierung von Grundwasserströmung und Stofftransport in heterogenen Aquifersystemen.- PhD thesis, Verlag Schelzky & Jeep, pp. 143, Berlin.

Schafmeister, M.-Th. and H. Burger (1995): Merging quantitative and qualitative information as input to contaminant flow simulation: A case study.- Abstract Volume for the Technical Programs at the International Meeting of IAMG, Osaka, Japan, october 29- november 2, 1995, 43-45.

Schafmeister, M.-Th. & Gh. de Marsily (1994): The influence of correlation length of highly conductive zones in alluvial media on the transport behaviour.- in: Dracos & Stauffer (eds.): Transport and Reactive Processes in Aquifers, 171-176, Balkema, Rotterdam, ISBN 90 5410 368 X.

Sommer-v. Jarmersted, Chr. (1992): Hydraulische und hydrochemische Aspekte der Uferfiltration an der Unterhavel in Berlin.- PhD thesis, Berl. Geow. Abh. Reihe A, Band 140, pp. 149, Berlin.

NUMERICAL CHARACTERIZATION OF CONTAMINANT TRANSPORT IN NESTED TWO-SCALE POROUS MEDIA

FENG RUAN AND DENNIS MCLAUGHLIN

Ralph M. Parsons Laboratory, Rm 48-209
Department of Civil and Environmental Engineering
Massachusetts Institute of Technology
Cambridge, MA 02139, U.S.A.

Abstract. Natural hydrogeologic properties (e.g., log hydraulic conductivity) may include many scales of variability. This paper investigates transport in multi-scale media by numerically simulating solute plumes in nested two-scale velocity fields that have a typical scale disparity on the order of 10 [Adams and Gelhar, 1992; Hess et. al., 1992]. Two-scale velocity fields with specified covariance properties are generated from a multi-variate Fast Fourier Transform algorithm. Two dimensional solute plumes are accurately derived from a spline-based Eulerian-Lagrangian transport solver. The plumes are simulated over a travel distance on the order of hundreds of large log k scales and thousands of small log k scales. The initial size of the solute plume is close to that of the large log k scale, and the large time plume size is much larger than this scale. Spatial second moments are computed and compared to theoretical predictions. The spatial structure of the plumes is examined through the use of the dilution index and mean-square measure of concentration variability. Time histories of these measures illustrate distinctive evolutionary features of solute plumes traveling in a small-scale, a large-scale and a two-scale medium. Our results indicate that the large-scale and the two-scale plumes do not reach asymptotic state after a long travel distance. We conclude by discussing the important implication in performing two-scale solute plume simulation.

A. Soares et al. (eds.), geoENV I–Geostatistics for Environmental Applications, 177–187.

1. Introduction

Natural hydrogeologic properties (e.g., log hydraulic conductivity) include many scales of variability ranging from the scale of individual pores (cm) to the scale of regional geologic units (km). Multi-scale media have been observed at several field sites in the United States (e.g., the Cape Cod site in Massachusetts and the Columbus site in Mississippi), where log hydraulic conductivity has been found to have two dominant scales with a scale disparity on the order of 10 [Adams and Gelhar, 1992; Hess et. al., 1992]. Solute plumes traveling through such two-scale aquifers appear to be controlled by the interplay between the small-scale and the large-scale [Gelhar, 1995]. In this study we assume log hydraulic conductivity is a multi-scale random field that has a (nested) two-scale correlation function [Gelhar, 1993]:

$$f(\mathbf{x}) = \sqrt{\alpha}(f_{Small}(\mathbf{x}) - \overline{f}) + \sqrt{1-\alpha}(f_{Large}(\mathbf{x}) - \overline{f}) + \overline{f} \qquad 0 \leq \alpha \leq 1 \quad (1)$$

where $f(\mathbf{x})$ is the log hydraulic conductivity field, α is a weighing factor, $\overline{f}(\mathbf{x}) = logK_G$ is the mean and K_G is the geometric mean. The small-scale log k field $f_{Small}(\mathbf{x})$ and the large-scale log k field $f_{Large}(\mathbf{x})$ are considered to be statistically independent. This yields $E[f(\mathbf{x})] = \overline{f}(\mathbf{x})$ and $\sigma_f^2 = \alpha\sigma_{f_{Small}}^2 + (1-\alpha)\sigma_{f_{Large}}^2$.

Existing stochastic theories of solute transport [e.g., Gelhar and Axeness, 1983; Gelhar, 1993; Dagan, 1987; Rajaram, 1991] apply to an ensemble of statistically similar solute plumes. Because we are concerned in particular with a single realization rather than the ensemble, the results of ensemble stochastic theories are applicable only when the ergodic hypothesis is valid and when a plume reached asymptotic state. This implies that the scale of the source size must be much larger than the characteristic scale of log k (the integral scale) and the plume must travel many log k correlation scales. We examine this ergodic assumption by observing solute plumes traveling in a small-scale, a large-scale and a two-scale medium.

In this paper we adopt a numerical approach to investigate the evolution of solute plumes in the presence of both one and two log hydraulic conductivity scales. It is difficult to simulate multi-dimensional transport through highly variable (random) velocity fields. We meet this challenge by using an accurate spline-based Eulerian-Lagrangian method on a large computational grid that resolves both small and large-scale velocity variations [Ruan and McLaughlin, 1996b].

Solute plumes are simulated in a synthetic two-scale aquifer over long distances (on the order of hundreds of the large log k scales). We consider three measures of spatial structure, namely the spatial second moment, the dilution index and a mean-square measure of concentration variability. We compute these measures during the simulation of solute plumes under small (λ_{Small}), large (λ_{Large}) and two-scale (λ_{Small} and λ_{Large}) random velocity fields. The initial source size is much larger than the small-scale and it is on the order of the large-scale. This allows a solute plume to sample many small correlation scales during the simulation. We are interested in whether a solute plume reaches asymptotic state and when the ergodic assumption is applicable [Dagan, 1990].

2. Approach

2.1. METHODS

The synthetic velocity fields required in our multi-scale transport simulations are generated directly, using a multi-variate Fast Fourier Transform algorithm [Ruan, 1996a]. An obvious alternative is to generate a synthetic log hydraulic conductivity field and then solve the associated groundwater flow equation (for hydraulic head) and the Darcy equation (for velocity). The primary disadvantage of this approach is the computational effort required to obtain a converged solution on a large computational grid (on the order of one million nodes) when log k is assumed to be moderately heterogeneous. Our direct generation approach is free from such constraint and is applicable to velocity fields of moderate log conductivity variance. These synthetic random velocity fields conserve mass and their spatial (co-)variances converge to the exact values [Graham and McLaughlin, 1989; Ruan, 1996a].

Solute transport simulations in one small log k scale have been performed using the particle tracking method [Bellin, et. al., 1994; Tompson and Gelhar, 1990]. This method is computationally expensive when the number of particles is made large enough to obtain accurate concentration solutions over large domains. As an alternative, we use a high-order spline-based Eulerian-Lagrangian method to solve the multi-dimensional transport problem. The Eulerian-Lagrangian method is suitable for convection-dominant transport problems (as in our case). The use of spline interpolation in our transport solver improves our ability to simulate sharp concentration fronts that are inherent under highly variable random velocity fields [Ruan and McLaughlin, 1996b].

2.2. SPATIAL MEASURES OF PLUME STRUCTURE AND CONCENTRATION VARIABILITY

The first and second spatial second moments are computed as functions of simulation time t over the spatial domain Ω:

$$M_{i0} = \frac{1}{M_0(t^n)} \sum_{k=1}^{N} (x_{1k} - \overline{x}_1(t))^i c_k^n \Delta x_1 \Delta x_2$$

$$M_{0j} = \frac{1}{M_0(t^n)} \sum_{k=1}^{N} (x_{2k} - \overline{x}_2(t))^j c_k^n \Delta x_1 \Delta x_2 \qquad i, j = 1, 2 \qquad (2)$$

where x_{1k} and x_{2k} are the coordinates of the kth nodal point, $\overline{x}_i(t), i = 1, 2$ is the position of the center of mass, $M_0(t^n)$ is the mass at time t^n, c_k^n is concentration at time t^n and Δx_1, Δx_2 are the spatial intervals of a rectangular computational grid that has a total of N nodes. The rates-of-change of the second moment are related to the dispersivities (macrodispersivity) as follows:

$$A_{11} = \frac{1}{2\overline{V}} \frac{dM_{20}}{dt} \qquad A_{22} = \frac{1}{2\overline{V}} \frac{dM_{02}}{dt} \qquad (3)$$

where A_{11} and A_{22} are, respectively, the longitudinal and the transverse dispersivities, and $\overline{V}$ is the mean velocity aligned to the longitudinal direction.

The dilution measure is computed as [Kitanidis, 1994]:

$$E(t^n) = \sum_{k=1}^{N} e^{-p_k^n log_e [p_k^n]} \Delta x_1 \Delta x_2 \qquad (4)$$

where $p_k^n = c_k^n / M_0(t^n)$ is the normalized concentration. The maximum dilution state $E_{MAX}(t^n)$ at time t^n is proven, via the maximum entropy analysis, to be a Gaussian plume with the same spatial second moment of a simulated plume [Kitanidis, 1994]. The relative dilution is used to measure the closeness of a plume's dilution state as compared to a perfectly diluted Gaussian plume. It is defined as [Kitanidis, 1994]:

$$I(t^n) = E(t^n) / E_{MAX}(t^n) \qquad (5)$$

A plume's concentration variability is measured by its mean-square distance from a asymptotic plume:

$$\|c(\mathbf{x}, t^n) - c_o(\mathbf{x}, t^n)\|_{L2} = \frac{1}{M_0(t^n)} \left[\sum_{k=1}^{N} (c_k^{t^n} - c_{k,o}^{t^n})^2 \Delta x_1 \Delta x_2 \right]^{\frac{1}{2}} \qquad (6)$$

where $c_{k,o}^{t^n} = c_o(\mathbf{x}, t^n)$ is Gaussian that has the same spatial second moment of the simulated plume.

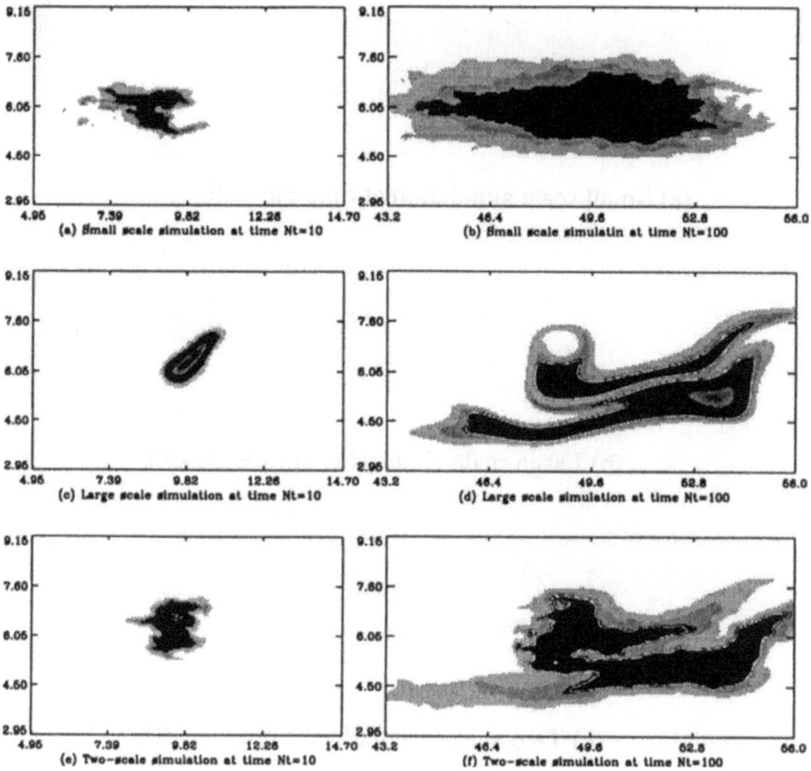

Figure 1. Simulated small-scale, large-scale and two-scale plumes at time $N_t = 10$ (left column) and at time $N_t = 100$ (right column). Plumes travel from the left to the right along the mean velocity field in the longitudinal direction, all length units are normalized by the large log k scale.

3. Results

3.1. THE SIMULATION SETUP

Both the one-scale (V_{Small}, V_{Large}) and two-scale (V_{Two} where α in Equation (1) is 0.5) random velocity fields are obtained on a rectangular computational domain of $4000\Delta x \times 250\Delta x$, where $\Delta x = \Delta x_1 = \Delta x_2 = 1.0$ m

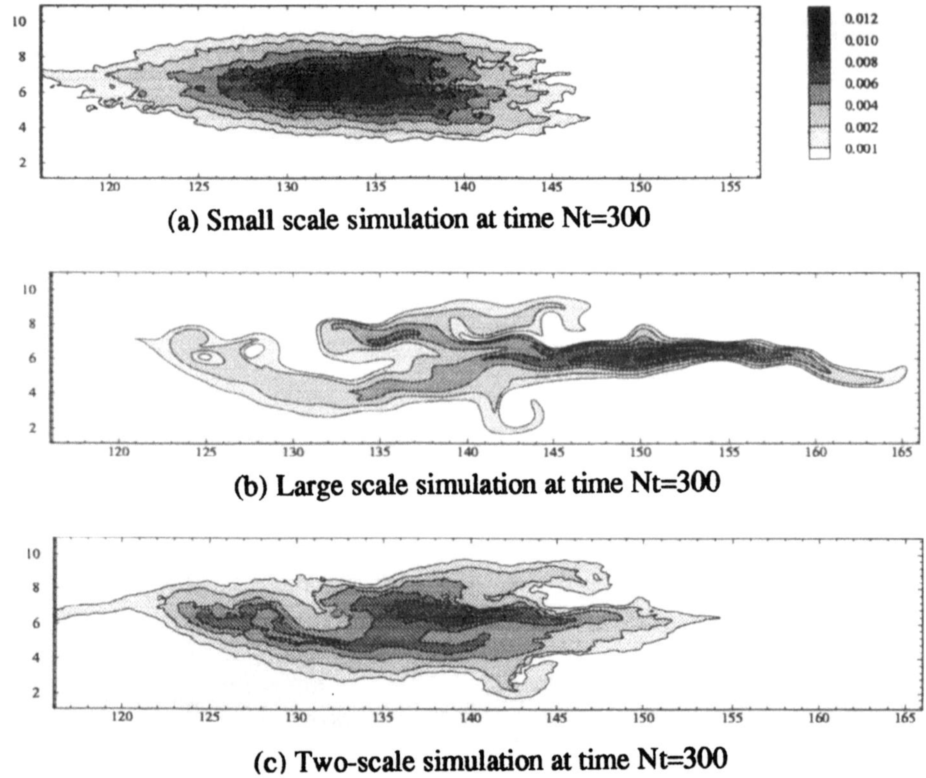

Figure 2. Simulated small-scale, large-scale and two-scale plumes at time $N_t = 300$

is the grid spacing. The log hydraulic conductivity is assumed to be Gaussian with unit variance and a geometric mean of 6.25 m/day. The size of the log k scales are $\lambda_{Small} = 2\Delta x$, $\lambda_{Large} = 20\Delta x$, and the scale disparity $\lambda_{Large}/\lambda_{Small}$ is 10. The computational domain spans over 200 large correlation scales (λ_{Large}) in the longitudinal direction, and $12.5\lambda_{Large}$ in the transverse direction. The mean velocity $\overline{V} = 0.41$ m/day is aligned with the longitudinal direction.

Our solute plumes are simulated from an instantaneous release of a square source close to the left boundary of the computational domain. The initial size of this source is $0.75\lambda_{Large} \times 0.75\lambda_{Large}$. This initial size is several times larger than that of the small correlation scale. The plume is propagated from

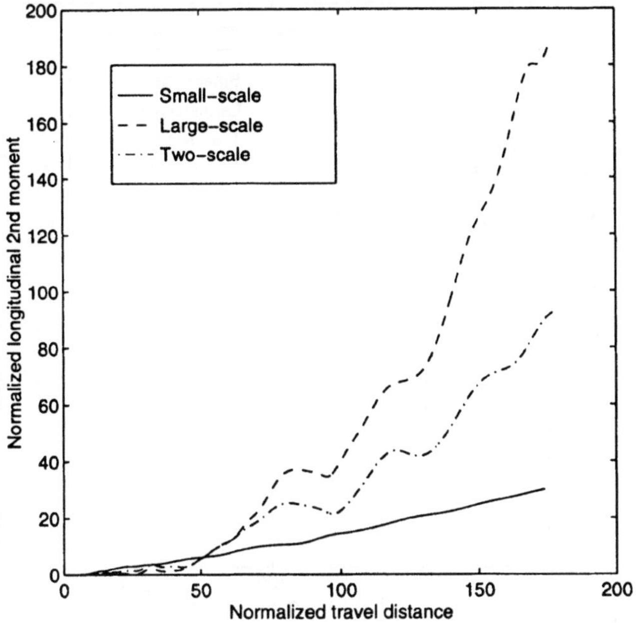

Figure 3. Computed longitudinal spatial second moment

the left to the right following the mean velocity. The local longitudinal and transverse dispersivities are assumed to be $\alpha_L = 0.1$ m and $\alpha_T = 0.1$ m respectively. The complete simulation time T is divided into $N_t = 400$ time steps. This results to a Peclet number of 10 and a Courant number of 8 derived from the mean velocity.

3.2. COMPUTED SPATIAL MEASURES

The simulated small-scale, large-scale and two-scale plumes are plotted at time $N_t = 10$ and at time $N_t = 100$ in Figure 1 and at time $N_t = 300$ in Figure 2. Mass is conserved for all simulated plumes. We note that (a) at the early time $N_t = 10$, concentration contours are scaled with a maximum value of 0.75. The two-scale plume demonstrates features somewhat similar to the small-scale plume (shown in Figure 1); (b) at the middle time $N_t = 100$, concentration contours are scaled with a value of 0.10. The two-scale plume exhibits spatial features similar to that of the large-scale plume. However spatial concentrations within dark blobs of higher

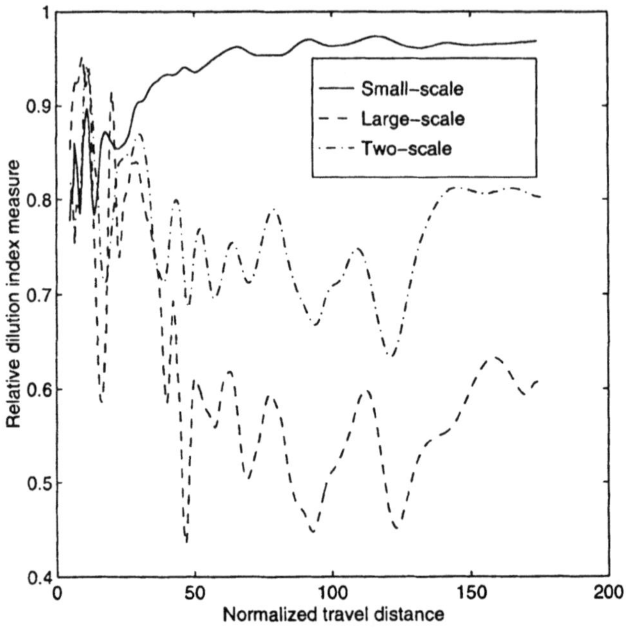

Figure 4. Computed relative dilution index measure

concentration threshold demonstrate mixing features similar to that of a small-scale plume. This indicates that small and large scales interact, the two-scale plume is both different from and similar to the small-scale plume and the large-scale plume; (c) at a late time $N_t = 300$, concentration contours are scaled with a value of 0.012. The two-scale plume indicates again both spatial features of the large-scale plume, and the well-diluted feature of the small-scale plume.

Computed spatial second moment M_{20} in the longitudinal direction are plotted in Figure 3 as functions of the plume travel distance (i.e., the displacement of center of mass), where the travel distance is normalized by λ_{Large} and M_{20} is normalized by λ_{Large}^2. The second moment of a small-scale plume grows with a constant slope, with an average of 0.17. This is larger than the Gelhar and Axness value [1983] of 0.10, but similar value has been observed from the numerical investigation of Tompson and Gelhar [1990]. The second moment of the large-scale and the two-scale plume does not reach a constant slope, they show significant variations or uncertainties

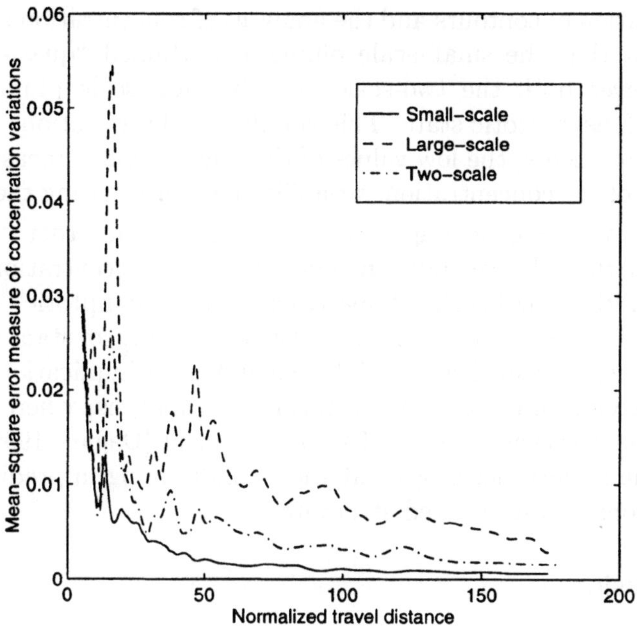

Figure 5. Computed mean-square error measure of concentration variations

after a travel distance of over 100 large correlation length.

The relative dilution measure is plotted in Figure 4 as a function of normalized travel distance. When a plume is perfectly diluted, its value should reach 1.0. Figure 4 shows that this index fluctuates rapidly for all three plumes when the travel distance is less than 50 large correlation length. With the further increase of the travel distance, the computed values for the small-scale plume reaches values close to 1.0. The index for the two-scale and the large-scale plumes fluctuates around a low value of 0.75 and 0.50, respectively. This indicates that these plumes do not reach a perfect dilution state during the entire simulation.

The concentration variability measure in equation (6) is plotted in Figure 5 as a function of normalized travel distance. The values of this measure fluctuate rapidly at a travel distance less than 50 λ_{Large}. It then decrease to lower values for all three plumes, where this measure is the lowest for the small-scale plume. It becomes larger for the two-scale plume and it is the largest for the large-scale plume.

The observation of contours and the analyses of concentration spatial structure suggest that the small-scale plume has attained "quasi-asymptotic" state. However both the two-scale and the large-scale plumes have not reached such asymptotic state. This is indicated by variations of their spatial second moments, the low values of the dilution index measures and the high values of the concentration variability measures during the simulation.
We note that it takes a long travel distance (on the order of 50 λ_{Large}) before both the relative dilution index and the concentration variability measure for the small-scale plume reach "quasi-asymptotic" state. Comparatively, its spatial second moment has attained a constant slope within a much shorter travel distance. This should have implication to the definition of asymptotic state in the theoretical analysis, where it has been based on the constant increase of second moment [Dagan, 1990]. This suggest also the importance of geostatistical conditioning in predicting spatial concentrations and its related structure.

4. Conclusion

Our results show that a solute plume traveling in a two-scale aquifer never reach the asymptotic state in the sense of Dagan [1990], and it exhibits features that are absent in both the small-scale and the large-scale plumes. The presence of two distinctive scales allows a solute plume to sample both large-scale non-uniformity and small-scale variations simultaneously. It becomes apparent that a two-scale plume can not be derived by ignoring the presence of either the large-scale or the small-scale.
The characterization of a solute plume in multi-scale aquifer is still an unsolved problem, where the constitutive relation (e.g., the Darcy's law) is non-local [Neuman, 1993; Deng et al., 1993]. The purpose of this work is to use state of the art numerical tools (e.g., the spline-based Eulerian-Lagrangian transport solver) to perform two-scale solute plume simulations, and supply these results as a basis for future development of analytical results that characterize solute plumes in two-scale (or multi-scale) porous media.

Acknowledgments

This research was supported by US National Science Foundation (grant EAR-9218602).

References

1. Adams, E. E., and L. W. Gelhar (1992) Field study of dispersion in a heterogeneous aquifer: spatial moments analysis, *Water Resour. Res.*, 28(12), 3293-3308.
2. Bellin A., Y. Rubin and A. Rinaldo (1994) Eulerian-Lagrangian approach for modeling of flow and transport in heterogeneous geological formations, *Water Resour. Res.*, 30(11), 2913-2924.
3. Dagan, G. (1987) Theory of solute transport by groundwater, *Ann. Rev. Fluid Mech.*, 19, 183-215.
4. Dagan, G. (1990) Transport in Heterogeneous Porous Formations: Spatial Moments, Ergodicity, and Effective Dispersion, *Water Resour. Res.*, 26(6), 1281-1290.
5. Deng, F.-W., J. H. Cushman and J. W. Delleur (1993) A Fast Fourier Transform stochastic analysis of the contaminant transport problem, *Water Resour. Res.*, 29(9), 3241-3247.
6. Gelhar, L. W. and C. L. Axeness (1983) Three-dimensional stochastic analysis of macrodispersion in aquifers, *Water Resour. Res.*, 19(1), 161-180.
7. Gelhar, L. W. (1993) *Stochastic Subsurface Hydrology*, Prentice Hall, Englewood Cliffs, New Jersery.
8. Gelhar, L. W. (1995) Perspectives on field scale applications of stochastic subsurface hydrology, *Second IHP/IAHS George Kovacs Colloquium, Subsurface Flow and Transport: The stochastic Approach*, UNESCO, Paris.
9. Graham, W., and D. McLaughlin (1989) Stochastic analysis of nonstationary subsurface solute transport, 1, Unconditional moments, *Water Resour. Res.*, 25(2), 215-232.
10. Hess, K. M., S. H. Wolf, and M. A. Celia (1992) Large-scale natural gradient tracer test in sand and gravel, Cape Cod, Massachusetts, 3, Hydraulic conductivity and calculated macrodispersivities, *Water Resour. Res.*, 19(1), 28(8), 2011-2027.
11. Kitanidis, P. K. (1994) The concept of dilution index, *Water Resour. Res.*, 30(7), 2011-2026.
12. Neuman, S. (1993) Eulerian-Lagrangian theory of transport in space-time nonstationary velocity fields: exact nonlocal formalism by conditional moments and weak approximation, *Water Resour. Res.*, 29(3), 633-645.
13. Rajaram, H. (1991) *Scale-dependent dispersion in heterogeneous porous media*, Ph.D. thesis, MIT, Cambridge.
14. Ruan, F. and D. McLaughlin (1996a) An efficient multivariate random field generator using the Fast Fourier Transform, *Advances in Water Resources*, in press.
15. Ruan, F. and D. McLaughlin (1996b) An Investigation of Eulerian-Lagrangian Methods for Solving Multi-Dimensional Heterogeneous Transport Problems, in preparation.
16. Tompson, A. F. B., and L. W. Gelhar (1990) Numerical simulation of solute transport in three-dimensional randomly heterogeneous porous media, *Water Resour. Res.* 26(10), 2541-2562.

INTERPOLATION OF GROUNDWATER QUALITY PARAMETERS USING ADDITIONAL INFORMATION

A. BÁRDOSSY
Institut für Wasserbau, Universität Stuttgart,
70550 Stuttgart, FRG.

U. HABERLANDT
National Hydrology Research Institute,
Saskatoon, Canada.

AND

J. GRIMM-STRELE
Landesamt für Umweltschutz Baden Württemberg,
76199 Karlsruhe, FRG.

Abstract. Groundwater quality parameters are usually more variable in space, than in time. Therefore measurements are made at a large number of locations, but only a few times a year. Experience shows that quite a bit of this variability can be explained from other factors, such as geology or land use. Therefore a straightforward interpolation based only on the measured data does not deliver plausible results. The goal of this paper is to present interpolation methods which can efficiently use additional information such as physical considerations (flow modelling), qualitative knowledge about the soil, land use or hydrogeological information. The method suggested for this purpose is a a simplified version of Markov-Bayes-Kriging called Markov-Bayes Updating. The problem of assessing prior distributions depending on the additional information is discussed. Estimations with and without consideration of the external data are compared statistically and visually. Last the problem of interpolating pollutants with very low concentrations often below the detection limit is also addressed. Extensive measurements of natural parameters pH, total hardness, and pollutants NO_3, athracin and tetrachlorethen made at more than 2600 locations in the state of Baden-Württemberg are used to illustrate the methodology on a regional scale.

A. Soares et al. (eds.), geoENV I–Geostatistics for Environmental Applications, 189–200.

1. Introduction

A central problem in environmental investigations is the estimation of pollution extent. This is usually done on the basis of point measurements performed at selected locations. These data are then used to assess values at unsampled locations. There are many methods available for this purpose. The goal of this paper is to present some geostatistical methods which can efficiently use available additional information for the estimation. The additional information can include physical considerations (flow modelling), qualitative knowledge about the soil structure, or other contamination-related information. Another frequent problem is the treatment of data below the detection limit. Indicator transformations and probability maps do offer a correct representation of these values. The methodology is demonstrated on a large regional scale, using groundwater pollution data. Without any modification the same estimation techniques can be applied at small scales, occurring usually in remediation problems.

2. Methodology

Among the various geostatistical interpolation methods that can be used, ordinary kriging is the best known one. Cokriging on kriging with an external drift are very efficient for taking into account additional variables in the estimation. The problem with these techniques is that they can only use numerical additional variables, and the relationship between the primary and the secondary variable is supposed to be linear. As for the present problem the most useful additional information is categorical, so that these methods could not be applied.

Instead different procedures, including simple kriging with a known mean, a simple updating procedure and Markov-Bayes kriging in an special form the Markov-Bayes updating were used. For the problem of considering values below the detection limit an indicator approach and its updating version is suggested. In the next section the methodology is shortly described: the third section contains results of the application to groundwater quality parameters.

2.1. ORDINARY KRIGING (OK)

Ordinary kriging is the most widely used geostatistical interpolation method (Matheron, 1971). The underlying statistical hypothesis is the so called intrinsic assumption, and no additional information is taken into account.

The weights of the linear estimator:

$$Z^*(x) = \sum_{i=1}^{n} \lambda_i Z(x_i) \tag{1}$$

are calcuated from the kriging system.

2.2. SIMPLE KRIGING (SK)

Simple kriging differs from OK that the mean is supposed to be known. Additional categorical information can be taken into account in this procedure by taking the means of the category as known.

The estimator is of the form:

$$Z^*(x) = m(x) + \sum_{i=1}^{n} \lambda_i [Z(x_i) - m(x_i)] \tag{2}$$

$m(x)$ and $m(x_i)$ are calculated from the values corresponding to the same category. A clear disadvantage of the method is that the quality of this mean value (depending on the sample size) is not taken into account.

2.3. SIMPLE UPDATING (SU)

Consider the situation where $Z(x)$ is complemented by a secondary variable $L(x)$ available at each point in the domain. This variable is discrete, and related to $Z(x)$ through a conditional expectation:

$$E[Z(x) \mid L(x) = l] = m_l \tag{3}$$

and a conditional variance:

$$\mathrm{Var}[Z(x) \mid L(x) = l] = \sigma_l^2 \tag{4}$$

In this case the first estimation of $Z(x)$ is that based on $L(x)$

$$Z'(x) = m_{L(x)} + \varepsilon_{L(x)}(x) \tag{5}$$

with $\varepsilon_{L(x)}(x)$ being a random error with 0 expectation and variance $\sigma^2_{L(x)}$. Using $Z'(x)$ for the estimation of $Z(x)$ combined with the observations one has:

$$Z^*(x) = \lambda_0 Z'(x) + \sum_{i=1}^{n} \lambda_i Z(x_i) \tag{6}$$

The estimation variance for this estimator is:

$$\mathrm{Var}[Z(x) - Z^*(x)] = -\sum_{j=1}^{n}\sum_{i=1}^{n} \lambda_j \lambda_i \gamma(x_i - x_j) +$$

$$+ 2\sum_{i=1}^{n} \lambda_i (1 - \lambda_0)\gamma(x_i - x) + \lambda_0^2 E[\varepsilon(x)^2] \qquad (7)$$

Minimizing this estimation variance with respect to the unbiasedness condition leads to the equation system:

$$\begin{aligned}
\sum_{j=1}^{n} \lambda_j \gamma(x_i - x_j) + \mu &= (1 - \lambda_0)\gamma(x_i - x) \;\; i = 1, \ldots, n \\
\sum_{j=1}^{n} \lambda_j \gamma(x - x_j) + \mu &= \lambda_0 \sigma_l^2 \\
\sum_{j=0}^{n} \lambda_j &= 1
\end{aligned} \qquad (8)$$

This is a linear equation system for the unkown weights λ_i, and the solution yields the estimator $Z^*(x)$. The additional variable $L(x)$ can be any discrete variable, for example soil type, landuse or geological code.

In practice we suppose that additional information in the form of a discrete index $(1, \ldots, M)$ $L(x)$ is available for every location in the field. For each class l the mean ant the variance can be calculated as

$$m_l = \frac{\sum_{i=1}^{n} Z(x_i) 1\{L(x_i) = l\}}{\sum_{i=1}^{n} 1\{L(x_i) = l\}} \qquad (9)$$

$$s_l^2 = \frac{\sum_{i=1}^{n} (Z(x_i) - m_l)^2 1\{L(x_i) = l\}}{(\sum_{i=1}^{n} 1\{L(x_i) = l\}) - 1} \qquad (10)$$

2.4. INDICATOR KRIGING (IK)

Very low pollutant concentrations values are difficult to interpolate. The main problem here is that there are measurements which are below the detection limit. To estimate the mean value of these parameters at unobserved locations assumptions about the values below the detection limits have to be made. There are several possibilities:

1. To consider these values as zero. This is a very optimistic assumption.
2. Ignore these data, which leads to an overestimation since all small values are removed.

3. To set all these values to a selected value between zero and the detection limit. This decision is arbitrary and replaces the uncertain data by exact values.

Instead of these approaches one can estimate the exceedence probabilities using indicator variables and indicator kriging (Journel; 1983).

The indicator variable $I(x,y)$ for a given threshold y is defined as:

$$I(x,y) = \begin{cases} 1 & \text{if } Z(x) \leq y \\ 0 & \text{if } Z(x) > y \end{cases} \tag{11}$$

The indicator variable $I(x,y)$ can also be regarded as the probability of $Z(x)$ being less than y:

$$I(x,y) = P\left[Z(x) \leq y\right] \tag{12}$$

If indicator coding is performed for each real value y then each observation is transformed into a step function. In practice a set of different y_k values $k = 1, \ldots, K$ is selected, and $Z(x)$ is transformed into the K dimensional vector $(I(x,y_1), \ldots, I(x,y_K))$.

2.5. MARKOV-BAYES UPDATING (MBU)

Cokriging and external drift kriging can embed additional information, but only in a numerical form and assuming a linear relationship. An estimation method to cope also with non-linear relationship is the Markov-Bayes kriging (MBK) described in Zhu and Journel (1993). The MBK method benefits from prior distributions at locations where additional information is available. In this application, a simplified form, called Markov-Bayes updating (MBU) is used. This differs from MBK in directly using the prior information available at the point for which the estimation has to be made. The reason for doing this was that additional information was available at each point of the catchment.

Formally MBU uses the assumption that additional information can be used for the assessment of prior distributions $U(x,y)$ at selected locations.

$$U(x,y) = P[Z(x) \leq y | \text{Additional information at } x] \tag{13}$$

If this $U(x,y)$ is different from the distribution $F(y) = P[Z(x) \leq y]$ then the additional information is useful for the estimation of $Z(x)$. The estimation is then performed by using the distribution $F(y)$, the local prior $U(x,y)$, the observations $Z(x_i)$ and the indicators defined in (11). These indicator values $I(x,y)$-s are estimated using a simple cokriging approach:

$$I(x,y) = \lambda_0 F(y) + \sum_{i=1}^{n} \lambda_i I(x_i,y) + \nu U(x,y) \tag{14}$$

Using the interpolated indicator values the estimator of the variable can be calculated using:

$$Z^*(x) = I^*(x, y_0)y_0 + \sum_{k=0}^{K}(I^*(x, y_{k+1}) - I^*(x, y_k))\bar{Z}_k \;, \tag{15}$$

with

$$\min z(x_i) = y_0 < y_1 < \ldots < y_K = \max z(x_i)$$

and $\bar{Z}_k$ being the mean of the measurement values between y_k and y_{k+1}

$$\bar{Z}_k = \frac{\sum_{i=1}^{n}(I(x_i, y_{k+1}) - I(x_i, y_k))Z(x_i)}{\sum_{i=1}^{n}(I(x_i, y_{k+1}) - I(x_i, y_k))} \tag{16}$$

In fact MBU combines three possible approaches:

1. assigning the same mean to the whole domain (λ_0)
2. using spatial interpolation ($\lambda_i, i = 1, \ldots, n$)
3. assigning the same value according to the corresponding class of the additional variable (ν)

The spatial dependence, the configuration of the observation points and the usefulness of the additional information are influencing the role of the above factors. For the calculation of the weights λ_i and ν one needs the covariance function of $I(x, y)$ and $U(x, y)$ and their cross-covariance function. A formulation of the equations by using variograms is also possible, but the covariance based form is simpler in this case. Lehmann (1995) used this method for the interpolation of soil moisture measurements. His results show that MBU is clearly superior to OK.

3. Application

The above described interpolation methods were applied to interpolate groundwater quality parameters measured at more than 2600 locations in Baden-Württemberg in 1994. Figure 1 shows the locations of the measurement points.

As a first step experimental variograms are calculated for the different variables. The experimental variograms were then fitted with theoretical curves. Figure 2 shows the experimental and the corresponding theoretical variograms for pH.

Table 1 shows the parameters of all theoretical variograms. One can see that the range of NO_3 is significantly lower than those of the natural parameters total hardness and pH. The reason for this is that increased NO_3 levels are caused by human activity, mostly agriculture, in resulting in an uneven distribution over the entire domain.

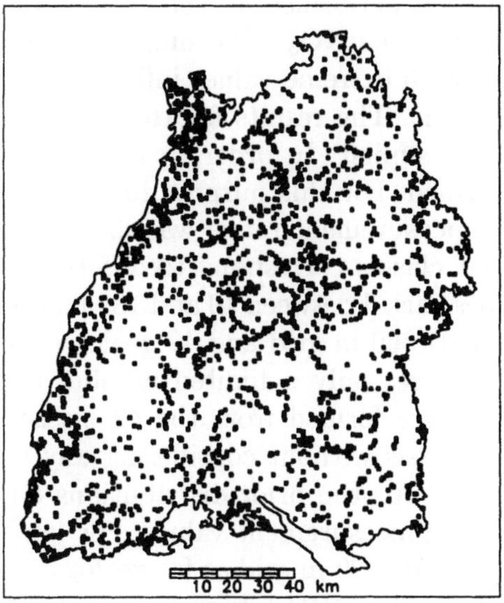

Figure 1. Measurement points of the groundwater quality parameters

Figure 2. Experimental and theoretical variogram for pH

Variable	Type	nugget	eff. range (m)	sill
NO_3	spherical	120.	4000.	510.
pH	spherical	.050	26000.	.120
Total hardness	spherical	1.50	27500.	2.75

TABLE 1. Variogram parameters

Ordinary kriging was performed for the three variables. Figure 3 shows the interpolation results for NO_3. The map does not reflect the opinion of the experts. Sites with extreme values influence their neighborhood and yield circular patterns. The results for the two other parameters, total hardness and pH, are similar. In order to account for geology and land use MBU was performed. For this purpose indicator variograms were calculated. The entire measurement range was divided into equally frequent classes, meaning that classes correspond to different measurement ranges. The number of classes varied between 10 and 18. The geological information was available on a 300×300 m grid and 42 geological classes were defined. As for quite a few classes only a limited number of measurements were available. 12 groups were formed from the 42 classes. The grouping was done on a geological basis without considering any statistical parameters. For each of these groups the prior distributions $U(x, y)$ was calculated from the corresponding measurement values. For land use 16 classes were available on the same 300×300 m grid as for geology. From these, 13 groups were formed. Results of the MBU interpolation for NO_3 using land use as additional information are shown on Figure 4. Compared to the OK this map looks more reasonable and corresponds better to the expert opinion. The extremes are not influencing their neighborhood as much as in the OK case. Figure 5 shows the results obtained using simple updating and landuse as additional information. The maps obtained by MBU and SU are very similar, which can be explained by the use of the same additional information. The SU procedure however does not reduce the maxima as strong as MBU does due to the indicator coding.

The performances of the different interpolation methods are compared using a cross validation technique, namely for each measurement point an estimation using the other measurements only was performed. Then the calculated and the measured values were compared, and different statistics were calculated. First the possible bias of the method was investigated using:

$$E_1 = \frac{1}{n}\sum_{i=1}^{n}(Z^*(x_i) - Z(x_i)) \tag{17}$$

For all estimation techniques and groundwater parameters the value of E_1 was close to zero, meaning that there was no bias.

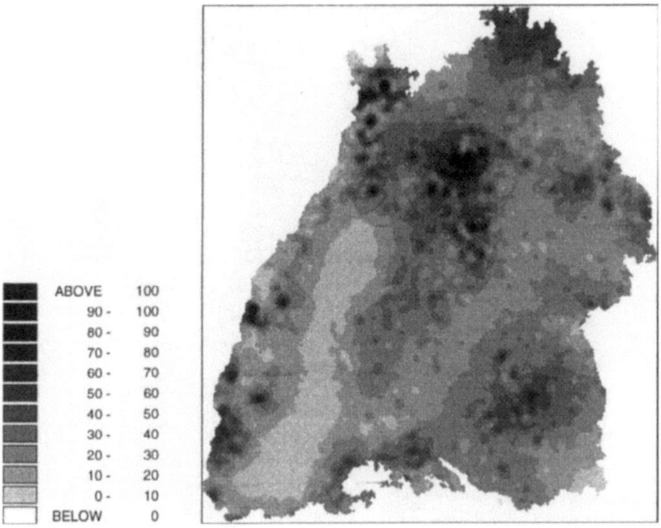

Figure 3. Interpolated map for NO_3 with OK $[mg/l]$

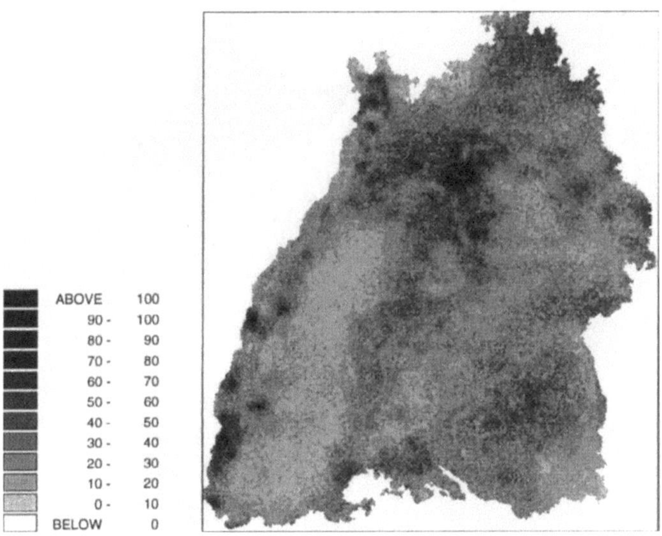

Figure 4. Interpolated map for NO_3 with MBU (land use) $[mg/l]$

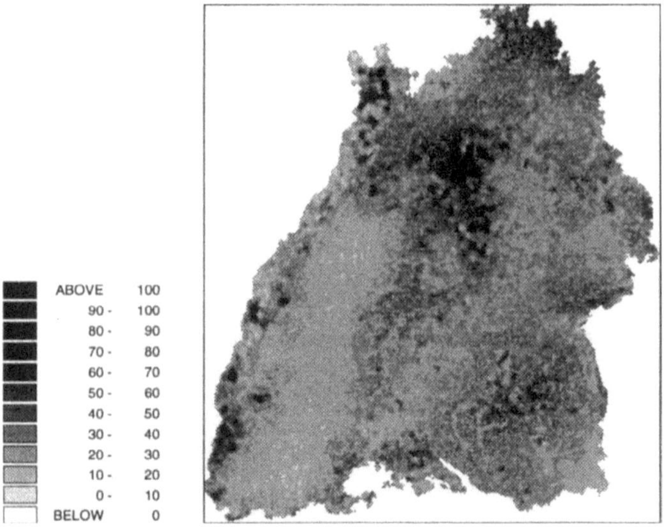

Figure 5. Interpolated map for NO_3 with SU (land use) $[mg/l]$

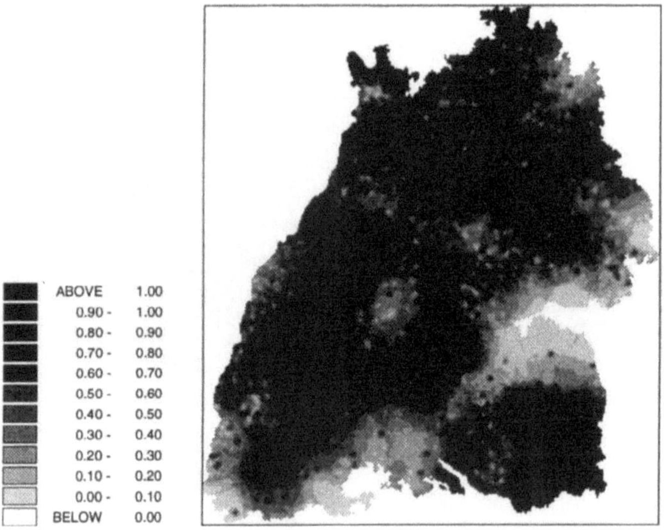

Figure 6. Probabilities for athracin being below $0.01\mu g/l$

The mean squared error was calculated to measure the accuracy of the estimation:

$$E_2 = \sqrt{\frac{1}{n}\sum_{i=1}^{n}(Z^*(x_i) - Z(x_i))^2} \tag{18}$$

Results for the different methods and parameters are given in Table 2.

Parameter	OK	SU (Geology)	SU (Landuse)	MBU (Geology)	MBU (Landuse)
NO_3	374	372	363	451	385
pH	0.089	0.086	0.091	0.086	0.097
Total hardness	2.04	1.97	2.07	1.95	2.04

TABLE 2. Cross validation results E_2)

For the two low concentrations, indicator variograms were calculated using different y levels. These levels correspond to different detection limits, to that the number of points considered increased with y. For example if a measurement was below the detection limit of 0.05 μg/l and the selected y level was 0.01 μg/l then this measurement could not be considered for the interpolation of $I(x, y)$. The reason for this is that one cannot decide wether the value is above or below this y level. On the other hand the same measurement was considered for the y level 0.05 μg/l, as it is clearly below this value. Figure 6 shows a map of exceedence probabilities for atracin corresponding to the y level 0.01 μg/l.

4. Discussion and Conclusions

In this paper different estimation methods have been described and compared. These methods can be used at different scales and for different parameters. The two dimensional evaluations used herein can as well be used in three dimensional cases.

Concluding points are as follows:

- The experimental variograms show that the investigated groundwater parameters exhibit a spatial dependency.
- Variogram ranges are small for anthropogenic pollutants such as nitrate, atrazin or tetrachlorethene. The natural parameters have substantially longer ranges.

- The use of external information improves the quality of the estimation, and delivers plausible maps. However it has to be noted that the improvement is very slight. For natural parameters the geology was the most useful additional variable and for nitrate land use.
- The two procedure MBU and SU produce very simmilar results, but the computational efforts for SU are substantially lower.
- Additional information in the form of numerical model results can also be considered to improve the interpolation.
- Indicator kriging is a method to overcome the problem of detection limits for pollutants with extremely low concentrations. Here estimation of the exceedence probabilities makes more sense than an attempt to estimate mean values.

References

Journel, A. G. (1983): Non parametric estimation of spatial distributions. Mathematical Geology, 15, 445-468.

Lehmann, W. (1995): Anwendung geostatistischer Verfahren auf die Bodenfeuchte in ländlichen Einzugsgebieten. Mitteilungen des Instituts für Hydrologie und Wasserwirtschaft, Nr.52, Universität Karlsruhe.

Matheron, G. (1971): The Theory of Regionalized Variables and its Applications. Les Cahiers du Centre de Morphologie Mathématique, Fasc. 5, Fontainebleau.

Zhu, H. and Journel, A.G., (1993): Formatting and Integrating Soft Data: Stochastic Imaging via the Markov-Bayes Algorithm. In *Geostatistics Tróia'92*, (ed. A. Soares), Kluwer Academic Publishers, Dordrecht, pp. 1-12.

A NEURAL NETWORK AIDED FILTERING ALGORITHM FOR DIAGNOSING AND PREDICTING NONLINEAR CONTAMINANT TRANSPORT DYNAMICS

M. NONAKA[1] and N. H. THOMAS[2]
[1]*Field of Geosystem Engineering, Graduate School, The University of Tokyo, Tokyo 113, Japan*
[2]*School of Chemical Engineering, The University of Birmingham, Birmingham B15 2TT, England*

Abstract

To enhance the estimation rate of the SEEK(Switching mode Enhanced Extended Kalman) filter algorithm a recurrent neural network is embedded within the SEEK algorithm. The SEEK-FIND(SEEK For Initialisation of Neural Descriptor) algorithm is efficient to estimate state variables in multi-modal nonlinear systems with slowly varying states. When a nonlinear system has regularities, i.e. slowly varying states, and irregularities, i.e. quickly varying states such as bifurcations, the recurrent neural network is applied to estimate the regular states instead of the SEEK filter after trained by using one-step delayed estimates and the current estimates from the SEEK filter as the input signals and the teacher signals, respectively. The SEEK filter is activated again when the system shows irregularities such as bifurcations. The SEEK filter and the neural network are automatically switched by monitoring the maximum estimation error. As a demonstration here, we offer the dual-states Lorenz system with long regular states and the three vortices flow system with bifurcations after taking long term regularities. The latter is simulated by the discrete vortex method as a simple example of multi-modal transport processes in high Reynolds number flows. It has been confirmed that the state variables are accurately estimated by smoothly switching the SEEK filter to the trained neural network, and vice versa. The estimation rate is enhanced in the SEEK-FIND algorithm by a few tens times compared with the original SEEK filter algorithm.

1. Introduction

The SEEK(Switching mode Enhanced Extended Kalman) filter algorithm[1,2] has been

A. Soares et al. (eds.), geoENV I–Geostatistics for Environmental Applications, 201–211.

developed so as to maximise the convergence rate of the estimation error covariance matrix by optimally selecting the observation matrix. The SEEK algorithm has been proved to achieve the excellent performance in estimating the state variables even in complicated chaotic systems such as a dual-states Lorenz system[2] controlled by the switching parameter. However, the algorithm seems to be redundant in the case where the system is in a regular state, i.e. the state variables are not abruptly varied but slowly varied. Too frequent switching of the observation matrix is inefficient in estimating regular state variables from a view-point of saving the estimation time in on-line filtering. The SEEK-FIND(SEEK For Initialisation of Neural Descriptor) algorithm is inspired by the motive that the estimation rate might be enhanced when the SEEK filter algorithm is replaced by a trained artificial neural network in a regular state to avoid the calculation of the optimal observation matrix in each time step. Regular-irregular multi-modal states are sometimes recognised in contaminant transport processes due to environmental inhomogeneity. For example, the parameters involved in the state space model of the transport process have to be adapted when the natural bounds such as temperature, wind and flow velocities, etc. are abruptly varied, which possibly results in a regular-irregular transition.

The combination of the extended Kalman filter and a neural network recently appears not only in estimating state variables and searching for optimal control parameters but also in modifying the training algorithm for the neural network. Simutis, Havlik & Luebbert[3] have estimated the states of a biotechnical process using the fuzzy aided extended Kalman filter and neural networks. Watanabe, Fukuda & Tzafestas[4], Iiguni, Sakai & Tokumaru[5], Ishida et. al.[6], Puskorius & Feldkamp[7] and Ruck et. al.[8] have proposed Kalman filter assisted algorithms for neural network training to enhance the learning speed instead of the conventional error back-propagation algorithm.

It should be essential for the environmental protection to diagnose the current state and to predict the future state of a contaminant transport system. Contaminant transport processes have been numerically analysed by using the advection-diffusion model[9-13], Lagrangian modelling[14] or Monte Carlo simulation[15,16]. Stochastic equations have also been used for modelling pollutant diffusion processes[17-19]. Moreover, the time series models have been applied to predict the contaminant transport behaviour in the atmosphere[20] and in the lake[21]. Although these methodologies may be useful to evaluate the pollution state in a localised area, the more sophisticated model has to be developed to diagnose and predict the contaminant transport behaviour in a large scale geosystem such as the ocean or the atmosphere. A state space model has been derived from extending discrete vortex simulations for horizontal contaminant dispersion[22], in which the dispersion area and the dispersion rate are regarded as the state variables and a few parameters have to be adaptively identified by referring the current observation. The SEEK-FIND algorithm should play a role to estimate contaminant transport dynamics using the state space model.

In this article, the neural network aided algorithm is proposed to enhance the estimation rate in multi-modal nonlinear systems with slowly varying states. The basic structure of the SEEK-FIND algorithm is explained in the second section. In the third section the SEEK-FIND algorithm is implemented and applied to estimate the state variables in the dual-states Lorenz system with long regular states and in the three vortices system with bifurcations after taking long term regularities. The concluding remarks are summarised in the final section.

2. SEEK-FIND algorithm

The SEEK filter algorithm[1,2] was inspired by the motive of enhancing the convergence rate of the estimation error covariance matrix by optimally selecting the observation matrix. However, it needs a long calculation time to estimate state variables because the optimal observation matrix has to be determined by evaluating the nonlinear objective function in each time step. We outline here the basics of the SEEK filter algorithm.

In the SEEK filter algorithm applied to a state space model composed of a nonlinear state equation and a linear observation equation, the switching mode observation matrix is defined by

$$H(t) = \mathrm{diag}[\mu_1, \mu_2(1-\mu_1), \cdots, \mu_{n-1}\prod_{i=1}^{n-2}(1-\mu_i), \prod_{i=1}^{n-1}(1-\mu_i)] \tag{1}$$

where one of the parameters μ_i $(i = 1, \cdots, n-1)$ is constrained to take 1, whilst others take 0, so as to maximise the convergence rate of the error covariance matrix. Here, we focus attention on the continuous Riccati-type matrix differential equation to evaluate the convergence rate of the estimation error covariance matrix. The nonlinear Riccati-type matrix differential equation is given by

$$\dot{P} = [F]P + P[F]^{\mathrm{T}} + [G]Q[G]^{\mathrm{T}} - PH^{\mathrm{T}}R^{-1}HP \tag{2}$$

where P is the error covariance matrix, $[F]$ is the Jacobian matrix derived from the nonlinear state equation for which the current estimates are substituted, $[G]$ is the current driving matrix, Q is the covariance matrix of the system noise, R is the covariance matrix of the observation noise and the superscript T denotes the transpose of the matrix. Namely, Equation (2) is evaluated in each time step when one of the observation variables is updated.

In order to maximise the convergence rate we focus attention on the trace of the current Jacobian matrix of Equation (2) assuming R is the unit matrix[2]. Consequently, the current objective function to be maximised is written as

$$J \underline{\underline{\Delta}} \,\mathrm{abs}\left[\begin{array}{l} \sum\limits_{i=1}^{n} a_{ii} - \{\mu_1^2 \hat{P}_{11} + \mu_2^2 \hat{P}_{22}(1-\mu_1)^2 + \cdots \\ \qquad + \mu_{n-1}^2 \hat{P}_{(n-1)(n-1)} \prod\limits_{i=1}^{n-2}(1-\mu_i)^2 + \hat{P}_{nn}\prod\limits_{i=1}^{n-1}(1-\mu_i)^2\} \end{array}\right] \tag{3}$$

where $a_{ii}\,(i = 1,\cdots,n)$ is the diagonal component of the matrix $[F]$, $\hat{P}_{ii}$ denotes the estimate of the diagonal component of the error covariance matrix. The conjugate gradient algorithm proposed by Fletcher & Powell[23] is applied to maximise Equation (3) in the SEEK filter algorithm.

In order to avoid the calculation of the optimal observation matrix in each time step when the system is in a regular state, i.e. slowly varying state, and to enhance the estimation rate we propose the SEEK-FIND algorithm, in which the SEEK filter is replaced by the trained artificial neural network when the system is in a regular state. In the training stage the SEEK filter returns the accurate estimates by optimally selecting the observation matrix, and the one-step delayed estimates are utilised as the input signals to the neural network, whilst the current estimates, i.e. one-step forward estimates, from the SEEK filter are regarded as the teacher signals for the recurrent neural network as illustrated in Figure 1. The errors between the teacher signals and the output signals from the neural network are back-propagated and the link-weight matrices between the input, hidden and output layers are decided so as to minimise the estimation error. The SEEK filter and the neural network are activated parallel in the training stage.

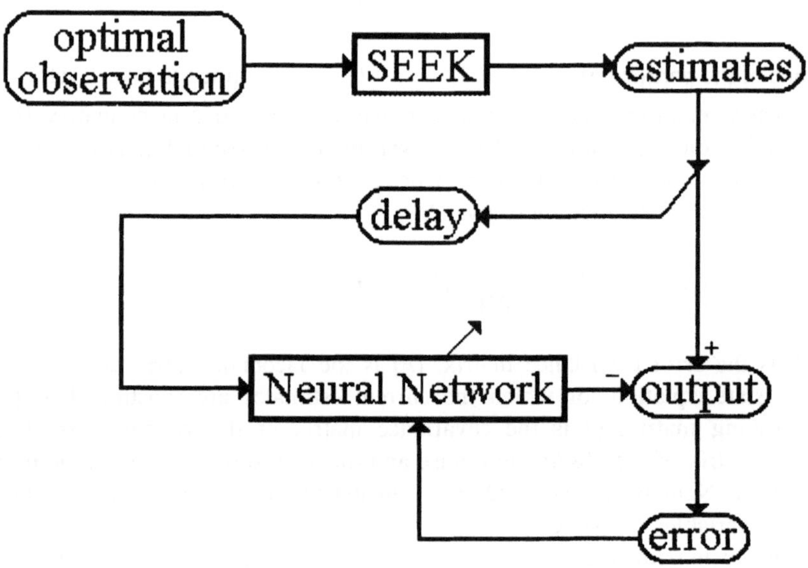

Figure 1. Training algorithm for the incorporated neural network in the SEEK-FIND algorithm

The SEEK filter is switched off after the training has been completed and then, the trained recurrent neural network replaces the SEEK filter to return the regular state estimates using the one-step delayed outputs as the input signals, as illustrated in Figure 2. At the first time step the input signals are given by the current estimates from the SEEK filter. The SEEK filter is switched on and another training of the recurrent neural network is activated when the estimated state variables lose track of true behaviour compared with the observed trajectory initialising the state estimates and the estimation error covariance matrix. The training and estimation of the incorporated neural network are automatically switched by monitoring the maximum estimation error in the SEEK-FIND algorithm. It is expected that the estimation rate might be enhanced when the regular state of a multi-modal system is long enough to train the incorporated neural network despite the training consumes another calculation time.

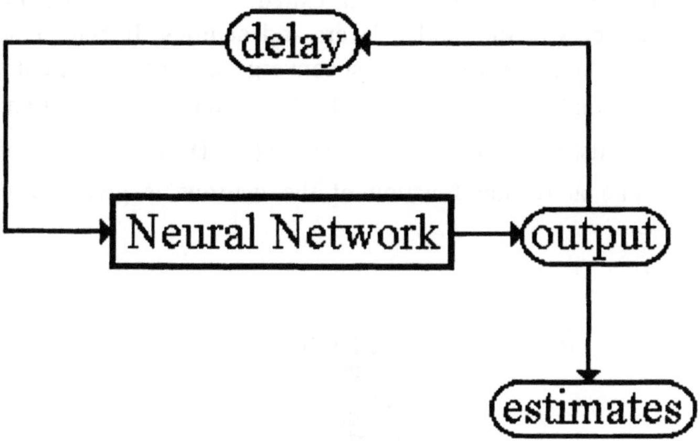

Figure 2. Estimation algorithm for the trained neural network incorporated in the SEEK-FIND algorithm

3. Application of the SEEK-FIND algorithm

3.1 DUAL-STATES LORENZ SYSTEM

As a model illustration here, we offer the following Lorenz equation[24] without noise ;

$$\begin{bmatrix} \dot{x}(t) \\ \dot{y}(t) \\ \dot{z}(t) \end{bmatrix} = \begin{bmatrix} -\sigma & \sigma & 0 \\ r & -1 & -x \\ y & 0 & -b \end{bmatrix} \begin{bmatrix} x(t) \\ y(t) \\ z(t) \end{bmatrix} \tag{4}$$

where the Lorenz parameters σ, r and b are positive definites. The aim of the demonstration is to evaluate the switching performance between the SEEK filter and the incorporated neural network and to examine the estimation performance of the recurrent neural network. Therefore, the system noise is not embedded within the state equation. We have focused attention on the dual-states parameter sets ($\sigma=10$, $b=8/3$, $r=28$) and ($\sigma=10$, $b=8/3$, $r=112$) when the strange attractors are activated[25], i.e. r is switched between 28 and 112 by a fixed time interval. In the dual-states system the state starting from $x=5$, $y=10$ and $z=20$ with $r=28$ is switched to the state controlled by $r=112$ starting from $x=10$, $y=20$ and $z=40$, and vice versa. Equation (4) is currently linearised by using the Taylor expansion and transformed to the difference equation to which the discrete Kalman filter is applied.

The structure of the neural network embedded within the SEEK-FIND algorithm is illustrated in Figure 3. The numbers of the neurons are 3 in the input layer, 9 in the single hidden layer and 3 in the output layer, called 3-9-3 ANN. It is not easy to search for the optimal numbers of the hidden layers and neurons. In this demonstration we select the 3-9-3 ANN after confirming a fast convergence and a sufficient accuracy. The one-step delayed estimates $\hat{x}(t)$, $\hat{y}(t)$ and $\hat{z}(t)$ from the SEEK filter are taken as the input signals, whilst the estimates $\hat{x}(t+1)$, $\hat{y}(t+1)$ and $\hat{z}(t+1)$ are used as the teacher signals. The activation function of the neurons is given by the sigmoidal function.

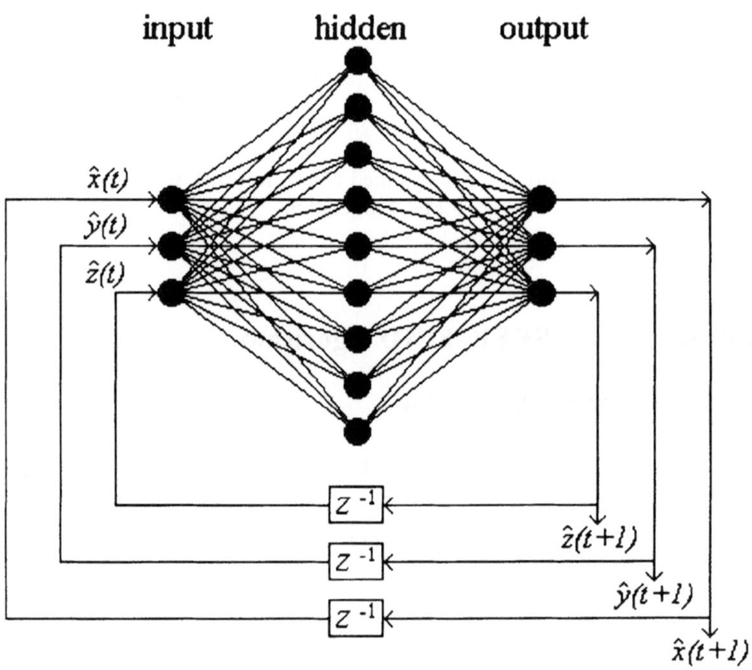

Figure 3. Structure of the embedded neural network within the SEEK-FIND algorithm to estimate the dual-states Lorenz system with long regular states

The link-weight matrices between the input and hidden layers and between the hidden and output layers are optimised by the error back-propagation method[26]. The leaning speed is enhanced by using the adaptive momentum algorithm[27]. The training of the recurrent neural network is completed for the Lorenz system with r=28 when the sum of the squared errors is lower than a desired value. Then the trained neural network replaces the SEEK filter to return the estimates $\hat{x}(t+1)$, $\hat{y}(t+1)$ and $\hat{z}(t+1)$ using $\hat{x}(t)$, $\hat{y}(t)$ and $\hat{z}(t)$ as the input signals until the maximum estimation error is increased over a desired value. It is expected that the 3-9-3 ANN is initialised and the training starts again when the state with r=28 is switched to the state with r=112, and vice versa.

The estimation performance of the SEEK-FIND algorithm applied to the dual-states Lorenz system is shown in Figure 4. Left-hand three illustrations show the time series behaviour of the estimates, whilst right-hand three illustrations represent the concealed trajectories calculated by the Runge-Kutta-Gill method. As can be seen, the SEEK filter is smoothly switched to the ANN at time *T1* when the sum of the squared errors is decreased to a desired value, and the 3-9-3 ANN returns the estimates of the regular state controlled by the Lorenz parameter r=28 until it fails to accurately estimate the state variables just after the state is switched at *T2* when the maximum estimation error is increased over the limitation. Then the 3-9-3 ANN is initialised and trained again to be adapted by another regular state controlled by the Lorenz parameter r.=112.

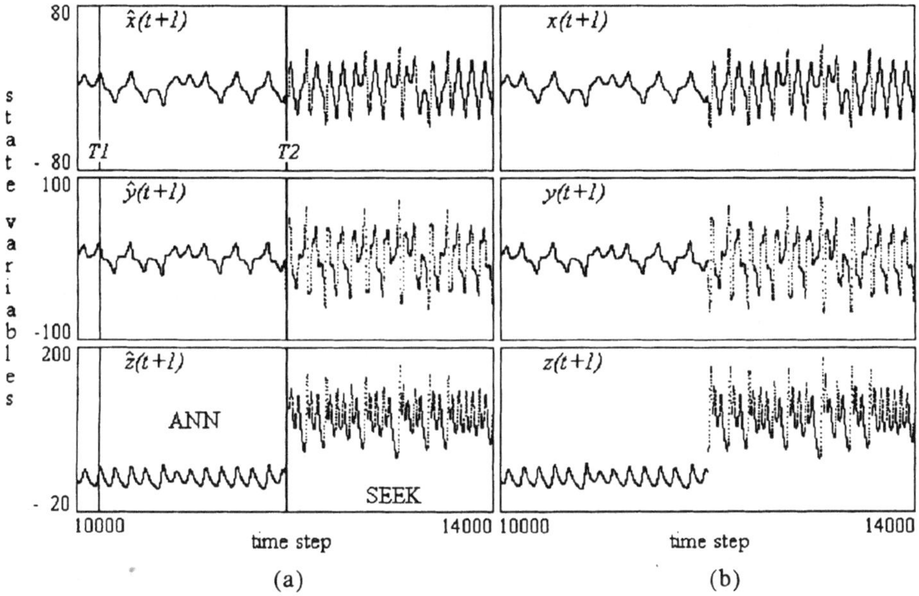

Figure 4. Estimated(a) and real(b) trajectories of the dual-states Lorenz system(The real trajectories are numerically calculated by using the Runge-Kutta-Gill method.)

After about 1.5 times longer training period the trained neural network replaces the SEEK filter to return the estimates of the regular state. Consequently, the estimation rate is enhanced by about twelve times when the SEEK algorithm is replaced by the trained neural network. On the other hand, it needs about 0.2 % longer calculation time for training of the neural network compared with the estimation by only the SEEK filter. Namely, the total estimation time by the SEEK-FIND algorithm is only 8.5 % of that by the SEEK filter.

3.2 THREE VORTICES SYSTEM

As a simple model of horizontal contaminant transport, we focus attention on the three vortices behaviour simulated by the two-dimensional discrete vortex method[2]. The transport behaviour in the ocean or in the atmosphere is dominated by the flow field with a large Reynolds number, in which the vortex element analysis should be useful according to the Kelvin-Helmholtz theory. We offer the three vortices system as one of the simplest examples of bifurcation behaviour.

The two-dimensional vortex behaviour is described by

$$\dot{x}_i = U + \frac{1}{2\pi} \sum_{\substack{j=1 \\ j \neq i}}^{M} \Delta\Gamma_j \frac{y_i - y_j}{\left(x_i - x_j\right)^2 + \left(y_i - y_j\right)^2} \tag{5}$$

$$\dot{y}_i = V - \frac{1}{2\pi} \sum_{\substack{j=1 \\ j \neq i}}^{M} \Delta\Gamma_j \frac{x_i - x_j}{\left(x_i - x_j\right)^2 + \left(y_i - y_j\right)^2} \tag{6}$$

where (x_i , y_i) is the space co-ordinates of the i-th vortex element, U and V are the flow velocities of the field, $\Delta\Gamma$ is the vortex strength and M is the total number of vortices. We have to account for viscous diffusion and inhomogeneity of the vortex strength in real flow systems. These disturbances are regarded as a system noise to be embedded within the state equation in the discrete vortex simulation[2]. However, it will be disregarded here to simply demonstrate the performance of the SEEK-FIND algorithm. Equations (5) and (6) are linearised by the Taylor expansion and transformed to the difference equations to which the discrete Kalman filter is applied.

The neural network has 6 neurons in the input layer, 18 neurons in the single hidden layer and 6 neurons in the output layer. The number of the neurons in the hidden layer was decided to keep a fast convergence and a sufficient accuracy. The sigmoidal function is used as the activation function of the neurons. The 6-18-6 ANN is trained by taking the one-step delayed estimates from the SEEK filter as the input signals, whilst the current estimates, i.e. one-step forward estimates, from the SEEK filter are regarded as the teacher signals for the ANN. The errors are back-propagated to seek for the optimal link-weights between the input and hidden layers and between the hidden and output layers.

The three vortices start from the space co-ordinates of (-1.0,0), (0,0) and (0.8,0) with respective vortex strength of 1.0, -0.5 and 1.4 assuming U=0 and V=0. The estimated trajectories are illustrated in Figure 5(a) as compared with the concealed trajectories calculated by the Runge-Kutta-Gill method(see Figure 5(b)). We can recognise the SEEK filter is smoothly switched to the 6-18-6 ANN at time *T3* and the ANN returns the accurate estimates until time *T4* when one of the bifurcations is activated. Then the SEEK filter is switched on, whilst the ANN is initialised and trained again parallel with the SEEK filter. Consequently, the trained recurrent neural network enhances the estimation rate by about twenty four times and the training consumes about 0.1 % longer calculation time compared with the original SEEK filter algorithm. Namely, the estimation rate by the SEEK-FIND algorithm is only 4.3 % of that by the SEEK filter.

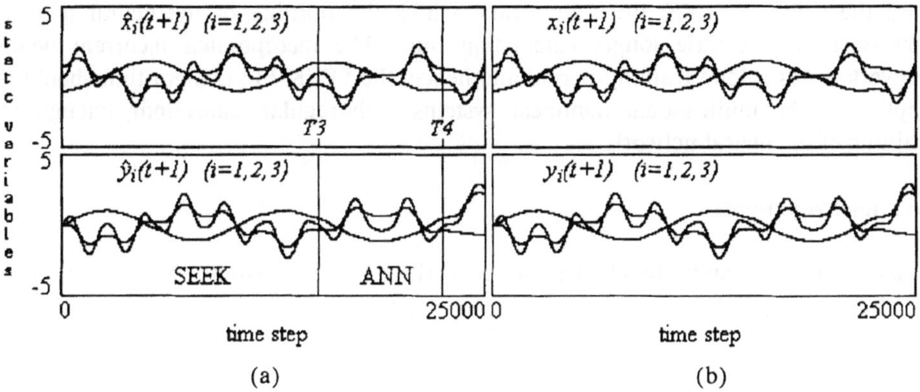

Figure 5. Estimated(a) and real(b) trajectories of the three vortices system with bifurcations(The real trajectories are numerically calculated by using the Runge-Kutta-Gill method.)

In this demonstration we incorporated the neural network with a single hidden layer to keep a fast training speed. However, the neural network still needs a long training period before the link-weight matrices converge to the desired values. Therefore, the SEEK-FIND algorithm should be applied to regular-irregular switching nonlinear systems in which the slowly varying period is long enough to train the neural network.

4. Conclusions

The SEEK-FIND(Switching mode Enhanced Extended Kalman filter For Initialisation of Neural Descriptor) algorithm has been proposed to enhance the estimation rate of the state variables in multi-modal nonlinear systems with slowly varying states. In a regular state, i.e. slowly varying state, the incorporated neural network is trained by taking the one-step delayed estimates from the SEEK filter as the input signals, whilst the current estimates, i.e. one-step forward estimates, are regarded as the teacher signals. The estimation errors are back-propagated to optimise the link-weights between the input and hidden layers and between the hidden and output layers. The SEEK filter is switched off and the trained recurrent neural network is switched on to return the state estimates when the link-weights converge to the desired values. The SEEK filter is activated again when the trained neural network comes across an irregular state such as a bifurcation and the estimation error is increased over a desired value. To demonstrate the performance of the SEEK-FIND algorithm we focus attention on the dual-states Lorenz system with long regular states and the three vortices system with bifurcations after taking long term regularities. We have recognised the SEEK filter is smoothly switched to the recurrent neural network, and vice versa, to return the accurate state estimates with a fast estimation rate in the multi-modal nonlinear systems. The estimation rate is enhanced by a few tens times in the SEEK-FIND algorithm compared with the original SEEK filter algorithm, whilst the training of the neural network consumes only a little longer calculation time. The incorporated recurrent neural network needs a long training period and hence, the SEEK-FIND algorithm should be applied to the multi-modal nonlinear systems with regular states long enough for training of the neural network.

Acknowledgements

The authors are grateful to The British Council for funding. Nittetsu Mining Co Ltd is also acknowledged for its additional financial support.

References

1. Nonaka, M. and Thomas, N. H.: Kalman filter method for statistical modelling of contaminant transport : An outline of the methodology, in *Proc. of the Sixth Int. Conference on Stochastic Hydraulics*, pp. 647-654, Taipei, 1992.

2. Nonaka, M. and Thomas, N. H.: Enhanced Kalman filter method for estimating statistical fields of transport dynamics-1. Mathematical basis, *J. Min. Material Process. Inst. Japan*, **112**(1996), 89-93.

3. Simutis, R., Havlik, I. and Luebbert, A.: A process state estimation and prediction in a production-scale beer fermentation using fuzzy aided extended Kalman filter and neural networks, in *IFAC Symposium Series No. 10*,.pp. 95-100, Pergamon Press, New York, 1992.

4. Watanabe, K., Fukuda, T. and Tzafestas, S. G.: Learning algorithm of layered neural networks via extended Kalman filter, *Int J Systems Science*, **22**(1991), 753-768.

5. Iiguni, Y., Sakai, H. and Tokumaru, H.: A real-time learning algorithm for a multilayered neural network based on the Extended Kalman filter, *IEEE Trans Signal Process*, **40**(1992), 959-966.

6. Ishida, R. et. al.: Superquick neuron training by extended Kalman filter(2nd Report), *Nippon Kikaigakkai Ronbunshu C*, **59**(1993), 2312-2317.

7. Pukorius, G. and Feldkamp, L. A.: Neurocontrol of nonlinear dynamical systems with Kalman filter trained recurrent networks, *IEEE Trans Neural Network*, **5**(1994), 279-297.

8. Ruck, D. W. et. al.: Comparative analysis of backpropagation and the extended Kalman filter for training multilayer perceptrons, *IEEEPAMI*, **14**(1992), 686-691.

9.. Lomen, D. O.: A model of diffusion and convection of pollutants from a point source, in Vansteenkiste, G. C., ed., *Modeling, Identification and Control in Environmental System*, pp. 361-378, North-Holland, Amsterdam, 1978.

10.. Zlatev, Z., Berkowicz, R. and Prahm, L. P.: Three dimensional advection-diffusion modeling for regional scale, *Atmospheric Environment*, **17**(1983), 491-499.

11.. Pozlewicz, A.: Modelling of pollutant diffusion in a stratified flow field, in Falconer, R. A., Goodwin, P. and Matthew, R. G. S. eds., *Hydraulic and Environmental Modelling of Coastal, Estuarine and River Waters*, pp. 381-389, Gower Publishing, Aldershot, England, 1989.

12.. Ozmidov, R. V.: *Diffusion of Contaminants in the Ocean*, Kluwer Academic publishers, Dordrecht, The Netherlands, 1990.

13. Nihoul, J. C. J.: Aplication of mathematical modelling to the marine environment, in Zannetti, P. ed.,.*Environmental Modeling*, Vol. 1, pp. 75-140, Computational Mechanics Publications, Southampton, England, 1993.

14. Muller, K. H.: A Lagrangian model of air pollutant dispersion, *Applied Mathematical Modelling*, **11**(1987), 104-109.

15. Horie, T.: Hydraulic investigation on seawater flow and substance dispersion in estuarine and coastal region, *Technical Note of The Port & Harbour, Research Institute of Japan*, **360**(1980), 1-222.

16. Brusasca, G., et. al.: Development of a portable FORTRAN 77 code for Monte Carlo particle modeling of atmospheric diffusion(MC-CAGPAR II): Validation against analytical solutions and tracer experiments, in *Computer Techniques in Environmental Studies*, pp. 431-450, Computational Mechanics Publications, Southampton, England, 1988.

17. Monin, A. S. and Yaglom, A. M.: *Statistical Fluid Mechanics*, Vol. 1,2, MIT Press, Cambridge, MA, 1975.

18. Seinfeld, J. H.: *Atmospheric Chemistry and Physics of Air Pollution*, Chap. 13,15, Wiley Interscience, New York, 1986.

19. Omatsu, S., et. al.: Estimation of nitrogen dioxide concentrations in the vicinity of a roadway by optimal filtering theory, *Automatica*, **24**(1988), 19-29.

20. Sawaragi, Y., et. al.: Statistical prediction of air pollution levels using non-physical models, *Automatica*, **5**(1979), 441-451.

21. Tanaka, M. and Katayama, T.: A robust identification of a linear system with observation outliers by the EM algorithm, *Trans. Institute of Systems, Control and Information Engineers*, **1**(1988), 117-126.

22. Nonaka, M. and Thomas, N. H : State space modelling of contaminant transport processes by extending discrete vortex simulations-1. Horizontal dispersion, *J. Min. Material Process. Inst. Japan*, **112**(1996), 225-230.

23. Fletcher, R. and Powell, M. J. D.: A rapidly convergent descent method for minimization, *Comput J*, **6**(1963), 163-168.

24. Lorenz, E. N.: Deterministic nonperiodic flow, *J. Atmos. Sci.*, **20**(1963), 130-141.

25. Sparrow, C.: *The Lorenz Equations : Bifurcation, Chaos and Strange Attractors*, Springer-Verlag, New York, 1982.

26. Rumelhart, D. E., Hinton, G. E. and Williams, R. J.: Learning internal representations by error propagation, in *Parallel Distributed Processing*, pp. 318-362, MIT Press, 1986.

27. Qiu, G., Varley, M. R. and Terrell, T. J.: Accelerated training of backpropagation networks by using adaptive momentum step, *Electron. Lett.*, **28**, 377-379.

HYDROLOGICAL MODELLING OF DISTRIBUTED RAINFALL / RUNOFF IN THE WATERSHED OF GUADIANA

Fernanda Gomes[1], Manuel Lacerda[1]
Maria Lurdes Duarte[2], José A. Almeida[2]

[1] *INAG - Instituto da Água, Direcção de Serviços Recursos Hidricos*
Av. Almirante Gago Coutinho, 30 - 1000 Lisboa Codex
Portugal
Tel. 351.1.8470080/93 - Fax. 351.1.8494787
[2] *Instituto Superior Técnico*
Av. Rovisco Pais, 1096 Lisboa Codex
Portugal
Tel. 351.1.8417247 - Fax. 351.1.8417442
E-Mail: ncmrp@alfa.ist.utl.pt

Abstract

This paper presents a rainfall / runoff model of the watershed of Guadiana river which is located in the South of Portugal. The methodology combines the geostatistical spatial characterisation of hydrological parameters and the posterior simulation of the surface runoff.

Both models - geostatistical and surface flow simulation - were developed and integrated in a Geographic Information System - GIS -, constituting an important tool for the hydrological management of the watershed.

The methodology can be resumed in the following sequential steps:
i) in a first step, the spatial distribution of rainfall and temperatures are simulated in all area for homogeneous time periods corresponding to a given month, through a stochastic simulation technique - Sequential Gaussian Simulation.
ii) in a second step, a simulation of the surface runoff, based on the Temez model is implemented in the GIS. This is an appropriate model, for characterising the hydrological behaviour of large watersheds, whose input are the climate parameters - rainfall and potential evapotranspiration (derived from step i) -, type of soil and land cover.

The combined model - stochastic images of spatial dispersion of rainfall and temperature and the simulation of runoff - can be used to visualise extreme situations of the hydrological behaviour of the watershed and local critical areas.

A. Soares et al. (eds.), geoENV I–Geostatistics for Environmental Applications, 213–221.

1. Introduction

This paper presents a combined methodology to simulate the hydrological behaviour of large watersheds. A stochastic simulation technique is applied for the characterisation of the spatial dispersion of temperature and rainfall for a given period of time.

The set of equiprobable images, which gives the spatial uncertainty of both variables, will feed a deterministic point dispersion model to simulate the surface runoff of the whole watershed.

This combined methodology, totally implemented in a GIS, can be described by the following points *(Figure 1)*:

i) in a first step, the spatial dispersion of rainfall and temperature are simulated in whole area in a given period of time by using the Sequential Gaussian Simulation method (Deutch and Journel, 1992). Thus, the information of watershed area is characterised by the following coverages in a GIS raster format: the DEM - Digital Elevation Model - regarding the topography of watershed; the type of soil and land cover coverage; and the set of stochastic images of rainfall and temperature which was transformed in potential evapotranspiration;

ii) in a second step a simulation model of surface runoff, based on the Temez model implemented in the GIS gives the different hydrological scenarios for the different stochastic images of climate parameters. An uncertainty measure regarding the water accumulation can be assessed in different sub-watersheds in all region. The final maps of the hydrological behaviour uncertainty are important tools for the planning of the hydrological resources of the region.

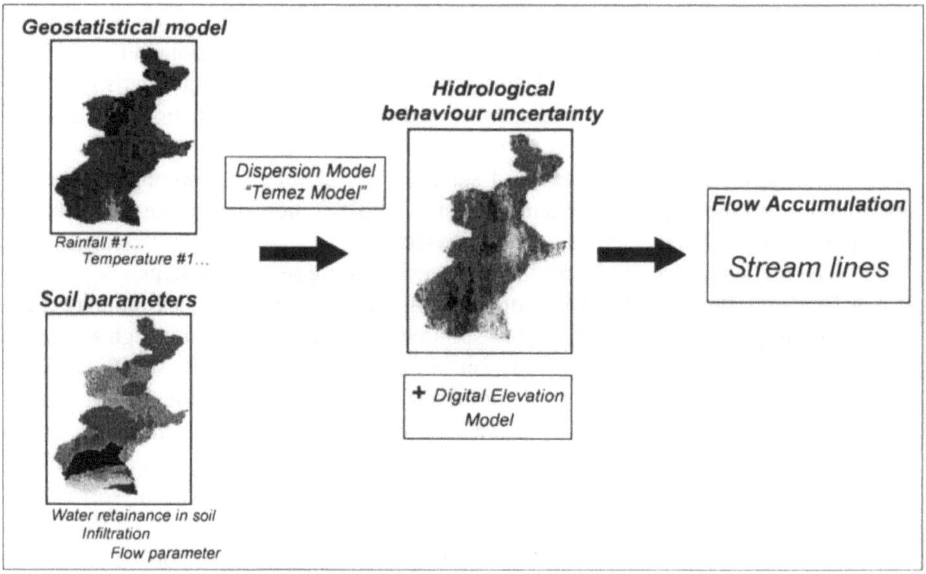

Figure 1 - Representation of combined methodology implemented in a GIS.

This two-step approach was implemented in the watershed of Guadiana located in the south of Portugal. The methodological sequence will be detailed in parallel with the case study of Guadiana watershed.

2. Guadiana Watershed

Located in Southeast border of Portugal with Spain, the Guadiana watershed[1] has the following data coverages in GIS raster format: i) DEM - The Digital Elevation Map regarding the topographic information *(Figure* 2); ii) the type of soil; iii) and land cover coverage.

The rainfall and the temperature are recorded by monthly measures in a network of monitoring stations of *figure 3*.

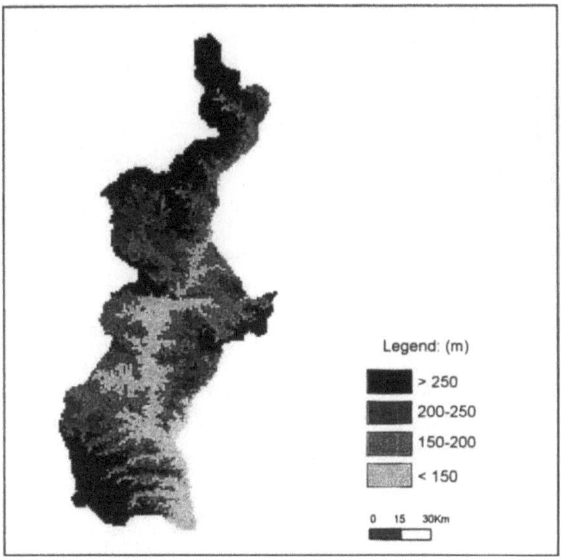

Figure 2 - Digital elevation model in Guadiana watershed.

3. Stochastic Simulation of Rainfall and Temperature

For this study, only one month period - April 1971 - has been selected for illustrative purposes.

Spatial continuity structures of rainfall are measured through spatial variograms calculated for this month period. The variogram used in this study was fitted by a anisotropic spherical model (figure 4), which exhibits a N-S main direction and a anisotropy ratio of 0.285.

---

[1] This study was conducted only for the portuguese part of the watershed.

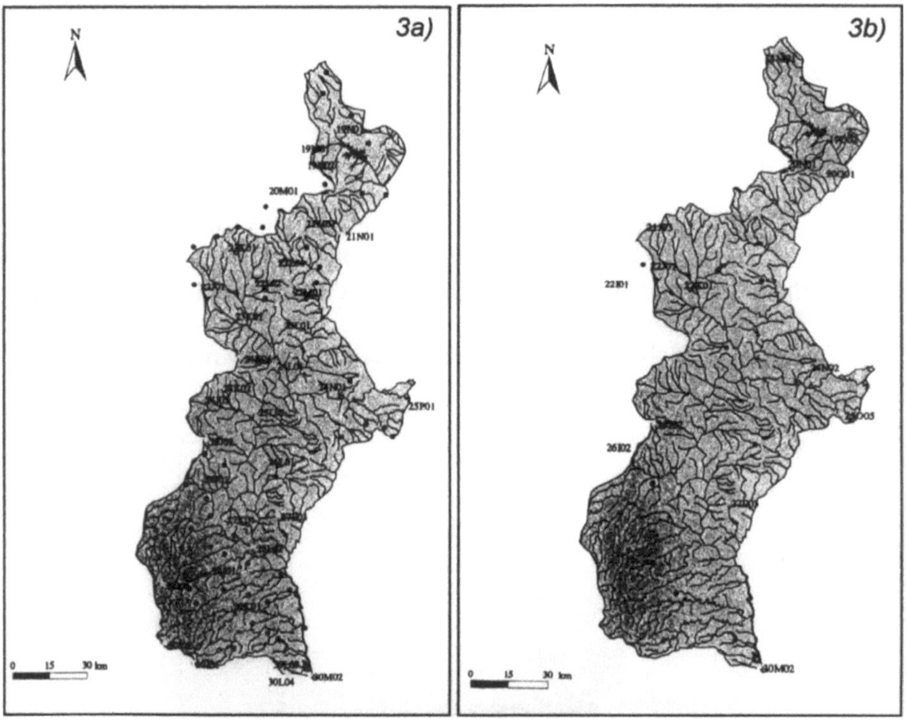

Figure 3 - Spatial location of the monitoring stations in Guadiana watershed: a) rainfall; b) temperature.

For the temperature, a very homogeneous variable in each season, an average variogram for a set of 30 years in April month was adopted: a isotropic spherical model was fitted to the experimental points (*see figure 5*).

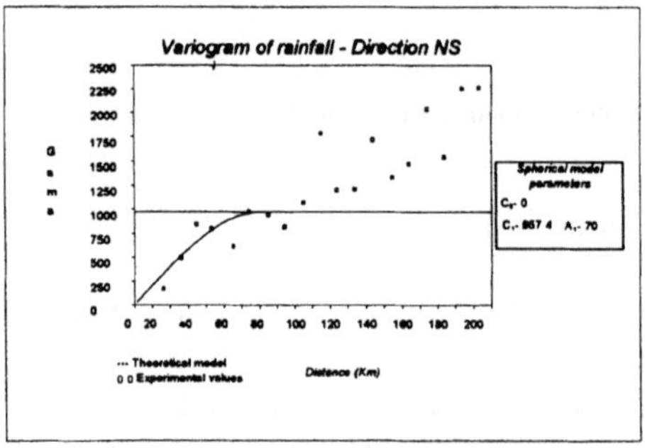

Figure 4 - Rainfall: experimental variogram and theoretical model.

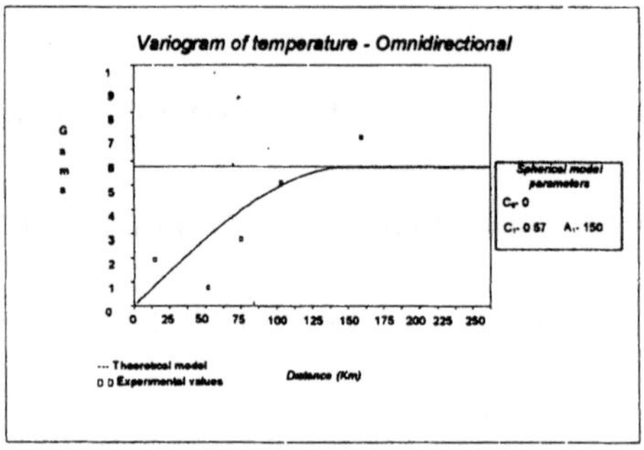

Figure 5 - Temperature: experimental variogram and theoretical model.

The Sequential Gaussian Simulation was applied to obtain the set of stochastic images of temperature (Figure 6) and rainfall (Figure 7). Each unit cell represents a square of 1 x 1 Km. The total area of the portuguese part of the Guadiana watershed is 11525Km^2.

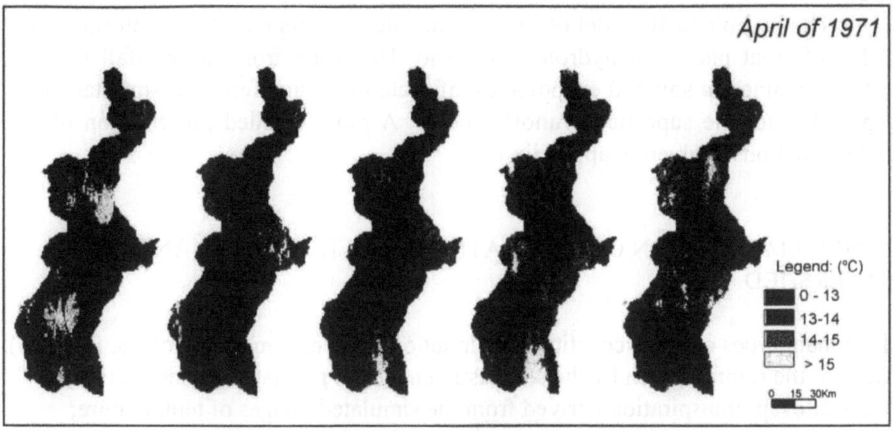

Figure 6 - Set of 5 simulated images of temperature.

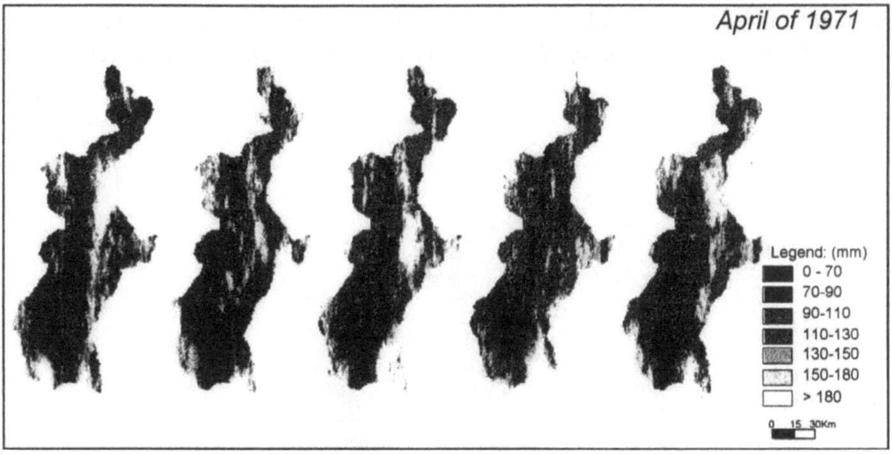

Figure 7 - Set of 5 simulated images of rainfall.

4. Surface Runoff Model

4.1 TEMEZ MODEL

In this study a surface runoff model adapted from the hydrological aggregate "Temez Model" (Temez, 1977) is proposed. Once the watersheds are divided in cells (raster format), each one has associated its own parameters. Thus, the equations of the hydrological model of Temez, which rule the evapotranspiration, infiltration and runoff process, are applied to each cell (Ruiz *el al,* 1994). The water flow of entire watershed results from the sum of the hydrological behaviour of the set of cells, given by the model.

The hydrological model of Temez simulate the essential of the waterflow process in the different phases of hydrological cycle. The water from the rainfall (except the portions retained in soil and evaporated) infiltrate in the aquifers and, simultaneously, is responsible for the superficial runoff process. A more detailed presentation of Temez model equations is given in appendix I.

4.2 IMPLEMENTATION OF SIMULATION MODEL TO GUADIANA WATERSHED

Three main types of maps constitute the input of the Temez model (Gomes, F., 1996):
.Maps of the rainfall given by the stochastic images of geostatistical simulation;
.Maps of evapotranspiration derived from the simulated images of temperature;
.Maps of main soil parameters: maximum capacity of water retainance in soil, maximum capacity of infiltration into the aquifers and flow parameter (see Appendix I).

These parameters are considered as constants in each sub-watershed. Results of the runoff model are visualised in the following maps: the real evapotranspiration *(Figure* 8), the soil infiltration *(Figure* 9) and the superficial runoff is represented in *figure 10.*

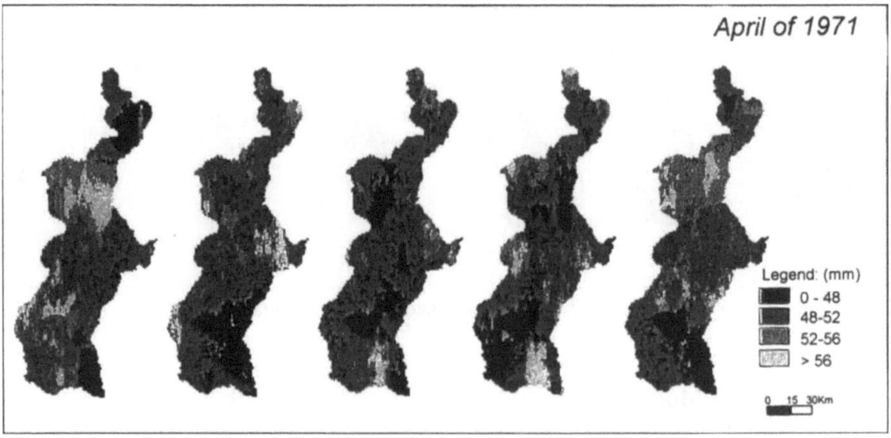

Figure 8 - Set of *5* images of real evapotranspiration (Temez model output).

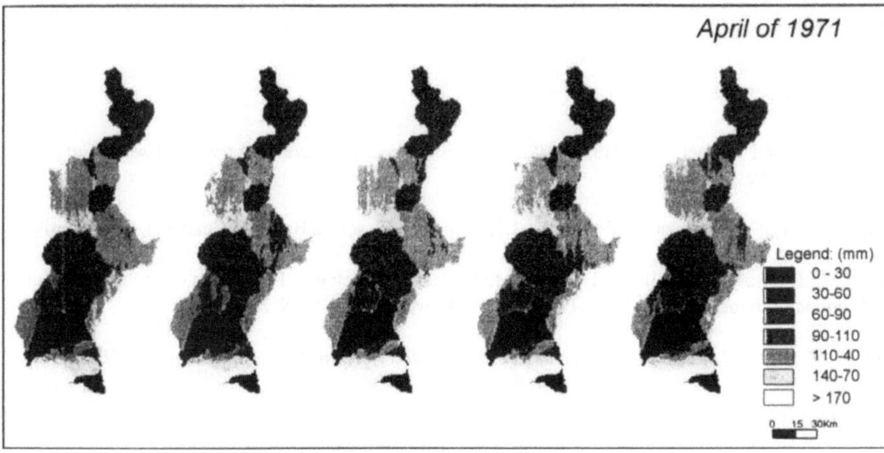

Figure 9 - Set of *5* images of water in the soil (Temez model output).

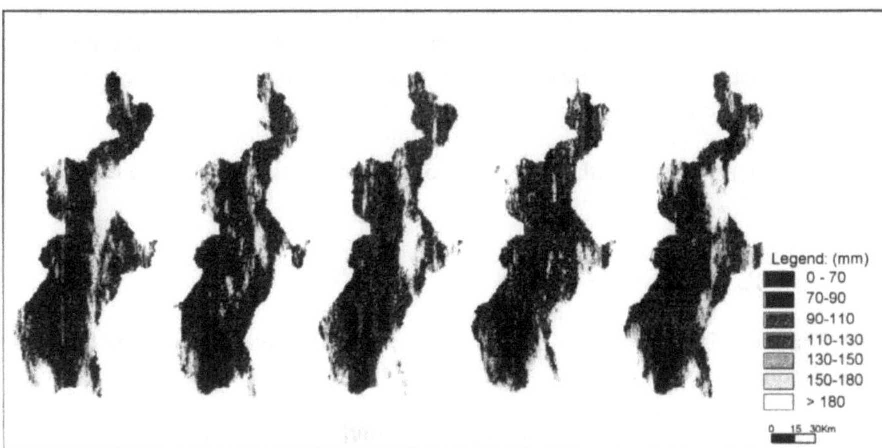

Figure 10 - Set of *5* images of Runoff (Temez model output).

5. Conclusions

This paper presents the methodological sequence of the simulation of hydrological behaviour of the Guadiana watershed. Based on the stochastic imaging of watershed runoff, an important tool for its hydrological planning can be build. In fact, after the calibration of Temez model with real data of watersheds, local distribution and extreme scenarios of superficial runoff can be predicted for given meteorological conditions.

For some periods, a relation between the topography and the rainfall seems to exist. Thus, the incorporation of the external effect of topography in the stochastic simulation is being taking into account in further developments of the proposed approach.

6. References

Deutch, C.V. and Journel, A.G. (1992) *GSLIB: Geostatistical Software Library and User's Guide*, Oxford University Press, Stanford.

Gomes, M.F. (1996) Modelação Hidrológica Distribuída: Aplicação à Bacia do Guadiana, MSc Thesis.

Ruiz, J.M., Estrela, T. and Quintas, L. (1994) Modelizacion Hidrologica Distribuida en el Proyecto Guadiana. El modelo Simpa 1.0 (Simulation Precipitation-Aportacion). CEDEX, Madrid, Spain.

Temez, J.R. (1977) *Modelo Matemático de Transformación Precipitación-Aportación*, ASINEL.

Appendix I

I.1) THORNTHWAITE METHOD

The ETP_i (potential evapotranspiration) is calculated from the temperature using the Thornthwaite method, by establishing correlation between the temperature and the potential evapotranspiration. Based on two indexes, I_i and J, respectively month and annual temperature index:

$$I_i = (T_i)^{1.5} / 5 \qquad \text{with } T_i \text{ - average of the temperature during one month (°C)}$$

and

$$J = \sum I_i \quad , i=1,\dots 12$$

The potential evapotranspiration in the equator and given the average temperature during a month T is:

$$ETP_o = 1.6 \frac{[10T]^\alpha}{J} \quad \text{with } \alpha = 0.49 + (17900\,J - 77.1\,J^2 + 0.675\,J^3) \times 10^{-6}$$

For other latitude, the potential evapotranspiration is given by

$$ETP_l = K\, ETP_o$$

where K is a constant dependent from the month and the latitude.

I.2) TEMEZ MODEL

The objective of the Temez model (Temez, 1977) is to quantify the water transport during the different phases of the hydrological cycle, by including transfer laws between the different involved phases, respecting the total water mass balance. This model assumes the division of the soil in two vertical zones: the undersaturated soil level and the saturated zone, that works as an aquifer with superficial recharges. The rainfall water, excluding the storage of water in the soil and the evapotranspiration, constitutes a surplus: part of it infiltrates and the other part constitutes the superficial runoff.

This surplus of water can be calculated using the following expressions:

$$T_i = 0 \quad \text{if} \quad P_i \leq P_0 \ \text{or}$$

$$T_i = (P_i - P_0)^2 / (P_i + \delta - 2\,P_0) \quad \text{otherwise}$$

with

$$\delta = H_{max} - H_{i-1} + ETP_i$$

$$P_0 = C\,(H_{max} - H_{i-1})$$

P_i - total rainfall during the period *i-1* to *i* (mm)
T_i - surplus of water in the period *i-1* to *i* (mm)
H_{max} - maximum capacity of water storage in the soil (mm)
H_{i-1} - water in the soil in time slice *i-1* (mm)
ETP_i - potential evapotranspiration in the period *i-1* to *i* (mm)
C - flow parameter

Once the surplus of water is calculated, the water in the soil (H_i) in time slice *i* is given by

$$H_i = \max\,(0, H_{i-1} + P_i - T_i - ETP_i) \quad \text{(mm)}$$

and the real evapotranspiration (ET_i) for the same time[2]

$$ET_i = \min\,(H_{i-1} + P_i - T_i\,,\ ETP_i) \quad \text{(mm)}$$

This model assumes the following infiltration law (I_i), function of the surplus of water and the maximum infiltration parameter (I_{max}):

$$I_i = I_{max}\,(T_i / T_i + I_{max})$$

The total flow of water (A_i) is given by the difference between the surplus of water and the infiltration:

$$A_i = T_i - I_i$$

[2] The maximum value for the real evaporation is the potential evapotranspiration.

SPATIAL CORRELATIONS OF HYDROCHEMICAL PARAMETERS

LAURA GUADAGNINI
Sezione Geologia Applicata - D.S.T.M.
Politecnico di Milano, P.zza L. da Vinci, 32, 20133 Milano (Italy)

Abstract

A first series of field measurements have been performed during a 1-week period in June 1995 in a region of the alluvial phreatic aquifer near Salsomaggiore, northern Italy. The study area is about 15 km^2 wide and the Ghiara and Stirone streams form a junction in its centre. Hydrochemical data are measured at some wells and at selected locations of the two streams. The collected data are presented and a geostatistical interpretation of their spatial distributions is given. To select the best among different variogram models, we used formal model discrimination criteria based on the Maximum Likelihood Cross Validation (MLCV) approach coupled with a generalised least square drift removal approach. To analyse how well the physical system is represented by a given model, different model identification criteria are adopted and the results are compared.

1. Introduction

Accurate information on the origin, age, geochemistry and migration paths of groundwater is needed to characterise the groundwater resource. In most cases these data are difficult and expensive to obtain. Furthermore, the interpretation of data taken from semiconfining, discontinuous alluvial sediments of the type discussed in this paper is not very simple: this is mainly due to the fact that the available well logs and water levels measurements do not allow to define a clear aquifer separation and the data often represent spatial averages of random variables over variable supports. The main motivation of this work is the need to obtain a more detailed knowledge of the spatial distribution of hydrochemical variables in a 15 km^2 wide area located in the southern part of the Po Valley, in the proximity of the Appennine chain (Province of Parma-northern Italy). Therefore, in this paper we describe the preliminary analyses of sampled data of pH, total salinity (TDS), electrical conductivity (EC) and chloride (Cl). The spatial covariance structure of these variables is evaluated by an adjoint state maximum likelihood cross validation (ASMLCV) method (Samper and Neuman, 1989a, b, c) and their spatial variability is estimated by kriging. We investigate the

A. Soares et al. (eds.), geoENV I–Geostatistics for Environmental Applications, 223–234.

utility of model identification criteria to select the best among different variogram models. We try to verify if it is possible to test geostatistically with the available data some hypotheses about the hydrochemical evolution of the investigated area.

2. Geology, hydrogeology and geochemistry of the basin

Detailed descriptions of the regional geology and hydrogeology of the studied area are outlined by Bernini and Papani (1987), Bigi et al. (1975) and Petrucci et al. (1975).

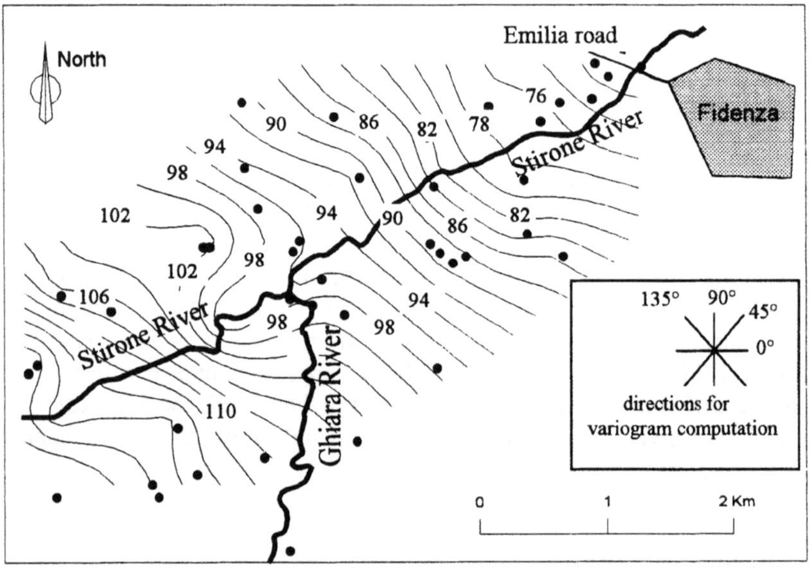

Figure 1: Location of the study area with sampling points and groundwater level contours (water levels in meters above mean sea level)

The area under study is part of the high alluvial plain laying upstream the city of Fidenza (*Figure* 1). The thickness of the sediments is defined by the top of the clayey Plio-pleistocenics marine formations, which lay more than 100 m below the ground surface. The prequaternary formations make up a fold system with synclines and anticlinals bends, often with faults and overthrusts. The northern limit of this complex system (Brè, Parma and Collecchio structures) is represented by the front of the Appennine chain buried beneath the Po Plain (External Thrust Front), while its southern limit forms the Pedeappennine margin (Pedeappenninic Thrust Front) (Bernini and Papani, 1987). The hydrogeological setting is characterised by alluvial deposits that make up the main aquifer and form lenses of different thickness, mean grain size and hydraulic conductivity, elongated in the N-S direction. Therefore, the main aquifer is a lithologically complex alluvial sequence (gravel with silt-clayey

intercalations) that define an undifferentiated to compartmental monostratum system with free to confined groundwater.
On the basis of the measurements, it is observed that in this area the main aquifer is located between 10 and 40 m below the ground surface. The groundwater system to the south of the city of Fidenza is mainly a phreatic one; as we move to Fidenza, a confined portion may be distinguished, too. *Figure* 1 shows the main pattern of the groundwater flow. Hydraulic heads were measured at 28 wells at the same time that the hydrochemical variables and they have been plotted by contour lines of constant head. The structural analysis of heads was performed with an adjoint state maximum likelihood cross validation (ASMLCV) method (Samper and Neuman, 1989 a, b) and kriging. As the variable exhibited a trend in the SW-NE direction, a polinomial drift was detrended from the original values, to yield the residuals. The application of ASMLCV to the piezometric head will be described elsewhere. Here we notice that groundwater mainly flows in the SW-NE direction and the contour lines seem to point out a hydraulic connection between the surface water of the Ghiara and Stirone rivers and the groundwater. On the right hand side of the Ghiara river a recharge area is recognisable. The Stirone river partly recharges and partly drains the groundwater system.
The different sources of chemicals release were investigated in situ and samples were collected at selected cross-sections in the two rivers. From the results of the laboratory analysis it was noted that the quality of the surface water noticeably changed downstream the junction of the rivers. For example, the chloride concentration in the Stirone river increased from 16 to 181 mg/l when moving downstream the junction. This is mainly due to the mixing with water from the Ghiara river, where the chloride concentration was 2040 mg/l close to the junction and it was 11000 mg/l further upstream. These data are somehow consistent with the findings of Alifraco et al. (1995). It is expected that the two rivers have different impacts on the local quality of the groundwater.

3. Data set, geostatistical analysis and discussion

Water samples were collected from some actively pumped wells used for domestic and agricultural purposes and at selected sections of the two streams during a 1-week period in June 1995; the location of the sampling points is shown in *Figure* 1. We measured in situ the pH, EC and TDS, while the chloride (Cl) concentration was determined by laboratory photometric analysis. EC is measured in microsiemens per centimetre, TDS and Cl in milligrams per litre. These parameters have been subjected to a geostatistical analysis based on the following steps: (a) computation of sample statistics, (b) computation of sample variogram γ, (c) evaluation of spatial drift, (d) variogram estimation, (e) kriging and (f) discussion of the results. Table 1 lists the sample statistics of the hydrochemical data. A base 10 logarithmic transformation was applied to EC, Cl and TDS. This choice is motivated by the fact that the resulting sample statistics correspond more closely to those typical of the Gaussian distribution

than would otherwise be the case. The behaviour of the resulting sample variograms is also improved by considering the log-transformation. Another advantage related to a log-transformation is that it allows to work with positive defined variables. The hypothesis of normality is accepted for the Log(Cl), Log(EC) and Log(TDS) data at the common 95% confidence level, based on the Kolmogorov-Smirnov and Chi-square tests. With regard to the pH, both Normal and Log-Normal hypotheses were rejected; this fact and its low variability explains why it was not transformed logarithmically.
For the variogram estimation, to select the best among different models we used formal model discrimination criteria based on the Maximum Likelihood Cross Validation (MLCV) approach of Samper and Neuman (1989 a, b, c) coupled with the generalised least square drift removal approach of Neuman and Jacobson (1984). Following this approach, the likelihood of the parameter vector of a given variogram model is maximised by minimising the Negative Log-Likelihood (NLL), as reported in Samper and Neuman (1989 a).
As MLCV assumes that the variogram model is known, the output variogram parameters are given without considering how well the physical system is represented by the adopted model. In order to answer this problem, earlier works by Carrera and Neuman (1986 a, b, c) and Samper and Neuman (1989 a, b, c) have compared four criteria, AIC (Akaike, 1974), MAIC (Akaike, 1977), HIC (Hannan, 1980) and KIC (Kashyap, 1982). In the following we adopt these model identification criteria and compare the obtained results. The kriging and error maps were computed by means of the GSLIB package (Deutsch and Journel, 1992). A detailed analysis of the Log(Cl), pH and Log(EC) is now presented. The variable TDS is directly related to EC and this is the reason why it was not analysed geostatistically. A discussion of the qualitative results is then presented.

TABLE 1. Sample Statistics of the hydrochemical data.

Variable	Number of data	μ Mean	m Median	σ^2 Variance	CV Coefficient of variation	Skewness	Kurtosis
pH	27	7.33	7.06	0.298	0.075	1.21	0.294
Log(EC)	27	3.14	3.00	0.112	0.106	1.72	6.384
Log(Cl)	26	1.97	1.73	0.64	0.406	0.84	2.960
Log(TDS)	27	2.92	2.82	0.051	0.077	0.37	1.947

3.1 CHLORIDE (Cl)

Values of Log(Cl) range from 1.04 to 4.068. The highest Cl values are measured close to the Ghiara river and lower values are detected near the Stirone river and in the area close to the city of Fidenza. In *Figure* 2a the raw variograms computed along the different directions listed in *Figure* 1 and with different angle tolerance are reported. A drift along the 90° e 135° directions seems to be recognisable. The number of sample pairs for the highest lags along directions with angle tolerance θ = 45° are below 10, so that the results are judged to be not very stable. It is felt that a higher number of data would be required to prove whether the detected drift is confirmed or it is an artifact of the data. To answer this problem, a more detailed sampling is currently being performed. For this reason, in the following the theoretical variogram model is evaluated only on the basis of the omnidirectional raw variogram. *Figure* 2b shows the omnidirectional raw variogram together with three models tested.

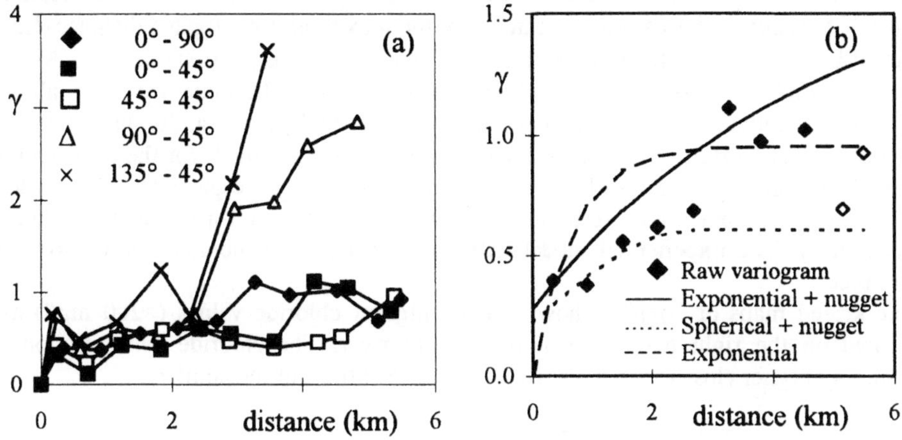

Figure 2a: Directional sample variograms of Log(Cl); directions are taken as reported in *Figure* 1; angle tolerance is 45° or 90° (omnidirectional).

Figure 2b: Omnidirectional raw variogram of Log(Cl) and the models fitted (empty diamonds are related to lags with less than 15 pairs).

TABLE 2. Identification of variogram structure of Log(Cl) measured.

Model	Exponential with nugget	Spherical with nugget	Exponential
NLL	58.0162	59.2015	59.6398
Number of parameters	3	3	2
Nugget	0.282147	0.222879	-
Sill	1.53396	0.382297	0.95783
Effective Range	14.79950	2.628929	2.01938
AIC	64.0162	65.2014	63.6398
Rank	2	3	1
MAIC	67.7904	68.9757	66.1559
Rank	2	3	1
HIC	65.1030	66.2883	64.3643
Rank	2	3	1
KIC	61.8681	68.8465	63.7135
Rank	1	3	2

The nugget associated with this variable appears to be larger than could be attributed to sampling and laboratory measures errors. It is felt that the main reason is the two dimensional nature of the adopted geostatistical analysis. Water samples have probably undergone mixing in wells which are in contact with the aquifer over different vertical intervals. This behaviour is observed also for the other studied variables.

The corresponding model parameters and the values of NLL and model identification criteria are listed in Table 2. It can be noted that all the tested models show a finite range. However, when a nugget is added to the Exponential structure, the result of MLCV estimation gives a range greater than the maximum investigated distance. Both the NLL and KIC rank the Exponential model with nugget as the best one; AIC, MAIC and HIC rank the Exponential model without nugget as the best one. However, all the results of the identification criteria are quite similar. In *Figures* 3a, 3b, 4a and 4b the kriged map and kriging variance of Log(Cl) are shown respectively for the Exponential model without nugget and for Exponential model with nugget respectively. The models yield quite different kriged (estimated) images of Log(Cl) and associated estimation uncertainty. The Exponential model without nugget gives generally a lower kriging variance.

The kriged maps of Log(Cl) show that the highest chloride values (>200 mg/l) are located on the right hand side of the Ghiara river. The chloride concentration is generally higher close to the rivers than at other locations of the aquifer.

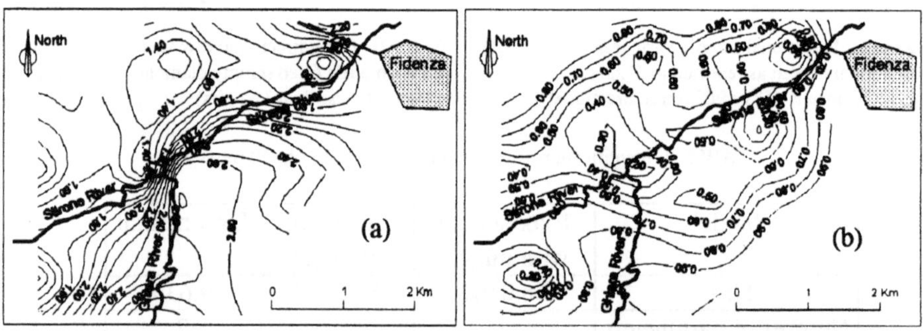

Figure 3: (a) Kriged map and (b) Kriging variance of Log(Cl) for the Exponential model without nugget.

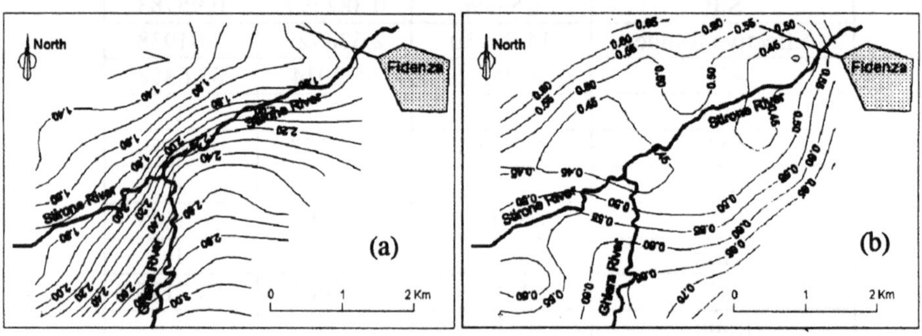

Figure 4: (a) Kriged map and (b) Kriging variance of Log(Cl) for the Exponential model with nugget.

3.2 pH

Values of pH range from 6.74 to 8.48 with a low variance and a small coefficient of variation. Sample variograms computed along the previously defined four directions showed a lack of anisotropy and no clear evidence of a drift. In *Figure* 5a the omnidirectional raw variogram is plotted together with four tested models. The corresponding model parameters and the values of NLL and model identification criteria are listed in Table 3.

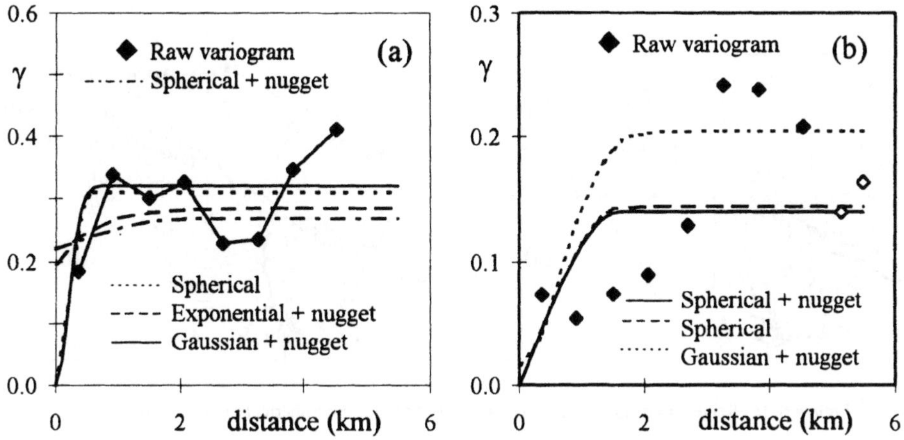

Figure 5: Omnidirectional sample variograms of (a) pH and (b) Log(EC) and the models fitted (empty diamonds are related to lags with less than 15 pairs).

TABLE 3. Identification of variogram structure of pH measured.

Model	Gaussian with nugget	Spherical with nugget	Exponential with nugget	Spherical
NLL	44.0	46.18	46.07	45.05
No. of parameters	3	3	2	2
Nugget	0.00028	0.22	0.19	-
Sill	0.32	0.048	0.086	0.31
Effective Range	0.478	2.0	1.71	0.59
AIC	50.0	52.18	52.07	49.03
Rank	2	4	3	1
MAIC	53.89	56.07	55.96	51.63
Rank	2	4	3	1
HIC	51.16	53.33	53.22	49.08
Rank	2	4	3	1
KIC	58.81	53.51	50.72	45.92
Rank	4	3	2	1

The lowest NLL value is associated to the Gaussian model with nugget. All the identification criteria rank the Spherical model without nugget as the best one. It is remarked that all the criteria give similar results; however, the KIC criterion seems to be very sensitive to the number of parameters of the adopted model. In *Figures* 6a, 6b, the kriged map and kriging variance of pH are shown for the Spherical model, while *Figures* 7a and 7b report the results of the estimation with the Gaussian model with nugget. Both models yield very similar kriged (estimated) images of pH but rather different measures of the associated estimation uncertainty. The Spherical model gives a lower kriging variance.

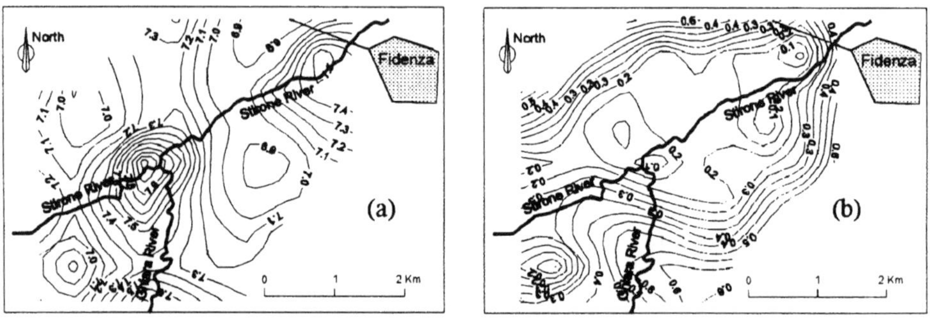

Figure 6: (a) Kriged map and (b) Kriging variance of pH for the Spherical model without nugget.

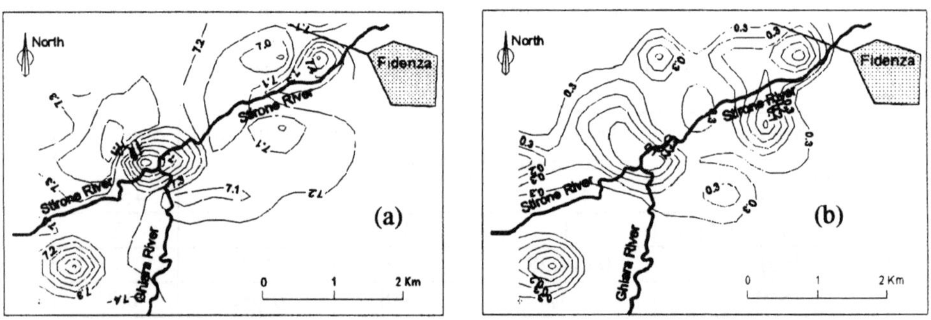

Figure 7: (a) Kriged map and (b) Kriging variance of pH for the Gaussian model with nugget.

3.3 ELECTRICAL CONDUCTIVITY (EC)

The results of the variogram analysis of Log(EC) along different directions are qualitatively similar to those of Log(Cl). *Figure* 5b shows the omnidirectional raw variogram together with the fitted models. Table 4 reports the corresponding model parameters and the values of NLL and model identification criteria. According to the

NLL value the model that fits the data best is the Gaussian one with nugget, while all four model discrimination criteria consistently rank the Spherical model without nugget, which is the most parsimonious one, as better. However, it is observed that the sill and range are very similar for the two models. The KIC criterion is the only one that gives values that are considerably lower than those obtained for the least acceptable models while AIC, MAIC and HIC are very similar. *Figures* 8a and 8b, 9a and 9b show the kriged map and kriging variance of Log(EC) respectively for the Gaussian model with nugget and for Spherical model without nugget. The models yield quite similar kriged (estimated) images of Log(EC) and associated estimation uncertainty. The Log(EC) spatial pattern is quite similar to that of the chloride. The highest values of Log(EC) are located in the region that lays to the east of the junction between the two rivers.

TABLE 4. Identification of variogram structure of Log(EC) measured.

Model	Gaussian with nugget	Spherical with nugget	Spherical
NLL	8.57203	8.74081	8.65660
Number of parameters	3	3	2
Nugget	0.016714	0.000117	-
Sill	0.187362	0.13931	0.143818
Effective Range	1.62320	1.610438	1.661055
AIC	14.57200	14.74080	12.65660
Rank	2	3	1
MAIC	18.45950	18.62830	15.24830
Rank	2	3	1
HIC	15.72800	15.89680	13.42720
Rank	2	3	1
KIC	28.35940	30.06870	19.34700
Rank	2	3	1

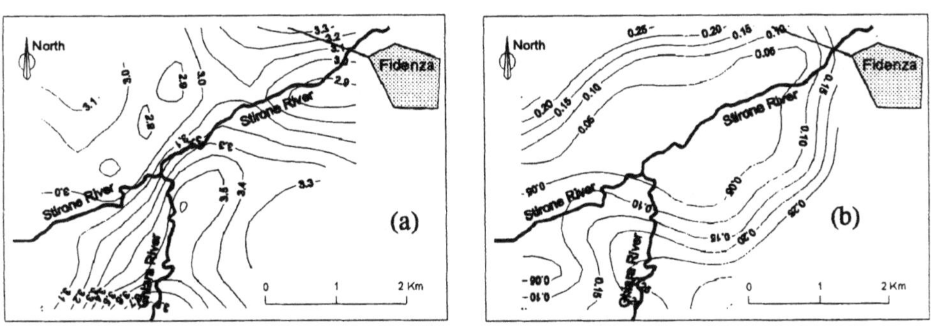

Figure 8: (a) Kriged map and (b) Kriging variance of Log(EC) for the Gaussian model with nugget.

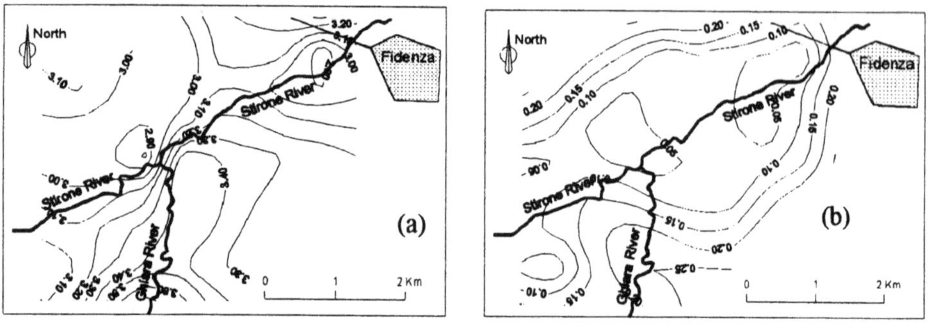

Figure 9: (a) Kriged map and (b) Kriging variance of Log(EC) for the Spherical model without nugget.

3.4 DISCUSSION

The analysis of the kriged maps of the chemical variables allows to recognise possible interactions between superficial waters and groundwater in the tested region. Water exchange between rivers and groundwater is enhanced by considering that the values of EC and Cl tend to increase locally in the proximity of the Ghiara and Stirone rivers. Samples collected at cross sections in the streams confirmed that superficial waters are characterised by Cl concentrations consistently higher than those measured in the groundwater system. The highest Cl and EC values are measured in areas where the pH is high. Superficial water usually shows higher pH levels than groundwater and the detected high pH in groundwater close to the streams may be attributed to a mixing process. Some areas may be distinguished, characterised by different qualitative behaviour:

- *Region on the right hand side of the Ghiara and Stirone rivers.* Chloride values are here very high: close to the river we detected Cl concentrations above 900 mg/l. In this area the aquifer is mainly recharged by the Ghiara river. The kriged maps reflect the basin-wide drift along the SW-NE direction with a high close to the Ghiara river and a monotonic descent toward the city of Fidenza. This supports the concept that groundwater is depleted in Cl as it moves SW-NE because of the influence of the Stirone river. Specifically, the Stirone river recharges the groundwater. Measured Cl contents in the Stirone water are much smaller than those detected in the Ghiara river. As a consequence of the mixing between the Stirone water and groundwater, the Cl concentration in the groundwater tends to decrease in the proximity of the Stirone river, over the area close to the city of Fidenza. The pH values in the groundwater are still high. It is evidenced geostatistically that the major source of aquifer pollution in the region is the Ghiara river, while water from the Stirone river has a local cleaning effect on the quality of groundwater.
- *Region on the left side of the Stirone river.* In this area a local groundwater recharge by the Stirone river is confirmed. Water with high Cl concentrations flowing from the Ghiara river has undergone mixing with the fresher Stirone water

downstream the junction. This explains the lower Cl values here detected. The highest pH values are always located close to the river.

- *Area between the Ghiara and Stirone rivers*. The available data does not allow a clear interpretation of the mixing process between surface water and groundwater. Anyway, it is felt that recharge areas are located close to the Ghiara river, while the Stirone river drains the groundwater system as it is suggested by the reproduced kriged maps.

In all the cases, the kriged maps are too smooth to reproduce local high and lows the existence of which could not be confirmed geostatistically in this study. Additional investigations, including multivariate statistical analysis techniques of various hydrochemical variables, are currently being developed in order to add further information to confirm the detected patterns and to improve the knowledge of the origin of groundwater in the tested system.

4. Conclusions

Hydrochemical parameters are suitable to geostatistical interpretation (e.g. Samper and Neuman, 1989 c; Shafer and Varljen, 1990; Guadagnini, 1996). The preliminary results of the geostatistical analysis of a series of field hydrochemical data that have been sampled during 1995 in a region of the aquifer near Salsomaggiore, northern Italy, is presented. The main subject of this experimental work is to test whether interactions between surface water and groundwater may or may not be detected geostatistically. As a first result, the analysed data show no discernible drift. However, this does not preclude the existence of spatial variations, according to expected patterns. The kriging and error maps of pH and the Log transformed Cl and EC have been computed. It has been confirmed that in the tested region there is evidence of an exchange of chloride from the surface water and groundwater. From this viewpoint, areas characterised by the hydrochemical influence of a river can be distinguished from others. One of the main conclusions of this preliminary study is that coupling geostatistical analysis with hydrogeological information allows to obtain useful suggestions about the evolution of chemical components in the groundwater. The above study area is currently being monitored and the data from different sampling periods are under investigation, to study the space-time evolution of the hydrochemical species detected.

5. References

Akaike, H., (1974) A new look at statistical model identification, *IEEE Trans. Automat. Control* **AC-19**(6), 716-722.

Akaike, H., (1977) An entropy maximization principle in *Applications of statistics*, Krishnaiah, P.R., North Holland-Amsterdam.

Alifraco, G., Beretta, G.P., Giovanardi G., Mozzanica, E., Munari, A., Pelosio, A., Pizzarotti, A., Tagliavini S., and Zavatti, A., (1995) Studi sull'acquifero parmense: il monitoraggio idrogeologico ed idrochimico e la cartografia automatizzata della vulnerabilità degli acquiferi, con un primo esempio relativo alla pianura fidentina (in Italian), Proc. of "Secondo convegno nazionale sulla protezione e gestione delle acque sotterranee: metodologie, tecnologie e obiettivi", *Quad. Tecn. Protez. Ambient.,3.103-3.122* Pitagora Ed., Bologna.

Bernini, M., and Papani, G., (1987) Alcune considerazioni sulla struttura del margine appenninico emiliano fra lo Stirone e l'Enza (e sue relazioni con il sistema del Taro) (In Italian), *L'Ateneo Parmense-Acta Naturalia*, **23**, 219-240

Bigi, B., Cavazzini R., Morestori, L., Panicieri, E., Pecorari M., Rossetti, G., and Petrucci, F., (1975) *Ricerche sulle acque sotterranee nella pianura parmense e piacentina: sezioni interpretative degli acquiferi (parte prima)* (in Italian), Grafiche STEP, Parma.

Carrera, J., and Neuman, P., (1986a) Estimation of aquifer parameters under transient and steady state condition: 1. Maximum likelihood method incorporating prior information, *Water Resour. Res.* **22**(2), 199-210.

Carrera, J., and Neuman, P., (1986b) Estimation of aquifer parameters under transient and steady state condition: 2. Uniqueness, stability, and solution algorithms, *Water Resour. Res.* **22**(2), 211-227.

Carrera, J., and Neuman, P., (1986c) estimation of aquifer parameters under transient and steady state condition: 3. Application to synthetic and field data, *Water Resour. Res.* **22**(2), 228-242.

Deutsch, C.V., and Journel, A.G., (1992) *Geostatistical software library and user's guide*, Oxford University Press, New York

Guadagnini, L., (1996) Analisi di strutture di correlazione spaziale di proprietà di un acquifero naturale, (in Italian) Proc. of V° Symposium G. I.A.S.T. - San Sepolcro 11-13 giugno 1996 (in press).

Hannan, E.S., (1980) The estimation of the order of an ARMA process, *Ann. Stat.* **8**, 197-181

Kashyap, R.L., (1982) Optimal choice of AR and MA parts parts in autoregressive moving average models, *IEEE Trans. Pattern. Anal. Mach. Intel PAMI* **4**(2), 99-104.

Neuman, S.P., and Jacobson E.A., (1984) Analysis of nonintrinsic spatial variability by residual kriging with application to regional groundwater levels, *Math. Geol.* **16**, 491-521.

Petrucci, F., Bigi, B., Morestori, L., Panicieri, E., and Pecorari M, Valloni, R., (1975) *Ricerche sulle acque sotterranee nella pianura parmense e piacentina. Pozzi idrici nelle provincie di Parma e Piacenza (destra Nure). Parte prima: da 1 a 1000* (in Italian), Grafiche STEP, Parma.

Samper, F.J., and Neuman, S.P., (1989a) Estimation Covariance Structures by Adjoint State Maximum Likelihood Cross Validation 1. Theory, *Water Resour. Res.* **25**(3), 351-361.

Samper, F.J., and Neuman, S.P., (1989b) Estimation Covariance Structures by Adjoint State Maximum Likelihood Cross Validation 2. Synthetic Experiment, *Water Resour. Res.* **25**(3), 363-371.

Samper, F.J., and Neuman, S.P., (1989c) Estimation Covariance Structures by Adjoint State Maximum Likelihood Cross Validation 3. Application to Hydrochemical and Isotopic data, *Water Resour. Res.* **25**(3), 351-361.

Shafer, J.M., and Varljen, M.D. (1990) Approximation of confidence limits on sample semivariogram from single realizations of spatially correlated random fields, *Water Resour. Res.* **26**(8), 1787-1802.

SEASONAL VARIATIONS ON SYSTEMS OF TRIPLE INFLUENCE

The Importance of Silicate Weathering, Soil Fertilization and Domestic Effluents to the Composition of Stream Water as Determined by Correspondence Analysis

A. M. P. ALENCOÃO, A. SOUSA OLIVEIRA, F. A. L. PACHECO
Geologist, Eng. Geologist, Eng. Geologist
Universidade de Trás-os-Montes e Alto Douro (Vila Real, Portugal)
E-mail: alencoao@utad.pt, soliveir@utad.pt, fpacheco@utad.pt

Abstract

The chemical composition of stream waters from the Pinhão river basin (North of Portugal) is attributed to three main influences (weathering, soil fertilization and urban pollution) which are described by factors obtained by Correspondence Analysis. The areal distribution of these influences is analyzed in six sampling epochs; differences between Autumn and Spring maps were generally attributed to changes in the water table.

1. Introduction

The composition of natural waters is determined by a number of processes, which include wet and dry deposition of atmospheric salts, evapotranspiration, water-soil and water-rock interactions. In inhabited areas, especially where there is intensive agriculture or industry, water composition is usually affected by fertilization and manuring, by leachates of solid waste, and by domestic and industrial effluents.

The contribution of each process affecting the water chemistry can be quantitatively assessed by mass balance models, as early recognized by Garrels & Mackenzie (1967) in a study of a pristine area (Sierra Nevada, USA) and lately by Pacheco & Van der Weijden (1996) in a study of an area with sizable anthropogenic input (Fundão, Portugal).

Factor Analysis and other eigenvector techniques are also used as methods for assessing the controls of water composition, especially in the study of huge hydrochemical data sets. Frequently the approach is strictly statistical, and although the analysis may identify the factors, the interpretation of these factors in terms of actual controlling processes is highly subjective.

Pacheco & Portugal Ferreira (1996) found significant correlations between factors obtained by Correspondence Analysis (C-vectors) and hydrochemical variables, for example the molar bicarbonate over silica ratio, which is characteristic

A. Soares et al. (eds.), geoENV I–Geostatistics for Environmental Applications, 235–246.

for particular water-mineral interactions. Based on the interpretation of these variables, they classified groundwater data sets from crystalline massifs as *systems of double influence* (silicate weathering and soil fertilization, under the background influence of atmospheric salt deposition). The discharge of domestic effluents directly into stream and river waters separates this type of contamination as an independent influence and, because of that, the samples collected in streams of the Pinhão river basin are here classified as a *system of triple influence.* Because samples were collected for three successive years (1993-1995), two times per year (April and October), seasonal variations could be analyzed.

2. Study area

The Pinhão river basin is located in the province of Trás-os-Montes (North of Portugal) and has an area of approximately 280Km^2 (Figure 1).

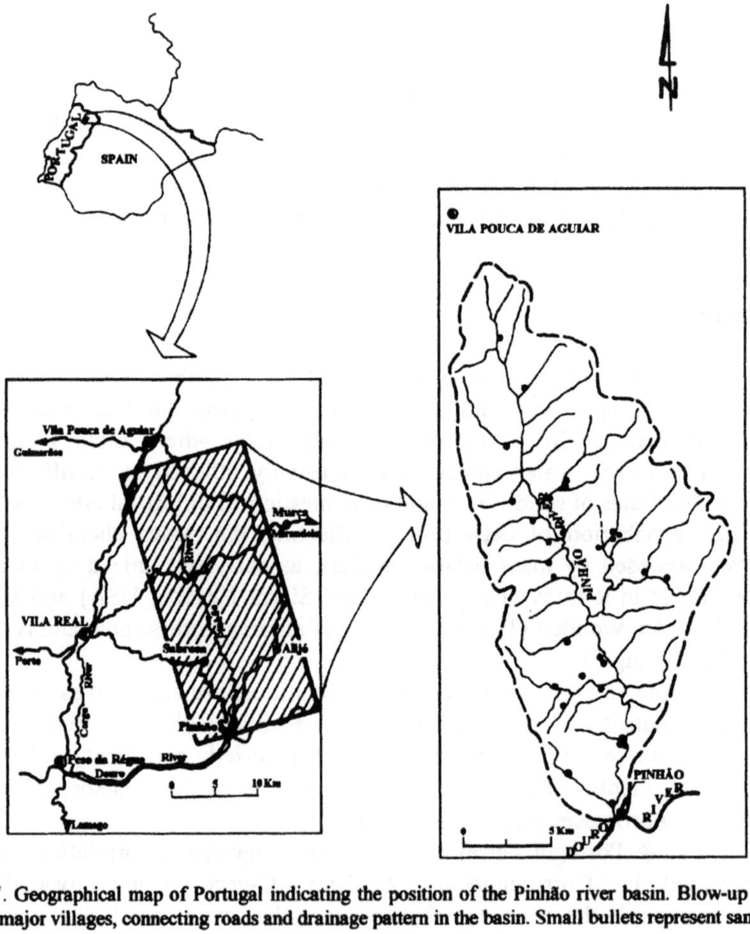

Figure 1. **Geographical map of Portugal indicating the position of the Pinhão river basin. Blow-up of the area, with the major villages, connecting roads and drainage pattern in the basin. Small bullets represent sampling sites.**

The northern area is dominated by hercynian granites and the southern area by paleozoic schists and graywackes (Figure 2). Granites are two mica, medium to coarse grained, sometimes with porphyroid texture; plagioclase varies in composition between albite and oligoclase/andesine. Soils developed from granites have variable amounts of kaolinite and gibsite (Alencoão *et al.*, 1996).

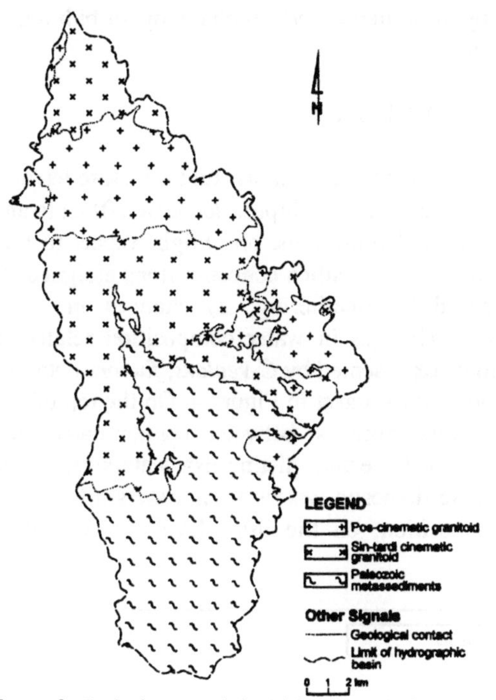

Figure 2. Geologic map of the Pinhão river basin; adapted from Sousa (1982) and Matos (1991).

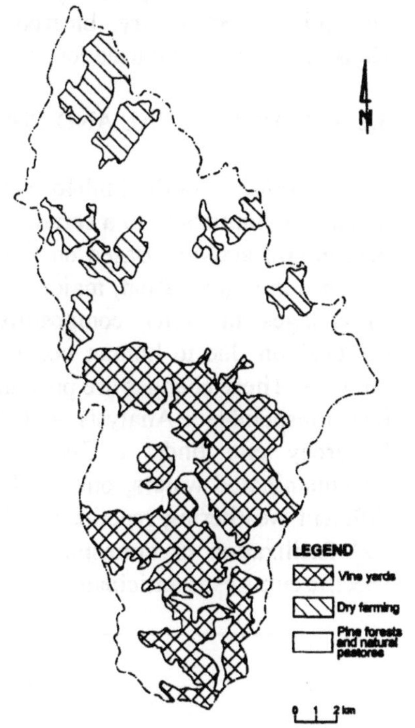

Figure 3. Land use map of the Pinhão river basin; adapted from Agroconsultores & Coba (1990).

Figure 3 is a simplified land use map of the Pinhão river basin. The northern area is occupied by pine forests where one can recognize small spots used for dry farming or natural pastures. The southern area is dominated by vine yards.

Human occupation is characterized by small settlements with no more than 200 people each, exceptionally reaching about 1000-1600 people in some small towns of the southern area (Sabrosa-1072, Favaios -1207, Sanfins-1555).

3. Materials and Methods

3.1. SAMPLING AND ANALYSIS

Stream water samples were collected for three successive years (1993-1995), two times per year (April and October), in the Pinhão river basin. 29 sampling sites were selected (cf. locations in Figure 1), 14 in granitic basins, 7 in schistose basins, and 8 in streams draining both granites and schists.

Chemical analyses were completed for major cations, major anions and dissolved silica and the results are given in Table 1. Sodium and potassium were determined by routine atomic emission spectrometry, magnesium and calcium by routine atomic absorption spectrometry, chloride, nitrate and silica by molecular absorption spectrometry; bicarbonate by titulimetry and sulphate by turbidimetry. House standards were used for calibration.

3.2. MATHEMATICAL AND CHEMICAL MODELING

Waters collected in the Pinhão river basin have drained granitic and schistose terrains and are contaminated to a certain extent by atmospheric input, domestic effluents and agricultural activity. Based on this background conditions, one might expect to find variance in major cation, major anion and dissolved silica concentrations assigned to: 1) changes in water composition caused by weathering; 2) changes in water composition due to human occupation; 3) changes in water composition related to land use. These changes are presumed to be represented by *C-vectors*, *factors* obtained by Correspondence Analysis, with the hierarchy suggest in Figure 4. On the top of this hierarchy one finds a C-vector that discriminates between contaminated and uncontaminated waters; one level below two C-vectors distinguish waters related to different weathering reactions and separate domestic waters from waters affected by soil fertilization or atmospheric salt deposition; on the third level the impact of specific cultures is anticipated.

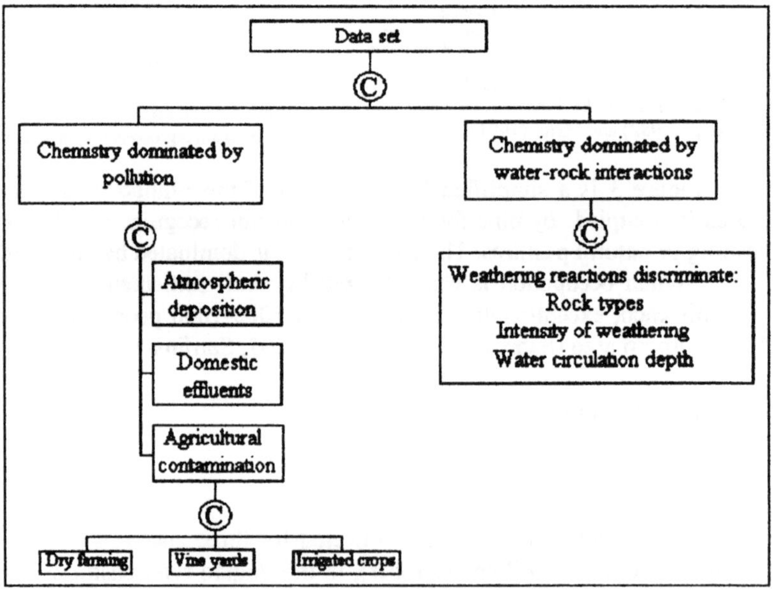

Figure 4: Dendritic structure of data sets; for stream water on crystalline massifs.

TABLE 1. Chemical composition of stream water samples collected in the Pinhão river basin. Concentration units in mg/l. (nd-not determined, vest-vestigial).

	P3	P4	P5	P6	P7	P8	P9	P10	P11	P12	P13	P14	P15	P16	P17	P20	P22	P23	P24	P26	P28	P29	P30	P31	P32	P33	P35	P36	P37
APRIL 1993																													
Ca^{2+}	0.5	0.4	1.9	0.5	1.7	2.4	1.3	2.3	9.6	2.3	9.2	6.0	1.5	2.4	0.5	1.4	1.8	8.8	9.0	5.2	8.7	7.3	0.8	4.1	0.6	0.5	0.7	0.3	0.9
Fe^{2+}	0.1	0.3	0.1	0.1	0.1	0.2	0.1	0.1	0.1	0.1	0.1	0.1	0.1	0.1	0.1	0.1	0.1	0.1	0.4	0.1	0.1	0.2	0.1	0.1	0.1	0.1	0.1	0.1	0.1
Mg^{2+}	0.4	0.3	0.6	0.4	0.9	1.0	0.8	1.3	5.2	1.2	4.8	3.2	0.6	1.4	0.5	0.8	1.0	6.4	5.8	3.5	4.5	4.5	0.5	1.6	0.5	0.4	0.4	0.4	0.6
K^{-}	5.0	vest	0.4	6.1	0.1	0.3	0.1	0.7	1.0	0.7	2.3	0.3	0.2	1.3	0.2	0.3	0.4	2.5	6.9	3.2	1.2	2.1	0.3	4.3	0.1	vest	vest	0.1	vest
Na^{+}	4.9	4.3	5.7	5.4	7.5	8.6	6.5	7.8	14.2	7.6	18.1	11.1	4.9	6.9	4.1	4.6	5.8	20.3	29.9	12.0	13.6	14.4	4.3	10.8	4.6	4.3	3.5	3.7	6.0
Cl^{-}	3.6	3.6	3.6	7.0	4.0	10.6	3.6	7.0	7.0	3.6	7.0	7.5	7.0	3.6	3.6	3.6	3.6	23.1	24.6	14.0	14.1	8.5	3.6	8.9	3.6	3.6	3.6	3.6	7.0
HCO_3^{-}	19.5	9.2	10.4	15.3	20.3	7.9	18.3	11.6	51.2	18.3	67.7	53.1	7.9	14.0	10.4	7.3	13.5	27.1	65.9	23.2	30.5	40.6	12.2	23.7	10.4	7.3	7.9	6.7	10.4
NO_3^{-}	nd	nd	1.3	nd	nd	nd	nd	1.1	4.1	1.1	0.9	0.4	nd	5.8	0.3	0.6	0.3	10.0	8.9	10.2	12.8	12.7	0.2	0.6	nd	nd	nd	nd	nd
SO_4^{2-}	nd	0.7	5.3	nd	0.7	7.8	1.3	8.9	19.4	7.2	18.0	1.5	0.9	6.7	nd	6.0	6.9	29.3	23.4	10.3	13.1	15.0	nd	13.2	0.7	1.2	nd	1.2	1.6
SiO_2	9.0	7.8	5.5	7.6	8.9	5.6	12.6	6.7	17.8	6.1	14.4	15.9	6.1	7.6	5.8	7.0	13.8	9.1	16.3	17.9	23.3	20.4	4.9	7.5	7.5	3.7	4.1	3.0	8.9
OCTOBER 1993																													
	P3	P4	P5	P6	P7	P8	P9	P10	P11	P12	P13	P14	P15	P16	P17	P20	P22	P23	P24	P26	P28	P29	P30	P31	P32	P33	P35	P36	P37
Ca^{2+}	1.6	0.8	9.6	0.8	2.8	3.1	1.4	2.2	12.9	1.8	12.4	7.9	7.6	4.5	5.0	3.0	3.1	9.8	13.1	8.1	12.4	10.2	1.0	4.0	0.8	1.2	0.7	0.8	1.2
Mg^{2+}	0.5	0.3	2.3	0.4	1.3	1.2	0.9	1.1	5.9	0.9	6.5	3.7	1.0	2.6	0.5	1.3	1.7	6.2	5.6	4.7	5.4	5.0	0.4	1.6	0.4	0.4	0.2	0.3	0.7
K^{+}	1.4	0.1	3.3	0.4	1.4	1.5	0.9	1.0	3.4	0.9	4.1	0.7	0.6	3.1	0.4	0.9	1.2	2.7	14.5	4.5	2.1	4.5	0.7	4.6	0.6	0.8	0.1	0.2	0.6
Na^{+}	4.7	3.2	9.2	5.2	7.7	7.4	6.0	5.0	14.3	4.7	17.8	10.6	5.0	12.9	3.8	6.7	7.0	19.0	39.8	12.0	12.5	15.3	3.3	9.0	4.0	5.0	3.0	3.3	6.2
Cl^{-}	6.4	4.2	13.6	5.2	11.3	12.1	4.3	5.9	7.9	6.1	14.4	9.1	5.7	18.4	4.9	7.1	9.2	20.4	35.0	15.5	13.6	15.9	4.1	13.5	5.0	7.0	3.0	4.4	7.8
HCO_3^{-}	7.9	3.1	4.9	8.5	11.0	4.9	15.9	6.7	62.2	3.7	75.6	38.4	29.9	6.1	17.1	2.4	6.1	43.9	106.8	30.5	32.3	35.4	6.1	12.9	6.1	6.1	4.9	3.1	7.3
NO_3^{-}	1.3	2.2	31.9	0.5	1.9	1.1	0.9	1.7	7.8	1.4	5.9	4.4	0.5	14.9	0.9	9.5	2.2	0.9	0.2	15.2	22.9	18.6	0.5	2.8	1.6	0.2	0.8	0.3	1.0
SO_4^{2-}	1.6	0.7	8.3	0.5	5.0	9.4	1.6	7.8	23.6	8.1	22.9	9.4	3.1	8.5	1.2	8.7	11.1	28.6	16.5	13.9	16.3	18.2	0.9	9.1	0.5	1.4	0.3	2.2	2.9
SiO_2	6.8	6.7	8.8	9.4	10.8	9.8	12.2	6.3	18.9	6.1	16.2	16.8	7.6	12.1	8.1	7.2	7.9	15.6	19.7	16.8	19.8	17.0	6.2	8.5	7.7	6.3	5.8	5.4	9.8
APRIL 1994																													
	P3	P4	P5	P6	P7	P8	P9	P10	P11	P12	P13	P14	P15	P16	P17	P20	P22	P23	P24	P26	P28	P29	P30	P31	P32	P33	P35	P36	P37
Ca^{2+}	1.0	1.5	2.5	1.0	3.0	2.0	1.5	3.0	9.0	3.0	8.5	5.0	1.0	1.5	1.0	2.0	2.0	7.5	8.0	4.5	8.0	6.0	1.0	2.5	1.0	1.0	1.0	1.0	1.0
Mg^{2+}	1.0	1.5	1.0	1.0	2.5	1.0	1.0	2.5	5.5	2.0	4.5	3.0	0.5	1.0	1.0	2.0	1.5	5.5	5.5	3.5	4.5	4.5	1.0	1.5	1.0	1.0	1.0	1.0	1.0
K^{+}	0.6	0.6	1.2	0.4	0.8	0.8	0.8	0.8	1.4	0.8	2.2	0.4	0.4	0.6	0.4	0.6	0.6	2.4	5.4	1.4	1.2	2.0	0.8	3.6	0.6	0.4	0.4	0.4	0.2
Na^{+}	1.5	1.5	3.5	2.5	6.5	4.5	4.0	6.0	13.0	6.0	15.0	8.0	2.0	2.5	2.0	1.5	4.5	15.0	24.0	8.0	11.5	11.0	1.5	6.5	2.0	1.5	1.0	1.0	3.0
Cl^{-}	3.9	3.8	7.8	4.7	7.5	7.2	6.8	8.5	12.4	8.7	13.4	8.5	6.1	6.5	5.3	6.0	7.3	12.9	18.8	10.8	9.0	11.7	5.4	10.2	5.1	4.2	4.0	3.8	6.1
HCO_3^{-}	6.7	12.2	5.5	7.3	4.3	6.1	6.1	11.0	32.3	8.5	40.3	29.3	10.4	4.3	5.5	4.9	4.3	22.0	25.0	15.9	37.8	20.7	4.3	8.5	4.3	6.1	5.5	4.3	5.5
NO_3^{-}	0.0	0.0	0.6	0.0	1.3	0.0	0.0	3.5	11.1	4.0	3.5	0.7	0.0	0.3	0.3	0.7	2.7	14.2	34.2	11.8	12.5	16.3	0.6	1.2	2.3	0.0	0.0	0.0	0.0
SO_4^{2-}	Vest.	Vest.	4.0	0.3	2.6	5.3	3.2	8.4	21.6	7.5	19.5	5.3	1.5	1.8	0.0	3.9	5.6	28.1	23.3	7.5	10.8	13.4	Vest.	8.7	0.0	0.1	Vest.	0.0	1.1
SiO_2	7.8	8.7	8.2	10.5	10.6	11.4	12.0	11.1	24.3	13.2	20.5	20.8	7.7	8.7	9.7	9.8	9.4	20.8	22.0	20.1	22.5	20.0	6.9	10.0	7.7	7.1	6.9	6.7	12.9

TABLE 1. Chemical composition of stream water samples collected in the Pinhão river basin. Concentration units in mg/l. (nd-not determined, vest-vestigial).

OCTOBER 1994

	P3	P4	P5	P6	P7	P8	P9	P10	P11	P12	P13	P14	P15	P16	P17	P20	P22	P23	P24	P26	P28	P29	P30	P31	P32	P33	P35	P36	P37
Ca^{2+}	1.5	1.0		1.0	2.0	3.0	2.0	4.5	10.0	4.5			3.6	6.0	1.8		8.2		10.5	5.0	8.5	8.0	1.5	3.5	1.0	1.5	1.0	1.0	1.5
Mg^{2+}	0.5	0.5		0.5	1.0	1.0	0.5	2.5	4.5	2.0			0.5	2.5	0.5		1.5		4.0	3.0	3.5	3.5	0.5	1.0	0.5	0.5	0.5	0.5	0.5
K^+	0.7	0.1		0.3	0.5	0.8	0.7	1.5	1.4	1.5			0.2	0.6	0.5		1.7		8.0	1.5	3.2	3.1	1.4	3.7	0.7	1.3	0.2	1.3	0.4
Na^+	4.0	3.5		4.5	6.5	7.0	6.0	9.0	13.0	9.0			6.2	11.7	5.1		13.1		51.0	9.5	11.5	15.5	8.0	7.5	3.5	4.5	3.0	3.0	5.0
Cl^-	3.9	3.9		5.8	4.6	8.8	7.0	9.6	10.8	8.7			7.1	7.7	6.1		6.7		27.8	9.0	9.9	14.3	7.5	10.3	5.5	8.6	3.4	4.0	1.7
HCO_3^-	9.5	6.1		8.5	6.1	14.0	9.8	17.1	46.4	10.4			14.6	7.3	5.8		15.2		92.7	20.7	28.1	24.4	15.9	12.2	7.4	6.1	6.1	6.8	6.1
NO_3^-	0.5	2.8		nd	0.7	0.0	nd	4.4	vest	4.2			0.9	22.6	1.0		28.6		vest.	7.9	14.3	18.4	0.6	2.3	0.0	vest	0.0	nd	0.4
SO_4^{2-}	nd	nd		nd	0.01	3.40	2.90	11.30	19.90	15.80			3.10	1.00	3.70		11.30		1.00	10.40	14.90	10.90	1.20	7.10	0.60	1.60	nd	0.50	0.01
SiO_2	2.6	2.9		3.3	8.9	5.1	4.7	3.8	5.9	3.3			2.5	10.2	2.5		2.8		5.7	5.0	5.6	5.2	1.5	2.3	2.3	1.6	1.8	1.6	7.5

APRIL 1995

	P3	P4	P5	P6	P7	P8	P9	P10	P11	P12	P13	P14	P15	P16	P17	P20	P22	P23	P24	P26	P28	P29	P30	P31	P32	P33	P35	P36	P37
Ca^{2+}	1.0	0.5	2.0	1.0	2.0	2.5	2.0	3.0	10.5	3.0	8.0	5.5	1.5	1.5	1.0	1.5	2.5	8.5	9.5	6.0	9.0	7.5	1.5	3.5	1.0	1.0	1.0	1.0	1.5
Mg^{2+}	0.5	0.5	0.5	0.5	1.0	1.0	1.0	1.5	5.5	1.5	4.0	3.0	0.5	0.5	0.5	1.0	1.5	5.5	1.5	4.0	4.5	4.5	0.5	1.5	0.5	0.5	0.5	0.5	0.5
K^+	0.2	0.2	0.6	0.2	0.4	0.4	0.4	0.8	1.4	0.8	1.6	0.4	0.4	0.6	0.4	0.6	0.6	2.4	5.6	1.6	1.4	1.8	0.6	3.8	0.4	0.2	0.2	0.2	0.2
Na^+	2.5	2.5	4.0	3.5	6.0	6.0	5.0	6.0	13.0	6.0	13.0	8.5	3.5	3.0	3.0	3.5	5.0	15.5	31.0	9.0	12.5	12.5	2.5	7.5	3.5	2.5	2.5	2.5	4.5
Cl^-	0.0	0.0	0.0	0.0	0.0	0.0	0.0	0.0	5.7	0.0	5.7	0.0	0.0	0.0	0.0	0.0	0.0	11.7	21.0	1.0	5.1	6.1	0.0	0.0	0.0	0.0	0.0	0.0	0.0
HCO_3^-	9.8	9.2	12.2	12.2	23.2	20.1	20.7	22.7	65.9	23.2	59.2	48.2	12.8	12.2	12.2	14.0	22.3	52.5	47.0	43.3	55.5	45.0	12.2	38.4	13.4	9.8	9.8	10.4	15.9
NO_3^-	0.2	0.4	3.4	0.3	0.7	0.4	0.3	3.8	9.8	3.7	1.9	2.2	0.0	0.6	0.2	1.5	2.4	13.3	26.7	11.8	14.2	15.7	0.5	1.0	0.0	0.1	0.0	0.1	0.6
SO_4^{2-}	1.9	1.8	3.3	1.7	2.9	4.1	2.9	4.6	8.1	4.7	8.1	3.7	2.9	3.3	2.0	3.7	4.2	9.3	11.6	5.2	6.5	6.7	2.0	3.7	1.8	2.0	1.8	1.8	2.3
SiO_2	6.4	10.0	8.3	10.4	10.5	10.8	9.8	8.6	15.1	8.8	18.8	13.2	8.0	8.8	7.8	10.2	7.7	18.4	19.3	15.6	10.1	17.0	5.6	9.8	7.6	6.6	5.5	5.2	11.7

OCTOBER 1995

	P3	P4	P5	P6	P7	P8	P9	P10	P11	P12	P13	P14	P15	P16	P17	P20	P22	P23	P24	P26	P28	P29	P30	P31	P32	P33	P35	P36	P37
Ca^{2+}	1.5	1.0	4.5	1.0	2.0	3.0	2.5	5.0	10.5	5.0			1.5	2.0	1.5		4.0		11.5	5.5	8.5	8.5	2.0	3.5	1.0		1.0	1.0	1.5
Mg^{2+}	0.5	0.5	1.5	0.5	1.0	1.5	1.0	5.0	5.5	3.0			0.5	1.0	0.5		2.5		6.0	4.0	4.5	4.5	0.5	1.5	0.5		0.5	0.5	0.5
K^+	0.8	0.4	5.4	0.6	0.6	1.0	1.0	1.8	1.8	1.8			1.0	1.6	0.8		1.6		12.0	2.2	2.0	3.8	1.8	4.0	1.0		0.6	0.4	0.6
Na^+	4.0	3.5	18.0	4.0	7.0	10.0	15.5	11.5	15.5	11.0			4.5	3.5	4.0		8.5		50.0	10.5	13.0	9.5	4.5	7.5	4.0		3.0	3.5	5.5
Cl^-	5.1	3.9	16.9	5.6	9.4	9.6	8.9	10.9	11.8	13.6			6.0	7.1	6.5		10.0		28.0	11.8	13.3	18.9	7.6	10.3	5.2		4.5	4.1	6.5
HCO_3^-	7.3	7.3	7.3	6.1	7.9	14.6	37.8	25.6	64.7	19.5			5.5	8.5	4.3		14.6		126.3	30.5	37.2	7.3	9.8	15.3	6.1		6.7	6.7	7.3
NO_3^-	0.5	0.4	28.0	0.5	0.2	0.7	0.9	3.6	0.0	3.9			1.2	1.3	1.7		3.8		0.3	8.1	11.9	16.8	0.6	3.9	0.7		0.2	0.2	0.6
SO_4^{2-}	1.7	1.1	9.1	1.3	5.3	10.5	1.8	12.0	16.5	12.4			4.3	1.0	1.7		11.1		19.6	8.8	11.5	11.6	1.9	6.8	1.2		1.1	1.1	3.9
SiO_2	4.4	5.0	6.9	7.0	7.9	10.8	9.5	6.7	14.0	6.2			5.4	10.4	5.5		6.7		15.5	11.4	12.4	12.6	3.9	5.1	4.5		4.2	3.4	7.0

Variance distribution is assumed to be hierarchic because the top-level C-vector is predicted to be the vector with the highest percentage of variance, the bottom-level C-vector is predicted to be the vector with the lowest percentage of variance, and so forth. If changes in water composition produced by one of the above selected causes are not significant, the corresponding C-vector is foreseen to be absent; for example, the top-level C-vector is not expected to incorporate the model if huge levels of contamination affect the whole sampled area.

The conceptual vector model of Figure 4 is checked in three consecutive steps:

1) C-vectors obtained as output of Correspondence Analysis are used as independent input variables of a linear multiple regression model, which has as dependent variable one relevant hydrochemical variable.
2) If regression parameters (the *b*'s) are low for all C-vectors but one, this vector is represented by the hydrochemical variable if the corresponding *b* is significant and the coefficient of multiple determination (or *goodness of fit*), R^2 is close to 100%
3) Because there are four C-vectors, Multiple Regression Analysis is applied four times with four different hydrochemical variables. Full validation of the vector model is obtained when each C-vector is represented by a different variable.

Hydrochemical variables are attached to C-vectors as explained bellow. In every cases concentrations are expressed in the *eq/l* scale.

1) The top-level C-vector in Figure 4 is attached to the percentage of pollution referred in Sousa Oliveira & Portugal Ferreira (1996) as

$$\% - \text{Pollution} = \frac{[Cl]+[SO_4]+[NO_3]}{[Cl]+[SO_4]+[NO_3]+[HCO_3]} \text{x}100 \qquad (1)$$

Because chloride, sulphate and nitrate concentrations in water, are usually derived from pollution sources and the bicarbonate concentration is essentially related to the hydrolysis of silicates, samples with %-pollution scores higher than 50% are marked as contaminated, while those with scores bellow 50% are labeled as uncontaminated.

2) The two C-vectors at the second level are attached to the percentage of chloride in pollution

$$\%\text{C}l = \frac{[Cl]}{[Cl]+[SO_4]+[NO_3]} \text{x}100 \qquad (2)$$

and to the molar bicarbonate over silica ratio ($[HCO_3]/[SiO_2]$).

The higher is the %Cl more important is the contribution of domestic effluents (and/or atmospheric salt deposition) to pollution; the lowest %Cl scores characterize the impact of soil fertilization, because sulphate and nitrate are important anion constituents of commercial fertilizers.

Garrels (1967) showed that dissolved HCO_3, dissolved SiO_2 and its ratios is a good diagnostic parameters for particular water-mineral interactions.

3) The bottom-level C-vector in Figure 4 is attached to the $[SO_4]/[NO_3]$ ratio because fertilizers used in dry farming have $[SO_4]/[NO_3]\approx0$ while those used in irrigated crops and vine yards have ratios around 3 and 6, respectively.

4. Results and Discussion

Correspondence Analysis (CA) is performed and hydrochemical variables are calculated from data in the Table 1. Multiple Regression Analysis (MRA) is applied as explained above and the results are presented in Table 2.

The variance absorbed by C-vectors is always higher than 89%. In all the cases but one ($[SO_4]/[NO_3]$ in April 1993), it is clearly recognized which is the C-vector correlative to a given hydrochemical variable because the regression parameter in that C-vector is high, when compared with those in the other C-vectors. In 14 of 19

TABLE 2. Results of Multiple Regression Analysis (MRA). C-vectors (C_x) are used as independent variables and hydrochemical variables as dependent variables. N_t is the total number of samples in the data set and N_m the number of samples used in the MRA calculations; N_t-N_m represent the number of anomalies in each hydrochemical variable. R^2 is the coefficient of multiple determination. Relevant regression parameters are represented in bold face. The program used to perform C.A. was adapted from Bencrezi (1973), the MRA - model was calculated by STATISTICA software.

April 1993: $C_1+C_2+C_3+C_4$=39.9+19.6+17.2+15.7=92.4% of system variance

	N_t	N_m	C_1	C_2	C_3	C_4	R^2(%)
%-pollution	29	29	**0.91**	0.36	0.41	0.22	98.9
%Cl [a)]	15	15	0.15	**0.64**	0.40	0.30	84.8
$[HCO_3]/[SiO_2]$	29	29	0.14	0.26	**0.90**	0.28	71.9
$[SO_4]/[NO_3]$ [b)]	17	12	0.32	0.37	0.41	**0.49**	79.2

a) group of samples with %-Cl<80%
b) 12 samples with NO_3=0

October 1993: $C_1+C_2+C_3$=52.8+23.0+13.3=89.1% of system variance

	N_t	N_m	C_1	C_2	C_3	R^2(%)
%-pollution	29	29	**0.78**	0.47	0.38	98.8
%Cl	29	26	0.18	**0.71**	0.43	93.5
$[SO_4]/[NO_3]$	29	23	0.47	0.17	**0.90**	85.1

April 1994: $C_1+C_2+C_3$=56.8+20.3+14.5=91.6% of system variance

	N_t	N_m	C_1	C_2	C_3	R^2(%)
%Cl	29	29	**0.90**	0.14	0.17	95.4
$[HCO_3]/[SiO_2]$	29	29	0.24	**0.90**	0.02	89.7
$[SO_4]/[NO_3]$ [a)]	20	20	0.29	0.18	**0.75**	58.9

a) 9 samples with NO_3=0

October 1994: $C_1+C_2+C_3$=52.5+23.4+14.2=90.1% of system variance

	N_t	N_m	C_1	C_2	C_3	R^2(%)
%Cl	24	24	**0.83**	0.38	0.14	95.8
$[HCO_3]/[SiO_2]$	24	24	0.36	0.23	**0.68**	63.2
$[SO_4]/[NO_3]$ [a)]	18	16	0.19	**0.87**	0.26	68.7

a) 6 samples with NO_3=0

April 1995: $C_1+C_2+C_3$=66.9+16.7+8.2=91.8% of system variance

	N_t	N_m	C_1	C_2	C_3	R^2(%)
%-pollution	29	29	**1.01**	0.34	0.16	94.3
%Cl [a)]	7	7	0.40	0.02	**0.87**	93.4
$[HCO_3]/[SiO_2]$	29	29	0.29	**0.76**	0.14	86.5

a) 22 samples with Cl=0

October 1995: $C_1+C_2+C_3$ = 60.6+19.7+10.5 = 91.1% of system variance

	N_t	N_m	C_1	C_2	C_3	R^2(%)
%-pollution	24	24	**0.93**	0.49	0.23	95.5
%Cl	24	23	0.19	0.32	**0.72**	92.0
$[HCO_3]/[SiO_2]$	24	21	0.47	**0.80**	0.07	93.0

cases the goodness-of-fit of the MRA-model (R^2) is higher than 84%. %Cl is always a significant hydrochemical variable; $[HCO_3]/[SiO_2]$ is absent in one sampling period and the other variables in two periods.

Figure 5 shows the spatial distribution of %-pollution scores in six sampling epochs. Contamination is higher in the southern area probably because viniculture produce more impact on stream water chemistry than dry farming (see also Figure 3); besides, population is more concentrated in the South. Except in the year of 1994, %-pollution increases when samples are collected after a dry season (sampling in October) and decreases when sampling was made after a humid season (April). In our opinion, that is a direct consequence of the hydrologic and land use cycles: a) In the area, the fertilizers are applied in Spring or early Autumn and stay in the soil until they became leached by Autumn rains; this produces the increase in the October scores. b) Continuous leaching during Winter cleans-up the soils and, consequently, stream waters in April are less contaminated; that is particularly visible in the period October 1994 - April 1995. The unexpected high scores of April 1994 are interpreted as the result of an early soil - washing caused by sudden occurrence of precipitation

Figure 5. Spatial distribution and seasonal variation of %-pollution. We used SURFER software to perform the surface maps, adopting kringing as interpolation method.

after seed-sowing, although distribution of these scores is not statistically significant as %-pollution is not assigned to any C-vector (cf. Table 2).

Distribution of %Cl values are shown in Figure 6; because 22 of 29 samples collected in April 1995 have anomalous zero chloride concentrations, the corresponding %Cl iso-contour map is not presented. Scores are high in the northern area and low in the southern area. Changes in climatic conditions from April to October seem to have little effect on the pattern of distribution maps, mainly in the southern area.

High %Cl scores in the northern area are attributed to atmospheric contamination, while low scores in the southern area are ascribed to agricultural activity (viniculture). The small spots of high %Cl surrounded by areas where %Cl is much lower are identified as the areas where domestic contribution is predominant; for example, contamination produced by Sabrosa domestic effluents is clearly foreseen in the small elliptical zones with %Cl between 60 and 70% located in the southern part of the October 1994 and October 1995 maps.

Seasonal changes in the pattern of %Cl dispersion maps are probably related to the type of soil uses practiced in the northern and southern areas: a) the non

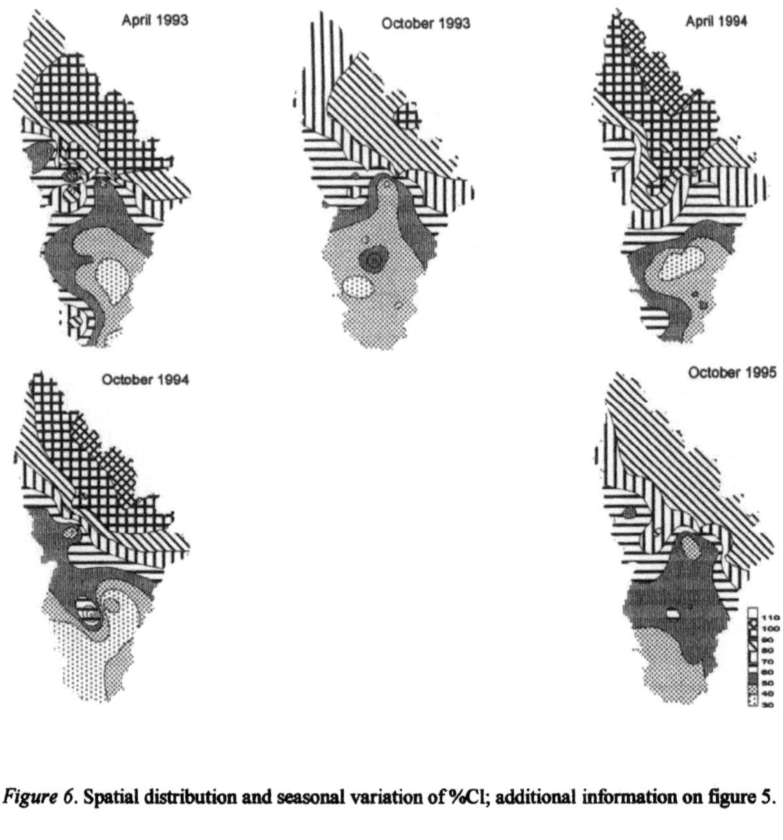

Figure 6. Spatial distribution and seasonal variation of %Cl; additional information on figure 5.

existing changes in the southern area are assigned to the fact that vine yards farming imposes a long standing signature on the soils, with its conservative influence over stream water composition; b) changes in the northern area are generally characterized by a decrease in the %Cl of samples collected in October (except October 1994). This reduction is probably caused by an increase in sulphate and nitrate concentrations related to fertilization of "non-perennial" cultures like wheat and rye; these cultures are normally sowed in late September and, as a consequence of top-dressing fertilization, one might indeed envisage changes in the chemistry of stream waters when sampling is made in October.

The $[HCO_3]/[SiO_2]$ ratio changes with time as it is illustrated in Figure 7; all maps are statistically significant, except that one of October 1993 (cf. second block of Table 2).

With the exception of April 1994, the $[HCO_3]/[SiO_2]$ scores are higher in the schists. A possible cause for this difference lies on the fact that weathering in the granites is controlled by plagioclase, whereas in the schists the biotite hydrolysis might determine the chemistry of groundwater.

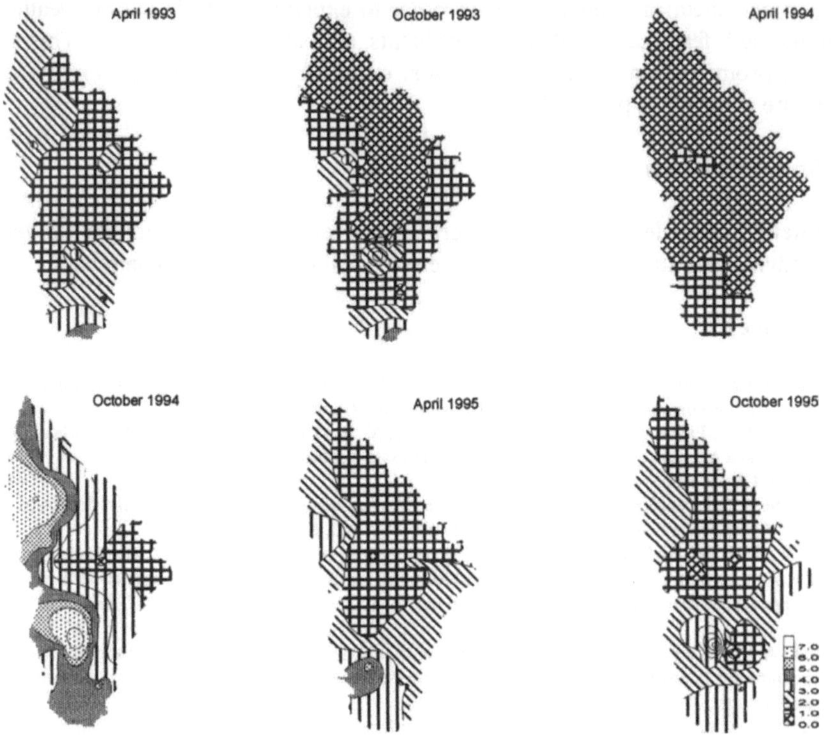

Figure 7. Spatial distribution and seasonal variation of $[HCO_3]/[SiO_2]$; additional information on figure 5.

Alencoão *et al.* Established, for the Pinhão river basin, a significant correlation between the $[HCO_3]/[SiO_2]$ ratio and the mean precipitation occurred four months before sampling. Based on these mean values of precipitation, sampling epochs were defined as *after dry* (April 1993, October 1994 and October 1995) or as *after humid* (October 1993, April 1994 and April 1995) seasons.

The lowest scores, to be found in the granites after wet periods, range between the theoretical values found for the hydrolysis of albite-oligoclase into kaolinite and gibsite, meaning that weathering is represented by reactions occurred in the soil. The highest scores, which are characteristic for the dry periods, probably result from water-rock interactions, naturally processed at more significative depths .

The information regarding the $[SO_4]/[NO_3]$ ratios is so scarce and highly heterogeneous that any sound explanation for their distribution is out of question.

5. Final Remark

It is to be remarked that combined application of Correspondence Analysis, Linear Multiple Regression Analysis and Hydrochemical Modeling might be used with stream water chemical compositions in order to explain its affiliation on weathering reactions, soil fertilization, domestic effluents and atmospheric inputs. This essay produced promising and convincing results, especially to evaluate the importance of each of the controlling processes.

6. Acknowledgments

We grateful acknowledge Prof. M.R. Portugal Ferreira, from the Coimbra University, for his advice, and to the anonymous reviewers for the constructive comments.

7. References

Agroconsultores & Coba (1990) *Carta de Solos, Carta do Uso Actual da Terra, e Carta de Aptidão da Terra do Nordeste de Portugal*, Universidade de Trás-os-Montes (PDRITM), Vila Real.

Alencoão, A. M. P., Pacheco, F. A. L. & Portugal Ferreira, M. R. (1996) Variação da razão bicarbonato/sílica, na água, e sua relação com a reacção de alteração em solo e subsolo granítico, *3° Congresso da Água e VII Silubesa*, **III**, 29-35.

Garrels, R. M. & Mackenzie F. T. (1967) Origin of the chemical compositions of some springs and lakes, In W. Stumm, *Equilibrium Concepts in Natural Water Systems*, Am. Chem. Soc.

Matos, A. V. (1991) *A Geologia da Região de Vila Real*. Tese de doutoramento, UTAD, Vila Real.

Pacheco, F. A. L. & Portugal Ferreira, M. R. (1996) Sistemas hidroquímicos com dupla influência: sua definição e interpretação no contexto da análise de correspondências, *3° Congresso da Água e VII Silubesa*, **III**, 177-185.

Pacheco, F. A. L. & Van der Weijden, C. H. (1996) Contributions of water-rock interactions to the composition of groundwater in areas with a sizable anthropogenic imput. A case study of the waters of the Fundão area, central Portugal, *Water Resouces Research* (in press).

Sousa, M. B. (1982) *Litostratigrafia e Estrutura do Complexo Xisto-Grauváquico ante-Ordovícico - Grupo do Douro (Nordeste de Portugal)*, Tese de doutoramento, Universidade de Coimbra.

Sousa Oliveira, A. & Portugal Ferreira M. R. (1996) A hidroquímica das águas meteóricas da região de Pedras Salgadas - Aplicação da análise de correspondências com vista à sua caracterização, *3° Congresso da Água e VII Silubesa*, **III**, 131-138.

KRIGING VS STOCHASTIC SIMULATION FOR RISK ANALYSIS IN SOIL CONTAMINATION

P. GOOVAERTS
Université Catholique de Louvain (Unité BIOM)
Place Croix du Sud, 2 Bte 16,
1348 Louvain-la-Neuve, BELGIUM

Abstract. This paper compares the performances of kriging vs simulation algorithms for delineating areas potentially contaminated by lead, and assessing the economic impact of declaring these areas safe (possible ill health). The comparison of predictions with actual topsoil lead concentrations at test locations shows that ordinary kriging underestimates large concentrations and the corresponding health costs. Better results are obtained using sequential indicator simulation which allows assessment of the uncertainty about health costs. Accounting for soft information as provided by a geological map is shown to reduce the financial loss attached to a wrong classification of zones as contaminated or safe.

1. Introduction

A common approach for delineating areas targeted for remediation consists of estimating the pollutant concentrations using kriging or some other interpolation algorithm, then declaring contaminated locations where the estimate exceeds a critical threshold. Unfortunately, least-squares interpolation algorithms tend to smooth out local details of the spatial variation of the attribute, hence small values are typically overestimated while large values are underestimated. Such conditional bias is a serious shortcoming when one aims at detecting large pollutant concentrations.

Stochastic simulation allows the generation of maps of pollutant concentrations which reproduce the spatial variability of the data without smoothing effect. Moreover, one may generate multiple realizations that all match reasonably the same sample statistics. The set of alternative realizations

A. Soares et al. (eds.), geoENV I–Geostatistics for Environmental Applications, 247–258.

can serve as input to complex transfer functions such as remediation processes. The resulting distribution of response values, e.g. remediation costs, provides a measure of response uncertainty that can be used in subsequent decision-making.

The "kriging" and "simulation" approaches are illustrated using an environmental data set related to soil contamination by lead in the Swiss Jura. The economic impact of decision-making based on estimated vs simulated lead concentrations is assessed and compared to the reference values provided by a validation set.

2. Setting the Problem

Consider the problem of assessing the economic impact of declaring a 14.5 km^2 region of the Swiss Jura safe with respect to lead. The information available consists of measurements of topsoil lead at 259 locations depicted at the top of Figure 1. A detailed description of the sampling, field and laboratory procedures is given in Atteia *et al.* (1994) and Webster *et al.* (1994).

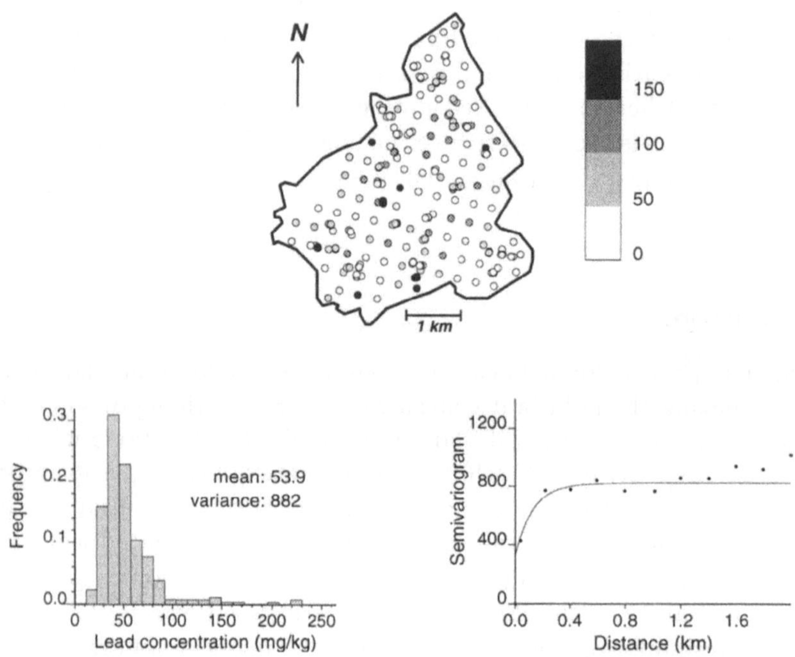

Figure 1. Location map of Pb data, and the corresponding sample histogram and semivariogram. The critical threshold for remediation is 50 mg kg^{-1}.

Figure 1 (bottom graphs) shows the sample histogram and semivariogram of lead. Note the positive skewness of the distribution of lead and the short-range structure of its semivariogram. According to the Swiss Federal Government (FOEFL, 1987) the tolerable maximum for healthy soil is 50 mg kg^{-1}, a value exceeded by 42.1% of the observations. That regulatory threshold applies to a 10 m $\times$ 10 m area small enough to be considered as equivalent to the sample support (1 m^2) when compared to the size (14.5 km^2) of the region.

The cost associated with classifying the region $\mathcal{A}$ as safe can be computed as the sum of local costs defined at N locations $\mathbf{u}'_j$ discretizing $\mathcal{A}$:

$$\mathcal{C} = \sum_{j=1}^{N} C(\mathbf{u}'_j) \tag{1}$$

where $C(\mathbf{u}'_j) = 0$ if the actual metal concentration at $\mathbf{u}'_j$, $z(\mathbf{u}'_j)$, is no greater than the critical threshold z_c (correct classification). If the guide value of 50 mg kg^{-1} is exceeded, consequences of soil contamination for human and animal health are likely to depend on whether the land is agricultural or forest. Some 17% of the study area is under forest and is shown light grey on the map of Figure 2. The remaining 83% is agricultural land, of which 80% is permanent grass and 3% cultivated for barley. To account for the greater risk associated with contamination of agricultural land, two different loss (cost) functions are considered, see Figure 2 (right graph):

$$C(\mathbf{u}'_j) = L_1(z(\mathbf{u}'_j)) = \begin{cases} 0.04 \cdot [z(\mathbf{u}'_j) - z_c]^2 & \text{(agricultural land)} \\ z(\mathbf{u}'_j) - z_c & \text{(forest)} \end{cases} \tag{2}$$

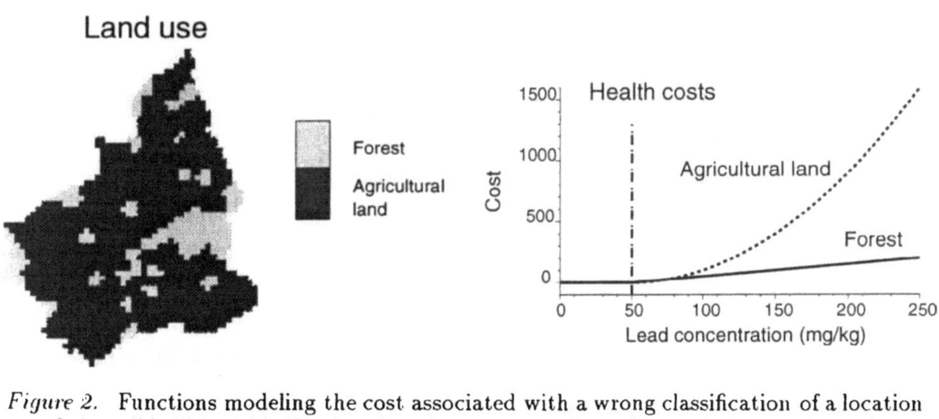

Figure 2. Functions modeling the cost associated with a wrong classification of a location as safe (possible ill health). The loss functions are specific to the particular land use (forest or agricultural land) at each place.

The computation of the health cost $C(\mathbf{u}'_j)$ calls for the actual metal concentration at $\mathbf{u}'_j$, which is unknown at any unsampled location. The unknown value $z(\mathbf{u}'_j)$ in expression (2) can be replaced by an estimate $z^*(\mathbf{u}'_j)$ or a simulated value $z^{(l)}(\mathbf{u}'_j)$.

3. Estimation vs Simulation

3.1. ORDINARY KRIGING

Let $\{z^*(\mathbf{u}'_j), j = 1, \ldots, N\}$ be the set of ordinary kriging estimates of attribute z over the study area $\mathcal{A}$. Each estimate $z^*(\mathbf{u}'_j)$ taken separately, i.e. independently of neighboring estimates $z^*(\mathbf{u}'_i)$, is "best" in the least-squares sense in that the local error variance $\text{Var}\{Z^*(\mathbf{u}'_j) - Z(\mathbf{u}'_j)\}$ is minimum. The map of such best local estimates, however, may not be best as a whole. As shown in Figure 3 (left column) ordinary kriging yields a smooth map of estimated lead concentrations: the variance of kriging estimates is six times smaller than the sample variance $\hat{\sigma}^2 = 882$. Application of the loss functions (2) to the kriged map yields the smooth map of health costs shown on Figure 3 (left bottom graph) with a global cost $\mathcal{C}$ of 19281.

3.2. SEQUENTIAL INDICATOR SIMULATION

Instead of a map of local best estimates, stochastic simulation aims at generating a map or a realization of z-values, say $\{z^{(l)}(\mathbf{u}'_j), j = 1, \ldots, N\}$ with l denoting the lth realization, which reproduces statistics deemed most consequential for the problem in hand. Typical requisites for such simulated map are:

1. data values are honored at their locations:

$$z^{(l)}(\mathbf{u}) = z(\mathbf{u}_\alpha), \quad \forall\ \mathbf{u} = \mathbf{u}_\alpha, \ \alpha = 1, \ldots, n$$

   the realization is then said to be conditional (to the data values).
2. the histogram of simulated values reproduces closely the sample histogram, in particular the proportion of large z-values.
3. the semivariogram model $\gamma(\mathbf{h})$ is reproduced.

Among the ever growing repertory of simulation algorithms, sequential indicator simulation (Alabert, 1987; Journel and Alabert, 1988; Gómez-Hernández and Srivastava, 1990) is retained here. This algorithm does not require any Gaussian hypothesis and allows one to account for class-specific patterns of spatial continuity. The algorithm proceeds as follows (Deutsch and Journel, 1992, p.146-151):

- Discretize the range of variation of z into $(K + 1)$ classes using K threshold values z_k. Then, transform each datum $z(\mathbf{u}_\alpha)$ into a vector

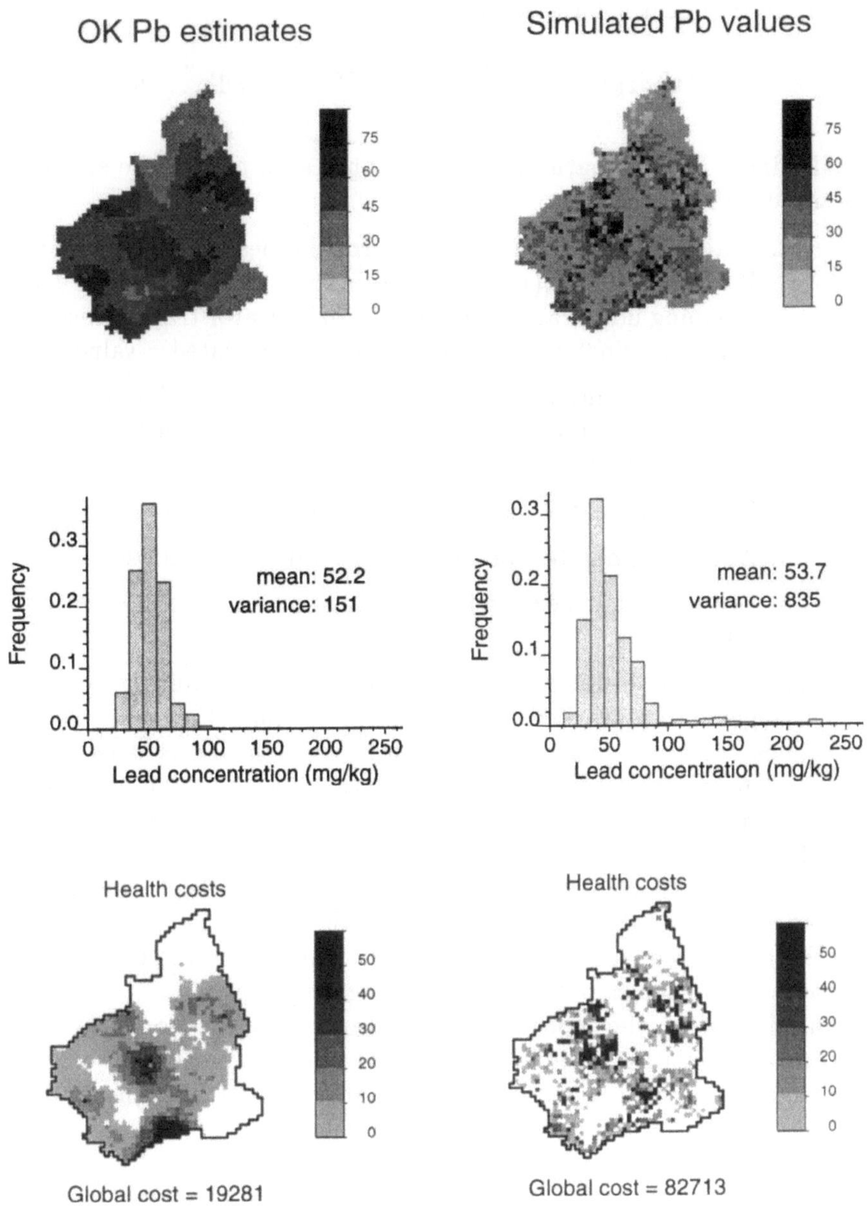

Figure 3. Ordinary kriging estimates and simulated values of Pb concentrations over the study area; middle graphs show the corresponding histograms. Bottom graphs show the maps of health costs computed using the loss functions of Figure 2. Note that the smoothing effect of kriging leads to a predicted cost much smaller than that provided by the simulated map.

of hard indicator data defined as:

$$i(\mathbf{u}_\alpha; z_k) = \begin{cases} 1 & \text{if } z(\mathbf{u}_\alpha) \leq z_k \\ 0 & \text{otherwise} \end{cases} \qquad k = 1, \ldots, K$$

- Define a random path visiting only once each node to be simulated.
- At each node $\mathbf{u}'$:
  1. Determine the K "posterior" probabilities $[F(\mathbf{u}'; z_k|(n))] = \text{Prob}\{Z(\mathbf{u}') \leq z_k|(n)\}$ using ordinary indicator kriging. The conditioning information (n) consists of indicator transforms of neighboring original z-data and previously simulated z-values.
  2. Correct for any order relation deviations, then build a complete conditional cumulative distribution function (ccdf) $F(\mathbf{u}'; z|(n))$ using interpolation/ extrapolation algorithms.
  3. Draw a simulated value $z^{(l)}(\mathbf{u}')$ from that ccdf.
  4. Add the simulated value to the conditioning data set.
  5. Proceed to the next node along the random path, and repeat steps 1 to 4.

One realization of the spatial distribution of lead values over the study area was generated using sequential indicator simulation and five threshold values corresponding to the 1st, 3rd, 5th, 7th and 9th deciles of the marginal distribution (cdf) of 259 lead data. The resolution of the discrete ccdf was increased by performing a linear interpolation between tabulated bounds provided by the sample cdf (Deutsch and Journel, 1992, p.131-135). Figure 3 (right column) shows the map of simulated lead values with the corresponding histogram and map of health costs. The sample histogram is well reproduced and the global cost is four times larger than the cost computed from the smooth map of kriging estimates.

In addition to the reproduction of sample statistics, one advantage of stochastic simulation over kriging is that one may generate several realizations that all match reasonably the same sample statistics and conditioning data. The set of alternative realizations $\{z^{(l)}(\mathbf{u}'_j), j = 1, \ldots, N\}$, $l = 1, \ldots, L$, provides a measure of uncertainty about the spatial distribution of attribute values. One hundred maps of simulated lead values were generated using sequential indicator simulation, yielding at each grid node $\mathbf{u}'_j$ a series of L=100 values $z^{(l)}(\mathbf{u}'_j)$. Figure 4 (bottom graphs) shows the maps of the mean (E-type estimate) and variance of the distribution of simulated values. While the maps of kriging and E-type estimates are similar, the maps of kriging variance and variance of simulated values are completely different, see Figure 4 (right column). The map of OK variance indicates greater uncertainty in the extreme west corner of the study area where data are sparse, while the uncertainty is smallest near data locations. Elsewhere the kriging variance is about the same whatever the surrounding data values. In contrast, the spread of the distribution of simulated values measured by the E-type variance indicates that the uncertainty is larger in the high-valued parts of the study area.

Application of the loss functions (2) to the set of 100 realizations yields 100 alternative maps of health costs that provide a measure of uncertainty[1] about the economic impact of taking no remedial measure. The uncertainty is here depicted by mapping the interquartile range of the local distributions of costs, that is the difference between the upper and lower quartiles of the distribution of health costs at each location, see Figure 5 (right graph). As for topsoil lead, the uncertainty about the health costs is larger where lead concentrations are large. In the low-valued zones the interquartile range is zero.

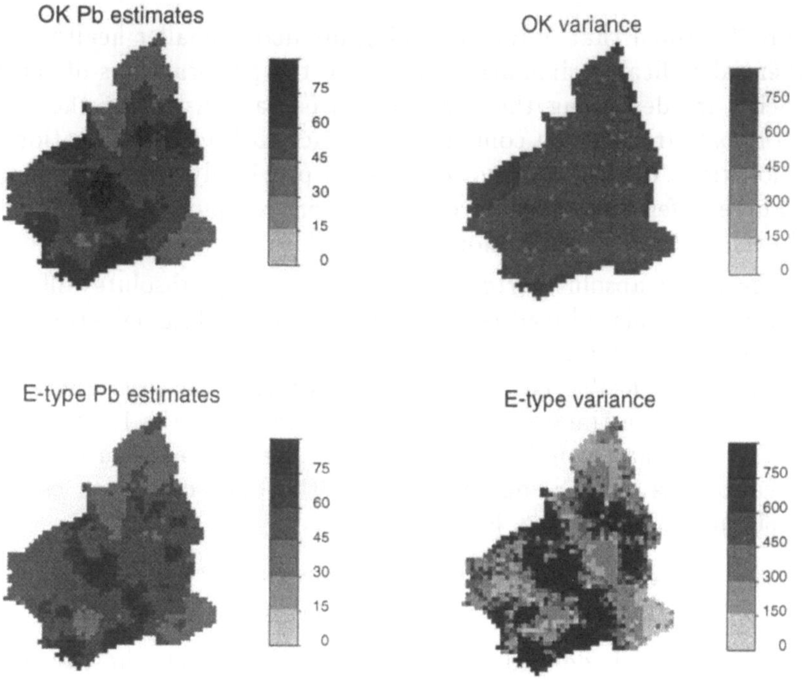

Figure 4. Ordinary kriging estimates of topsoil lead and the corresponding estimation variance (top graphs). Bottom maps show the mean (E-type estimate) and variance of 100 realizations generated using sequential indicator simulation.

[1]Actually, only the uncertainty resulting from our imperfect knowledge of the distribution in space of lead concentrations is accounted for. There is also uncertainty about the loss functions, and this could be taken into account by considering alternative models for the health costs.

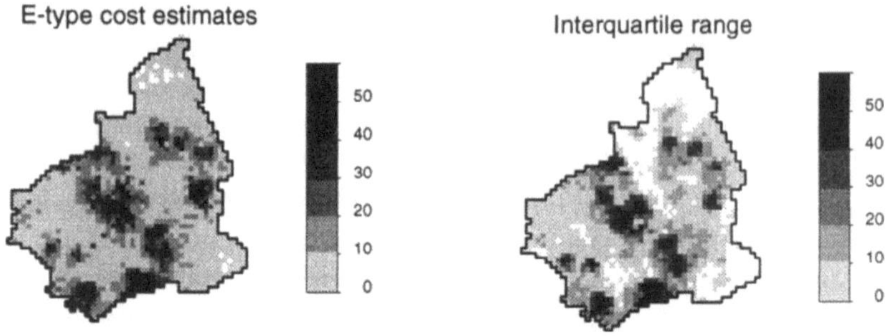

Figure 5. Maps of the mean and interquartile range of the local distributions of health costs obtained by applying the loss functions of Figure 2 to 100 simulated values at each grid node.

4. Performance comparison

Figure 3 showed that ordinary kriging predicted smaller health costs than sequential indicator simulation. To check the performances of the two approaches in identifying the hazardous areas and assessing the associated health costs, results are compared with actual lead concentrations at 100 test locations $\mathbf{u}_t$ that were not considered previously (Figure 6, top graph).

Three different performance criteria were computed for the kriging map and each of the 100 realizations:

1. the mean absolute error, that is the average absolute difference between the actual lead concentrations and the kriging estimates or the simulated values,
2. the global health cost obtained by applying the loss functions (2) to the kriging estimates or the simulated values at the 100 test locations,
3. the cost that would result from a wrong classification of the test locations as safe or contaminated on the basis of kriging estimates or simulated values, e.g. for location $\mathbf{u}_t$:

$$C'(\mathbf{u}_t) = \begin{cases} L_1(z(\mathbf{u}_t)) & \text{if } \mathbf{u}_t \text{ is classified as safe} \\ L_2(z(\mathbf{u}_t)) & \text{otherwise} \end{cases}$$

where $L_1(.)$ is defined as in (2), and $L_2(.)$ models the cost associated with assigning a place as contaminated and is depicted in Figure 6 (left bottom graph):

$$L_2\{z(\mathbf{u}_t)\} = \begin{cases} 0 & \text{if } z(\mathbf{u}_t) > z_c \\ 12.5 & \text{if } z(\mathbf{u}_t) \leq z_c \text{ (agricultural land)} \\ 25.0 & \text{if } z(\mathbf{u}_t) \leq z_c \text{ (forest)} \end{cases} \tag{3}$$

If $\mathbf{u}_t$ is actually contaminated, then the classification is correct and there is no loss. If $\mathbf{u}_t$ is actually safe then any restriction or remediation is actually unnecessary and will incur unnecessary costs. If remediation involved removing the upper layer of soil or cleaning it in some way the cost might be independent of the concentration of pollutant. Let it be a constant value that depends on land use: it is more expensive to clean forest soil than the more accessible farmland. The total misclassification cost is the sum of the local costs $C'(\mathbf{u}_t)$ over the 100 test locations.

Figure 6 shows, for each criterion, the probability distribution obtained by post-processing the 100 realizations generated using sequential indicator simulation. Ordinary kriging yields, in average over the 100 test locations, better predictions than the simulation algorithm in that the mean absolute prediction error is smaller for kriging estimates than for simulated values. However, the underestimation of large pollutant concentrations by ordinary kriging (conditional bias) leads to underestimating the global health cost: 700 instead of 1449. For this criterion, simulation performs better: the median of the distribution of 100 costs is 1345. The last criterion indicates that a classification based on smooth kriging estimates costs more than any of the classifications based on simulated values, i.e. the OK misclassification cost is larger than the maximum of the probability distribution, see Figure 6 (right bottom graph).

5. Accounting for soft information

Precise or "hard" measurements of lead concentration are supplemented by indirect or "soft" information provided by the geological map shown in Figure 7 (left top graph). Table 1 gives, for each threshold value z_k, the proportion of sites where z_k is not exceeded for each of the five geological formations. Most of the small values are located on Argovian rocks, and so it might be worth incorporating this information in sequential indicator simulation. Goovaerts and Journel (1995) presented several ways of taking into account soft information when estimating the posterior probabilities. In this paper the most straightforward approach is retained, that is simple indicator kriging of the hard indicator data, $i(\mathbf{u}_\alpha; z_c)$, using the soft indicator data (prior probabilities) of Table 1 as local indicator means (Goovaerts *et al.*, 1997).

One hundred realizations of the spatial distribution of lead concentrations were generated using sequential indicator simulation and the local means as provided by the geological map. The average simulated value (E-type estimate) at each grid node is mapped in Figure 7 (right top graph). Accounting for soft information yields smaller E-type estimates on the Argovian formation, compare with Figure 4. The impact on large values is negligeable because large concentrations are evenly distributed on all rock types and so are poorly discriminated by geology. Consequently, the his-

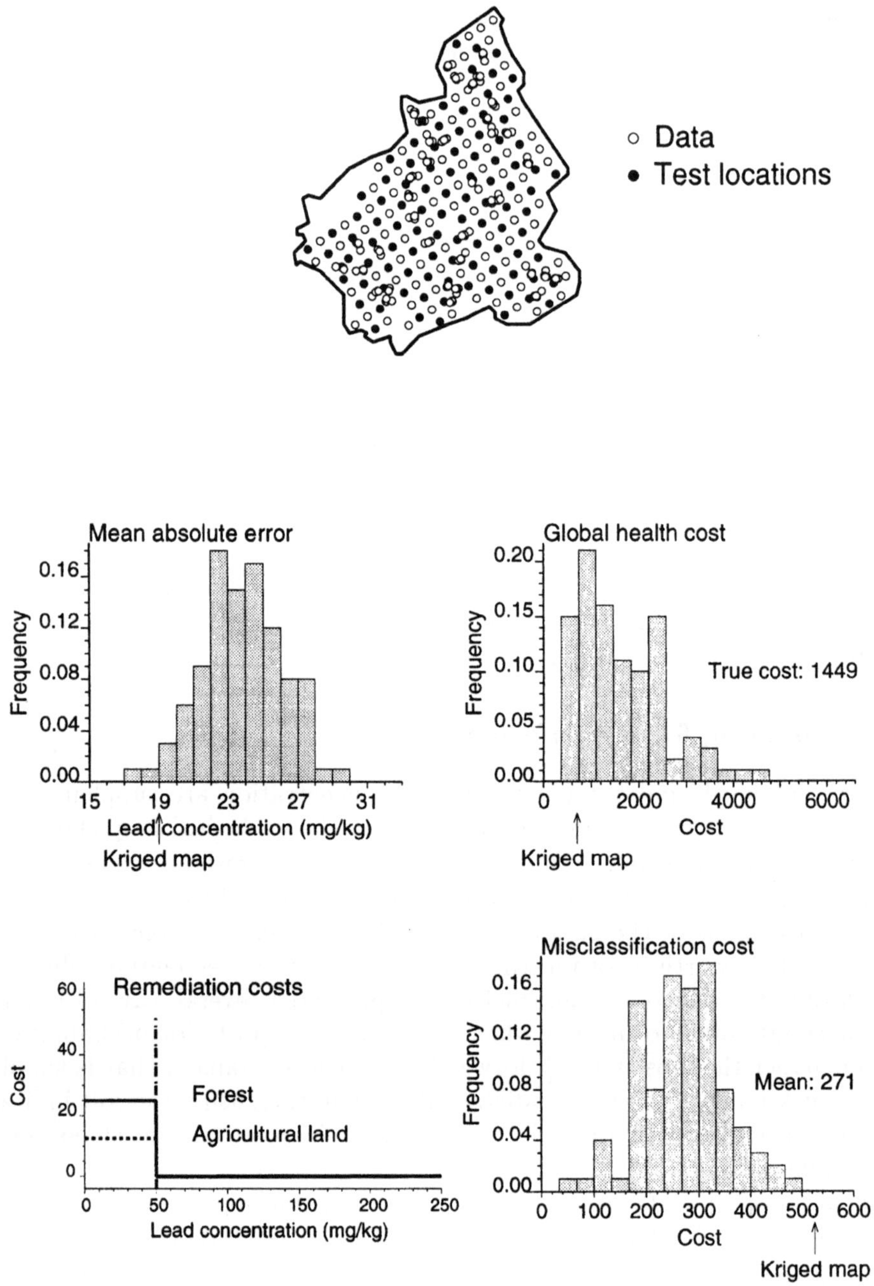

Figure 6. Histograms of three performance criteria computed by comparing actual lead concentrations at 100 test locations (top graph, black dots) with simulated values. For the mean absolute error and misclassification cost, best results are zero, and the actual health cost for these 100 test locations is 1449.

TABLE 1. Proportions of sampled sites where the lead concentration is no greater than the threshold value z_k for each formation separately.

Formation	Threshold z_k (mg kg^{-1})				
	29.92	37.56	46.0	56.4	80.4
Argovian	0.23	0.58	0.70	0.91	0.98
Kimmeridgian	0.01	0.21	0.44	0.64	0.91
Sequanian	0.03	0.19	0.44	0.63	0.83
Portlandian	0.00	0.33	0.67	0.67	1.00
Quaternary	0.18	0.27	0.45	0.67	0.89

togram of global health costs shown in Figure 7 (left bottom graph) is fairly similar to that obtained ignoring soft information. The better prediction of low-valued areas leads, however, to smaller misclassification costs, compare the histograms shown at the right bottom of Figures 6 and 7.

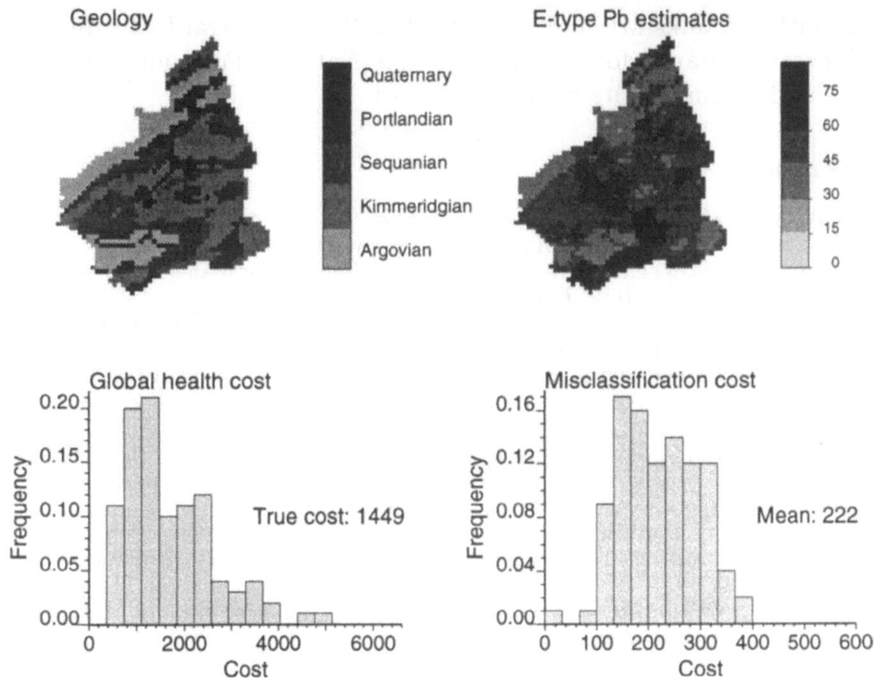

Figure 7. Mean (E-type estimate) of 100 realizations generated using sequential indicator simulation and a calibration of the geological map. Bottom graphs show the histograms of health costs and misclassification costs computed from these realizations at 100 test locations.

6. Conclusions

Delineation of areas targeted for remediation should not be based on smooth interpolated maps which may lead to a severe underestimation of the economic impact of the contamination. Stochastic simulation allows the generation of several maps that reproduce the spatial variability of the data and provide an assessment of the uncertainty about the spatial distribution of pollutant concentrations. The impact of such uncertainty on prediction of health costs can be investigated by post-processing the set of alternative realizations using loss functions, resulting in a distribution of possible costs.

In the case-study, sequential indicator simulation is shown to result in much smaller misclassification costs than ordinary kriging. These costs are further reduced by accounting for soft information provided by the calibration of a geological map.

7. Acknowledgement

The author thanks Mr. J.-P. Dubois of the Swiss Federal Institute of Technology at Lausanne for the data and the National Fund for Scientific Research (Belgium) for its financial support.

References

Alabert, A. (1987) *Stochastic Imaging of Spatial Distributions Using Hard and Soft Information.* Master's thesis, Stanford University, Stanford, CA.

Atteia, O., Dubois, J.-P., and Webster, R. (1994) Geostatistical analysis of soil contamination in the Swiss Jura, *Environmental Pollution* **86**, 315-327.

Deutsch, C.V. and Journel, A.G. (1992) *GSLIB : Geostatistical Software Library and User's Guide.* Oxford University Press, New-York.

FOEFL (Swiss Federal Office of Environment, Forest and Landscape) (1987). *Commentary on the Ordinance relating to Pollutants in Soil (VSBo of June 9, 1986).* FOEFL, Bern.

Gómez-Hernández, J. and Srivastava, R. (1990) ISIM3D: an ANSI-C three-dimensional multiple indicator conditional simulation program, *Computers & Geosciences* **16**, 395–440.

Goovaerts, P. and Journel, A.G. (1995) Integrating soil map information in modelling the spatial variation of continuous soil properties, *European Journal of Soil Science* **46**, 397-414.

Goovaerts, P., Webster, R., and Dubois, J.-P. (1997) Assessing the risk of soil contamination in the Swiss Jura using indicator geostatistics, *Environmental and Ecological Statistics* **4**, in press.

Journel, A.G. and Alabert, F. (1988) Focusing on spatial connectivity of extreme valued attributes: stochastic indicator models of reservoir heterogeneities, SPE paper # 18324.

Webster, R., Atteia, O., and Dubois, J.-P. (1994) Coregionalization of trace metals in the soil in the Swiss Jura, *European Journal of Soil Science* **45**, 205-218.

A FRAMEWORK FOR OPTIMIZATION OF SOIL SAMPLING STRATEGY AND SOIL REMEDIATION SCENARIO DECISIONS USING MOVING WINDOW KRIGING

C.F. VAN TOOREN & M. MOSSELMAN
Public Works Rotterdam, Environmental Engineering Department,

PO-Box 6633, 3002 AP Rotterdam, the Netherlands
E-mail: imfrap@gw.rotterdam.nl.

Abstract

At heavily polluted sites a reliable determination of the nature and extent of the contaminants is required for the choice between soil remediation scenarios. Sampling protocols generally prescribe the quantity of sampling rather than anticipating on the information needed for the choice of a remediation scenario.

In this paper, we present a framework for an iterative procedure of sampling, model calculations and optimized further sampling. 3D moving window kriging is used to account for mean and covariance non-stationarity of the concentration levels caused by different sources of contamination at old industrial sites.

The interpolation results are used to calculate the remediation costs and environmental benefit for different remediation alternatives, the remediation error and the evaluation of the need for further sampling. Optimal further sampling takes place at those locations where the largest reduction of the remediation error is expected for the remediation with the best balance between remediation cost and environmental benefit.

1 Introduction

After the first discoveries of soil contamination in the Netherlands in 1980, the Dutch government formulated the goal to clean up these contaminated sites to multifunctional soils within one generation. Multifunctional soil is defined as soil where any possible type of landuse is not inhibited by the presence of soil contamination, which is only guaranteed by remediation up to the natural background level of the soil.

A. Soares et al. (eds.), geoENV I–Geostatistics for Environmental Applications, 259–270.

However, the problem of soil contamination appeared to be far more widespread than was expected in the early eighties. In 1991, it was estimated that according to the policy of multi-functionality 110,000 sites should be remediated at a total cost of 50 billion dutch guilders (RIVM, 1991). Nearly all (former) industrial sites are involved. At present, the policy goal of remediation to natural background level of all locations within one generation is considered to ambitious. Therefore, it is necessary to set priorities in soil remediation and to pay attention to the balance between the cost of remediation and the environmental benefit that is achieved by the remediation.

Planning departments need insight in the remediation scenario options, the spatial distribution of the costs involved and the associated financial risks, caused by the uncertainty of the expected contamination pattern. A remediation scenario is defined by the choice between threshold concentration levels used for remediation and the choice between remediation techniques like excavation or in situ treatment.

In this paper, a Soil Sampling Strategy and Cleanup Decision Support (SACLEDS) framework is presented, to optimize sampling and remediation scenario decisions.

2 Problem definition

At heavily polluted sites a reliable 3D assessment of the nature and extent of the contaminants is required for determination of:

- possible soil remediation scenarios
- cost of soil remediation scenarios
- benefit of soil remediation scenarios
- actual environmental risks

Sampling is not 3D continuous, so interpolation is needed to get a continuous 3D model of the pollution. Kriging and stochastic simulation can be used to determine 3D grid estimates of contamination including reliability estimates.

Over the years industrial areas have been contaminated heterogeneously due to various sources from different industrial processes. The heterogeneous character of the soil contamination is intensified by transport of the contaminants and human activities. As a result mean and covariance non-stationarity of the concentration levels can be expected at these sites. Division of these sites into seperate subregions with mean and covariance stationarity is often difficult. A specific interpolation method is needed to account for these problems.

At present, Dutch site investigation protocols prescribe the quantity of sampling for soil remediation studies without anticipation on these specific aims of the soil investigation. However, sampling strategy can be optimized by anticipating on the remediation scenario decisions to be made. For this anticipation, an iterative procedure of sampling, interpretation and modelling and evaluation of the need of further sampling is needed.

3 3D moving window kriging

One approach to allow for non-stationarity of mean and covariance is the division of the site into subregions that identify different populations based on different contamination sources and/or different lithologies. In the investigations of contaminated sites these specific regions are often difficult to determine, because different populations are indistinguishly superimposed on each other and the spatial extent of these populations is generally not well known. If these subregions can be assigned, and the covariance functions can be estimated in these regions, kriging and stochastic simulation results may show discontinuities at the boundaries of these regions.

Another approach to allow for non-stationarity is the use of a moving window technique. The rationale for why a moving window scheme is effective with non-stationary processes is derived from Journel and Huijbregts' (1978) definition and discussion of quasi-stationarity. A non stationary process defined over a large region will exhibit smaller absolute trend and a more homogeneous covariance structure over a subregion relative to the entire region. In other words the process will be quasi-stationary within this subregion (Haas, 1990a,b). The moving window approach is suitable to overcome the problem of not being able to determine regions with identifiable different populations. The moving window method may be combined with the identification of subregions.

On the one hand for a spatial process with expected first and second order non-stationarity a small as possible window size is preferred to achieve quasi-stationarity. On the other hand, the window needs to contain enough data points for reliable semivariogram determination. In a cross-validation procedure, the window size is optimized so that just enough samples are available for the semivariogram modelling.

Cross-validations of soil contamination data show that the heterogeneous characteristics of varying concentration levels and reliability estimates are averaged out using single window methods instead of using a moving window method (Van Tooren & Haas, 1993; Van Tooren, 1993).

A drawback of the moving window approach is that the use of an automated procedure for semivariogram fitting is unavoidable. The fitting of these semivariograms is difficult to control, due to the amount of fitted semivariograms.

We chose in the SACLEDS framework to use kriging in stead of stochastic simulation, because most of the site remediations take place by excavation. For the set up and execution of an excavation, the smoothed or averaged results of kriging are suitable.
For the design of a remediation using groundwater flushing, one is mainly interested in the heterogeneity of the soil contamination. In that case, the results of a stochastic simulation in a moving window procedure would be more suitable. Unfortunately, no software is available to perform stochastic simulation in a moving window procedure.

We use the computerprogram KRIG (Haas, 1996a) for moving window kriging. In KRIG, small circular or elliptical windows consisting of a predefined number of samples are centred at each estimation grid node. Only the samples within the window are used for automatic determination of the window specific semivariogram and solution of the kriging system. Because semivariogram modelling and kriging are performed over small windows, mean and covariance stationarity are assumed.

For the 3D modelling of the contamination pattern we use the spatio-temporal extension of the moving window residual kriging (Haas, 1995), replacing time by depth. The estimation process consists of a two stage generalized regression estimate of the local trend in space and depth at the estimation location followed by a kriging estimation of the residual process at that location.

4 Application of the SACLEDS framework on a former gasworks site.

The SACLEDS framework consists of an iterative procedure of:
- sampling
- 3D moving window cross-validation and kriging
- calculations of remediation costs and remediation error for different remediation scenarios
- evaluation of the need for further sampling based on remediation error for a given remediation scenario

Application of the framework will be illustrated below for the investiagation of a former gasworks site in a harbour area in the city of Rotterdam (The Netherlands).

4.1 SITE DESCRIPTION

The area of the former gasworks itself is about 120,000 m^2, but the total study area has an area of 180,000 m^2. The gasworks operated from 1913 until 1967. The production of the coal gas and the subsequent purification and storage of the gas served as individual overlapping sources of contaminants. Moreover the different contaminated waste products often were used as material for raising of the location. Currently several industrial activities are present at the former gasworks location.
The geology consists of a 15 m thick layer of holocene clay and peat deposits above pleistocene sand deposits. The present terrain surface is not the same as during the operation of the gasworks. As the holocene deposits have a low bearing capacity the terrain has been raised with a layer of sand and junk with an average thickness between 2 and 4 metres.

4.2 INVENTARISATION OF PRIOR INFORMATION AND SAMPLING

At large heavily polluted sites, usually one or more prior investigations have been performed before the decision is taken to investigate the extent of the contamination in detail for remediation scenario decisions. New sampling takes place after inventarisation and analysis of all prior data.
There are conflicting concepts in the choice of a sampling strategy. For kriging interpolation itself sampling on a regular triangular grid is preferable. For the modelling of the semivariogram some samples at relatively small distances apart are needed. In the case of moving window kriging, each window should contain such a cluster of samples, which needs careful planning of the sampling.

At the former gasworks site, 17 prior soil investigations took place up to 1993. These prior investigations covered only parts of the location and in many cases their aim was proving the presence of soil contamination rather than mapping the extent of the contamination. Samples with an expected high degree of contamination, due to a strange colour or smell, had been analyzed relatively more frequently (preferential analyzing). Often, from one borehole no clean samples were available above or below a polluted sample. As a result these prior data were biased.

In 1994, it was decided to integrate all investigations at the former gasworks site. Additional sampling was done up to a data density of a horizontal grid of minimally 32 by 32 metre over the whole site.

4.3 CORRELATION STUDIES AND CHOICE OF KEY CONTAMINANTS

The moving window kriging procedure requires a large computational effort. Therefore, the interpolations have to be focused on the key contaminants. Correlations between the occurrence of different contaminants were studied. Furthermore, the choice of key contaminants depends on different types of treatment of the contaminated soil. The possibilities to treat soil that is contaminated with heavy metals differ from those for organic compounds.

Typical contaminants at former gasworks sites are PAH, cyanide, benzene and mineral oil. For the former gasworks site under study, cyanide and PAH were chosen as key organic contaminants. In a later stage, mineral oil had to be added to these key contaminants as in parts of the location high concentrations of mineral oil appeared to be measured without the occurrence of high concentrations of PAH or cyanide. A part of the former gasworks site appeared to be polluted with heavy metals. Lead and zinc were chosen as key heavy metal contaminants.

4.4 3D MOVING WINDOW CROSS-VALIDATIONS

A combination of the cross-validation measures mentioned in Van Tooren (1993) and in Persicani (1995) was determined for different window sizes and horizontal and vertical lag sizes. For the interpolations, window and lag sizes were chosen that scored well for most cross-validation measures.
Problems may arise in the comparison of the Normalised Mean Error and Mean Square Deviation Ratio for different window sizes when the kriging standard deviation is near zero at small window sizes containing mainly equal (below detection limit) concentration levels.

For soil remediation studies, classification with respect to certain threshold levels is important. In the cross-validation procedure was also evaluated how many samples are misclassified with respect to these threshold levels.

For the former gasworks site, 60-100 data points per window, were found for optimal results. This agrees with our experience at other gasworks sites.

4.5 3D MOVING WINDOW KRIGING

For the former gasworks site, we used an interpolation grid of 7.07 * 7.07 * 0.5 metre up to a depth of 6 metre below terrain level. Every interpolated grid point represents a block of soil with a volume of 25 m^3. Smaller block sizes would have given more detail but would have no practical value for the excavation process.

In some cases the concentration levels determined in the laboratory are correlated with other contaminants (quantitative) or with qualitative data like the presence of junk in the soil, smells observed in the field or lithology. If the correlations are strong between the sparsely analyzed contaminants and other measurements, and there is also a good physical argument for it, cokriging can be used to improve both the estimates and the estimated kriging standard deviation. Haas (1996b) describes cokriging in the presence of non-linear trend and covariance non-stationarity using only the within window samples to determine correlations between dependant variables.

The results of the interpolations showed that, especially at larger depths, preferential analyzing locally resulted in overestimation of the concentration levels. The limited number of samples at these larger depths also created extrapolation problems. As a result the estimates at those depths need to be considered carefully.

For most pollutants horizontal semivariogram ranges of 40-70 meters were found. The vertical range for most pollutants varied between 0 and 1.5 meters. When clusters of samples are absent in a windows, the nugget parameter for this window's semivariogram is extrapolated from sample variogram values at larger lags. For most of the pollutants an average of about 30% of the variation of the concentration levels was explained at the mean sampling distance.

4.6 DETERMINATION OF EXCEEDANCE PROBABILITIES FOR DIFFERENT REMEDIATION LEVELS

Insight in the reliability is needed to evaluate the need for further sampling and the risks involved with misclassification of the soil. In the evaluation of the reliability of the kriging estimates, the reliability with respect to certain threshold levels is important rather than the absolute value of the unreliability of the estimates.

For every block of soil, probabilities of exceeding specified remediation concentration levels were calculated (Press e.a., 1986) by combining the estimate and the associated standard deviation and given threshold (remediation) levels: the dutch S-, ½(S+I)- and I-levels that have replaced the former ABC-levels. The S-level is defined as the natural background level. In the Dutch Soil Protection Law, concentrations above the I-level are considered a "severe case of pollution".

The calculation of exceedance probabilities is based on the assumption that the estimation error is normally distributed. For correct estimation of the error variance all input data have to be transformed to approximate a normal distribution. For the former gasworks site we used a logarithmic transformation. The observed variable Y is modeled as a trend plus error. Mathematical statistics states that the result of adding a deterministic trend (nonrandom) to a normally distributed error variable results in a normally distributed observed variable. Hence, Y is normally distributed with a nonconstant mean (the trend) and the error is normally distributed with a constant mean of zero.

The aim of further sampling is reduction of the remediation error (Van Tooren, 1993; Haas et al, 1995) for a given remediation scenario. The remediation error, which results from a misclassification of soils, is the wrong decision about either treating unduly clean soils or leaving behind contaminated soils.

A probability of 40% of exceeding the S-level means that the soil, on average, is misclassified as clean in four times out of ten. Sampling needs to focus on places with high probabilities of misclassification with respect to the threshold level that will be used in the soil remediation. This is the zone where the probability of exceeding a remediation level is between e.g. 40-60%.

Figure 1 shows maps of the degree of misclassification based on the estimates for cyanide with respect to the S-level and the I-level. The estimates for cyanide were based on 277 samples from 227 boreholes distributed more or less regularly over the site. Probabilities of misclassification for each depth traject of 0.5 metre were combined to one value for every grid square (column of soil). The dark areas have a relative high degree of misclassification. The difference between the two maps shows, that optimal further sampling mainly depends on the choice of the remediation threshold level. The choice of the remediation scenario in turn depends on the balance between remediation cost and environmental benefit. Therefore, indications of remediation cost and environmental benefit have to be determined before optimal further sampling can be determined.

4.7 CALCULATION OF REMEDIATION COST FOR DIFFERENT REMEDIATION LEVELS

A remediation cost calculation model determines the cost of different remediation scenarios yielding remediation cost maps in guilders/m^2. Thus far only excavation remediation scenarios have been considered. The model uses the 3DMWK results of concentration levels, an interpreted model of the lithology of the site and several remediation cost variables. The lithology is divided in two groups, depending on whether the soil can be remediated by extraction methods in a soil treatment plant. The main cost variables are the

Figure 1. Unreliability of the estimate for cyanide with respect to the S-level and I-level

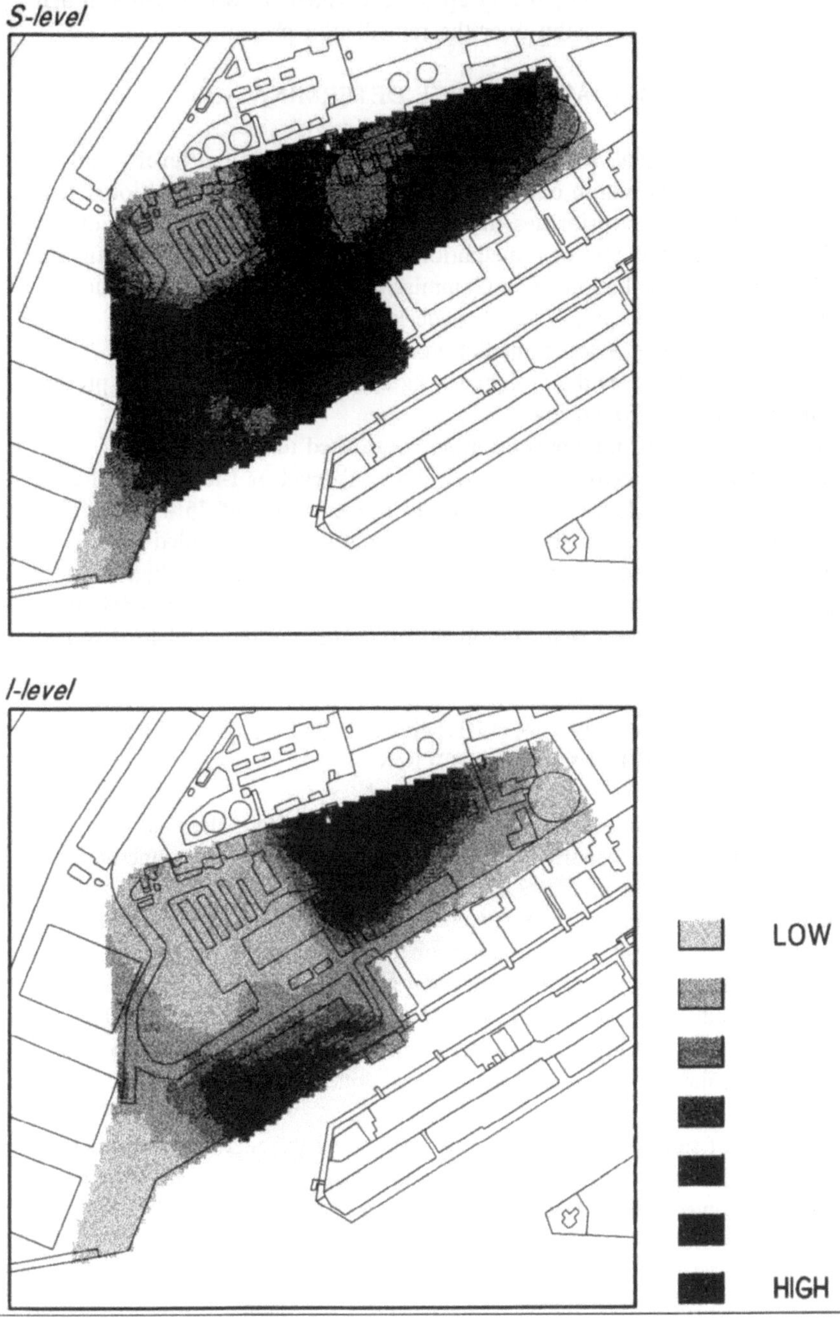

destiny of outcoming soil (cleanup treatment, waste dump or reuse elsewhere) and excavation costs. Depending on the remediation scenario, for the former gasworks site the treatment of the soil forms 50% to 75% of the total estimated remediation cost. Other cost parameters are transportation to a soil treatment plant or waste dump, filling up the excavated pit with clean soil and engineering cost like supervision over the remediation project.

4.8 OPTIMAL REMEDIATION LEVEL DETERMINATION

Currently, the policy goal of remediation to natural background level of all contaminated locations within one generation is considered too ambitious. It is therefore necessary to set priorities in soil remediation and to pay attention to the balance between the cost of remediation and the environmental benefit that is achieved by the remediation. The Commissie Welschen (1993) introduced the Removed Load Concept ("Vrachtbenadering") to quantify the environmental benefit of a soil remediation scenario. The load of contamination is defined as the weight (in kg's) of contaminants that is present above the natural background level.
The kriging estimates of the concentrations were used to calculate the removed load when excavating to S-level, ½(S+I)-level or I-level. For the former gasworks site, when excavating up to I-level a load of 75% of the contamination is removed for 35% of the cost that would be needed to remediate the location up to natural background level. For a remediation of the soil above (S+I)/2 the results are 80% of the contamination for 45% of the cost. By definition, at a remediation up to S-level the removed load is 100 %, but this does not take into account the remediation error.

4.9 EVALUATION BETWEEN ADDITIONAL SAMPLING VERSUS REMEDIATION ERROR REDUCTION

Once the optimal remediation scenario has been determined, the cost of further sampling and the benefit of the expected reduction of the remediation error for this remediation scenario are compared. The remediation error has to be expressed in a remediation error cost as well, but objective criteria to express the remediation error in remediation error cost still have to be developed. Sampling has to continue as long as the reduction of remediation error costs exceeds the cost of additional sampling. Sampling needs to take place in the zone with a high probability of misclassification with respect to the chosen remediation level. In figure 1 these zones are the dark areas.

The procedure of sampling, interpolations and model calculations continues iteratively until sampling costs exceed the reduction of remediation error costs that will be achieved with further sampling for a chosen remediation level.

At the former gasworks site, the remediation error is especially high for regions with extrapolation problems in the vertical direction. The need for further sampling is there still apparent. Further sampling will take place in those regions where the unreliability of the estimates with respect to ½(S+I)-level and I-level are highest.

5 Discussion, conclusions and further steps

The SACLEDS framework presents a systematic approach for a sampling strategy, which is focused on gathering information needed for the remediation scenario in which the balance between cost and environmental benefit is optimal.

In the use of geostatistics mean and covariance non-stationarity of concentration levels can be a problem. At old industrial sites it is difficult to devide the site in more stationary subregions. 3D moving window kriging can be used to account for non-stationarity, still allowing the use of some subregions.

A critical step in the moving window kriging approach is the choice of the amount of samples per window. In our opinion the use of, relative to ordinary single window kriging, limited amounts of samples and therefore suboptimal determined semivariograms, is to be preferred above not being able to allow for non-stationarity. Unfortunately, in a moving window approach the use of an automated procedure for semivariogram fitting is unavoidable.

Careful planning of the sampling is needed to obtain enough clusters of samples in each window, to avoid the negative impact of preferential analyzing and to avoid extrapolation in both the horizontal and vertical direction.

The remediation error can be used for the evaluation of the need for further sampling. Additional sampling has to take place at those locations where the largest reduction of the remediation error is expected for the remediation with the best balance between cost and environmental benefit. Optimal sampling depends on the chosen remediation concentration level.

Kriging is suitable for the determination of the remediation cost and its unreliability, using excavation as remediation option. For the design of in situ remediations, where representation of heterogeneity is important, moving window stochastic simulation will be more suitable.

Visualisation by map of the spatial distribution of the remediation costs of different alternatives, supports decisionmaking by planning departments and policymakers.

6 Acknowledgements

The authors wish to thank Dick Engelhardt of the Port of Rotterdam for supporting the development of the framework financially, Frits Apeldoorn for the preperation of the maps and Tim Haas for his helpful comments on the article and for the KRIG modifications he made for us.

7 Literature

Commissie Welschen (1993) Saneren zonder stagneren. Eindrapport van de Werkgroep bodemsanering. Den Haag, november 1993.

Haas, T.C. (1990a) Kriging and automated semivariogram modelling within a moving window. A tmospheric Environment Vol. 24A, no. 7, pp. 1759-1769.

Haas, T.C. (1990b) Lognormal and Moving Window Methods of Estimating Acid Deposition. Journal of the American Statistical Association Vol. 85, No. 412. pp. 950-963.

Haas, T.C. (1995) Local prediction of a spatio-temporal process with an application to wet sulfate deposition. Journal of the American Statistical Ass. Vol. 90, No. 432, applications and case studies, pp 1189-1199.

Haas, T.C., C.F. van Tooren & V. Brettschneider (1995). Optimization of a cleanup using moving window kriging. Contaminated Soil '95, pp 205-206. Kluwer Academic Press

Haas, T.C. (1996a) KRIG manual

Haas, T.C. (1996b) Multivariate spatial prediction in the presence of non-lineair trend and covariance non-stationarity. Environmetrics, Vol. 7, pp. 145-165.

Journel, A. & C.J. Huijbregts (1978) Mining Geostatistics. Academic Press, New York, 1978.

Persicani, D., (1995) Evaluation of soil classification and kriging for mapping herbicide leaching simulated by two models. Soil Technology, 8, pp. 17-30.

Press, W.H., B.P. Flannery, S.A. Teukolsky & W.T. Vetterling (1986) Numerical recipes, the art of scientific computing. Cambridge University Press

RIVM, 1991 Nationale milieuverkenning 2 1990-2010. Bilthoven, 1991. 550 pp.

Tooren, C.F. van & T.C. Haas (1993) A site investigation strategy using moving window kriging and automated semivariogram modelling. Contaminated Soil '93, pp. 609-622, Kluwer Academic Press.

Tooren, C.F. van (1993) Sampling strategies and mapping of soil contamination; possibilities of kriging interpolation. TNO report IMW R 93/001, 108 pages.

ANALYSIS OF SOLUTE REDISTRIBUTION IN HETEROGENEOUS SOIL:

I. Geostatistical approach to describe the spatial scaling factors

D. JACQUES[1], J. VANDERBORGHT[1], D. MALLANTS[1], B.P. MOHANTY[2] AND J. FEYEN[1]

[1]: *Institute for Land and Water Management, K.U.Leuven*
Vital Decosterstraat 102, B-3000 Leuven, Belgium

[2]: *U.S. Salinity Laboratory*
450 W. Big Springs Road, Riverside, CA 92507-4617, USA

1. Abstract

The soil water retention curve (WRC) and the saturated hydraulic conductivity (K_s) were measured in three overlying horizons in a vertical soil profile. The spatial variability in WRC and K_s was described with two separate sets of scaling factors, i.e., $\alpha_{r,\psi}$ and $\alpha_{r,K}$ for WRC and K_s, respectively. Both sets of scaling factors were found to be lognormally distributed. The $\log_e$-transformed scaling factors exhibited stationarity in their variance but showed non-stationarity in their median, both in vertical and horizontal direction. The trends or large scale deterministic variations were successfully removed using a median polish algorithm. The large scale variations contributed significantly to the total observed variability for the $\alpha_{r,\psi}$ parameters of the C2 horizon and for the $\alpha_{r,K}$ parameters of the Ap horizon. Spatial structure for scaling parameters was only observed in these two cases, whereas a pure nugget was observed in all other cases.

2. Introduction

Mathematical models will continue to play an increasing role in the management and the remediation of contaminated soils. An important factor impeding the use of simulation models is the inherent spatial variability of both physical and chemical soil properties which significantly influences water flow and solute transport in the vadose zone. As a result, there have been considerable research efforts to predict solute transport considering the spatial distribution of the governing parameters (Destouni, 1992, Russo et al., 1994, among others). Several numerical studies revealed significant effects of the spatial correlation structure, boundary conditions, and medium heterogeneity on flow and transport phenomena (Tseng and Jury, 1994, Roth, 1995).

In this study, we analyzed the spatial variability of the scaling parameters for the water retention characteristic and the saturated hydraulic conductivity. The WRC was described

A. Soares et al. (eds.), geoENV I–Geostatistics for Environmental Applications, 271–282.

with the analytical parametric function of van Genuchten (1980). The parameters of this function can be viewed as regionalized random variables and consequently, the statistical properties can be characterized using geostatistical tools (e.g., Mallants et al., 1996c). In order to simplify the description of the spatially variable soil hydraulic functions, scaling was invoked. The scaling factors pertaining to the WRC and K_s were analyzed within a robust geostatistical framework to investigate their spatial correlation structure in a multi-layered soil profile. The probability density function (pdf) and the spatial correlation structure of the scaling factors were used to generate two-dimensional random fields which served as input to a numerical experiment designed to investigate the effects of parameter variability on solute transport under different boundary conditions (Vanderborght et al., 1996, this issue).

3. Materials and Methods

3.1. SAMPLING AND LABORATORY MEASUREMENTS

Soil samples were collected along a 31-m long transect on an experimental field located at Bekkevoort, Belgium. The soil is defined as a well-drained Eutric Regosol (Udifluvent). A dye-study revealed that large amounts of macropores are present across the Ap-C1 horizon, i.e., 6 % at 13 cm depth to 0.37 % at 55 cm depth (Timmerman et al., 1996). The soil profile exhibits three horizons: Ap (0-25 cm) and C1 (25-55 cm) which both developed in colluvial material, and a C2 (55-100 cm) which is a textural B layer. In each soil horizon, 60 undisturbed soil cores (height 5.1 cm, diameter 5 cm) were taken with an alternating spacing of 10 and 90 cm at 10, 50 and 90 cm depth (Mallants et al., 1996b).
The drainage curve of the water retention characteristic was determined for each soil core at 0, 0.5, 1, 1.5, 2, 2.3, 2.8, 3.4, and 4.2 pF (pF = $\log_{10}(-\psi)$, where ψ is pressure head in cm water). In addition, the saturated hydraulic conductivity was measured on the same cores using a constant head permeameter. In previous applications, the WRC was described by means of the analytical function of van Genuchten (1980) and the spatial variability of its parameters was characterized by the multivariate distribution (Mallants et al., 1996b) and the variograms of each parameter (Mallants et al., 1996c). In this study, however, the spatial variability in hydraulic properties will be characterized by using the univariate distribution of the scaling factors.

3.2. SCALING OF SOIL HYDRAULIC PROPERTIES

In this study, the linear component of the spatial variability of the hydraulic soil properties is described using scaling factors which cause a linear transformation of the variables of interest (Vogel et al., 1991). Based on physical relationships and using the concept of similar media, i.e., media with identical internal geometries and porosities but different characteristic lengths, λ, Miller and Miller (1956) derived the relation between the pressure head, ψ (cm), of a medium with the pressure head of a reference medium with the same internal geometry. They also derived an analogous relation for the hydraulic conductivity, K. These original relations are given by:

$$\psi_r = (\lambda_m/\lambda_r)\,\psi_m \tag{1}$$

$$K_r = (\lambda_r/\lambda_m)^2\,K_m \tag{2}$$

where the subscript m denotes the reference medium and the subscript r the scaled medium. The scaling factor α_r is defined as λ_r / λ_m.

In the present study, we adopt the modifications to the original concepts introduced by Warrick et al. (1977). First, the assumption of equal porosities is eliminated by replacing the measured water content θ by the degree of saturation, $S = \theta / \theta_s$, with θ_s the saturated water content. Second, instead of using physical laws to derive scale factors (dimensional analysis, Tillotson and Nielsen, 1984), scale factors are obtained by relating the soil hydraulic function at the different locations to an average or reference soil hydraulic function. The latter procedure, the so-called functional normalization (Tillotson and Nielsen, 1984), assumes functional similarity (Simmons et al., 1979). In the present study, two sets of scaling factors are obtained: (1) $\alpha_{r,\psi}$ for the WRC and (2) $\alpha_{r,K}$ for the K_s data. The WRC data are scaled using the method of Warrick et al. (1977) and Hopmans (1987). In an iterative procedure, the sum of squares,

$$SS = \sum_{r=1}^{R}\sum_{i=1}^{I(r)}\left[\psi_m(S_{r,i}) - \alpha_{r,\psi}\psi_r(S_{r,i})\right]^2 \tag{3}$$

is minimized using the constraint that the mean scaling factor must be 1. During the scaling procedure, the reference curve, $\psi_m(S)$, is described by a third degree function in S. It is first fitted to the unscaled ψ-S data and subsequently to scaled data. This procedure was carried out using the WRC-data for the three horizons separately and for all data combined (pooled data). Only the latter results are discussed in this paper. The van Genuchten function was fitted through the scaled data and served as the reference curve for the simulations done by Vanderborght et al. (1996, this issue). The scaling factors for the hydraulic conductivity are obtained by $\alpha_{r,K} = K_{s,r} / K_{s,m}$ where $K_{s,m} = \exp(\langle Y\rangle)$ and $\langle Y\rangle$ is the arithmetic average of $\log_e(K_s)$. In the following discussion we will compare results in terms of the mean squared error, MSE = SS/df, with df being the degrees of freedom.

3.3. GEOSTATISTICAL PROCEDURES

The spatial correlation structure of the scaling factors is investigated with the experimental semivariograms of the regionalized random variable, $Z(\mathbf{x})$ with $\mathbf{x}$ the vector containing the spatial coordinates. It is assumed that $Z(\mathbf{x})$ obeys the intrinsic hypothesis (Matheron, 1963) that assumes stationarity in mean and that the covariance for each pair $[Z(\mathbf{x}), Z(\mathbf{x}+\mathbf{h})]$ is only function of the separation vector $\mathbf{h}$. The estimator of the semivariogram, $\hat{\gamma}(h)$, is given as (Matheron, 1963):

$$\hat{\gamma}(\boldsymbol{h}) = (1/2N(\boldsymbol{h}))\sum_{i=1}^{i=N(h)}\left[z(\boldsymbol{x}_i + \boldsymbol{h}) - z(\boldsymbol{x}_i)\right]^2 \tag{4}$$

where N(h) the number of pairs for the lag h, $z(\mathbf{x}_i)$ the measured data at location $\mathbf{x}_i$ and

$z(\mathbf{x}_i+\mathbf{h})$ the data at location $\mathbf{x}_i+\mathbf{h}$. The experimental semivariogram is described with the spherical semivariogram model, $\gamma(h)$, and is defined as:

$$\gamma(h) = C_0 + C_s\left[(3h/2a) - (h^3/2a^3)\right] \qquad 0 < h \leq a \tag{5}$$

$$\gamma(h) = C_0 + C_s \qquad h>a \tag{6}$$

The parameters of the spherical model are the nugget C_0, the sill-value, $C_0 + C_s$, and the range of spatial dependence, a.

4. Results and Discussion

4.1. SCALING AND DISTRIBUTION OF SCALE FACTORS

Preliminary analysis of the scaling procedure using the complete ψ-S data set for the entire pressure range (from saturation to 4.2 pF) gave poor results. Better results were obtained when data for the two driest pressure steps (3.8 and 4.2 pF) were omitted. Because the behavior of the retention curve for these dry conditions are less important for our purposes, we continued the analysis with the reduced data set.

The ratio of the minimum MSE after scaling to the MSE before scaling was 0.877. In Figure 1a, the scaled S values are plotted together with the fitted van Genuchten function. This curve is considered as the reference or mean soil hydraulic function and the MSE, now based on the difference between scaled and estimated S values using van Genuchten's function, is reduced by 31 % in comparison with the fit through the non-scaled S data. The fitted retention curve overpredicts the observations near saturation as well as in the dry range, but gives a fairly good description of the data in the pressure range 1-3 pF. These discrepancies may be due to the presence of a structural pore-size distribution close to saturation. It was previously shown by Mallants et al. (1996b) that the retention behavior of this macroporous soil could be better described when a linear superposition of two or

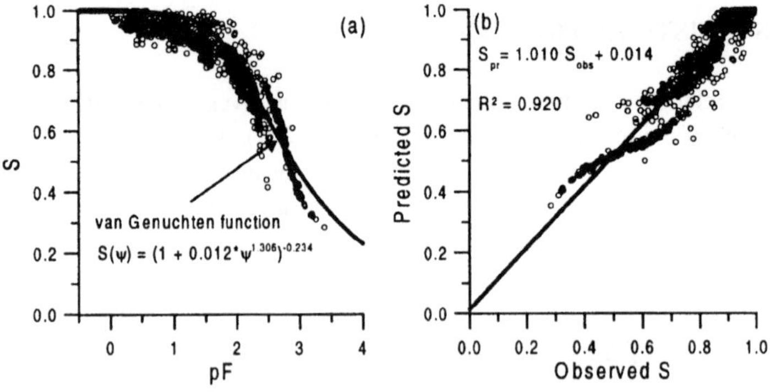

Figure 1: (a) Scaled WRC data and mean van Genuchten function and (b) scatter plot of fitted versus observed degree of saturation, S.

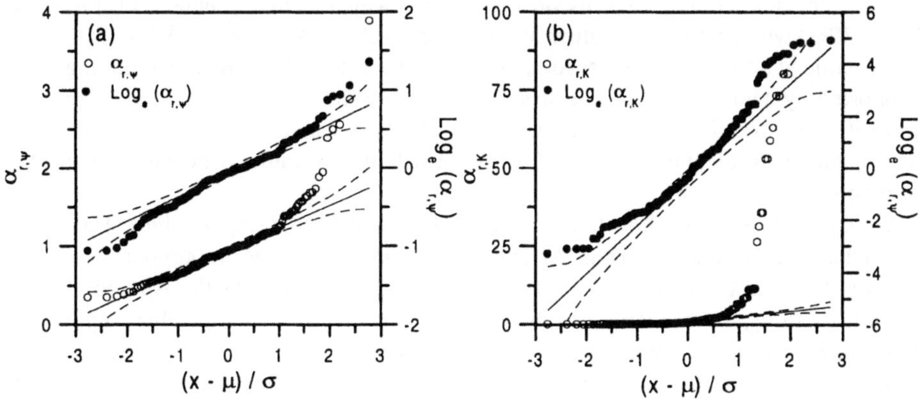

Figure 2: Normal probability plots of the original and $\log_e$-transformed scale factors of (a) the WRC and (b) the K_s.

even three van Genuchten functions was considered. The scatter plot (Figure 1b) of predicted versus observed S values does not indicate biased estimations (high R^2 and a slope coefficient of 1). However, part of the observed variability in the WRC-data cannot be explained by invoking scaling. Vogel et al. (1991) defined this part as nonlinear variability (e.g., the n-parameter in the van Genuchten function), whereas the other part of the total variability which can be explained by scaling is refered to as the linear variability. The probability plots of both the untransformed and $\log_e$-transformed scaling factors are given in Figures 2a for $\alpha_{r,\psi}$ and 2b for $\alpha_{r,K}$. The theoretical function (solid line), together with an approximated 95 % confidence interval (dashed line), is estimated based on robust estimates, i.e., using the median for the mean and the interquartile-range for the standard error (Chambers et al., 1983). Both $\alpha_{r,\psi}$ and $\alpha_{r,K}$ are better described with a lognormal distribution than with a normal distribution. This is in agreement with the results of Warrick et al. (1977), Hopmans (1987), and Mallants et al. (1996a).

Statistics of the $\log_e$-transformed scaling factors are given in Table 1. The ensemble average, $<\alpha_{r,\psi}>$, decreases with depth. Therefore, we can stated that the top layers consist

Table 1: Statistics of $\log_e$-transformed scale factors.

	$\mu_{\ln\alpha}$	$\sigma^2_{\ln\alpha}$	$<\alpha>$	Stdev	CV	N
			$\alpha_{r,\psi}$			
Pooled	-0.076	0.144	0.996	0.393	0.39	178
Ap	0.068	0.194	1.179	0.545	0.46	60
C1	-0.003	0.101	1.049	0.342	0.33	59
C2	-0.295	0.068	0.770	0.205	0.27	59
			$\alpha_{r,K}$			
Pooled	≈ 0	3.561	5.932	34.69	5.85	176
Ap	0.054	3.671	6.617	40.95	6.19	59
C1	-0.305	2.471	2.536	8.35	3.29	59
C2	0.255	4.511	12.313	116.85	9.49	58

$<\alpha>=\exp(\mu_{\ln}+0.5\sigma_{\ln}^2)$, Stdev$=(\exp(2\mu_{\ln}+\sigma_{\ln}^2)(\exp(\sigma_{\ln}^2)-1))^{0.5}$, CV$=(\exp(\sigma_{\ln}^2)-1)^{0.5}$, N=number of observations of more coarse-textured materials since a location with a high $\alpha_{r,\psi}$-value is characterized by a WRC with an air entry values at a lower pF-values. An observed WRC with an air entry value at a low pF is typical for a sandy or coarse-textured soil. These findings were supported by data on particle size distribution, which showed percentage sand decreased from 40% in Ap and C1 to 20% in C2. The variability in scaling factors also decreases with depth. The coefficient of variation (CV) is consistent with values previously reported by Mallants et al. (1996a).

The $<\alpha_{r,K}>$ values do not show the same trend with depth as the $<\alpha_{r,\psi}>$ values. The C2 has the highest $<\alpha_{r,K}>$, followed by the Ap and C1 horizon. Since $K_{s,r}$ is proportional to $\alpha_{r,K}$ (see section 3.2), the K_s for the C2 horizon is higher than the K_s of the Ap and C1. The variability in $\alpha_{r,K}$ is very large and the CVs are much larger than the values reported by Mallants et al. (1996a), although they were determined on the same soil. However, Mallants et al. (1996a) scaled the unsaturated hydraulic conductivity, whereas in this study only the saturated hydraulic conductivity was scaled. In general, the latter is much more variable than the unsaturated hydraulic conductivity. Furthermore, the large variability in $\alpha_{r,K}$ may also be attributed to the 100 cm^3 small soil cores used in the determination of K_s.

If the similar medium concept is valid for this soil, the scaling factors of the WRC, $\alpha_{r,\psi}$, should be equal to $(\alpha_{r,K})^{-0.5}$ (eqs. 1, 2) and also the $Var(\alpha_{r,\psi}) = Var(\alpha_{r,K}^{-0.5})$. However, this is not the case for this soil which is obvious from Figure 3. Moreover, the $Var(\alpha_{r,\psi})$ is almost 5 times smaller than the $Var(\alpha_{r,K}^{-0.5})$, for which two reasons may be given. Firstly, results from the study of Mallants et al. (1996a) on the same soil indicated that the soil is geometrically dissimilar since the scaling factors obtained from scaling the WRC and from scaling the unsaturated hydraulic conductivity were different. Secondly, the measurement of K_s is mainly determined by the presence or absence of macropores in the soil samples, whereas the WRC is determined by the whole range of pore sizes. Therefore, there may be a very strong geometric non-similarity between anchor-pores for these two hydraulic soil properties. Furthermore, the occurrence of macropores in a 5-cm diameter soil sample is highly susceptible to spatial variability which in turn causes the large variability in K_s and $\alpha_{r,K}$.

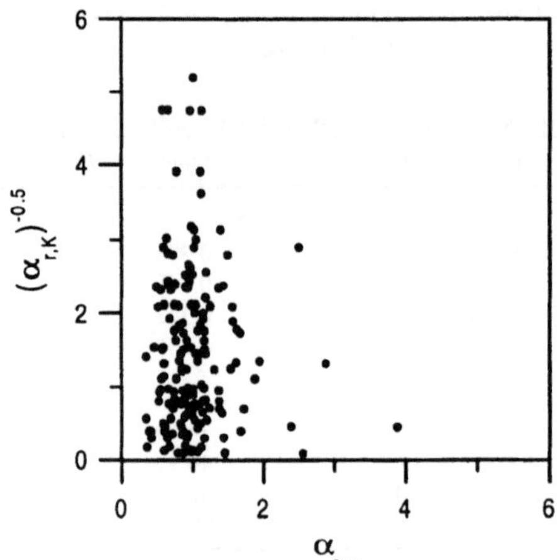

Figure 3: Comparison between $\alpha_{r,\psi}$ and $\alpha_{r,K}$.

4.2. ANALYSIS OF NON STATIONARITY

Prior to the estimation of the variogram, the assumptions of the intrinsic hypothesis were investigated using a robust approach which had been previously used by Cressie and Horton (1987), Mohanty et al. (1991), and Mohanty and Kanwar (1994). The stationarity of the variance is checked using median versus interquartile-range squared (IQ^2) plots. The transect was divided into 10 equal windows of each 3 m. For each window and depth, the median and the IQ^2 was calculated for the untransformed and the $\log_e$-transformed $\alpha_{r,\psi}$ and $\alpha_{r,K}$. The results for $\alpha_{r,\psi}$ are shown in Figure 4. For the untransformed data, a slight increase in IQ^2 is observed with increasing values of the median. The $\log_e$ transformation stabilizes the non-stationarity of the variance, except for 3 outliers in the Ap. Similar variance stabilizing effects were obtained for $\alpha_{r,K}$ (results not shown). The further geostatistical analysis of the scaling factors will therefore be performed with $\log_e$ transformed data.
In Figure 5, the individual values of the $\log_e$-transformed scale factors are plotted along the transect for the 3 horizons. To detect possible trends in a more efficient way, the medians of the above defined windows are also plotted. The straight dashed line represents the median value for the specific horizon. There is a linear increase of $\alpha_{r,\psi}$ along the transect in the Ap and C2 horizon, whereas no such trend is observable in the other cases. However, there is an apparent cyclic component present in the $\alpha_{r,K}$ values of the Ap horizon, and the C2 horizon.
Table 1 further reveals a nonlinear trend with depth. Moreover, Figure 5 reveals some indications of non-stationarity in the mean along the transect. Thus, we can decompose the random variable, $Z(\mathbf{x})$, into a deterministic large-scale variation component $\mu(\mathbf{x})$ and a stochastic small scale variation component, $\varepsilon(\mathbf{x})$, where $\mathbf{x}$ denotes the location vector. In this study, we can decompose the $\mu(\mathbf{x})$ into an overall mean μ, a depth effect, δ_i (i=1,2,3), and a horizontal effect, η_j (j=1,2,...,30). Those deterministic components will be estimated with the median polish algorithm (Cressie, 1993, p. 186; Mohanty and Kanwar, 1994) resulting in:

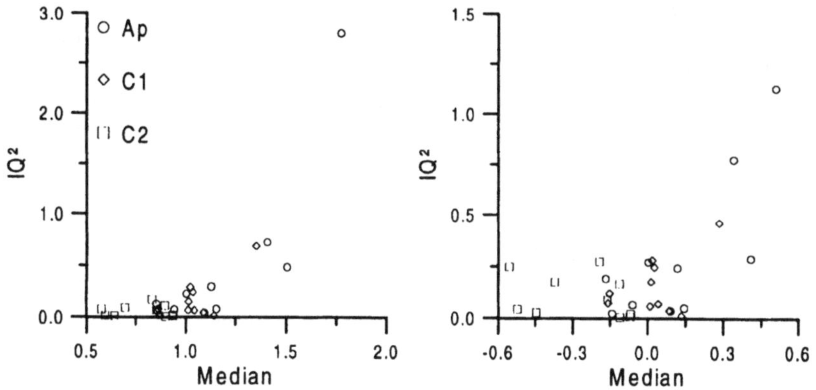

Figure 4: Interquartile-range squared versus median for the original (left) and the $\log_e$-transformed (right) $\alpha_{r,\psi}$.

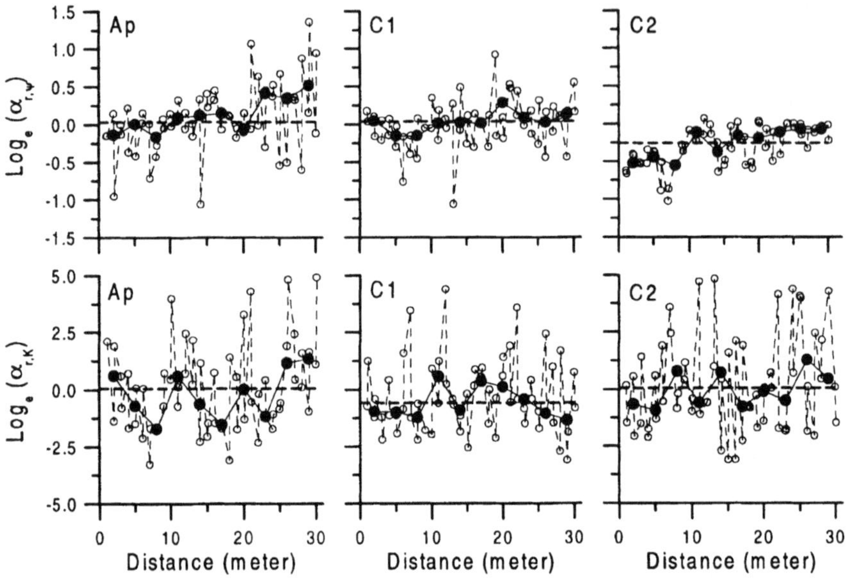

Figure 5: Spatial variability of $\log_e$-transformed scale factors of the WRC (top) and of K_s (bottom) for the Ap (left), C1 (middle) and C2 (right) horizon. Open circles indicate the individual scale factors, closed circles are the medians of the windows and the dashed line represents the median for the whole data set.

$$Z(\boldsymbol{x}) = m + d_i + h_j + R(\boldsymbol{x}) \tag{7}$$

where m, d_i and h_j are the robust estimates of μ, δ_i and η_j, respectively. The variograms will be estimated based on the residuals, R(**x**), which represent a stationary process. The median polish algorithm is invoked here since it is more robust with respect to outliers than mean-based methods (Mohanty et al., 1991). In an iterative median polish procedure (Mohanty and Kanwar, 1994), the different deterministic components of the $\log_e$-transformed $\alpha_{r,\psi}$ and $\alpha_{r,K}$ were removed.

A final check of the histograms of the residuals (for $\log_e(\alpha_{r,\psi})$ in Figure 6) reveals normality without outliers. Also note that the $\log_e$-transformation improves the data normality and results in variance stationarity. A similar good performance of the median polish algorithm was found by Mohanty and Kanwar (1994).

4.3. SEMIVARIOGRAM ANALYSIS

In Figure 7, the experimental variograms are plotted for both the original $\log_e$-transformed scaling factors and the residuals of the median polish procedure. In general, the spatial correlation structure and the variogram estimates are smaller for the median-polished data than for the untransformed data. Moreover, the variograms based on the median polished data are less erratic.

Variograms of the original $\log_e(\alpha_{r,\psi})$ and the median polished data of the Ap and C1-horizon

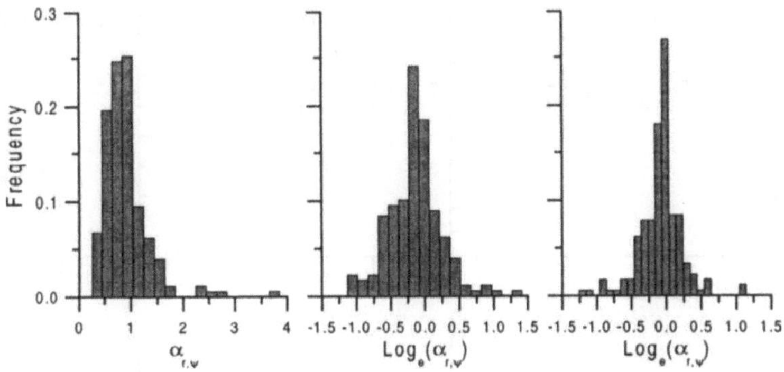

Figure 6: Histograms of original (left), $\log_e$-transformed (middle) and median polished (right) scaling factors of the WRC.

reveal pure nugget, whereas the variogram of the C2-horizon exhibits weak spatial correlation for the original $\log_e(\alpha_{r,\psi})$ and pure nugget for the median polished data. The largest difference between two variograms before and after median polishing is found in the C2-horizon, indicating that a considerable part of the observed variability is due to the large scale deterministic variation and that the small-scale variation has only a minor contribution to the total variability. The spherical model (5) was fitted to the variogram of the C2 horizon of the original data (not shown here). The range of spatial correlation was approximately 3 m with a sill-value of 0.59. Ignoring the linear trend in the data leads to an overestimation of the spatial correlation structure, as was shown in the theoretical study of Russo and Jury (1987). In the Ap-horizon, the large scale variation contributes less to the total variability in the data, whereas the variability in the C1 horizon is only caused by small scale variation without any spatial correlation.

The variograms of the original $\log_e(\alpha_{r,K})$ data for the three horizons exhibit a hole-effect due to the apparent cyclic trend (Figure 5). The variograms based on the median polish data also reveal this hole-effect, except for the C1-horizon. The failure of the median polish algorithm to remove the hole effect is explained by the fact that the vertical medians were close to zero. This is because the cyclic trend in the Ap was the reverse of the cyclic trend in the C2. The data for the C1 horizon do not exhibit a specific trend (Figure 5). Therefore, the corrections with the median polish algorithm are quite small.

The variograms of the C1 horizon reveal pure nugget. Both variograms of the C2 horizon are very erratic and are pure nugget. In the variograms of the Ap horizon, weak spatial correlation can be detected. Variograms are described with the spherical model, ignoring the presence of the possible hole-effects. For the original $\log_e(\alpha_{r,K})$, the nugget is 2.4, the sill-value 3.7 and the range 4 m. Fitting the variogram for the median polish data revealed a nugget of 2.3, a sill value of 3 and a range of 2.4 m. Again, the influence of the large-scale variation on the spatial correlation structure is clear. In contrast to the $\log_e(\alpha_{r,\psi})$ data, the contributions of the large scale variation to the total variability of the $\log_e(\alpha_{r,K})$ data is larger for the Ap and C1 horizon than for the C2 horizon.

The WRC is primarily determined by the texture of the soil material. Because the C2 horizon consists mainly of undisturbed material it has maintaned its spatial structure, probable due to a depositional pattern, and large scale deterministic variations. However,

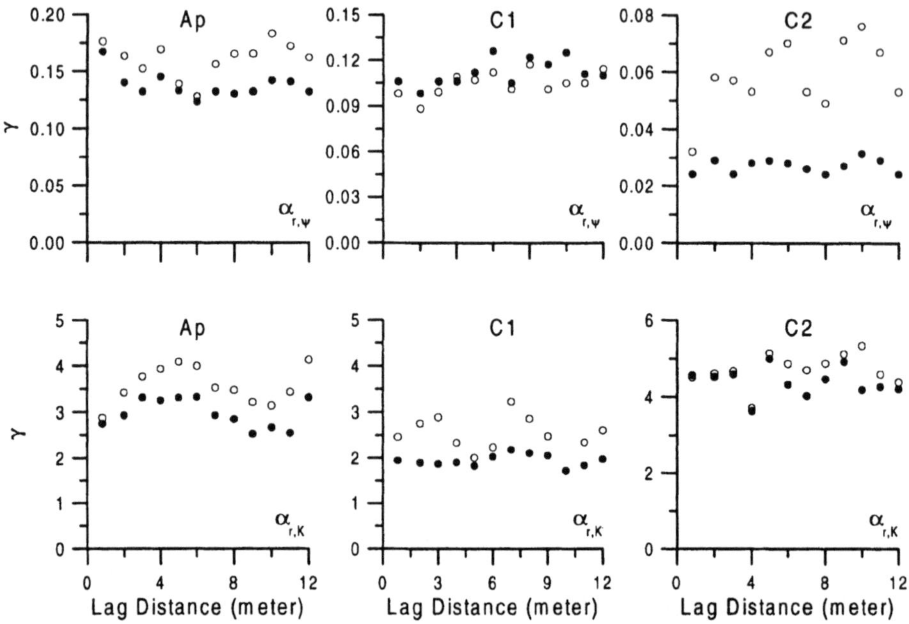

Figure 7: Experimental semivariograms of the scale factors of the WRC (top) and K_s (bottom) for the Ap (left), C1 (middle) and C2 (right). The semivariogram of the $\log_e$-transformed data are plotted as open circles. The semivarograms of the median polished data are plotted as closed circles.

at this moment, no detailed analysis is performed to investigate the reason for this spatial structure. In contrast, the Ap and C1 developed in younger colluvial material are affected by external processes such as climate and tillage. As a consequence, the $\alpha_{r,\psi}$-data are not characterized by a distinct spatial structure. The K_s and $\log_e(\alpha_{r,K})$ data are mainly determined by macropores formed by earthworms and tree roots. These factors have a major influence in the top soil, causing the spatial structure observed in the Ap-horizon.

5. Conclusions

The purpose of the present spatial analysis of the scale factors of two hydraulic properties was to obtain input parameters for random field generator and simulation models. Those models may help to improve our understanding of solute transport through the vadose zone at the field-scale. We were especially interested in the statistical parameters of scaling factors, such as the degree of heterogeneity, the large scale variation and the spatial correlation. Using scaling factors in the analysis of flow and transport problems is very attractive since it reduces the number of regionalized random variables and makes computational procedures simpler and more efficient.

For this study, scaling factors were obtained separately for the WRC and the K_s data. Furthermore, different trends in the data for vertical and horizontal directions and in the

spatial correlation structure of the two sets of scaling factors were found.
The two sets of scaling factors have been analyzed within a robust geostatistical framework. Scaling factors were $\log_e$-transformed to obtain stationarity in variance and a median polish algorithm was used to remove large-scale variation in both vertical and horizontal direction. It is important to carry out such a detailed data analysis in an explorative stage to detect the presence of large-scale variation since the occurrence of trends has a large influence on the spatial correlation estimates.
Large scale variation has an important influence on the variability of the $\log_e(\alpha_{r,\psi})$ data of the C2 horizon and of the $\log_e(\alpha_{r,K})$ data of the Ap and C1 horizons. Spatial correlation structures were only observed for the $\log_e(\alpha_{r,\psi})$ of the C2-horizon and for the $\log_e(\alpha_{r,K})$ of the Ap-horizon. The scaling factors exhibited non-stationarity in both vertical and horizontal direction and have different spatial correlation structures with depth. In the accompanying paper (Vanderborght et al., 1996), the statistics of the scaling factors obtained here will be used to generate two-dimensional random fields of scaling factors. Those random fields will be used in a numerical flow and transport experiment designed to study effects of medium heterogeneity on transport of chemicals through soils.

6. Acknowledgement

The first author acknowledges the financial support of the scholarship of the Flemisch Institute for the Encouragement of Scientific-Technological Research in the Industry (IWT).

7. References

Chamber, J.M., Cleveland, W.S., Kleiner, B., and. Tukey, P.A (1983) *Graphical methods for data analysis*, Belmont, CA, Wadsworth publishing company.

Cressie, N.A.C. (1993) *Statistics for spatial data*, John Wiley, NY.

Cressie, N.A.C., and Horton, R. (1987) A robust-resistant spatial analysis of soil water infiltration, *Water Resources Research* **23**, 911-917.

Destouni, G. (1992) The effect of vertical soil heterogeneity on field scale solute flux, *Water Resources Research* **28**, 1303-1309.

Hopmans, J. (1987) A comparison of various methods to scale soil hydraulic properties, *Journal of Hydrology* **93**, 241-256.

Mallants, D., Jacques, D., Tseng, P.-H., van Genuchten, M.Th., and Feyen, J. (1996a) Comparison of three hydraulic property measurement methods, *Journal of Hydrology*, accepted.

Mallants, D., Jacques, D., Vanclooster, M., Diels, J., and Feyen, J. (1996b) A stochastic approach to simulate water flow in a macroporous soil, *Geoderma* **70**, 299-313.

Mallants, D., Mohanty, B.P., Jacques, D., and Feyen, J. (1996c) Spatial variability of hydraulic properties in a multi-layered soil profile., *Soil Science* **161**, 167-181.

Matheron, G. (1963) Principles of geostatistics, *Economical Geology* **19**, 129-149.

Miller, E.E., and Miller, R.D. (1956) Physical theory for capillary flow phenomena, *Journal of Applied Physics* **27**, 324-332.

Mohanty, B.P., and Kanwar, R.S. (1994) Spatial variability of residual nitrate-nitrogen under two tillage systems in central Iowa: A composite three-dimensional resistant and exploratory approach, *Water Resources Research* **30**, 237-251.

Mohanty, B.P., Kanwar, R.S., and Horton, R. (1991) A robust-resistant approach to interpret spatial behavior of saturated hydraulic conductivity, *Water Resources Research* **27**, 2979-2992.

Roth, K. (1995) Steady state flow in an unsaturated, two-dimensional, macroscopically homogeneous, Miller-similar medium, *Water Resources Research* **31**, 2127-2140.

Russo, D., and Jury, W.A. (1987) A theoretical study of the estimation of the correlation scale in spatially variable fields 2. Nonstationary fields, *Water Resources Research* **23**, 1269-1279.

Russo, D., Zaidel, J., and Laufer, A. (1994) Stochastic analysis of solute transport in partially saturated heterogeneous soil. 1. Numerical experiments, *Water Resources Research* **30**, 769-779.

Simmons, C.S., Nielsen, D.R. , and Biggar, J.W. (1979) Scaling of field-measured soil water properties, *Hilgardia* **47**, 77-154.

Tillotson, P.M., and Nielsen, D.R. (1984) Scale factors in soil science, *Soil Science Society of American Journal* **48**, 953-959.

Timmerman, A., Vanderborght, J., Mallants, D., Jacques, D., and Feyen, J. (1996) Detecting macropore flow in the field with TDR and staining patterns, *European Geophysical Society, XXI General Assembly*, The Hague, The Neterlands, 6-10 May, p. 327.

Tseng, P.-H. and Jury, W.A. (1994) Comparison of transfer function and deterministic modeling of area-averaged solute transport in a heterogeneous field, *Water Resources Research* **30**, 2051-2063.

Vanderborght, J., Jacques, D., Mallants, D., Tseng, P., and Feyen, J. (1996) Analysis of solute redistribution in heterogeneous soil: II. Numerical simulation of solute transport, *First European Conference on Geostatistics for Environmental Applications*, 20 - 22 November, Lisbon, Portugal, this issue.

van Genuchten, M.Th (1980) A closed-form equation for predicting the hydraulic conductivity of unsaturated soils, *Soil Science Society of American Journal* **44**, 892-898.

Vogel, T., Cislerova, M., and Hopmans, J.W. (1991) Porous media with linearly variable hydraulic properties, *Water Resources Research* **27**, 2735-2741.

Warrick, A.W., Mullen, G.J., and Nielsen, D.R. (1977) Scaling field-measured soil hydraulic properties using a similar media concept, *Water Resources Research* **13**, 355-362.

ANALYSIS OF SOLUTE REDISTRIBUTION IN HETEROGENEOUS SOIL:
II. Numerical Simulation of Solute Transport

[1]J. VANDERBORGHT, [1]D. JACQUES, [1]D. MALLANTS,
[2]P.- H. TSENG AND [1]J. FEYEN
[1]*Institute for Land and Water Management, KULeuven*
Vital Decosterstraat 102, B-3000 Leuven, Belgium
[2]*Institute of Geophysics and Planetary Physics, UCR*
Riverside, CA 92521-0412, USA

1. Abstract

Field-scale solute dispersivities are derived in a macroporous sandy-loam soil for unsaturated and saturated flow conditions from unconditional numerical simulations of solute transport. Heterogeneous soils are generated using geostatistical information on soil hydraulic properties. The effect of the measurement technique, size and saturation degree of the soil sample on the geostatistical parameters of the soil hydraulic properties and on the simulated field-scale dispersivities is investigated. A comparison of field-scale dispersivities derived from simulated and measured solute concentrations reveals that for unsaturated flow conditions, the heterogeneity of hydraulic conductivities measured under unsaturated conditions and on small soil samples gives the best results. For saturated flow conditions, the simulated dispersivities largely underpredict the measured dispersivities. This is because flow and transport heterogeneity due to macropore activation cannot be described by the heterogeneity of macroscopic hydraulic properties.

2. Introduction

To assess the risk of ground water contamination or to remediate polluted sites, models that describe solute transport processes in the vadose zone are of utmost importance. Transport of inert solutes at the field-scale depends largely on the heterogeneity of macroscopic soil hydraulic properties which results in extremely variable local water flux or solute dispersion. Using geostatistical information about the soil hydraulic properties, the field-scale solute dispersion can be derived. Unfortunately, the spatial variations and correlation structures of relevant soil hydraulic properties (i.e., the hydraulic conductivity) depend largely on both the measurement methods (Mallants et al., 1996b; Mohanty et al., 1994) and the scale of the soil sample on which these properties are determined (Mallants et al., 1996a). In addition, the saturation degree also

A. Soares et al. (eds.), geoENV I–Geostatistics for Environmental Applications, 283–295.

influences the variability of the hydraulic conductivity considerably. This is especially true in highly structured and macroporous soils. When close to saturation, the variability of the conductivity can increase drastically with increasing degree of saturation.

The objectives of this study are to evaluate the hydraulic property heterogeneity in terms of measuring scales and saturation degree, and to investigate its effect on field-scale solute transport in a macroporous sandy-loam soil. The field-scale solute dispersion in heterogeneous soils will be simulated under both saturated and unsaturated conditions and will be compared to measured field-scale dispersion to assist in future development of improved characterisations of solute transport in heterogeneous soils.

3. Materials and Methods

3.1 CHARACTERISATION OF SOIL HYDRAULIC PROPERTIES

Data on soil hydraulic properties and the spatial variability of these properties were collected at an experimental site located in Bekkevoort (Belgium). The soil was characterised as a macroporous well-drained sandy loam (Udifluvent or Eutric Regosol). Water retention curves, $\theta(h)$, with θ the volumetric water content and h the pressure head, were determined on 5.1-cm long and 5-cm diameter soil samples (Kopecky rings). Saturated hydraulic conductivity, K_{sat}, was measured on both Kopecky rings and on larger 20-cm long and 20-cm diameter samples (crust rings). Hydraulic conductivity curves, $K(h)$ or $K(\theta)$, were derived from θ and h measurements during a drainage experiment in 100-cm long and 30-cm diameter soil columns (large columns). All soil samples were taken along a transect with a regular spacing between each sample. For a detailed description of the experimental setup and the data sets, the readers are referred to Mallants et al. (1996b). At two nearby plots, field-scale leaching experiments were carried out under unsaturated (Jacques et al., 1996a) and saturated (Mallants et al., 1996c) flow conditions.

To characterise the heterogeneity of the hydraulic properties, the linear scaling technique was used (cf. accompanying paper by Jacques et al., 1996b). Spatially varying $\theta(h)$ and $K(h)$ curves were linearly scaled to deterministic reference curves, $\theta(h^*)$ and $K(h^*)$, using scaling factors α_h and α_k, respectively:

$$h = \alpha_h h^* \quad (1)$$

$$K(h) = \alpha_k K^*(h^*) \quad (2)$$

The deterministic $\theta(h^*)$ was described by the van Genuchten water retention curve, WRC (van Genuchten, 1980):

$$\theta(h^*) = \theta_r + \frac{\theta_s - \theta_r}{(1 + |\alpha h^*|^n)^m} \qquad h^* < 0 \quad (3a)$$

$$\theta(h^*) = \theta_s \qquad h^* \geq 0 \quad (3b)$$

where θ_s and θ_r denote the saturated and residual volumetric water content respectively and α, n and m ($m = 1 - 1/n$) are shape factors. Note that the linear scaling technique is only applicable when the shape factors n and m are deterministic. In addition, since we did not scale the saturated water contents, it was also assumed that θ_s and θ_r are deterministic. The assumption of deterministic n, θ_s, and θ_r parameters was considered to be reasonable since these parameters did not vary considerably (Mallants et al., 1996d).

The conductivity curve, $K^*(h^*)$, was derived from the WRC (Eq. 3) using Mualem's pore size distribution model (Mualem, 1976):

$$K^*(h^*) = \frac{K^*_{sat}\left\{1 - |\alpha h^*|^{nm}\left[1 + |\alpha h^*|^{n}\right]^{-m}\right\}^2}{\left[1 + |\alpha h^*|^{n}\right]^{ml}} \qquad h^* < 0 \qquad (4a)$$

$$K^*(h^*) = K^*_{sat} \qquad h^* \geq 0 \qquad (4b)$$

where K^*_{sat} is the reference saturated conductivity and l is a pore connectivity parameter. When l is fixed, the shape of the $K^*(h^*)$ curve is entirely determined by the shape of the $\theta(h^*)$ curve. As for the $\theta(h)$, the linear scaling technique assumes that the shape factors n,m and l of the local $K(h)$ curves are deterministic. In Table 1, the parameters θ_s,θ_r,α,n,K^*_{sat}, and l for $\theta(h^*)$ and $K^*(h^*)$ are shown.

TABLE 1. Parameters used for $\theta(h^*)$ and $K^*(h^*)$ characteristics

θ_s	θ_r	α (cm^{-1})	n	K^*_{sat} (cm d^{-1})	l
0.4	0.001	0.0109	1.288	38	20

The heterogeneity of the soil hydraulic properties was quantified in terms of $\sigma^2\ln(\alpha_h)$ and $\sigma^2\ln(\alpha_k)$, the variance of the log-transformed scaling factors. The spatial correlation of a log-transformed scaling factor was characterised by an exponential correlogram.

Based on the data used to characterise the heterogeneity of soil hydraulic properties, we considered 4 different cases. The first case assumed that only the data on the variability of $\theta(h)$ were available. Using the concept of geometrically similar pore spaces (Miller and Miller, 1956), the scaling factor of the $K(h)$ curve can be derived from the scaling factor of the $\theta(h)$ curve:

$$\alpha_k = 1/\alpha_h^2 \qquad (5)$$

From Eq. (5) it follows that:

$$\sigma^2 \ln(\alpha_k) = 4\, \sigma^2 \ln(\alpha_h) \qquad (6)$$

For the second case, scaling factors α_k were derived from K_{sat} measurements on crust rings (crust ring case):

$$\alpha_k = K_{sat}/K^*_{sat} \quad (7)$$

$$K^*_{sat} = \exp(\langle \ln(K_{sat}) \rangle) \quad (8)$$

where $\langle \ln(K_{sat}) \rangle$ is the expected value or the ensemble average of the log transformed saturated conductivity. From Eq. (7), it follows that:

$$\sigma^2 \ln(\alpha_k) = \sigma^2 \ln(K_{sat}) \quad (9)$$

The third case used K_{sat} measurements on Kopecky rings to define the scaling factors α_k (Kopecky ring case). Since for the first, second, and third cases no K(h) characteristics where measured on these soil samples, l in Eq. (4) was chosen to be 0.5, a representative value suggested by Mualem (1976) for various soil types. For the fourth case, K(h) curves that were determined from drainage experiments in the large columns were scaled to a reference K^*(h*) curves while α_k was determined according to Eq. (2) (large column case). The variances of the log transformed scaling factors together with the horizontal, γ_{x1}, and vertical, γ_{x2}, correlation lengths are shown in Table 2 for the four cases.

TABLE 2. Geostatistical parameters for the four cases

	$\sigma^2 \ln(\alpha_h)$	$\sigma^2 \ln(\alpha_k)$	γ_{x1} (m)	γ_{x2} (m)
similar field case	0.169	0.676	1.5	0.5
crust ring case	0.169	1.55	1.5	0.5
Kopecky ring case	0.169	3.6	0.1	0.1
large column case	0.169	2.55	0.1	0.1

As indicated in Table 2, $\sigma^2 \ln(\alpha_k)$ varies considerably between different cases and is the largest when based on K_{sat} measurements on the smallest samples (Kopecky ring case). On the other hand, $\sigma^2 \ln(\alpha_k)$ becomes the smallest when characterised by the variability of the θ(h) curve. This indicates that the variability of the hydraulic conductivity due to variations in soil texture, which define the variability of the θ(h) curve, is much smaller than the variability of the hydraulic conductivity due to variations in soil structure, which define the variability of K_{sat}. For unsaturated conditions, the effect of variable soil structure on the hydraulic conductivity is smaller than for saturated conditions as is indicated by the lower $\sigma^2\ln(\alpha_k)$ for the large column case than for the Kopecky ring case. For all cases, there was no information available for the vertical correlation lengths, γ_{x2}. The values of γ_{x2} were thus chosen in accordance with those used in other studies on solute transport in heterogeneous fields (Tseng and Jury, 1994). For the Kopecky ring and large column cases, the experimental horizontal variograms of α_k showed a pure nugget effect (Mallants et al. 1996d). This nugget effect can be explained by small scale variations of soil structure which also explain the drastic decrease of $\sigma^2\ln(\alpha_k)$ with increasing soil sample scale (Kopecky vs. crust ring cases). Therefore, correlation lengths were chosen to be of the same order of magnitude as the scale of the soil volume over which the variables that were used to derive the hydraulic conductivity (water contents, pressure heads or water fluxes) were averaged by the measurement device.

3.2 GENERATION OF RANDOM FIELDS

Unconditional 2-D random fields of 512 by 512 nodes representing a vertical cross-section were generated using spectral techniques in combination with the fast Fourier transform, FFT (Gutjahr et al., 1995). From these generated fields, smaller fields of 200 by 200 nodes were selected. The horizontal grid size, Δx_1, was determined large enough to allow sufficient heterogeneity within the field. In the literature (Tompkins et al., 1994), a field width of 20 horizontal correlation lengths is recommended. On the other hand, the grid size should be small enough in order not to smooth out small-scale heterogeneity. A minimum of 5 nodes per correlation length is suggested (Tompkins et al., 1994). The vertical grid size, Δx_2, was determined small enough to avoid oscillatory behaviour of the numerical solution when solving the solute flow equation.

TABLE 3. The scales of the simulation domains and the grid sizes

	Δx_1 (m)	Δx_2 (m)	width (m)	depth (m)
similar field and crust ring cases	0.15	0.01	14.85	1.99
Kopecky and large column cases	0.01	0.01	1.99	1.99

3.3 2-D SIMULATION OF WATER FLOW AND SOLUTE TRANSPORT

It is assumed that water flow in a rigid, variably saturated porous medium can be described by the Richards equation:

$$\frac{\partial \theta}{\partial t} = \frac{\partial}{\partial x_i}\left[K(h)\left(\frac{\partial h}{\partial x_i} + \delta_{i2}\right)\right] \tag{10}$$

where δ_{ij} is the Kronecker delta, and x_i (i =1, 2 for planar flow in a vertical cross-section) are the spatial co-ordinates with x_2 the vertical co-ordinate which is positive upward. The mass flow of an inert dissolved solute is assumed to be described by the 2-D convection-dispersion equation (CDE):

$$\theta\frac{\partial C}{\partial t} = \frac{\partial}{\partial x_i}\left[\theta D_{ij}\frac{\partial C}{\partial x_j}\right] - J_{wi}\frac{\partial C}{\partial x_i} \tag{11}$$

where C is the concentration of the solute in the water phase, J_{wi} is the water flux in direction i and D_{ij} is the ij entry of the dispersion coefficient tensor $\boldsymbol{D}$. Neglecting molecular diffusion, D_{ij} is given by (Bear, 1972):

$$\theta D_{ij} = \lambda_T |J_w| \delta_{ij} + (\lambda_L - \lambda_T)\frac{J_{wi}J_{wj}}{|J_w|} \tag{12}$$

where $|J_w|$ is the magnitude of the water flux vector J_w, and λ_L and λ_T are the longitudinal and transverse hydrodynamic dispersivities respectively. For all cases, λ_L and λ_T were assumed to be 1cm and 0.01 cm, respectively.

Eq. (10) and (11) were solved numerically by the SWMS_2D code (Šimůnek et al., 1994) using the Galerkin finite element method. For the unsaturated flow simulations, a uniform water flux of 2.8 cm d^{-1} at the inlet, a free drainage bottom boundary condition and no flow lateral boundary conditions were implemented. For the saturated flow simulations, constant heads were defined at the top and bottom boundaries. The initial conditions, h(x_1,x_2) at t = 0 were determined from the relation between the random processes of the hydraulic properties and pressure heads in the frequency domain (Yeh et al., 1985). After back transformation to the spatial domain, an approximation of the steady state pressure head distribution is obtained. Before a solute pulse was applied to the soil surface, it was checked to verify that the field was under steady state condition. In general, it took approximately 40 days of continuous water application to reach this condition. To solve the solute flow equation, an initial solute free soil profile, a solute flux top boundary, no flux lateral boundary, and a no gradient lower boundary conditions were imposed. When Eq. (11) is solved, volume averaged or resident concentrations, C^r, are obtained. For some practical applications, solute fluxes and flux-averaged concentrations, C^f, are more relevant. For multi dimensional transport, Sposito and Barry (1987) defined C^f as the ratio of the projection of the solute mass flux in the direction of the water flux to the magnitude of the water flux. However, since (i) the direction of water flux in soils is difficult to measure and (ii) the solute flux across a plane perpendicular to the vertical component of water flow is of interest or is measured, we define C^f as the ratio of the vertical solute flux to the vertical water flux:

$$C^f = C^r - \frac{\theta}{J_{w2}}\left[D_{i2}\frac{\partial C^r}{\partial x_i}\right] \tag{13}$$

with J_{w2}, the vertical component of the water flux.

3.4 CALCULATION OF FIELD AVERAGED FLUX CONCENTRATIONS AND FIELD-SCALE DISPERSIVITY, λ_{eff}.

Breakthrough curves, BTCs, of local concentrations were withdrawn at 10 selected depths in the profile (0.05, 0.1, 0.2, 0.3, 0.4, 0.5, 0.6, 0.7, 0.8, 0.9, and 1.0m). To calculate field-scale averaged flux concentrations, $\hat{C}^f$, local flux concentrations were weighted by the local vertical water flux:

$$\hat{C}^f = \frac{< C^f J_{w2}>}{<J_{w2}>} \tag{14}$$

For each BTC of $\hat{C}^f$, the field-scale dispersivity, λ_{eff}, was derived. The solution of the

1-D convection-dispersion equation (CDE) for a first type inlet boundary condition which describes flux concentrations in a homogeneous 'effective' soil column (Parker and van Genuchten, 1984) was fitted to the BTC using non-linear least squares optimisation:

$$\frac{\partial \hat{C}^f}{\partial t} = v_{eff}\,\lambda_{eff}\frac{\partial^2 \hat{C}^f}{\partial x_2^2} - v_{eff}\frac{\partial \hat{C}^f}{\partial x_2} \tag{15}$$

with v_{eff} the field-scale pore water velocity.

4. Results and Discussion

4.1 VARIABILITY OF WATER TRANSPORT PARAMETERS

For the crust ring and Kopecky ring cases, the generated random fields and the vertical water fluxes for the unsaturated flow simulations are shown in Fig. 1. Water flow seems to be channelled in "stream tubes" which are formed close to the inlet surface. The vertical water flow is much more heterogeneous for the Kopecky ring case. For the crust ring case, the stream tubes are predominantly vertical and continuous along the simulation domain indicating that water flow for this case is mainly vertical. The tortuous nature of the stream tubes for the Kopecky ring case indicates considerable horizontal water flow. This illustrates the effect of the horizontal correlation scale (1.5 m for the crust ring case and 0.1 m for the Kopecky ring case) on the horizontal water flow. When the horizontal correlation scale is large as compared to the range over which capillary forces act in the soil, water flow is predominantly one dimensional (Yeh et al., 1985). In Table 4, mean (μ), variance (σ^2) and coefficient of variation (*CV*) of the water transport parameters are shown. The variability of both $|J_{w1}|$ (absolute value of horizontal flow) and J_{w2} increase (i) with increasing variability of the hydraulic conductivity, and (ii) from unsaturated to saturated conditions. The larger horizontal water flow, $|J_{w1}|$, for the Kopecky ring, large column, and saturated flow cases reduces the variability of local pressure heads, h, as compared to the crust ring case although the variability of the hydraulic conductivity for the latter case is smaller than that for the former cases. The higher mean pressure head for the large column case is explained by the different $K^*(h^*)$ curve, due to a different ℓ parameter.

TABLE 4. Variation of the water transport parameters

	$\|J_{w1}\|$ (cm d^{-1})			J_{w2} (cm d^{-1})			θ			h (cm)		
	μ	σ^2	*CV%*	μ	σ^2	*CV%*	μ	σ^2	*CV%*	μ	σ^2	*CV%*
similar field case	0.14	0.015	90	2.8	0.06	9	0.375	0.00023	4	-35	51	20
crust ring case	0.63	0.9	151	2.8	1.06	37	0.375	0.00035	5	-41	1057	79
Kopecky ring case	1.25	3.46	149	2.8	12.38	125	0.375	0.00032	5	-41	485	57
large column case	1.03	2.03	138	2.8	8.23	102	0.387	0.00007	2	-20	93	48
saturated flow case (Kopecky rings)	15.6	645	162	37.9	2889	142	-	-	-	0	397	-

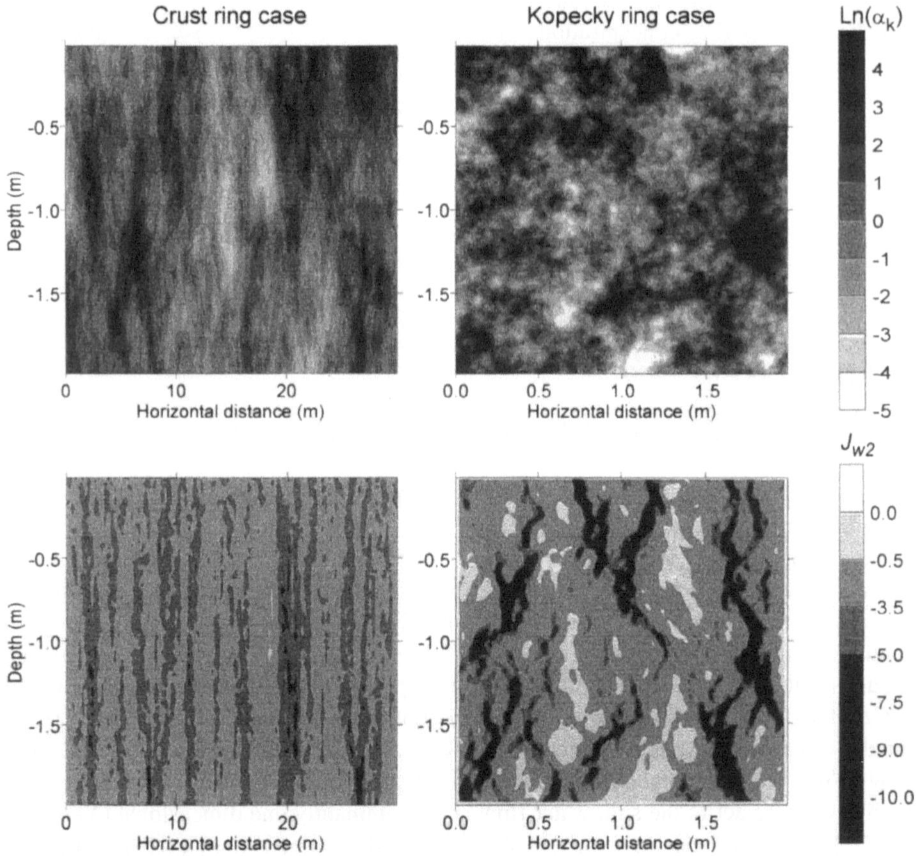

Figure 1. **Generated random fields of the log-transformed scaling factor ln (α_k) and simulated vertical water flow, J_{w2} (cm d^{-1}), for unsaturated flow conditions.**

4.2 CONCENTRATION PROFILES AND BTCS OF $\hat{C}^{f}$

In Fig. 2, concentration profiles for unsaturated flow conditions 12.5 days after the application of the solute pulse are shown. For the similar field case, the solute concentration profile does not vary much along the transect indicating relatively homogeneous solute flow. For the crust ring case, the concentration profile varies considerably along the transect. Analogue to the water flow, solute flow is mainly vertical resulting in unimodal depth profiles of concentrations at nearly all horizontal locations along the transect. For the Kopecky ring and large column cases, the concentration profiles are very irregular and reflect both the variability and two-dimensionallity of the water flow for these two cases. Due to horizontal solute flow, bi- or multi-modal depth profiles are observed at several locations along a horizontal transect.

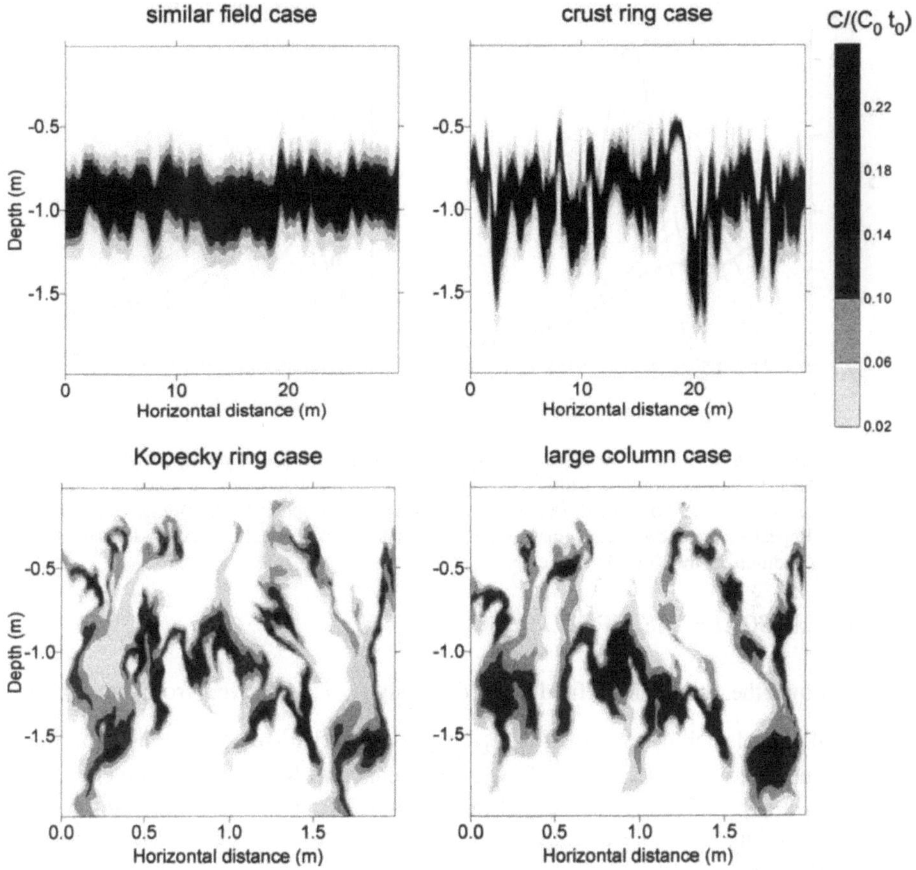

Figure 2. Simulated concentration profiles 12.5 days after application of a solute pulse (C_0 and t_0 are the pulse concentration and pulse duration, respectively)

In Fig. 3, BTCs of $\hat{C}^f$ calculated at 0.9-m below the input surface are plotted as a function of pore volumes, $T = q\, t / < \theta >$, with q the applied water flux (2.8 cm d^{-1}). Also shown in the figure is the standard deviation of the local flux concentrations. From Fig. 3, it is clear that (i) the dispersion and skewness of the BTCs, and (ii) the spatial variability of the concentrations increase with (i) increasing heterogeneity of the soil hydraulic properties and, (ii) from unsaturated to saturated flow conditions.

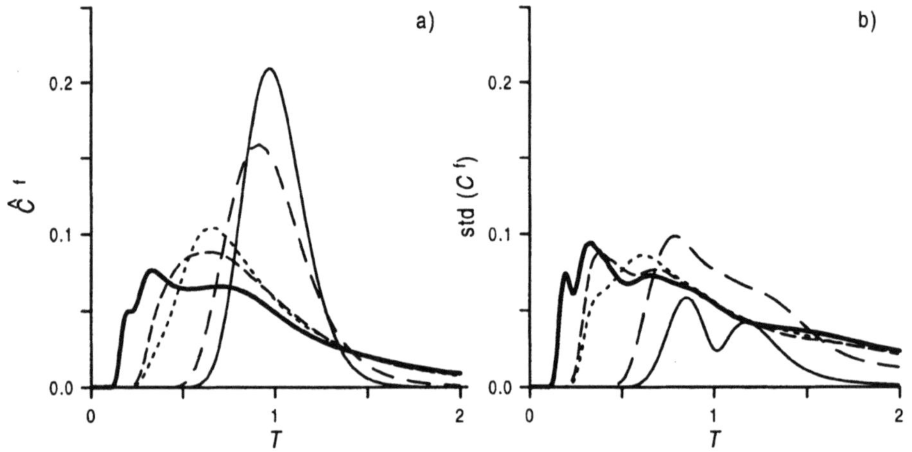

Figure 3. Simulated (a) BTCs of field-scale averaged flux concentrations, $\hat{C}^f$, and (b) standard deviation of local flux concentrations, std(C^f), at 90 cm below the input surface as a function of the applied amount of pore volumes, *T*. Decreasing dash lengths correspond to the similar field, crust ring, Kopecky ring, and large column cases in this order. Thick line corresponds to the saturated flow condition for the Kopecky ring case.

4.3 FIELD-SCALE DISPERSIVITY, λ_{eff}

Fig. 4 shows the "simulated" field-scale dispersivity, λ_{eff}, derived from time series of simulated $\hat{C}^f$. Also shown in Fig. 4 is λ_{eff} derived from field-scale leaching experiments. For both unsaturated and saturated flow experiments, the "measured" λ_{eff} increases with increasing depth indicating that lateral mixing of solute is not yet complete enough to allow each solute molecule to explore all of the different flow paths at the observation scale. Under this condition, the solute transport process is a stochastic-convective process and is characterised by a linear increase of λ_{eff} with depth (Simmons, 1982). Under unsaturated flow conditions, the following cases are discussed. For the similar field case, the simulated λ_{eff} remains fairly constant with increasing depth and underpredicts the measured λ_{eff} considerably. Since the simulated λ_{eff} is nearly identical to the local-scale dispersivity, λ_T (1 cm), the effect of heterogeneity of soil hydraulic properties on solute transport is minimal for this case. For the crust ring case, the simulated λ_{eff} increases with depth but the measured λ_{eff} is still underpredicted especially at greater depths. This increasing λ_{eff} with depth is explained by the vertical continuity of water flow channels shown in Fig. 1. However, the simulated variability of water flow is not large enough to predict the measured λ_{eff}. The Kopecky ring case results in an increasing λ_{eff} with depth but the rate of increasing diminished at greater depths. This is explained by the transverse mixing of fast and slow moving solutes due to tortuous and piecewise continuous local flow pathways. The simulated λ_{eff} for this case overpredicts the measured λ_{eff} especially at shallower depths indicating that water flow heterogeneity is overpredicted. Finally, the best agreement between measured and simulated λ_{eff} for unsaturated flow conditions is obtained for the large column case.

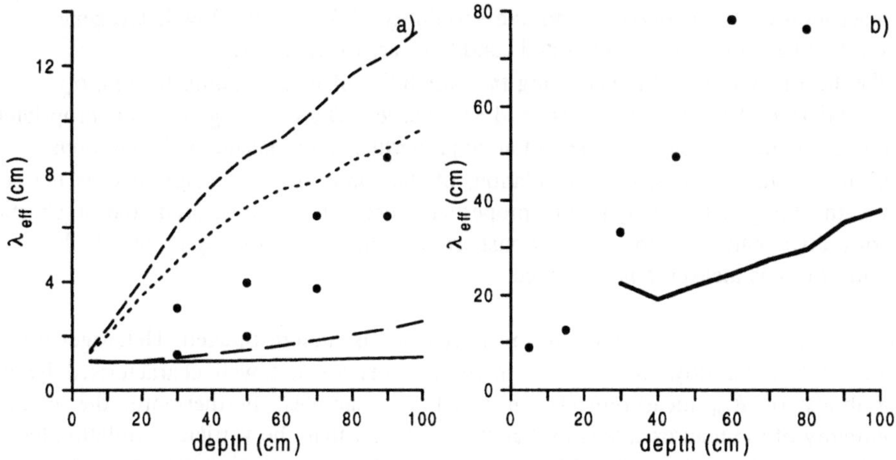

Figure 4. Measured (closed circles) and simulated (lines) field-scale dispersivities, λ_{eff}, for (a) unsaturated and (b) saturated flow conditions. Decreasing dash lengths correspond to the similar field, crust ring, Kopecky ring, and large column cases in this order.

However, at shallower depths, the simulated λ_{eff} still somewhat overpredicts the measured λ_{eff} indicating that water flow heterogeneity close to the surface boundary is not well represented.

For the saturated flow conditions, both measured and simulated λ_{eff} are larger than their counterpart under unsaturated conditions. The simulated λ_{eff} underpredicts the measured λ_{eff} considerably indicating that water flow heterogeneity was underestimated even if K_{sat} measurements on the Kopecky rings, exhibiting the largest variability, were used to determine σ^2 ($\ln\alpha_k$).

5. Conclusions

For the unsaturated water flow conditions, the following conclusions are drawn:

- Using the variability of the θ(h) curve to predict the variability of the *K*(h) curve based on the similar medium assumption leads to considerable underestimation of the solute flow heterogeneity and the field-scale dispersivity, λ_{eff}.
- Using the variability of K_{sat} determined on 20-cm long and 20-cm i.d. soil cores (crust rings) also results in an underestimation of λ_{eff} indicating that the essence of water and solute flow heterogeneity within these soil cores has not been characterised appropriately.
- Using the variability of K_{sat} determined on 5.1-cm long and 5-cm i.d. soil cores (Kopecky rings) results in an overestimation of λ_{eff}. This can be explained in part by soil macroporosity due to which the soil hydraulic conductivity is much more heterogeneous under saturated than under unsaturated conditions. As a result, when the variability of the soil hydraulic conductivity under unsaturated conditions in a

macroporous soil is derived from the variability of K_{sat}, water flow heterogeneity and solute dispersion under unsaturated conditions are overpredicted.

- The best results are obtained using the variability of linear scaling factors, α_K, which were derived from locally determined K(h) curves. These scaling factors appropriately reflect the heterogeneity of the hydraulic properties under unsaturated conditions.
- Using a small-scale spatial correlation, of the same order of magnitude as the soil volume on which the hydraulic properties were actually measured, the stochastic-convective nature of the solute transport over the range of depths at which solute transport was measured is preserved.

For saturated flow conditions, the measured λ_{eff} is underestimated. Heterogeneity of water and solute flow due to flow in macropores are not well characterised by the variability of K_{sat} measurements on small soil samples. Furthermore, the vertical continuity of the macropores is expected to be larger than the vertical correlation length used in this study. It is, however, questionable whether macropore structure can be described using second order stationary random fields. Modelling of water and solute flow in macropores may therefore require a different approach.

6. Acknowledgements

The authors would like to thank the Belgian National Fund for Scientific Research (N.F.W.O.). The first author is a research assistant of the Belgian National Fund for Scientific Research (N.F.W.O.).

7. References

Bear, J. (1972) *Dynamics of fluids in porous media*, 764 pp., Elsevier, New York.

Jacques, D., Kim D.-J., Diels. J., Vereecken, H., and Feyen, J. (1996a) Analysis of steady state chloride transport through two heterogeneous soils. *Water Resour. Res.* (Submitted)

Jacques,D., Vanderborght,J., Mallants,D., Mohanty,B.P., and Feyen J. (1996b) Analysis of solute redistribtution in heterogeneous soil: I Geostatistical approach. *First European Conference on Geostatistics for Environmental Applications*, 20-22 November, Lisbon, this issue.

Gutjahr, A. L., Hatch, S., and Bullard, B. (1995). *Random field generation, Conditional simulation, and flow modelling using the fast fourier transform.* N.M. Inst. of Min. and Technol. , Socorro, NM, USA.

Mallants,D., Mohanty, B.P., Jacques,D., and Feyen, J. (1996d) Spatial variability of hydraulic properties in a multi-layered soil profile. *Soil Science*, **161**, 167-181.

Mallants, D., Mohanty, B.P.,Vervoort, A., and Feyen, J. (1996a) Spatial analysis of the saturated conductivity of a soil with macropores. *Soil Tech.* (in press)

Mallants, D., Jacques, D., Tseng, P.-H., van Genuchten, M.Th., and Feyen, J. (1996b) Comparison of hydraulic property measurement methods using three sizes of soil cores. *J. Hydrol.*(accepted)

Mallants, D., Toride, N., Vanclooster, M., van Genuchten, M.Th., and Feyen, J. (1996c) Steady-state solute transport through large undisturbed soil columns. *Soil Sci. Soc. Am. J.* (submitted)

Mohanty, B.P., Kanwar, R.S., and Everts C.J. (1994) Comparison of saturated conductivity measurement methods for a glacial till soil. *Soil Sci. Soc. Am. J.*, **58**, 672-677.

Miller, E.E., and Miller, R.D. (1956) Physical theory for capillary flow phenomena. *J. Appl. Phys.*, **27**, 324-332.

Mualem, Y. (1976) A new model for predicting the hydraulic conductivity of unsaturated porous media. *Water Resour. Res.*, **12**, 513-522.

Parker, J.C., and M. Th. van Genuchten (1984) Flux-averaged and volume-averaged concentrations in continuum approaches to solute transport. Water Resour. Res., **20**, 866-872.

Simmons, C.S. (1982) A stochastic-convective transport representation of dispersion in one-dimensional porous media systems. *Water Resour. Res.*, **18**, 1193-1214.

Šimůnek, J., Vogel T., and van Genuchten, M.Th. (1994) *The SWMS_2D code for simulating water flow and solute transport in two-dimensional variably saturated media.* Research Report No. 132, U.S. Salinity Lab., ARS, USDA, Riverside, CA, USA.

Sposito, G., and Barry, D.A. (1987) On the Dagan model of solute transport in groundwater: foundational aspects. *Water Resour. Res.*, **23**, 1867-1875.

Tompkins, J.A., Gan, K.C., Wheater, H.S., and F. Hirano (1994) Prediction of solute dispersion in heterogeneous porous media: effects of ergodicity and hydraulic conductivity discretisation. *J. Hydrol.*, **159**, 105-123.

Tseng, P.-H., and Jury, W. (1994) Comparison of transfer function and deterministic modeling of area-averaged solute transport in a heterogeneous field. *Water Resour. Res.*, **30**, 2051-2063.

van Genuchten, M.Th. (1980) A closed form equation for predicting the hydraulic conductivity of unsaturated soils. *Soil Sci. Soc. Am. J.*, **44**, 892-898.

Yeh, T.-C. J., Gelhar,L.W., and Gutjahr, A.L. (1985) Stochastic analysis of unsaturated flow in heterogeneous soils. 1. Statistically isotropic media. *Water Resour. Res.*, **21**, 447-456.

SIMULATING LOCATIONAL ERROR IN FIELD-BASED MEASUREMENTS OF REFLECTANCE

P.M. ATKINSON
Department of Geography
University of Southampton
Highfield
Southampton
SO17 1BJ, UK

Tel: [01703] 594617
FAX: [01703] 593295
E.mail: pma@soton.ac.uk

ABSTRACT

Measurement error may be a large component of the total variation encountered in environmental variables. Therefore, for environmental analyses measurement error in sample data should be acknowledged and quantified. Further, it is important to understand *how* the variogram is affected by different kinds of measurement error. In this paper, a survey of field-based reflectance was undertaken in such a way as to allow the estimation of both the sample variogram $\gamma_Z(\mathbf{h})$ and the component of the variogram due to measurement error $\gamma_e(\mathbf{h})$. The underlying variogram $\gamma_u(\mathbf{h})$ of true values $u(\mathbf{x})$ was represented with a simple function. Then sequential Gaussian simulation was undertaken to introduce, via a locational error ($\delta\mathbf{y}$), a measurement error $e_s(\mathbf{x})$ into the underlying true values $u_s(\mathbf{x})$ to give a set of 'observed' values $z_s(\mathbf{x})$. The results demonstrated the effects of locational error on the sample variogram $\gamma_{zs}(\mathbf{h})$, and in particular a negative cross-correlation between the underlying values and measurement error that results from a locational error. In addition, it is shown that the component of the sample variogram of reflectance due to measurement error $\gamma_e(\mathbf{h})$ could not be explained solely by a locational error $\delta\mathbf{y}$.

A. Soares et al. (eds.), geoENV I–Geostatistics for Environmental Applications, 297–308.

1. Introduction

Geostatistics has been developed primarily in the mining industry (David, 1977; Journel and Huijbregts, 1978; Isaaks and Srivastava, 1989) and more recently in soil survey (Webster and Oliver, 1990). The measurement errors in the variables encountered in such disciplines may be a relatively small component of the overall variation and in many instances the errors are so small as to be *treated* as negligible. Consequently, measurement errors have received relatively little attention in geostatistics.

Geostatistics is now being applied in a wide variety of environmental disciplines including ecology (Rossi *et al.*, 1992), pollution monitoring and remote sensing (for example, Dungan *et al.*, 1994, Rossi *et al.*, 1994). The errors encountered in certain environmental data may be a large component of the overall variation. This is important because measurement error can have a large effect on the results produced from the analysis of such data and their interpretation. Consequently, it is important that where measurement errors are expected to be a large component of the overall variation (i) their presence is acknowledged, (ii) the sources of the errors are identified, and (iii) the magnitudes of the errors are quantified. The need for such information on errors in environmental data has been recognised for some time (for example, Goodchild and Gopal 1989; Davis *et al.*, 1991, Lunetta *et al.*, 1991).

For geostatistics, measurement error affects kriging and conditional simulation through the actual values of the sample data, but it also may affect the sample variogram (the term 'sample' is used in place of the term 'experimental' since there is nothing experimental about the variogram) and, therefore, subsequent geostatistical analyses carried out using that sample variogram. Therefore, for geostatistical analyses it is important to (i) understand the way that measurement error contributes to the sample variogram (particularly where different sources of measurement error may affect the variogram differently) and (ii) represent measurement error in the sample variogram explicitly with the fitted model.

This paper deals with measurement error using an example involving *in situ* measurements of reflectance made with a field radiometer (Milton 1987), and discusses how measurement error contributes to the sample variogram.

2. Theory

Readers will be familiar with the basic theory and so only a brief summary is given here.

2.1 variogram

The semivariance is defined as half the expected squared difference between the random functions, RFs, $Z(\mathbf{x})$ and $Z(\mathbf{x+h})$, at a particular vector distance and direction of separation or lag, $\mathbf{h}$. The variogram, $\gamma(\mathbf{h})$ (defined as a parameter of the RF model), is then the function that relates the semivariance to lag:

$$\gamma(\mathbf{h}) = 1/2 \; E[\{Z(\mathbf{x}) - Z(\mathbf{x+h})\}^2]. \tag{1}$$

2.2 Geostatistical simulation

Kriging estimates optimally by drawing the mean value from the cumulative distribution function (CDF) of the estimate conditional upon the variogram, and the values and configuration of the n local data surrounding the estimate. Kriging also estimates the estimation variance as a by-product of the kriging process. Therefore, kriging identifies the mean and variance of the conditional CDF (CCDF) of the estimate. If the CDF of the original data is multivariate Gaussian normal then kriging identifies the full CCDF of the estimate. This principle is used in sequential Gaussian (SG) simulation to provide a CCDF for every node to be simulated. Random values drawn from each CCDF will be such as to ensure that the spatial dependence in the original variable (represented by the variogram) will be reconstituted in the simulated data.

Given a set of sample observations, $z_v(\mathbf{x}_i)$, i=1,2,...,n, the simulated value is drawn at random from the CCDF of the estimate, conditioned by the n data:

$$F(N)\,(z_1,...,z_N\;(n)) = \text{Prob.}\;\{Z_i < z_i,\; i=1,...,N\;(n)\}. \tag{2}$$

SG simulation, as the name implies, involves simulating each location one at a time, whilst updating the prior probabilities at each step to include the previously simulated value. That is, the simulated values are treated as data and used to condition subsequent simulations. Not all the data are necessary for each simulation since closer values screen the influence of more distant values. Nevertheless, to reproduce the variogram up to the range a of the function may require a large set of conditioning data and so it may be necessary to perform a two-stage simulation, the first on a coarse grid and the second on a fine mesh that fills the spaces between the nodes of the coarser grid (Deutsch and Journel, 1992).

A second technique for simulation, known as indicator simulation, provides freedom from the assumption of a Gaussian CCDF and is becoming more widely applied than SG simulation. However, SG simulation is adopted here because it is conceptually simpler and is sufficient for the present purpose.

3. Measurement error and the sample variogram

3.1 A MODEL OF MEASUREMENT ERROR

Measurements of properties such as reflectance are rarely made perfectly and, therefore, a component of the observed spatial variation is likely to be due to measurement error. The true values of the property of interest (that is, the measured variable with the measurement error removed) are referred to as the underlying variable.

The following model of a RF, $Z_v(\mathbf{x})$, is appropriate where the measurement errors are large:

$$z_v(\mathbf{x}) = u_v(\mathbf{x}) + e_v(\mathbf{x}), \tag{3}$$

where $u_v(\mathbf{x})$ is a realization of the RF, $U_v(\mathbf{x})$, the spatially dependent underlying variation, and $e_v(\mathbf{x})$ is a realization of the RF, $E_v(\mathbf{x})$, the measurement error. This model has appeared previously (for example, Yakowitz and Szidarovszky 1985, De Kwaadsteniet 1990; Deutsch and Journel, 1992), but it has not been applied widely.

Measurement error (defined in attribute space) may also arise from a locational error (defined in geometric space). In these circumstances it is common for the error to be modelled as follows:

$$z_v(\mathbf{x}) = u_v(\mathbf{x}+\delta\mathbf{y}), \tag{4}$$

where, $\delta\mathbf{y}$ is the locational error.

3.2 MEASUREMENT ERROR IN THE SAMPLE VARIOGRAM

The sample variogram may be estimated for supports of size $|v|$ by

$$\hat{\gamma}_v(\mathbf{h}) = 1/2P(\mathbf{h}) \sum_{(i,j)\in D(\mathbf{h})}^{P(\mathbf{h})} \{z_v(\mathbf{x}_i) - z_v(\mathbf{x}_j)\}^2, \tag{5}$$

where $D(\mathbf{h})$ is defined as $\{(i, j) \mid \mathbf{x}_i\text{-}\mathbf{x}_j\text{=}\mathbf{h}\}$, and $P(\mathbf{h})$ is the cardinality of $D(\mathbf{h})$. Where measurement error is a non-negligible component of the total variation in a sample, it may affect the sample variogram and should be accounted for.

3.3 TYPES OF MEASUREMENT ERROR

Atkinson *et al.*, (1996) suggest that it may be helpful, in the first instance, to distinguish between measurement error which is cross-correlated with the underlying variation and measurement error which is uncorrelated or orthogonal to the underlying variation, since they affect the variogram differently.

The locational errors reported by Chilès (1976) are an example of cross-correlated measurement error. Where the locational error is random, the observed values are essentially drawn from the underlying univariate CDF, and the CDF of the observed variable is expected to be the same as that of the underlying variable.

Locational error should result in an increase in semivariance at small lags only, with $\gamma_v(0^+)$ (where the superscript indicates some very small positive difference either in space or time) estimating measurement error in the sample variogram. The semivariance at lags larger than the range should not be affected since the underlying CDF is unaltered.

Measurement error which is uncorrelated with the underlying variation generally arises from some source external to the variation in the property of interest. The CDF of the observed variable is then the convolution of the CDFs of the underlying variable and of the measurement error. Uncorrelated measurement error results in an increase in semivariance at all lags except zero, and so the underlying variogram is simply shifted upwards.

3.4 THE RELATION BETWEEN UNDERLYING VARIATION AND MEASUREMENT ERROR

The variogram, $\gamma_{vZ}(\mathbf{h})$, of a property, Z, can be represented as the sum of, $\gamma_{vU}(\mathbf{h})$, the variogram of the underlying variation, $\gamma_{vE}(\mathbf{h})$, the variogram of the measurement error, and $\gamma_{vUE}(\mathbf{h})$ the cross-variogram between the two (Journel and Huijbregts 1978):

$$\gamma_{vZ}(\mathbf{h}) = \gamma_{vU}(\mathbf{h}) + \gamma_{vE}(\mathbf{h}) + 2\gamma_{vUE}(\mathbf{h}). \qquad (6)$$

It has already been noted (assuming that the variograms models are bounded or transitive) that for locational error the sills of $\gamma_{vZ}(\mathbf{h})$ and of $\gamma_{vU}(\mathbf{h})$ should be equal (that is, adding locational error to the underlying variogram does not result in a change to the sill). Where $h >> a$ and the sills of $\gamma_{vZ}(\mathbf{h})$ and $\gamma_{vU}(\mathbf{h})$ are equal, the relation of the cross-variogram with the variogram of the error is given by:

$$\gamma_{vE}(\mathbf{h}) = -2\gamma_{vUE}(\mathbf{h}), \qquad (\text{for } h >> a) \qquad (7)$$

Eq. (7) means that measurement error due to a locational error is negatively cross-correlated with the underlying variation, and that the cross-variogram $\gamma_{vUE}(\mathbf{h})$ has a negative sill with a magnitude equal to one half of the sill of the variogram of the measurement error $\gamma_{vE}(\mathbf{h})$.

For uncorrelated measurement error Eq. (6) implies that $\gamma_{vU}(\mathbf{h})$ and $\gamma_{vE}(\mathbf{h})$ are orthogonal and, therefore, that $\gamma_{vUE}(\mathbf{h})$ is zero.

4. Example

In the following example $\gamma_v(0^+)$ is estimated directly by measuring repeatedly the reflectance of a pasture with a field radiometer.

4.1 FIELD SITE AND SAMPLE DATA

A field site was chosen at the Wenallt situated on the northern outskirts of Cardiff, UK. The area is some 200 m above sea level, approximately 8 km north of the coast, and is composed of a well drained soil underlain by old red sandstone. The terrain is generally undulating with slopes of up to 30°. A field of pasture consisting primarily of rye grass (*Lolium perenne* and *Lolium multiflorum*) was chosen for analysis.

The reflectance of the field of pasture was measured using a field radiometer. Since the measurements are not destructive it was possible to measure at the same location, **x**, over the same support, v, more than once. The reflectance of the grass canopy was measured from 12:00 to 14:00 hours GMT on 16 February 1992 using a Milton multiband radiometer (MMR) (Milton 1980). The MMR was positioned at a height of approximately 2 m above the ground giving a support with a diameter of approximately 0.5 m.

One hundred measurements of radiance in wavebands in both the red (0.6 μm-0.7 μm) and near-infrared (0.8 μm-1.1 μm) wavelengths were made on approximately level ground in one direction along a 25 m transect with the observations spaced 0.25 m apart. The survey was then repeated in the reverse direction giving one hundred pairs of observations in both wavebands at a lag of just greater than zero. A Kodak grey card of known reflectance in the two wavebands was measured simultaneously using a second sensor enabling the estimation of reflectance using:

$$\text{Ref(target)} = \quad \text{Ref(card)} \,.\, \text{Rad(target)} \,/\, \text{Rad(card)}, \tag{8}$$

where, Ref is reflectance and Rad is radiance. The normalized difference vegetation index, NDVI, was computed from the reflectance data using:

$$\text{NDVI} = (\text{NIR-Red})/(\text{NIR+Red}), \tag{9}$$

where NIR is near-infrared reflectance and Red is red reflectance

4.2 SAMPLE VARIOGRAMS

The sample variogram of the NDVI was computed using Eq. (4) and all two hundred measurements (Fig. 1). The quantity $\hat{\gamma}_v(0^+)$ was computed as 30.5 x 10^{-6} units^2 (where the units are of NDVI) using the one hundred co-located pairs of observations of NDVI. This estimate represents variation in the NDVI due to measurement error only. It does not represent spatial variation since each observation in a given pair is supposed to represent the same place.

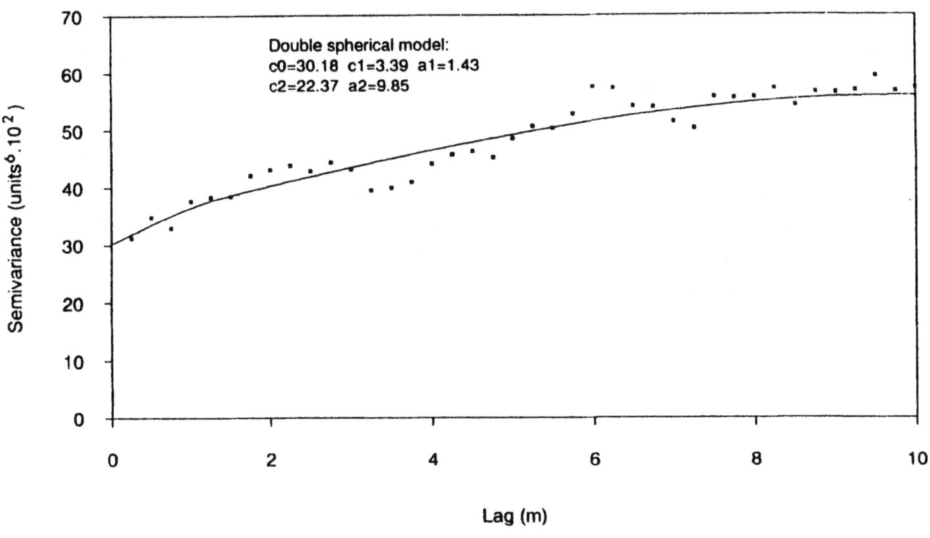

Figure 1. Sample variogram of NDVI fitted with a double spherical model

Several models were fitted to the sample variogram of NDVI (including $\hat{\gamma}_v(0^+)$) up to a lag of 43 observations (each observation being 0.5 m in diameter and spaced 0.25 m from its neighbours) by weighted least squares approximation. Of the models authorized in two dimensions the spherical model provided the best fit. However, the double spherical model provided a better fit at the smallest lags (Fig. 1). The nugget variance of the double spherical model, 30.18 x 10^{-6} units^2, was close to $\hat{\gamma}_v(0^+)$ (30.5 x 10^{-6}

units^2).

4.3 SOURCES OF MEASUREMENT ERROR IN FIELD-BASED REFLECTANCE DATA

There are many possible sources of measurement error in field-based reflectance data, each of which may have a different effect on the variogram.

In terms of field-based reflectance, locational error amounts to pointing the sensor at the wrong place, and in practice this error may be large because the operator must judge from a distance where the sensor is pointing.

Examples of uncorrelated measurement error in reflectance data are sensor noise, human error in recording the values, the limit to the precision of the measuring device, and subsequent errors in processing the data, for example, on a computer. This component of measurement error is likely to be very small for the reflectance of a grass canopy measured with a field radiometer.

There are many other possible sources of error. For example, the support might be larger or smaller than the defined support, and moreover the geometry of the support might be affected by the angle-of-view of the sensor. An important source of error is variation in the property of interest over time (Atkinson *et al.*, 1996). For example, as one measures reflectance over a period of say one hour the soil moisture and moisture in the vegetation may change, wind may induce large changes in reflectance, the solar geometry may change and so on.

5. Simulating locational error

From section 5.3, one may hypothesize that a large proportion of the measurement error in NDVI is due to a locational error. Therefore, the objectives in this section are to show how measurement error (due to a locational error) contributes to the sample variogram (and, in particular, that it is negatively cross-correlated with the underlying variation), and to use this information to infer whether the error in Fig. 1 may be attributable solely to locational error.

5.1 SIMULATING LOCATIONAL ERROR

In the first instance, a simple variogram model was chosen to represent the underlying variation. A spherical model with no nugget variance, a structured component of 10 units and a range of 10 units (defined as metres to aid interpretation) was specified.

This model was used in SG simulation (Deutsch and Journel, 1992) to produce a set of some 400 by 400 simulated nodes on a square grid with a spacing of 0.05 m (that is, 20 by 20 1 m^2 cells, each with 20 by 20 sub-cells). A two stage simulation was undertaken to ensure that the long range (as well as the short range) spatial dependence was accurately recreated. In the first stage, a grid of 20 by 20 nodes was simulated, and in the second stage, (treating the first set of simulated nodes as data) the gaps between the nodes were simulated to form a grid of 400 by 400 nodes covering 20 m by 20 m. The underlying variogram, $\gamma_u(\mathbf{h})$, was estimated using Eq. (5) with lag spacings of 1,2,3...,12 m from this data set (Fig. 2).

To add locational error, it was necessary to modify the realization drawn from the CCDF of each simulated point. This modification amounted to displacing the spatial position of the *i*th RV by an amount δ_i (where δ_i is drawn at random from a second CDF representing the spatial dimension of the locational error). In the present example, a Gaussian two-dimensional error was used with a mean of zero and a standard deviation of 1 m. The six steps involved in the simulation process are given in Table 1. In addition to that for the underlying variation, variograms were estimated for the observed variation (underlying variation plus measurement error) and for the measurement error, and also, the cross-variogram was estimated between the underlying variation and the measurement error (Fig. 2).

Table 1. Steps used in simulation

1.	Define an underlying variogram model $\gamma_u(\mathbf{h})$
2.	Simulate on a square grid with locations $\mathbf{x}_j$, for j= 1,2,...,m and grid spacing much less than the spacing between sample data
3.	Simulate Gaussian locational error ($\delta \mathbf{y}_i$) for i = 1,2,...,n locations
4.	Extract data sets for perfect locations ($\mathbf{x}_i$) to give $u_s(\mathbf{x}_i)$ and 'observed' locations ($\mathbf{x}_i + \delta \mathbf{y}_i$) to give $z_s(\mathbf{x}_i)$
5.	Subtract $z_s(\mathbf{x}_i)$ from $u_s(\mathbf{x}_i)$ to give the error $e_s(\mathbf{x}_i)$
6.	Estimate $\gamma_s(\mathbf{h})$ for underlying $u_s(\mathbf{x}_i)$, 'observed' $z_s(\mathbf{x}_i)$ and error $e_s(\mathbf{x}_i)$ variables

5.2 RESULTS

Fig. 2 shows that introducing locational error results in a change to the form of the variogram which is generally similar to that predicted from Eqs. (6) and (7). For example, the variogram increases at small lags, while it does not increase at all at large lags. Further, the measurement error appears to be spatially uncorrelated, while the cross-semivariance is negative and decreases with lag to a negative sill which is equal in magnitude to about one half of the sill of the variogram of measurement error. Finally, the range of the variogram of the observed variation is greater than that of the variogram of the underlying variation by an amount that is determined directly by the locational error δ_i.

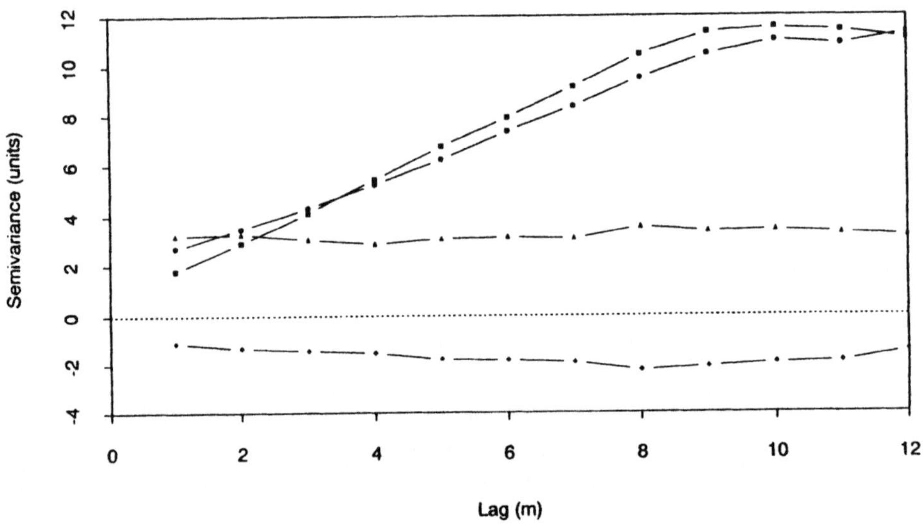

Figure 2. Variograms estimated from simulated data for underlying variable $u(\mathbf{x}_i)$ (squares), 'observed' variable $z(\mathbf{x}_i)$ (circles) and measurement error $e(\mathbf{x}_i)$ (triangles) and the cross-variogram between $u(\mathbf{x}_i)$ and $e(\mathbf{x}_i)$ (diamonds)

However, there is one important feature which was not predicted, and it relates to the semivariance at a lag of $h = 0^+$. It was expected that $\gamma_v(0^+)$ would be determined by and equal to the semivariance in the variogram of the error. However, it is clear from

Fig. 2 that this is not the case. The semivariance for the 'observed' variogram at a lag of 1 m is less than the equivalent value for the variogram of the error. It appears that the intercept at the ordinate of the variogram of the observed variation will be even less than this.

Importantly, to achieve the variogram of measurement error $\gamma_e(\mathbf{h})$ in Fig. 2, it was necessary to make the standard deviation of the locational error $\delta\mathbf{y}$ equal to 1 m. This value is much larger than might reasonably be expected for field-based measurements of reflectance (in the example, the support is 0.5 m and the spacing between observations along the transect is 0.25 m), and yet the variogram of the simulated error $\gamma_{es}(\mathbf{h})$ is a smaller proportion of the overall variance than is the variogram of the observed error $\gamma_e(\mathbf{h})$. It may be concluded that only a small proportion of the semivariance due to measurement error (and, in particular, $\hat{\gamma}_v(0^+)$) is due to a locational error. It is likely that much of the remaining proportion is due to errors created by differences in the timing of the observations, although this has yet to be established.

6. Conclusions

Simulation using the variogram represents a powerful tool for investigating the nature of spatial data. In this paper, a distinction is drawn between cross-correlated (and, in particular, locational) and uncorrelated measurement error. The NDVI was measured with a field-based radiometer, and the sample variogram of NDVI estimated. Sequential Gaussian simulation was then used to demonstrate how locational error contributes to the sample variogram. It was then possible to deduce that the semivariance due to measurement error in the variogram of field-based reflectance (and, in particular, $\hat{\gamma}_v(0^+)$) (Fig. 1) could not be due solely to a locational error. Further research is required to evaluate more fully the effects of locational error and other sources of error on the sample variogram.

Acknowldegement

The research for this paper was initiated and the field work was undertaken while the author held a Leverhulme Trust research grant at the University of Bristol.

References

Atkinson, P.M., R. Dunn, and A.R. Harrison, 1996, Measurement error in reflectance data and its implications for regularizing the variogram, *International Journal of Remote Sensing* (in press).

Chilès, J.P., 1976, How to adapt kriging to non-classical problems: three case studies. In *Advanced Geostatistics in the Mining Industry*, edited by M. Guarascio, M. David, and C.J. Huijbregts (Dordrecht: Reidel), p. 69-90.

David, M., 1977, *Geostatistical Ore Reserve Estimation* (Dordrecht: Elsevier).

Davis, F.W., Quottrochi, D.A., Ridd, M.K., Lam, N. S-M., Walsh, S.J., Michaelson, J.C., Franklin, J., Stow, D.A., Johannsen, C.J., and Johnston, C.A., 1991, Environmental analysis using integrated GIS and remotely sensed data: some research needs and priorities. *Photogrammetric Engineering and Remote Sensing*, **57**, 689-697.

Deutsch, C.V., and Journel, A.G., 1992, Geostatistical Software Library User's Guide (Oxford: Oxford University Press).

Dungan, J.L., Peterson, D.L., and Curran, P.J., 1994, Alternative approaches for mapping vegetation quantities using ground and image data. In *Environmental Information Management and Analysis: Ecosystem to Global Scales*, edited by W. Michener, S. Stafford and J. Brunt (London: Taylor and Francis), p. 237-261.

Goodchild, M., and Gopal, S., (Eds.), 1989, *Accuracy of Spatial Databases* (London: Taylor and Francis).

Isaaks, E.H., and Srivastava, R.M., 1989, *Applied Geostatistics* (Oxford: Oxford University Press).

Journel, A.G., and Huijbregts, C.J., 1978, *Mining Geostatistics* (London: Academic Press).

De Kwaadsteniet, J.W., 1990, On some fundamental weak spots of kriging technique and their consequences. *Journal of Hydrology*, **114**, 277-284.

Lunetta, R.S., Congalton, R.G., Fenstermaker, L.K., Jensen, J.R., McGwire, K.C., and Tinney, L.R., 1991, Remote sensing and geographic information system data integration: error sources and research issues. *Photogrammetric Engineering and Remote Sensing*, **57**, 677-687.

Milton, E.J., 1980, A portable multiband radiometer for ground data collection in remote sensing. *International Journal of Remote Sensing*, **1**, 153-165.

Milton, E.J., 1987, Principles of field spectroscopy. *International Journal of Remote Sensing*, **8**, 1807-1827.

Rossi, R.E., Mulla, D.J., Journel, A.G., and Franz, E.H., 1992, Geostatistical tools for modeling and interpreting ecological spatial dependence. *Ecological Monographs*, **62**, 277-314.

Rossi, R.E., Dungan, J.L., and Beck, L., 1994, Kriging in the shadows: geostatistical interpolation for remote sensing. *Remote Sensing of Environment*, **49**, 32-40.

Webster, R., and Oliver, M.A., 1990, *Statistical Methods for Soil and Land Resources Survey* (Oxford: Oxford University Press).

Yakowitz, S.J., and Szidarovszky, F., 1985, A comparison of kriging with nonparametric regression methods. *Journal of Multivariate Analysis*, **16**, 21-53.

APPLICATION OF GEOSTATISTICS TO 3D MODELLING OF CONTAMINATED SITES: A CASE STUDY

Michel Garcia and Roland Froidevaux
FSS International

Abstract: This paper presents, through an actual case study, a general approach for the application of geostatistics to the diagnosis of polluted sites. Diagnoses are preliminary work prior to deciding about remediation plans. The main objectives are to delineate the soil pollution for some concentration thresholds, to derive contaminated soil volumes, and to assess the uncertainty attached to them. Volumes can then be converted into treatment costs which provide support for decision making regarding remediation or complementary investigations.
For such a diagnosis task, geostatistics provide many tools that allow better integrating different types of information and give direct access to all kind of uncertainty measurements. In this study, the geostatistical approach relies on the following three-step procedure: 1) model the geometry of the site which here consists of infill materials delimited by a topographic surface and underlying alluvium; 2) fill in, through conditional simulation, the whole site domain with the different soil types which have been identified, each soil type being associated to a specific contamination degree (discrimination of concentration populations); 3) generate, from the simulated soil type images, probability maps that provide the local probability of exceeding given concentration thresholds.
In this case study, the available information consist of contaminant concentrations measured on soil samples, soil descriptions recorded on several drill-holes, and resistivity measurements along several profiles over the site. Resistivity measurements are not directly related to soil contamination. It is proven, however, that they are correlated to some specific soil types and allow detecting their presence or absence.

Keywords: geostatistics, soil contamination, probability map, conditional simulation, integrating soil resistivity, risk analysis

Introduction

The soil contamination observed in steelwork and coal processing sites often result from several sources of pollution over an activity history of several tens of years. Typical pollutants are metals (e.g. zinc, lead) and polycyclic aromatic hydrocarbon (PAH). The origin of the pollution may be linked to the industrial activity itself and

A. Soares et al. (eds.), geoENV I–Geostatistics for Environmental Applications, 309–325.

the sources be found at the location of former processing units. Very often, the pollution also comes former surface or underground storage tanks, or is attached to originally contaminated steelwork unit wastes used as backfill materials for construction work.

In such an industrial environment, soil contamination problems become complex non-stationary three-dimensional problems regarding large areas of very heterogeneous soils (i.e. a wide variety of backfill materials) contaminated by many pollutants from several, partly unknown, sources of contamination. The soil contamination diagnosis can then include different types of investigation surveys, from drill-holes aimed at identifying rocks or materials and providing samples for chemical analysis, to geophysical or more generally physical measurement surveys that allow providing, at a reasonable cost, rather high-resolution secondary information about soil contamination or materials.

At any step of the diagnosis process, data analyses are required to interpret the available data, to delineate polluted areas and derive volumes of contaminated soils, to assess the uncertainty about those volumes and locations, and more generally to help to decision making. The aim of this paper is to present, on an actual case study, how geostatistics provide tools to integrate all relevant data and thus reach such objectives.

The case study (Colin *et al.*, 1996) is from an industrial site in Lorraine (eastern France). Though several pollutants were identified, we shall focus on the PAH contamination. The available data include PAH concentrations measured on soil samples, soil (material) descriptions (logs) recorded on drill-holes, observed tar beds, geophysical (soil resistivity or conductivity) data and topographical information.

Ojectives and steps of the study

When geostatistics were called to help to modelling the soil contamination, several investigation surveys had been carried out and most of the data were already collected. After several months of investigation and unsuccessful contamination mapping, the industrial client required that the site characterization studies be completed and that soil contamination models be proposed so that remediation planning could begin.

As this often happens in actual case studies, the sampling strategy was designed by third party investigators who did not use geostatistics in their sampling design. Nevertheless, here were all the available data and it was the geostatistician's task and responsability to make use of them at best and to provide exploitable results for decision making. Specially, the aim was not to demonstrate that the sampling design was not appropriate, the number of samples insufficient, or the uncertainty on contaminated soil volumes very high (say a ratio of 20 between maximum and minimum contaminated soil volumes). Instead, the objective was to identify regions that could be classified with reasonable confidence as contaminated or clean, and to

recognize *transition zones*, i.e. unclassified zones, in between identified contaminated and clean regions, where the contamination limits are likely to exist. By looking at the volumes of transition zones, the associated treatment costs provide support for decision making regarding remediation or complementary investigation.
In this study, the geostatistical approach relied on the following three-step procedure: 1) model the geometry of the site which here consists of infill materials delimited by a topographic surface and underlying alluvium; 2) fill in, through conditional simulation, the whole site domain with the different soil types which were identified, each soil type being associated to a specific contamination degree (discrimination of concentration populations); 3) generate, on the simulated soil type images, probability maps that provide the local probability of exceeding given concentration thresholds.

The first step consisted in integrating the abundant elevation data to better model the complex site topography, which includes several hills and hollows, and to derive a reliable thickness map of backfill materials over the whole site. The second step allowed distinguishing contaminated and non-contaminated soil types so as to better evaluate volumes of contaminated soils. This was a helpful way of integrating available three-dimensional data, correlated to the soil contamination, in addition to sparse concentration data. In the third step, the probability maps were used to identify contaminated, clean and unclassified (transition) regions and to provide support for decision making.

Note that instead of estimating conditional probabilities, a more complete approach would have been to sequentially simulate the soil types and the PAH concentrations. This would have required, however, much additional human and computer effort — that may not be economically acceptable in such a soil contamination study where budgets are rather small (e.g. a total of 10 to 15 man-days is common) — and would have called for the analysis and modelling of the different sources of uncertainty. Specially, for each soil type, this would have required to characterize the uncertainty about the whole probability distribution function (histogram) of PAH concentrations, uncertainty which needs by itself running another kind of (sampling-related) stochastic simulations (see Englund and Heravi, 1992, or Journel and Xu, 1995).

Using this approach is possible to generate a series of equi-probable contamination scenarios (images) from which a much better assessment and quantification of different kinds of uncertainties, including the uncertainty about contaminated soil volumes for different critical thresholds, can be derived. Unfortunately, as this was mentioned above, when the contamination problem is very complex and the number of available data is rather small or the sampling layout not appropriate, the resulting uncertainty about contaminated soil volumes may become very high and its quantification does not help much to decision making (e.g. what decision regarding a maximum to minimum volume ratio of several tens on polluted soils).

Available data

Four types of data were available and used. They are briefly presented hereafter.

Structural data

The soil consists of a upper layer made of different types of backfill (non-geological) materials overlying geologically emplaced alluvial deposits. The backfill materials are comprised of industrial wastes, issued from ten's of years of former industrial activities (steelwork and coal processing), construction materials and natural lands. The total thickness of this upper layer ranges from 1 to 5 m over the site.

A number of 553 topographical data, which were collected for geophysical interpretation purposes, provided surface elevation values of the backfill material layer. The underlying alluvium was intersected by 11 drill-holes from which the alluvium top elevation was measured and the upper layer thickness derived. Not fully penetrating drill-holes (i.e. that did not reach the alluvial deposits) also provided minimum thickness values of the upper layer. Additional information came from the geologist who mentioned that the top surface of the alluvial deposits should be a rather flat, maybe slightly tilted, plane.

The diagnosis only concerned the backfill material layer, no significative contamination having been observed in the alluvium. The latter result was justified by the rather static pollution in the backfill materials and a within alluvion aquifer which is likely to transport incomig contaminants.

Soil type description information

Hereafter and throughout the paper, *soil types* will refer to backfill material types found in the upper (backfill material) soil layer.

Soil type descriptions (logs) were available on 169 drill-holes distributed over the site. Soil samples were initially classified in seven categories, and this classification was further refined by taking into account colours. The resolution of this classification proved to be too high for the purpose of the study. Indeed, some of the classes were under-represented (only observed in one or two holes), whereas others were associated to a too few number of concentration values to derive any concentration distribution. A simplified classification was obtained by gathering the soil types into three families hereafter denoted soil types 1 to 3. Two of the families (types 1 and 3) were proved not to be contaminated for PAH concentrations higher than a threshold of 200 ppm. The latter threshold will be the only critical threshold considered here to illustrate the proposed geostatistical approach. The relationship between soil types and PAH concentrations will be presented in a subsequent section.

PAH concentration data

The number of PAH concentrations measured on soil samples is 135, with up to five samples collected from a same hole at different depths. In most of the holes, however, no more than one sample was analysed.

In addition to concentration values, the locations of tar (in beds or in impregnation), observed in some of the holes, were reported. According to the geochemistry, tar could be associated to highly contaminated soils (PAH concentrations of several thousands ppm). In practice, this information about tar was interpreted as PAH concentrations higher, though unknown, than 200 ppm. Note that by so integrating tar information it was implicitely made the assumption that the same genetic (mass transport) phenoma applied to different forms of soil contaminations by PAH[1].

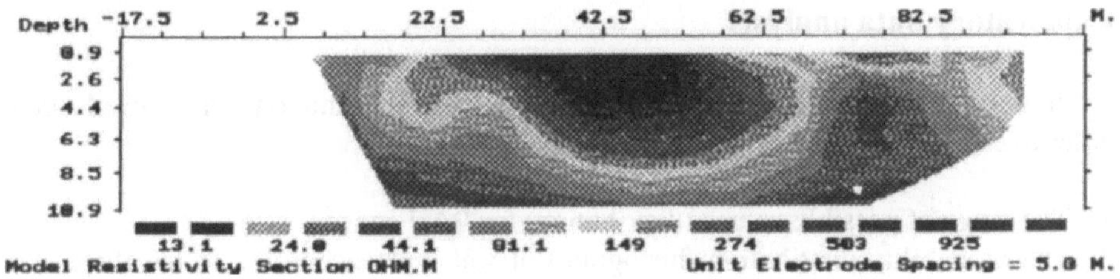

Figure 1. Resistivity profile (provided by Géostock)

Geophysical data

Two kinds of geophysical surveys were carried out on the site: electromagnetic surveys which provided soil conductivity measurements at the nodes of a 5 m × 5 m grid covering most of the site, and dipole-dipole electrical measurement surveys which allowed deriving soil resistivity profiles. A geostatistical approach, aimed at integrating conductivity information so as to directly map the contamination through a non-linear correlation between concentration and conductivity variables, was presented in Colin *et al.* (1996). This approach is applicable only if a meaningful relationship, albeit non-linear, can be established between the primary (concentration) and secondary (conductivity) variables. This condition, which would have required experimental measurements of soil material resistivities according to PAH concentrations, could not be satisfied and the approach was eventually not retained.

[1] Tars in bed (i.e. solid tar) were probably buried and may not contaminate surrounding soils. Conversely, tars in impregnation reflect the accumulation of liquid hydrocarbons above very low permeability materials that act as flow barriers. It can then be suspected that the overlying part of the soil, through which hydrocarbons probably flowed, is also highly contaminated. With the indicator formalism (Eq. 2) used to directly estimate the probability of exceeding a given PAH threshold, the integration of such an information is straightforward.

In the study here presented, only the 2200 m of resistivity profiles were integrated. A full inversion of dipole-dipole electrical measurements, that would allow deriving resistivity values versus depths, was not possible. Instead, the resistivities profiles (Figure 1) provided weighted averages of soil resistivity values over depth intervals. The resistivity is directly correlated to the soil nature. Though several different soil types (materials) may show same resistivities, the three-class description of soils allowed gathering within a same class (class 3) all high resistivity materials (resistivity greater than 200 ohm.m). Consequently, looking at maximum resistivity values along resistivity profile lines, it could be admitted that type 3 materials exist, at an unknown depth, where high resistivity values have been measured, and type 3 materials are absent over the whole backfill material thickness where the maximum resistivity is smaller than 200 ohm.m. As this will be presented later, such an information was used for conditional simulation of soil types.

Exploratory data analysis

An initial exploratory data analysis of the available information was carried out in order to establish the following data features.

1. The range of possible values (distribution) for PAH concentrations.
2. The probability distribution (histogram) of soil types according to the three-class description.
3. The relationships between soil types and PAH concentrations.
4. The range of resistivity values for each soil type (this was provided by the geophysicist).
5. The spatial continuity of PAH concentration and soil types.

Structural data about the surface topography and the top of alluvium were also analysed but will not be discussed in the present paper.

Univariate distributions

Building histograms (probability distributions) of the variables under study is, traditionally, the first step of any geostatistical work. Histograms allow detecting suspicious outlier values, checking on clustered data and identifying populations (multimodal distributions). They also provide information to derive a priori models for stochastic simulations.

PAH concentrations
Figure 2.a shows the experimental histogram and a statistics summary. As is often observed on contaminant concentration data, the distribution shows a strong positive skewness (the mean is 256 whereas the median is 25, and the variation coefficient is 3) with a few outlier values. Looking at the histogram with a logarithmic scale (Fig. 2.b), two populations can be distinguished. They partly overlap around a concentration

threshold of 100 ppm. Considering a critical PAH concentration threshold of 200 ppm to identify contaminated soils, the first population (on the left) would only represent non-contaminated soil samples. It will be shown that the three-class soil description allows discriminating such PAH populations.

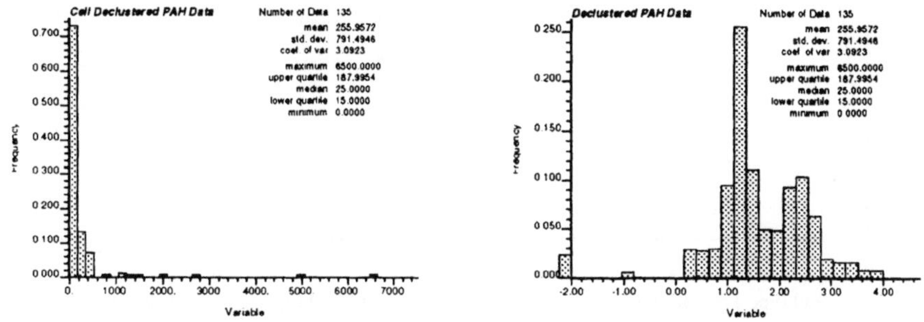

a) b)

Figure 2. Declustered histograms of a) PAH concentrations and b) logarithmic of PAH concentrations

Soil types

The experimental histogram and summary statistics of the three-class soil description are given in Figure 3. The corresponding soil type proportions (frequencies) are 15% for type 1 (clayey lands), 69% for type 2 (various steelwork wastes and construction materials) and 16% for type 3 (slag deposits). Those proportions determine the a priori histogram model that will be used for conditional simulation.

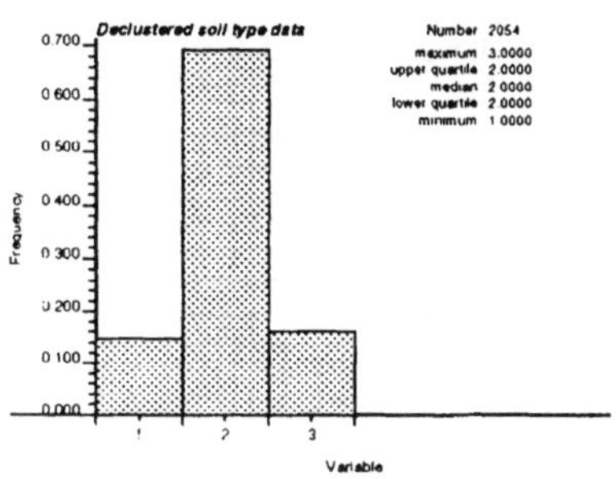

Figure 3. Histogram of soil types

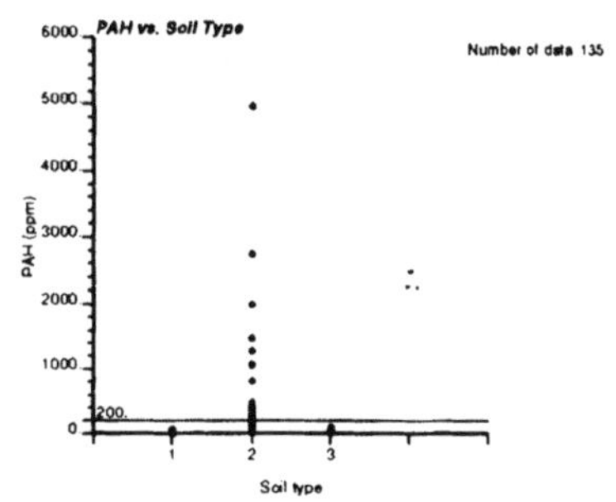

Figure 4. Relationships between PAH and soil types

Relationships between PAH concentrations and soil types

The next step of the data analysis consisted in checking whether the two modes observed on the PAH histogram could be associated to different soil types. Figure 4 exhibits a cross-plot of PAH concentrations versus soil types. It can be seen that all measured PAH concentrations on soil samples coming from the two soil types 1 and 3 are less than 100 ppm. This potentiality of not being contaminated was confirmed by the geologist. Indeed, on the one hand, the soil type 1 is made of recent (overlying) backfill lands that are posterior to the major (identified) contamination phenomena. On the other hand, the soil type 3 mostly includes very low permeability materials (slag deposits) that rather act as flow barriers.

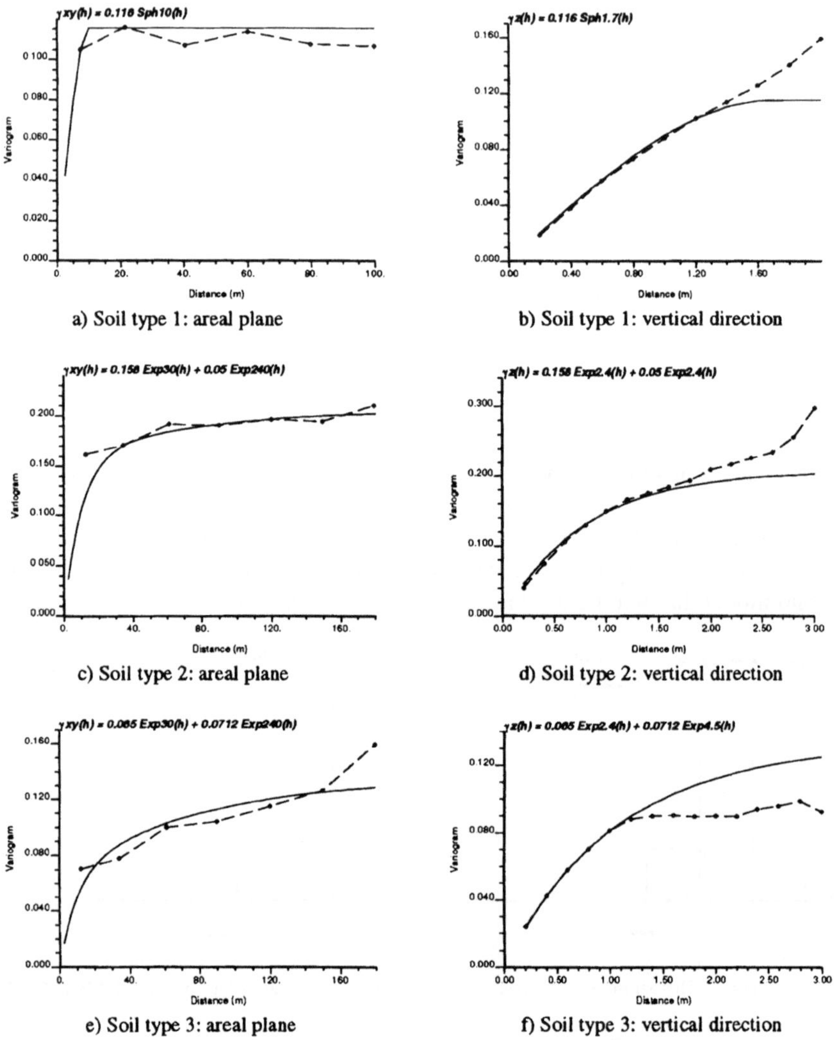

Figure 5. Experimental and theoretical variograms of soil types

Therefore, considering a 200 ppm value for the critical PAH concentration threshold, it can be admitted that the soil types 1 and 3 are clean. If there is no need anymore to assess the local probability that the concentration exceed 200 ppm (this probability is zero everywhere in soils of type 1 or 3), the spatial distribution of those two types of non-contaminated soils are required to correctly evaluate the volumes of contaminated backfill materials. Conditional simulations were carried out to generate soil type models. They are presented in a subsequent section.

Spatial continuity analysis

Variogram models are required to estimate or simulate variables at unsampled locations. For the three-step approach used in this study, several variogram models had to be inferred from the available data. Those models were for the following variables.

1. Domain topography (surface elevation) and residual of the top of alluvium (the residual was the difference between top elevation data and average elevation values from a least-square fitted plane).
2. Soil types (directional models were generated for each indicator variable).
3. PAH concentration and indicator variables for different thresholds including 200 ppm.

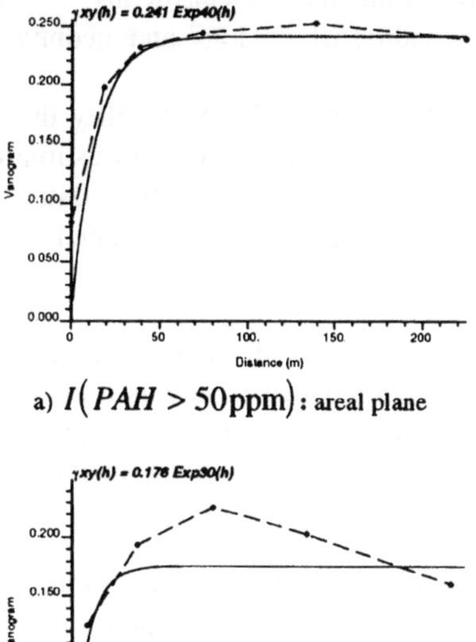

a) $I(PAH > 50\text{ppm})$: areal plane

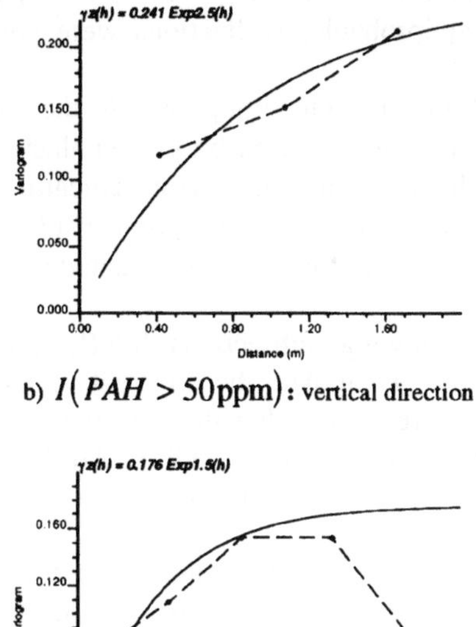

b) $I(PAH > 50\text{ppm})$: vertical direction

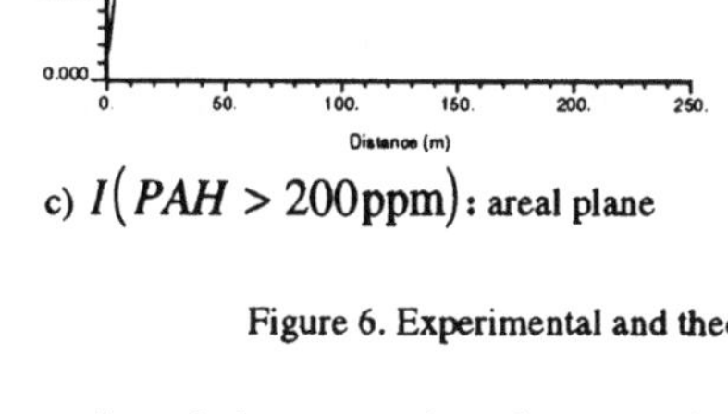

c) $I(PAH > 200\text{ppm})$: areal plane

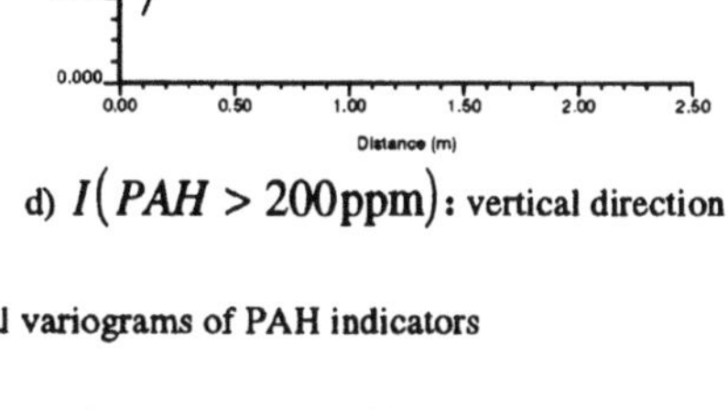

d) $I(PAH > 200\text{ppm})$: vertical direction

Figure 6. Experimental and theoretical variograms of PAH indicators

Figures 5 and 6 summarize the experimental and theoretical variograms for the soil types and the PAH variables[2]. For all the variables, there was no evidence of anisotropy in the horizontal plane. Regarding the PAH indicators, it can be noted the small (effective) ranges which do not exceed 40 m in the areal plane.

Three-step approach

A three-step approach was undertaken to successively model the backfill material layer (geometrical domain), the spatial distribution of soil types and the probability that the PAH concentration exceed 200 ppm. This approach can be summarized as follows.

1. Domain modelling. The domain, which just consisted of the upper (backfill material) layer, was modelled by estimating the site topography and the top of alluvial deposits from the available topographic and drill-hole data.
2. Soil type modelling. To generate realistic images of the spatial distribution of soil types (realistic here means images that reproduce a priori statistical and spatial variability models), the soil types were simulated stochastically. Several (equiprobable) realizations were generated conditional on logs and geophysical data.
3. Probability modelling. Within the backfill material layer, the probability that the PAH concentration be greater than 200 ppm was estimated. Such an estimation, which requires the prior indicator transform of PAH concentration data, was carried out in two ways, by considering all soil types equally contaminated and by accounting for non-contaminated soil types 1 and 3.

The following subsections briefly present the geostatistical methods applied to estimate or simulate the different variables. The geostatistical programs used in this work were from either the Geostatistical Toolbox (Froidevaux, 1990) or the GSLIB (Deutsch and Journel, 1992). All hereafter mentioned geostatistical methods are described in Deutsch and Journel, 1992.

Estimated or simulated values were computed either at the nodes of a two-dimensional (2D) regular grid, for structure modelling, or at the nodes of a three-dimensional (3D) regular grid for soil type and PAH modelling. The grid spacings were 5 m in the x and y-directions, and 0.2 m in the z-direction. The 3D grid was abutted on the top of alluvium and was comprised of 104×100×25 nodes, part of them being outside the domain's boundaries and thus inactive (not visited during estimation or simulation).

[2] It can be noted that the available PAH concentration data did not allow building reliable concentration variogram models for the different types of soils. Indeed, 112 data over 135 were attached to type 2 soils. Therefore, it was admitted that the inferred variogram models reflect the spatial continuity of PAH concentrations (or related indicator variables) within type 2 soils, and those variogram models were also attributed to the other soil types when required, i.e. when the soil type discrimination did not apply.

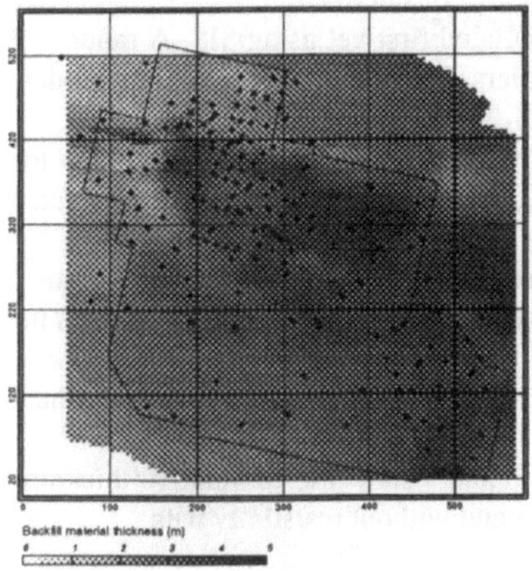

Figure 7. Estimated Backfill material thickness

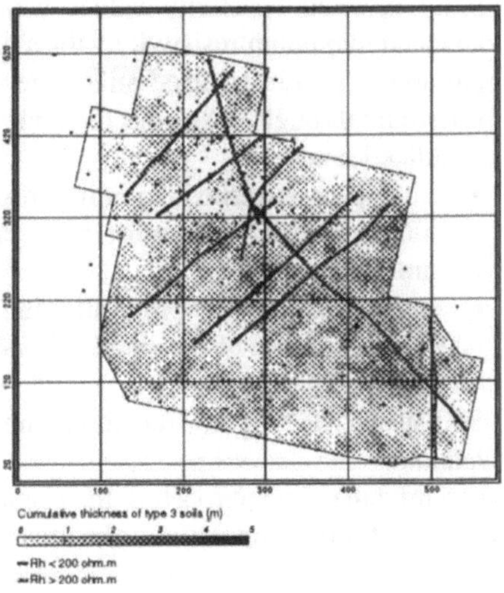

Figure 8. Conditional simulation of soils without integrating resistivity data

Domain modelling

The topography was estimated (interpolated) from the 553 elevation measurements through ordinary kriging with local neighbourhood (inference of a non-stationary random function model). Based on geological considerations, the a priori surface (average plane) of the top of alluvium was obtained by least-square fitting of the 11 top of alluvium elevation data. This a priori surface was then conditioned to the data through simple kriging with local mean. By subtracting everywhere surface and top of alluvium elevations, the total thickness of the backfill material layer was derived (Fig. 7). For consistency purposes, it was then checked that the depth of all drill-holes that do not intersect the alluvium was smaller than the local (estimated) backfill material thickness. The resulting total volume of backfill materials is 512,000 m^3.

Simulation of soil types: integrating resistivity information

The soil types were simulated by sequential indicator simulation (SIS). Two approaches were considered: conditioning the simulations to drill-hole data only and integrating resistivity profile information. In the latter case, the sequential simulation algorithm was adapted as follows to account for constraints on the existence or absence of type 3 soils according to maximum local resistivity values.

1. The 3D grid, which fills in the whole backfill material layer, can also be seen as a 2D (areal) regular grid of vertical *node-pillars*, each node-pillar being identified by two grid indices (i,j). By so doing, the resistivity profiles can be represented by a series of intersected node-pillars to which measured resistivity values are attached.

Fully inverted resistivity values being not available (i.e. resistivity vs. depth), attached maximum resistivity values only are considered.

2. All nodes from intersected pillars are simulated first yet using SIS. A random path is defined through all intersected node-pillars. For each pillar visited, the nodes are simulated randomly.
   - If the pillar-attached maximum resistivity value is smaller than 200 ohm.m: all simulated soil types must be 1 or 2, and those equal to 3 are rejected and simulated again.
   - Conversely, if the pillar-attached maximum resistivity is greater than 200 ohm.m: if none of the simulated soil types along the pillar is 3, all pillar's nodes are simulated again.
3. Still using SIS, all the remaining nodes are then simulated by visiting them at random.

To illustrate the two approaches, the Figures 8 and 9 show the cumulative thickness of type 3 soils for two simulations obtained with and without resistivity data.

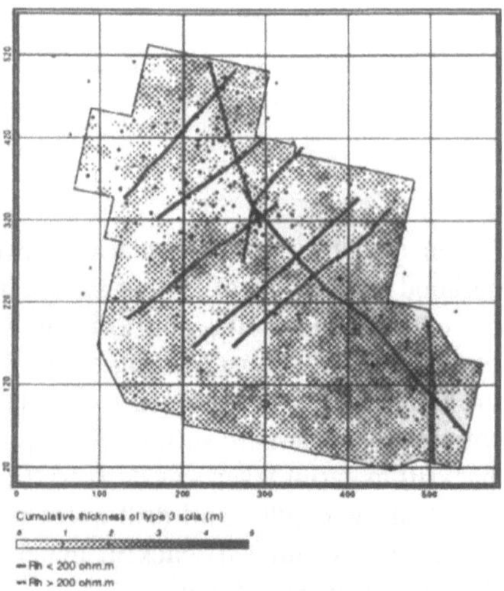

Figure 9. Conditional simulation of soils honoring resistivity data

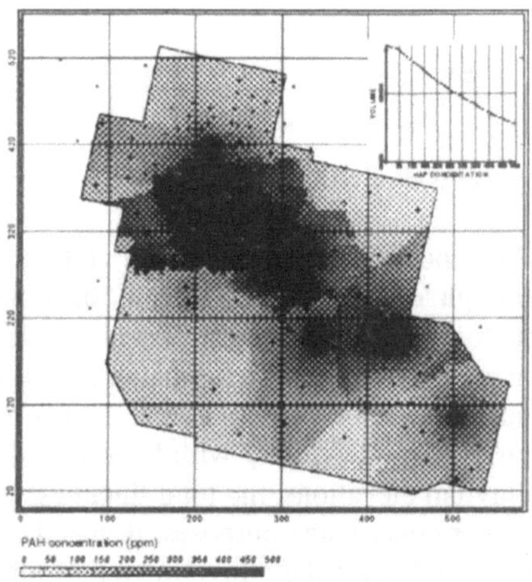

Figure 10. Estimated PAH concentrations

Probabilistic model of PAH concentration

As mentioned before, the objective of the study was to compute the probability of exceeding given concentration thresholds. Considering a concentration threshold of 200 ppm, this consists in estimating at any unsampled location **r** the probability,

$$Prob\{C(\mathbf{r}) > 200 \mid (n(\mathbf{r}))\}, \tag{1}$$

conditional to a set of $n(\mathbf{r})$ surrounding concentration data denoted $(n(\mathbf{r}))$, that the random concentration variable $C(\mathbf{r})$, measured in $\mathbf{r}$, be greater than 200 ppm. Using the indicator formalism, it can be demonstrated (Journel, 1984-a) that:

$$Prob\{C(\mathbf{r}) > 200 \,|(n(\mathbf{r}))\} = E\{I(\mathbf{r};200)|(n(\mathbf{r}))\} \quad (2)$$

where the right-hand side term is the conditional expectation, which can be estimated directly by kriging, of the random indicator variable $I(\mathbf{r};200)$ here defined as $I(\mathbf{r};200) = 1$ if $C(\mathbf{r}) > 200$ ppm (or tar are located in $\mathbf{r}$) and = 0 otherwise.

This approach was repeated for the following cases.

1. All soil types were treated together (no discrimination) by assuming that the pollution equally applies to all soils.
2. The probabilities were computed on type 2 soils only, soils of types 1 and 3 being assumed non-contaminated. Several soil type realizations were then considered so as to assess the uncertainty on soil volumes for different probability threshold (i.e. the uncertainty on the volume of soils where the probability is smaller than a given threshold).

The following sections present the results and the interpretation for decision making.

Probability (risk) map results and interpretation

Note on visualization of 3D grids

Looking at the probability values or more generally at estimated or simulated values as computed at the nodes of a 3D grid, several three-dimensional views or cross-sections are required generally to depict and understand how the phenomenon under study is spatially distributed. In some cases, synthesized 2D areal maps can be derived from 3D grids by only plotting one particular feature of value distributions in the vertical direction. Specially, it may be useful looking at maximum value maps that show, for each vertical grid-pillar, the maximum node value, or at thickness or proportion maps that give the cumulative thickness or the proportion of nodes for which the attached values are within a given class or interval (e.g. see Fig. 8-9).

Probability map results

Though the typical strong positive skewness of concentration distributions makes unsuitable direct estimation of pollutant concentrations (Journel, 1984-a, Isaaks and Srivastava, 1989), estimated (interpolated) concentration values often are computed to delineate soil contaminations. For the sake of comparison, estimation by ordinary kriging with local neighbourhood (accounting for non-stationarity) was carried out on

PAH concentrations (Fig. 10). The corresponding volume of contaminated soil (PAH concentration > 200 ppm) reaches 154,000 m^3.

The probability of exceeding 200 ppm was computed first without soil discrimination and second by just considering type 2 soils as potentially polluted. In the latter case, a series of previously simulated soil images was considered to better account for the uncertainty on the spatial distribution of soils. Figure 11 shows the maximum probability map without soil distinction and Figure 12 the maximum probability map estimated for one particular soil stochastic image. Table 1 gives a summary of soil volumes for different probability thresholds. Value ranges (i.e. mininum and maximum values) are provided for volumes computed on several simulated soil images.

TABLE 1. Soil volumes (m^3) for different probability thresholds of exceeding 200 ppm

Soil model	*Prob* > 0.2	*Prob* > 0.4	*Prob* > 0.6	*Prob* > 0.8
All soils together	236,300	61,590	12,025	765
Type 2 soils only	167,300 - 177,200	44,400 - 46,600	9,940 - 10,200	650 - 705

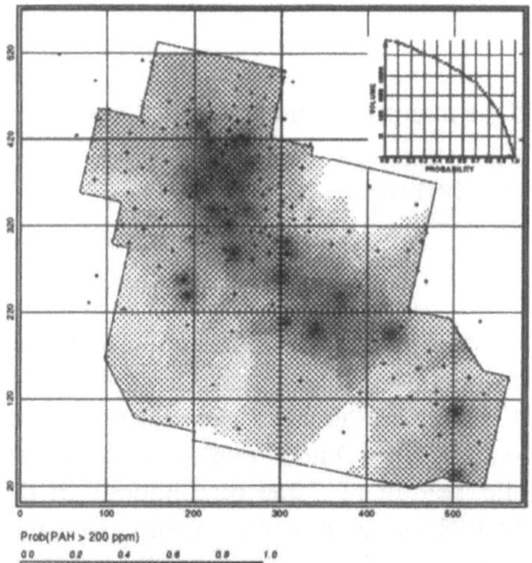

Figure 11. Maximum probability that PAH>200 ppm (all soils together)

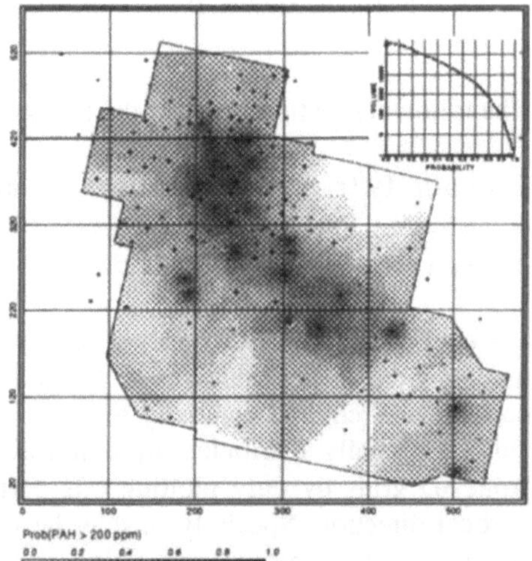

Figure 12. Maximum probability that PAH>200 ppm (Type 2 soils only)

The probability thresholds reflect the risk a project manager would take regarding the decision of what is clean and what is not. Comparing the volumes in Table 1 to the one obtained with direct estimation of PAH concentrations (Fig. 10), one will note that over the 154,000 m^3 of soils that estimation would identify as contaminated for a 200 ppm threshold, 12,025 m^3 (i.e. less than 8 %) would be more than 60 % chance contaminated, and only 765 m^3 (i.e. less than 0.5 %) would be more than 80 % chance contaminated. The comparison can also be done for non-contaminated soils. Whereas concentration estimation shows 358,000 m^3 of non-contaminated soils, the probability map approach reveals that 450,000 and 275,700 m^3 would be more than 40 % and 20 % chance, respectively, not to be contaminated. Considering only type 2 soils (bottom row in Table 1), the soil volumes become 8 to 30 % smaller according to the probability threshold. Those percentages are to be compared to the 31 % proportion of soils of types 1 and 3. As expected, the maximum probability maps (Fig. 11 and 12) reveal almost same low and high probability zones with and without discrimination on soil types. The following section proposes guidelines to interpret such probability results and use them for decision making.

Interpretation

The proposed approach aims at classifying regions as contaminated, non-contaminated or undefined (transition regions) according to the probability of exceeding a particular critical concentration threshold (not a full distribution). Decision criteria are associated with this region classification. This approach only assumes that acceptable probability thresholds (risk levels) are provided to classify soils. Practically, it consists of the following risk definition and decision making criteria.

1. Low risk regions: below a given minimum probability threshold (e.g. a 20% probability), the risk of finding contaminated soils is small enough to classify regions as clean. The decision is to leave those regions.

2. High risk regions: above a given maximum probability threshold (e.g. a 80% probability), there is little hope soils be clean and the regions can reasonably be classified as contaminated. The decision is to treat (clean) those regions.

3. Medium risk (unclassified or uncertain) regions: in between the two minimum and maximum probability thresholds, the probability that soils be contaminated is too small to systematically classify regions as contaminated with the risk of unduly cleaning non-contaminated materials, and too large to consider them as non-contaminated and take the risk of leaving toxic pollutants. Such unclassified regions can be seen as transition regions, located in between low and high risk regions or at their periphery, where contamination limits are likely to exist. For decision making, economical aspects can then be integrated to conclude about cleaning or complementary investigation[3]. One way to proceed consists in dividing those regions into disjointed continuous zones and to apply the following decision rules to each zone.

   – If the zone volume is large enough (see hereafter), it is worth investing in

[3] Note that here the two options are to clean or explore unclassified regions. This implicitly means that the zones, where the probability of exceeding a critical threshold is over the low risk probability, cannot be left.

additional samples and chemical analyses to better delineate locally the contamination and possibly avoid cleaning non-contaminated soils. An alternative decision would also to establish a concentration control procedure on digged out soils during the remediation work.
- Otherwise, complementary investigations are not justified economically and cleaning is recommended (see footnote 3).

In the above decision rules, the zone volumes can be converted into treatment costs and be compared to the costs of complementary sampling, analyses and modelling. If the latter costs do not exceed a given percentage (e.g. 10%) of the overall treatment cost, the first decision should be preferred; otherwise the second should prevail.

Regarding the complementary sampling, an optimization method was developed to determine the minimum number and locations of additional samples. The optimization objective is to minimize the within-zone risk of misclassifying soils, i.e. to minimize a global objective[4] function which is computed on all internal zone grid-nodes. This is obtained by first generating a series of (equiprobable) conditional simulations (images) of the PAH concentration. The same indicator variogram models applying to all PAH images, one kriging system only must be solved to estimate at an unsampled location the probability, which is actual and simulated data dependent, of exceeding 200 ppm. Given a number of additional samples, a suitable algorithm then allows modifying the sampling layout until the objective function is minimized. If the minimum reached is not small enough, the number of additional samples is increased and the optimization process repeated.

Conclusions

The aim of this paper was to show, on a case study, how geostatistics can be applied in practice to help to decision making for the remediation of a contaminated site. As in most actual soil contamination problems, the challenge was to clearly specify diagnosis objectives, to make good use of various types of data that are related or correlated diffe-rently to the pollution, to establish decision criteria that must be justified and often call for good sense, and also to meet financial and time constraints inherent to all studies.

In this case study, the available data included quantitative (concentration, soil resistivity, topographic) and categorical (soil type) information as well as expert judgements (about structural trends). A three-step approach was proposed to

[4] Basically, the objective function, computed on each node, is zero if the local conditional probability is less than the low risk probability or greater than the high risk probability (i.e. the classification is reliable), otherwise is equal to the absolute difference between the conditional probability and the closest of the two low and high risk probability thresholds. The global objective function then is calculated by taking the norm (any norm may apply according to the objectives) of all node objective function values.

progressively integrate the available information, to better characterize the uncertainty about the soil contamination (limits, volumes) and to provide relevant elements for decision making. The decision criteria were based on the interpretation and use of estimated conditional probabilities of exceeding a critical concentration threshold. The proposed approach can easily be implemented and provide valuable results.

References

Colin, P., Froidevaux R., Garcia, M. and Nicoletis, S. (1996) Integrating geophysical data for mapping the contamination of industrial sites by polycyclic aromatic hydrocarbons: a geostatistical approach, in R. M. Srivastava, S. Rouhani, M. V. Cromer, A. I. Johnson, Ed., *Geostatistics for Environmental Applications, ASTM STP 1238*, American Society fpr Testing Materials, Philadelphia.

Deutsch, C. V. and Journel, A. G. (1992) *GSLIB: Geostatistical Software Library and User's Guide*, Oxford University Press, 340 p.

Englund, E. J. and Heravi, N. (1992) Conditional Simulation: Practical Application for Sampling Design Optimization, in Amilcar SOARES, Ed., *Geostatistics Tróia '92*, Kluwer Academic Publishers (1992), pp 613-624.

Froidevaux, R. (1990) Geostatistical Toolbox Primer, version 1.30. FSS International, Geneva, Switzerland.

Isaaks, E. H. (1984) Risk qualified mappings for hazardous waste sites, a case study in distribution-free geostatistics, MSc Thesis, Applied Earth Sciences Dept., Stanford University.

Isaaks, E. H.and Srivastava R. M. (1989) *An Introduction to Applied Geostatistics*, Oxford University Press, 561 p.

Journel, A. (1984a) New Ways of Assessing Spatial Distributions of Pollutants, G. Schweitzer Ed., *Environmental Sampling for Hazardous Wastes*, Amer. Chem. Soc. Press, pp 109-118.

Journel, A. (1984b) Decision-making tools for toxic chemical sites, EPA report, Project No. CR-811893-01-0.

Journel, A. G. and Xu, W., (1995) Resampling from Stochastic Simulations, SCRF report N°8, Stanford University.

MAPPING CADMIUM CONCENTRATION AND THE RISK OF EXCEEDING A LOCAL SANITATION THRESHOLD USING INDICATOR GEOSTATISTICS

J. MOHAMMADI (*), M. VAN MEIRVENNE (*) & P. GOOVAERTS(**)
(*) *University Gent, Dept. Soil Management, Coupure 653, 9000 Gent, Belgium*
(**) *Université Catholique de Louvain, Unité BIOM, Place Croix du Sud 2, Bte 16, 1348 Louvain-la-Neuve, Belgium*

Abstract

In the north-east of Belgium an extensive area has been contaminated with cadmium as a result of past industrial activities. The performances of two geostatistical algorithms (ordinary kriging with a global trend model and indicator kriging) for estimating the topsoil Cd content at 276 test locations are compared. Better re-estimation scores are obtained for the "E-type" estimate, that is the mean of the conditional cumulative distribution function provided by indicator kriging. Besides an estimate of Cd content, indicator kriging allows one to assess the probability to exceed the sanitation threshold for vegetable gardens (most risky land use). That threshold varies across the region depending on local clay and organic matter contents. The comparison of estimated probabilities with actual Cd concentrations at test locations indicates a good assessment of the risk by indicator kriging.

1. Introduction

In north-east Belgium, three zinc-ore smelters, located within 10 km from each other, have emitted cadmium through their chimneys during about one century. Since the seventies, most of this Cd is being recuperated, so the Cd pollution of the soils of this region is of an historical nature. In the early eighties research started to investigate the extend of this Cd pollution. Mainly vegetable gardens were sampled, since consuming vegetables grown in Cd contaminated soil represents the most risky pathway (Chainey, 1990). Based on several sampling campaigns, hand drawn isoline maps were used to delineate the most heavily contaminated areas and inhabitants received recommendations about soil management.

More recently, Van Meirvenne *et al.* (1993) analysed the Cd contamination using geostatistical tools. They identified a deterministic trend and an anisotropic variogram due to the influence of the major wind directions of this area. However, they did not evaluate their inventory in respect to alternative interpolation methods and no effort was made to extend the inventory with the probability of exceeding the sanitation threshold.

The goals of this paper are twofold : (1) to compare the performances of different geostatistical algorithms for estimating the Cd concentrations in the soils of the study area

A. Soares et al. (eds.), geoENV I–Geostatistics for Environmental Applications, 327–337.

and (2) to map the probability of exceeding the official Flemish sanitation threshold for vegetable gardens that depends on the amount of clay and organic matter of the soil.

2. Sample data set

The sample data set consists of 1690 Cd analyses of the topsoil of vegetable gardens. The sampling locations displayed in Figure 1 are spatially clustered since the focus is only on vegetable gardens. Figure 2 shows the sample histogram and Table 1 summarises the most important descriptive statistics.

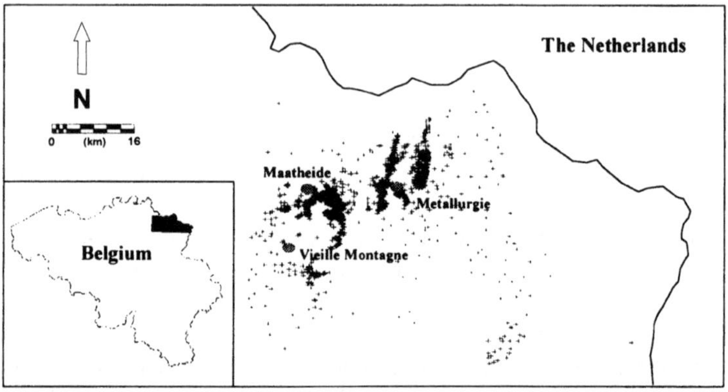

Figure 1. Localisation of the study area inside Belgium, the 1690 sample points (+) and the three zinc factories (greyish areas)

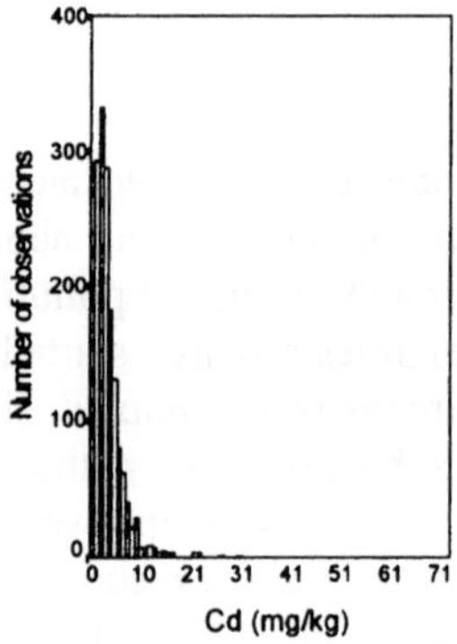

Figure 2. Sample histogram of the Cd data

According to the Flemish regulation, a Cd content of 0.8 mg kg^{-1} (for a "standard" soil with 10 % of clay and 2 % of organic matter) can be considered to be a "background" concentration found under natural conditions (Vlaamse Gemeenschap, 1996). Ninety two percent of the sampling locations exceed this value, which indicates the prevalence of soils enriched with Cd in the region.

TABLE 1. Descriptive statistics of Cd data set (n = 1690).

	Cd mg kg^{-1}
Average	3.86
Median	3.00
Minimum	0.01
Maximum	70.5
Variance	16.2
Skewness	5.68
Kurtosis	61.6

3. Performance comparison of interpolation methods

3.1 INTERPOLATION METHODS

The following two geostatistical algorithms were used to estimate the topsoil Cd content in the region :

3.1.1. Ordinary Kriging with a global trend model (OKt)

This algorithm proceeds in three steps :

(1) A deterministic trend is modelled as a function of : (i) the distance to the nearest zinc factory and (ii) the frequency of wind, classified in four directional classes. Figure 3 shows the year-average frequencies of wind blowing *from* a particular direction. This wind-rose was reversed to represent the wind blowing *to* a particular direction. The latter was grouped into 4 classes of direction angles (336°-79°, 79°-169°, 169°-281° and 281°-336°). Figure 4 shows the deterministic trend model for one of these classes. A power function was found to represent the trend of the data best for all wind classes.

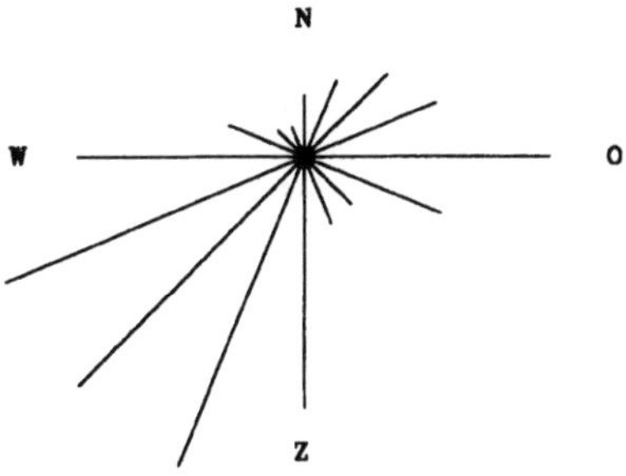

Figure 3. Yearly averaged frequencies of wind blowing from a particular direction, the longest line (SSW) represents 12.5 % of the days, (data from a meteorological station near to the study area ; N = north, O = east, Z = south and W = west).

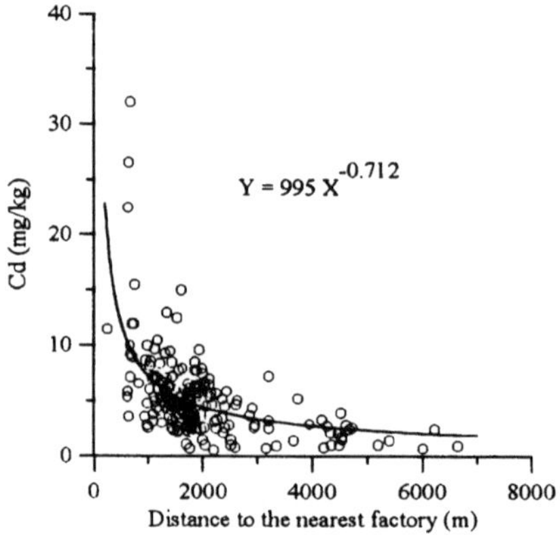

Figure 4. Fitted deterministic model for Cd data located in wind direction class 169°-281° (see fig. 3) as a function of the distance to the nearest factory.

(2) Trend values are subtracted from measured values, and the resulting residuals are interpolated using ordinary point kriging and the variogram of Figure 5.
(3) Final estimates of Cd concentration are obtained by adding the trend estimates to the estimates of the residuals.

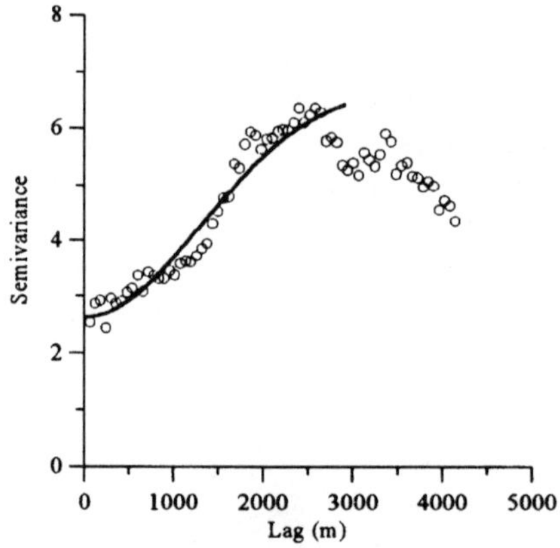

Figure 5. Isotropic variogram of the residuals (after trend removal). The curve is a gaussioan model with a nugget of 2.635, a sill of 6.749 and an apparent range of 3149 m, fitted to a lag of 2700 m.

3.1.2. "E-type" estimate provided by Indicator Kriging (E-IK)

Indicator kriging (Journel, 1983 ; Goovaerts, 1994) is but kriging applied to an indicator transforms of the data, i.e. each observation $z(\mathbf{u}_\alpha)$ is coded into an indicator $i(\mathbf{u}_\alpha; z_k)$ equal to 1 if $z(\mathbf{u}_\alpha) \leq z_k$ and 0 otherwise. Journel (1984) showed that the kriged estimate of the indicator at an unsampled location $\mathbf{u}_0$ is the estimated conditional probability that the z-value at $\mathbf{u}_0$ is no greater than the threshold z_k :

$$F(\mathbf{u}_0; z_k|(n)) = \text{Prob}\{Z(\mathbf{u}_0) \leq z_k|(n)\} \tag{1}$$

where the notation (n) expresses the conditioning to the surrounding n observations. Conditional probabilities estimated for a series of K threshold values z_k (k = 1,..., K) provide a discrete model of the conditional cumulative distribution function (ccdf) at $\mathbf{u}_0$. The mean of the ccdf, called the "E-type" estimate of z(**u**), is defined as :

$$[z(\mathbf{u})]_E^* = \int_{-\infty}^{+\infty} z\, dF(\mathbf{u}; z|(n)) \tag{2}$$

In practice, one uses the following approximation (Deutsch and Journel, 1992, p. 76) :

$$[z(\mathbf{u})]_E^* \approx \sum_{k=1}^{K+1} z'_k [F(\mathbf{u}; z_k|(n)) - F(\mathbf{u}; z_{k-1}|(n))] \tag{3}$$

where $z_0 = z_{min}$, $z_{K+1} = z_{max}$ are the minimum and maximum of the data. The conditional mean z'_k of each class was obtained by a linear interpolation between tabulated bound values provided by the sample cdf (Deutsch and Journel, 1992, p. 133), with the maximum set to 70 mg kg^{-1}.

Twelve thresholds were considered in this study (Figure 6), and Table 2 provides the parameters of the fitted variograms at each threshold. Note the decrease of the range as the threshold value increases, which indicates that small Cd concentrations are better structured in space than large concentrations.

TABLE 2. Parameters of the indicator variograms for twelve thresholds

Cd threshold (mg kg^{-1})	Global cdf	Variogram model	nugget	sill	range (†) (m)
0.9	0.058	Spherical	0.0306	0.0761	1700
1.2	0.111	Spherical	0.0493	0.1230	1562
1.8	0.240	Spherical	0.0609	0.1859	1412
2.2	0.312	Spherical	0.0878	0.2049	1322
2.6	0.411	Spherical	0.0996	0.2079	1062
3.1	0.519	Spherical	0.1043	0.1932	745
3.6	0.604	Spherical	0.1028	0.1682	671
4.4	0.710	Exponential	0.0806	0.1222	747
5.5	0.804	Exponential	0.0313	0.0865	510
7.4	0.901	Exponential	0.0155	0.0475	603
10	0.946	Exponential	0.0077	0.0279	310
15	0.975	Exponential	0.0000	0.0170	102

(†) Apparent range for the exponential models

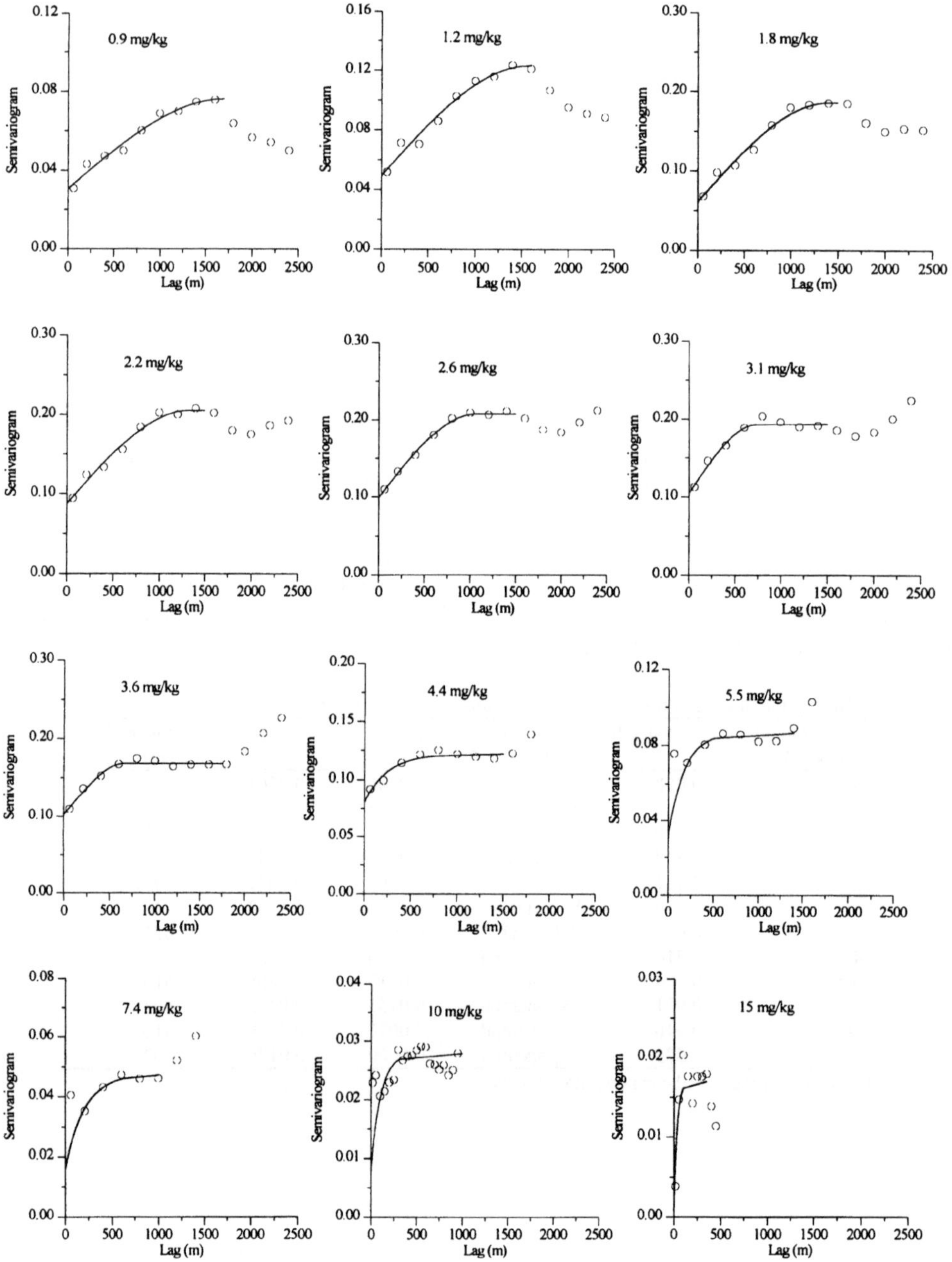

Figure 6. Isotropic indicator variograms for the 12 threshold values (see Table 5 for the models and their parameters).

3.2. ANALYSIS OF RE-ESTIMATION SCORES

The data set was split into a prediction set (1414 data) and a validation set including 276 randomly selected test locations. Cd concentration was estimated at each test location using the two interpolation algorithms. Prediction performances were assessed using the following criteria:

(1) The Mean Estimation Error (MEE) :

$$MEE = \frac{1}{N} \sum_{\alpha=1}^{N} [z(\mathbf{u}_\alpha) - z^*(\mathbf{u}_\alpha)] \tag{4}$$

(2) The Mean Squared Estimation Error (MSEE) :

$$MSEE = \frac{1}{N} \sum_{\alpha=1}^{N} [z(\mathbf{u}_\alpha) - z^*(\mathbf{u}_\alpha)]^2 \tag{5}$$

(3) The Pearson correlation coefficient r between the measured ($z(\mathbf{u}_\alpha)$) and estimated ($z^*(\mathbf{u}_\alpha)$) Cd values.

Table 3 shows that "E-type" estimation provided by IK performs best for the three criteria. Figure 7 shows a scatter diagram of the "E-type" IK estimate *vs.* the measured Cd concentration at the test locations.

TABLE 3. Criteria used to compare the two interpolation methods for the independent data set (N = 276)

Method	MEE	MSEE	r
OKt	0.153	8.93	0.669
E-IK	0.017	7.92	0.697

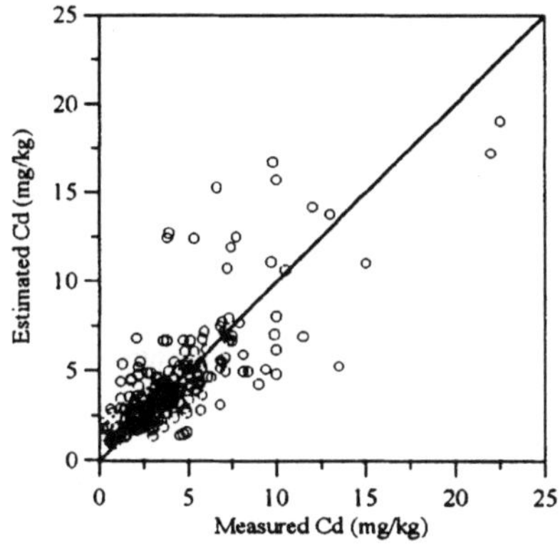

Figure 7. "E-type" estimate vs. the Cd content measured at the 276 test locations.

3.3. MAPPING THE Cd CONTENT

At every estimation location the ccdf values were determined using ordinary indicator kriging at the 12 threshold values listed in Table 2. Only 0.77 % of the estimated probabilities showed order relation problems such as negative probabilities or series of probabilities that are a decreasing function of the threshold value. All deviations are of a small magnitude (average deviation = 0.0003) and were corrected using the average of an upward and downward correction (Deutsch & Journel, 1992, p. 80). Figure 8 shows the map of the "E-type" estimate of topsoil Cd concentration in the region. It is clear that the Cd concentration increases strongly near the zinc factories, with important areas with an estimated Cd content exceeding 10 mg kg^{-1}.

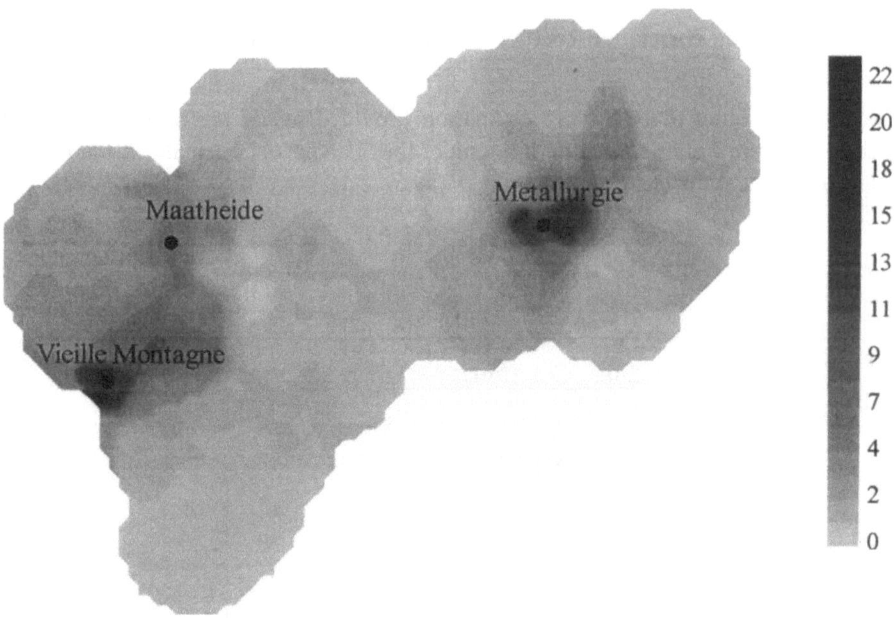

Figure 8. "E-type" estimate of Cd concentration (mg kg^{-1}).

4. Probability of exceeding a spatially variable threshold

4.1. MAPPING THE SANITATION THRESHOLD

The Flemish government has decided (Vlaamse Gemeenschap, 1996) that the threshold for soil sanitation depends on the land use and varies according to the soil composition (clay and organic matter). For the land use "vegetable gardens" the sanitation threshold is given by :

$$N(x,y) = 2(\frac{0.4 + 0.03x + 0.05y}{0.8}) \qquad (6)$$

with $N(x,y)$ the sanitation threshold for Cd in mg kg^{-1}, x the clay content (%) and y the % of organic matter. Since both soil properties vary in space, the threshold varies as well and must be mapped.

During the National Soil Survey, conducted some 30 years ago, clay content has been measured at 1253 locations in the study area. Descriptive statistics in Table 4 indicate that the soils in the area contain little clay ; the dominant soil texture class is sand.

TABLE 4. Descriptive statistics of topsoil clay and organic matter (OM).

	Clay %	OM %
Number of samples	1253	1378
Average	3.8	6.97
Minimum	0.7	1.55
Maximum	19.9	22.41
Variance	5.63	5.47
Skewness	2.88	1.48
Kurtosis	12.69	6.62

The organic matter content was measured for most of the samples taken from the vegetable gardens. Table 4 shows that, on average, the topsoil of the vegetable gardens of the study area are rich in organic matter. If the average clay and organic matter content are entered into Eq. 6, one finds an average sanitation threshold of 2.16 mg Cd kg^{-1} for land use "vegetable garden".

Both variables were estimated at the same locations as Cd content using ordinary point kriging. The sanitation threshold was then computed using Eq. 6.

4.2. MAPPING THE PROBABILITY OF EXCEEDING THE THRESHOLD

The probability that the Cd concentration exceeds the local sanitation threshold estimated in section 4.1 was retrieved from the ccdf model as estimated in section 3.3. The resulting probability map is shown in Figure 9. This figure shows a dominance of probability > 0.8 to exceed the threshold, mainly as two large patches extending generally N.E.- S.W. from the three factories. It is obvious that the spatial variability of the threshold causes the probability map to be more complex than the map of Cd estimates (Figure 8).

4.3. VALIDATION OF THE PROBABILITY MAP

To assess the performance of the probability map in identifying contaminated areas, the probability to exceed the sanitation threshold was computed at the 276 test locations. The test locations were then split into two groups according to whether the sanitation threshold is actually exceeded or not, and the average estimated probability was computed for each group. The average probability to exceed the sanitation threshold is 0.475 at the test locations below that threshold (true probability = 0), whereas it is 0.817 at the test locations where the threshold is exceeded (true probability = 1). Thus, the probability map allows for a fairly good discrimination between contaminated and safe locations.

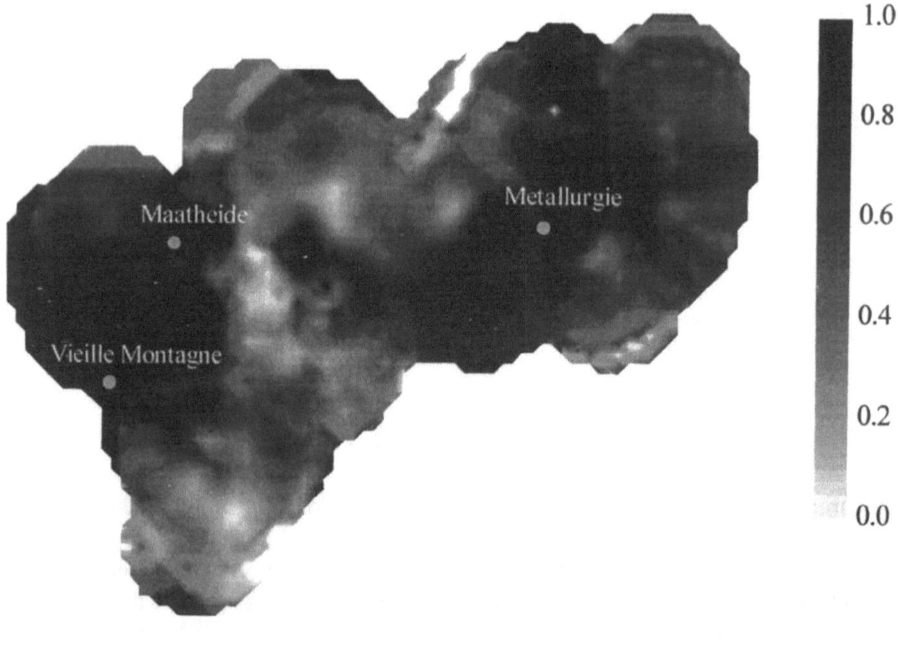

Figure 9. Probability that the Cd content exceeds the spatially variable sanitation threshold for land use "vegetable gardens".

Deutsch (1996) proposed to assess the goodness of probabilistic models by measuring their accuracy and precision. A probability distribution is said to be accurate if the fraction of true values falling in the p symmetric probability interval exceeds p for all p in [0,1], for example if at least 10 % of true values fall in the 10 % probability interval. Precision is a measure of the narrowness of the distribution: in the previous example the distribution is precise if the 10 % probability interval contains exactly 10 % of the true values. To assess the precision and accuracy of the ccdfs provided by indicator kriging at the 276 test locations we proceeded as follows: (1) for a series of probability values p in [0,1], we derive at each test location u_t the corresponding symmetric p-probability interval as: $[F^{-1}(u_t;(1-p)/2|(n)),\ F^{-1}(u_t;(1+p)/2|(n))]$, (2) for each probability p we calculate the proportion p* of locations u_t where the true value $z(u_t)$ is within the probability interval, and (3) we plot p versus p*, see Figure 10. The ideal case is where all points fall on the first bisector of the so-called accuracy plot. Figure 10 shows that the probability distributions are accurate and precise up to a probability interval of width p around 0.9. The small discrepancies at large probabilities p mean that the corresponding probability intervals contain less than a proportion p of true values. For example, the 95 % probability intervals contain around 90 % of true values.

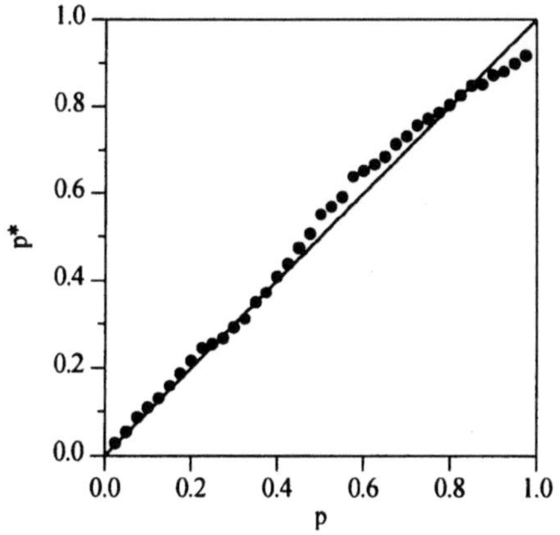

Figure 10. Proportion of test locations where the actual Cd concentration falls within a symmetric probability interval p derived from IK estimates versus the width of that interval.

5. Conclusions

An extended data set allowed us to compare the performance of two geostatistical algorithms for estimating topsoil Cd concentrations. Analysis of re-estimation scores at test locations indicates that the "E-type" estimate provided by indicator kriging performs best. Another goal was to map the probability to exceed any given threshold at a particular location which can be retrieved from the ccdf model estimated by IK. In this paper, IK was used to estimate the probability to exceed a sanitation threshold for "vegetable gardens" that varies locally according to the contents in clay and organic matter. The validation of the probability map indicates a good discrimination between locations that are below the sanitation threshold and those that are above that threshold.

6. References

Chaney, R.L. (1990) Twenty years of land application research, part I, *BioCycle* **31**, 54-59.

Deutsch, C.V. (1996) Direct assessment of local accuracy and precision. In *Geostatistics Wollongong 96* (ed. E. Baafi), Kluwer Academic Publisher. In press.

Deutsch, C.V., and Journel, A.G. (1992) *GSLIB Geostatistical Software Library and User's Guide*, Oxford University Press, New York.

Goovaerts, P. (1994) Comparative performance of indicator algorithms for modeling conditional probability distribution functions, *Mathematical Geology* **26**, 389-411.

Journel, A.G. (1983) Non-parametric estimation of spatial distributions, *Mathematical Geology* **15**, 445-468.

Journel, A.G. (1984) New ways of assessing spatial distributions of pollutants. In : G. Schweitzer (Ed.), *Environmental Sampling for Hazardous Wastes*, Amer. Chem. Soc. Press, pp. 109-118..

Van Meirvenne, M., Vercauteren, F., Hofman, G., and Ide, G. (1993) Geostatistical analysis of the cadmium pollution in north-west Limburg, Belgium, *Biometrie et Praximetrie* **33**, 45-56.

Vlaamse Gemeenschap (1996) Besluit van de Vlaamse regering houdende vaststelling van het Vlaams reglement betreffende de bodemsanering, *Belgisch Staatsblad*, 27.03.1996 : 7018-7058.

GEOSTATISTICAL METHODS APPLIED TO CHARACTERIZATION OF THE HETEROGENEITY OF THE MIGMATITE FORMATION AT OLKILUOTO, WESTERN FINLAND

EEVALIISA LAINE
Researcher
Helsinki University of Technology, Laboratory of Engineering Geology and Geophysics, 02150 Espoo, Finland

Abstract

Geostatistical methods were applied to study the character of the heterogeneity of the bedrock in the Olkiluoto area - a possible nuclear waste site in Finland. The structural model was deduced from the geological interpretations made earlier in the area and from the variogram analysis. Rock type, as a categorical variable, was mapped with sequential simulation and with the indicator principal component simulation (IPCS). In the Olkiluoto area, the latter method proved to produce more reliable images of the bedrock than the sequential simulation. The simulations indicate that the bedrock of Olkiluoto is rather heterogeneous. The accuracy of the simulations could be improved by using geophysical data and an advanced geological model incorporating the variation in the tectonic directions in the study area.

1. Introduction

Olkiluoto is one of the three possible high-level nuclear waste sites in Finland. The bedrock of the Olkiluoto area is Precambrian and a part of the Svecofennian schist belt in western Finland (Fig. 1).

The bedrock of this area consists mainly of mica gneiss and granite (e.g. Anttila et al., 1992). It could be interpreted as an enormous migmatite formation with mica gneiss as a paleosome and granite as a neosome. Eight boreholes, spaced about 500 metres apart, have been drilled into this migmatite formation (Fig. 2). They were drilled approximately perpendicular to the layering of the rock formation. The lithology of the drill cores have been studied. The rock properties between the boreholes are to be predicted from the drill core data and outcrop observations (Fig. 2). Until now, this has mainly been done deterministically (e.g. Saksa et al., 1993). In the present paper, statistical inference and simulation methods are applied to the data.

A. Soares et al. (eds.), geoENV I–Geostatistics for Environmental Applications, 339–350.

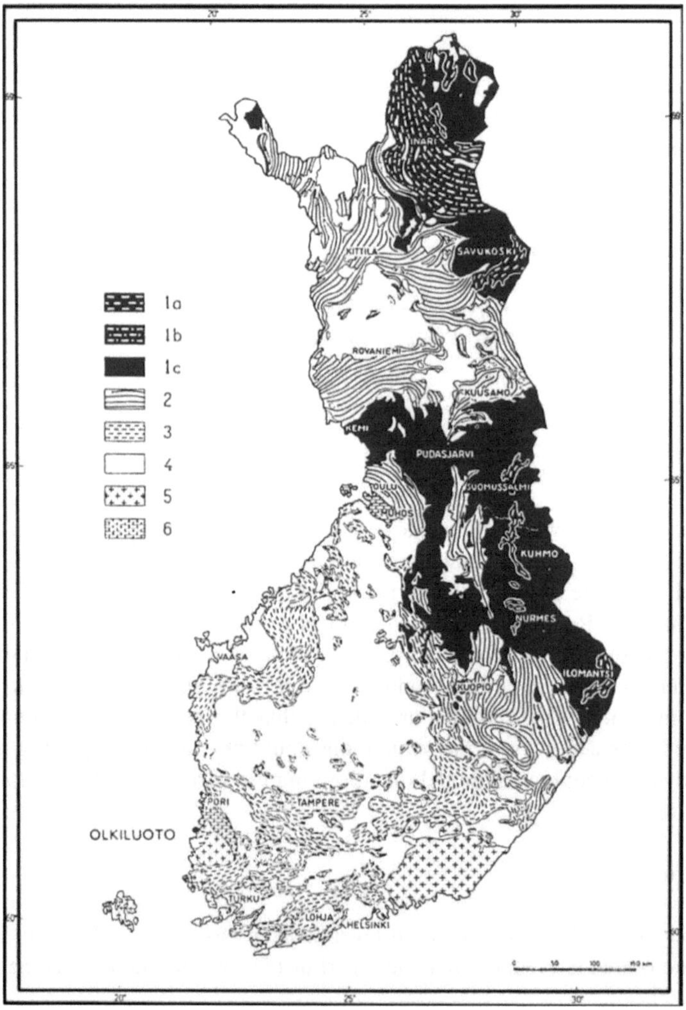

Fig. 1 Main structural units of the Finnish bedrock (Simonen, 1980); location of the Olkiluoto site. *Presvecokarelidic*: 1a, schist and paragneiss; 1b, granulite; 1c, orthogneiss. *Svecokarelidic*: Karelidic schist belt; 4, orogenic plutonic rocks. Postsvecokarelian; 5, rapakivi granites; 6, Jotnian sediments.

The aim was to simulate the lithology of the Olkiluoto bedrock in order to evaluate the heterogeneity of the bedrock. In nuclear waste disposal investigations it is important to predict the rock properties very exactly. However, this may be difficult if the bedrock is very heterogenous. Rock type as a categorical variable was simulated by sequential simulation (e.g. Alabert and Massonnat, 1990) and indicator principal component simulation (Suro-Perez, 1992). The geostatistical simulations were done for a volume of 1 200 m x 1 500 m x 300 m within the Olkiluoto study site (Fig. 2). The data used in this study were mainly from the data base of Posiva Oy and provided by Fintact Oy.

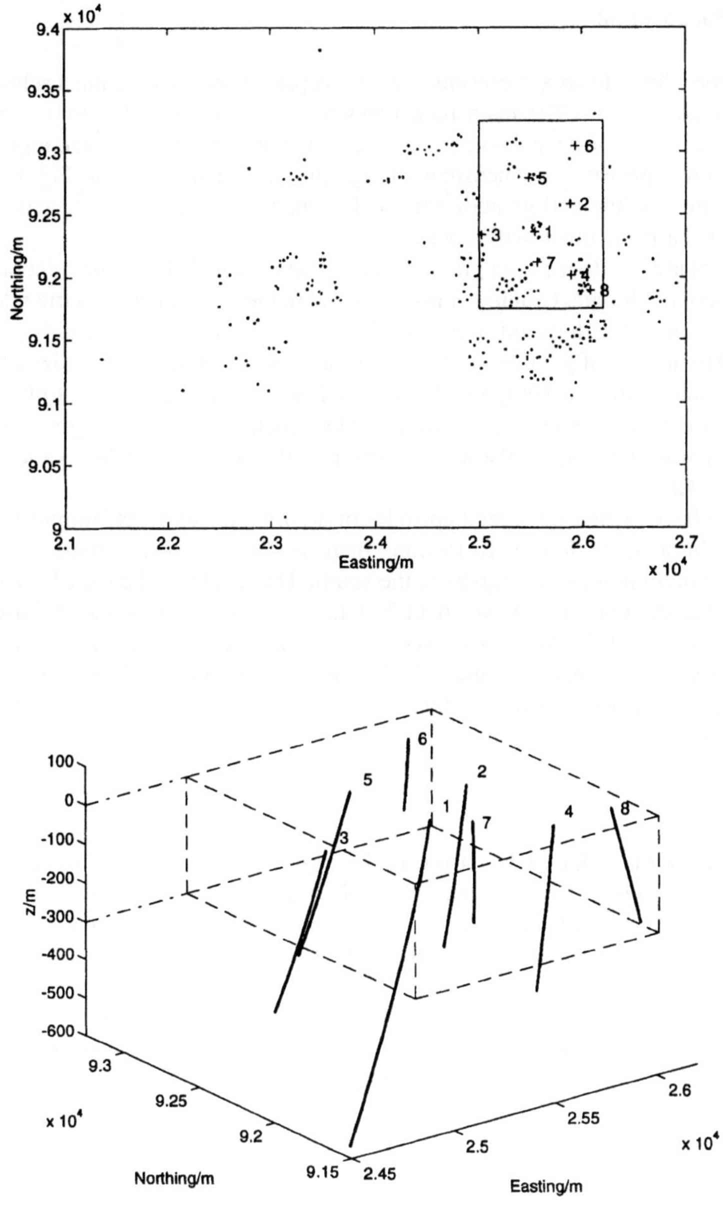

Figure 2. The location of the boreholes (+) and the outcrops (·). A is the upper surface of the detailed study volume V.

2. Geological Background

The bedrock of the Olkiluoto area is Precambrian and is part of the Svecofennian schist belt in western Finland (Fig. 1). The main rock types are mica gneiss and granite forming migmatites of different scales: the whole site area could be interpreted as a large migmatite formation with mica gneiss as a paleosome and granite as a neosome (e.g. Fig. 6E). In addition, there are tonalitic and granodioritic rocks, quartz feldspar gneisses, and minor occurrences of diabase and other mafic rocks.

Mica gneisses are fine to medium grained, banded and foliated. The main minerals are plagioclase, quartz and biotite. Granitic veins are found in the mica gneiss. Quartz feldspar gneiss occurs intercalated with the mica gneiss. The granite occurs as veins, mainly parallel to the foliation. The amount of granitic material determines whether the rock is migmatite or mica gneiss. Granite, often coarse grained, forms dykes conforming with the foliation or intersecting it. It contains nebulitic inclusions of older rocks. Tonalitic and granodioritic rocks form elongated lenses along the main foliation in the area. Their fabric varies from massive to gneissose.

The bedrock has experienced several episodes of deformation and metamorphism (e.g. Härme, 1960). In spite of these deformations there is a dominating, almost east-west schistosity plane dipping 45 - 65 degrees to the south. The length of the rock formations along the main strike direction vary from about 200 metres to thousands of metres based on the geological map (Fig. 6E). 200 m was used as the measure of the maximum continuity along the elongation of the rock formations in the structural model. The lack of data along the strike direction of the rock formation did not permit a rigorous geostatistical variogram range to be determined.

3. The Data Set

The data consisted of the drill core logs from all the eight boreholes KR1 - KR8 (Lindberg and Paananen, 1991, and POSIVA Oy's data base). The rock types in the outcrops were also determined (POSIVA Oy's data base). The lithological data were used to create a categorical variable describing the rock types in the Olkiluoto area (TABLE 1).

TABLE 1. The Rock Type Coding. N is the amount of observations.

Rock type	Code	N (Drill cores)	% (Drill cores)	N (Outcrops)	% (Outcrops)
Granite	1	632	16.8	64	22.0
Migmatite	2	210	5.6	44	15.1
Mica Gneiss	3	2384	63.4	121	41.6
Mafic Rocks	4	16	0.4	0	0.0
Tonalites and Granodiorites	5	401	10.7	62	21.3
Quartz Feldspar Gneiss	6	118	3.1	0	0.0

Note that the geological descriptions are approximate but are coded exactly. Furthermore, the outcrop mapping is not as detailed as drill core logging and could be biased as outcrops represent the most erosion resistant parts of the bedrock.

4. Variogram Analysis

Experimental variograms were calculated with rock type as a categorical variable (or exclusive class variables), e.g., set to 1 if a specific rock type prevails at a certain location, to 0 if not). The data are mainly aligned along the boreholes, almost perpendicular to the assumed main schistosity plain (dip direction 175/ dip 45). Thus, the variograms are practically directional variograms perpendicular to layering. Outcrop data were analysed separately in order to study the anisotropy of the occurrence of the rock types.

Experimental variograms for granite, migmatite, mica gneiss and mafic rocks are modelled using an exponential model (TABLE 2 and Fig. 3). On the other hand, the variograms describing the occurrence of quartz-feldspar gneisses and tonalitic-granodioritic rocks are modelled using a spherical model.

Table 2. The parameters for variograms of the class variables derived from the lithology. Exp - exponential and Sph - spherical model.

Rock type	Model	Nugget	Range	Sill
Granite	Exp	0	3.5	0.12
Migmatite	Exp	0	4.0	0.02
Mica gneiss	Exp	0	4.0	0.13
Mafic rocks	Exp	0	3.0	0.0044
Tonalite/granodiorite	Sph	0	23.0	0.035
Quartz-feldspar gneiss	Sph	0	23.0	0.011

The anisotropy, i.e. directions of maximum and minimum continuity, can be effectively studied by a variogram surface (Isaaks and Srivastava, 1989), on which each cell displays the value of the variogram for the corresponding separation distance or lag. The outcrop data was analysed by calculating variogram surfaces for the binary categorical variables

The binary categorical variable for granite shows a maximum continuity along the main strike (80 degrees from the north) of the rock formations (Fig. 4A). This is in accordance with geological interpretations. In addition, there is another continuity direction towards the nortwest (Fig. 4A). By the variogram surface analysis, the migmatites show a weak continuity along northeast-southwest direction, almost along the main strike of the rock formations (Fig. 4B). A binary categorical variable for mica gneiss has the direction of maximum continuity in the same direction (Fig. 4C), whereas tonalite shows a direction of maximum continuity to the northwest (Fig. 4D). This is not necessarily a contradiction with the geological interpretation of tonalites conforming to the main strike: the variogram surface describes the continuity of the occurrence of the different rock types, not the schistosity. The old tonalite intrusions may have deformed with surrounding gneisses, but the locations of the tonalite intrusions are not controlled by the gneisses. The anisotropies of the binary categorical variables for different rock types were used in sequential simulation of the lithology in the study volume V (Fig. 2).

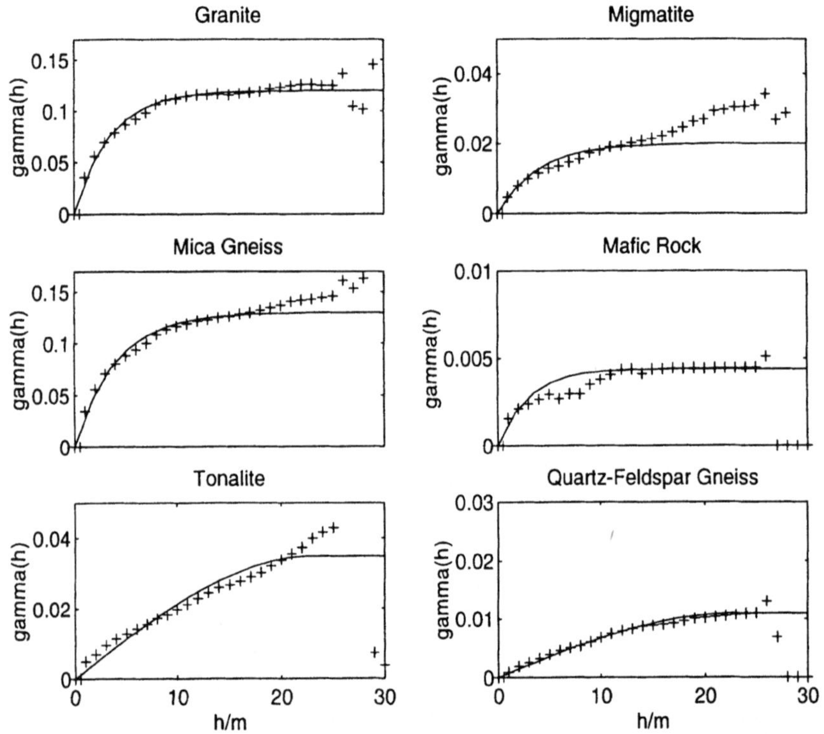

Figure 3. Experimental variograms for the binary categorical variables for the rock types and corresponding variogram models (solid lines).

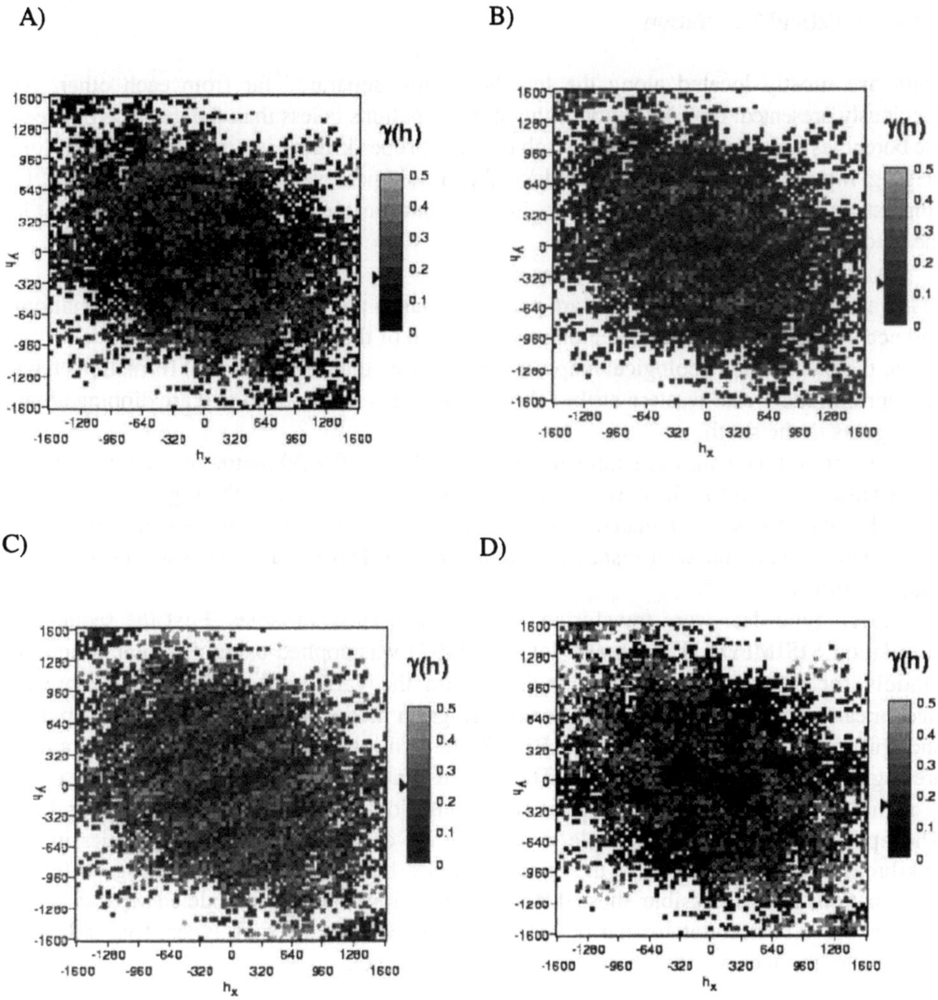

Figure 4. Variogram surface of the binary categorical variable for A) granite, B) migmatite, C) mica gneiss and D) tonalite/granodiorite. **h** (m) is separation distance. The pixels with no variogram value are white.

5. Geostatistical Simulation

Data are mostly located along the few boreholes, separared far from each other. As previously presented, the continuity of the rock formations is less than the distance between the boreholes and thus there is not enough data for proper kriging interpolation. In addition, kriging interpolation would not produce the lithological variability of the bedrock. In contrast, geostatistical realizations reproduce the data variability, an essential component in the study of rock heterogeneity.

A very rough structural model was applied:
1) It was assumed that the boreholes are perpendicular to the layering of the rock formations and accordingly they are aligned along the direction of the minimum continuity.
2) On the basis of the geological maps, it was assessed that the geological formations have an approximately east-western strike (about 80 degrees from north) and are dipping about 45 degrees to the south.

The rock type data were simulated using a 120 x 150 x 30 metre grid. Each grid cell was 10 m x 10 m x 10m. The corresponding surface area is shown in the Figure 2. The rock type distribution was simulated using sequential simulation and indicator principal component simulation with respective algorithms SISIMPDF and IPCSIM (Deutch and Journel, 1992).

Rock type data were transformed to exclusive class variables. First the sequential simulation SISIMPDF (Deutch and Journel, 1992) was applied using different structural models for different rock types implied from the variogram analysis of the binary categorical variables. The three-dimensional variogram model was based on the outcrop data and drill core logs (Figures 3, 4 and TABLE 2). In addition, it was assumed that the granites are aligned along the main schistosity. The second continuity direction to the northwest was disregarded (Fig. 4A). The resulting visualizations did not look very reliable (e.g. Fig. 6B). The applied model was too simple to describe the complicated bedrock tectonics in the Olkiluoto area. For example, the granitic rocks have been formed during different deformations and, if possible, these different generations should be modeled separately.

In order to account for the relationships between the six rock types, five indicator principal components of the indicator covariance matrix C_I

$$C_I = \begin{matrix} .14110 & -.01020 & -.10540 & -.00068 & -.01870 & -.00510 \\ -.01020 & .05640 & -.03720 & -.00024 & -.00660 & -.00180 \\ -.10540 & -.03720 & .23560 & -.00248 & -.06820 & -.01860 \\ -.00068 & -.00024 & -.00248 & .00398 & -.00044 & -.00012 \\ -.01870 & -.00660 & -.06820 & -.00044 & .09790 & -.00330 \\ -.00510 & -.00180 & -.01860 & -.00012 & -.00330 & .02910 \end{matrix}$$

at lag 0 were calculated. The rock type indicators constitute a closed system, and thus, only five principal components are needed to characterize the six rock types. The five principal components should be orthogonal for all separation distances for indicator principal component simulation (Deutch and Journel, 1994). The correlation between the obtained principal components was low, and, the assumtion of the orthogonality was reasonable.

A variogram analysis was performed for these components (TABLE 3 and Fig. 5). The indicator principal components were then used in the indicator principal component simulation IPCSIM (Deutch and Journel, 1992) with the assumption that the rock type dimensions are about 200 metres or larger along the main strike of the rock formations. The resulted realizations (e.g. Figures 6A and 6C) show similar features and seem to have similarities with the geological map (Fig. 6E), although, the simulations give more heterogenous and perhaps more detailed images of the bedrock. The reason for the more reasonable realizations compared to the sequential simulation case is probably due to taking account for the relationships between the rock units within the highly deformed bedrock. These images give, however, only a probable distribution of the rock types in the studied rock block (Figure 2); they are not geological maps. Based on fifty simulations, the probability of the occurrence of the migmatite at a certain location was calculated. The obtained probability map shows that the southern part of the study area is strongly migmatized and, thus, very heterogeneous.

Table 3. The parameters of the exponential variogram models for the indicator principal components along the boreholes.

	Nugget	Range	Sill
IPC1	0.03	5.5 m	0.19
IPC2	0.01	8.0 m	0.06
IPC3	0.00	6.0 m	0.03
IPC4	0.001	7.0 m	0.009
IPC5	0.0007	9.0 m	0.009

6. Conclusions

Geostatistical methods were applied to evaluate the heterogeneity of the bedrock in the Olkiluoto area - a possible nuclear waste site in Finland. The bedrock is Precambrian and thus, the rock formations have a complicated deformation history. The sequential simulation did not produce reliable looking visualizations. The model used was too simple to describe the tectonics of the bedrock. Better results was achieved by the indicator principal component simulation when imaging the lithology of the Olkiluoto area with more reasonable realizations of the bedrock. As a starting point for the geostatistical simulations, the relationships between the rock units seem to be more relevant than the simple tectonic model. According to the simulations, the bedrock of Olkiluoto is rather heterogeneous, especially the southern parts of the study area.The reliability of these geostatistical simulations could be improved by using geophysical data and an advanced geological model incorporating the variations of the tectonic directions. In order to get a measure of uncertainty of the simulations hundreds of simulations are needed.

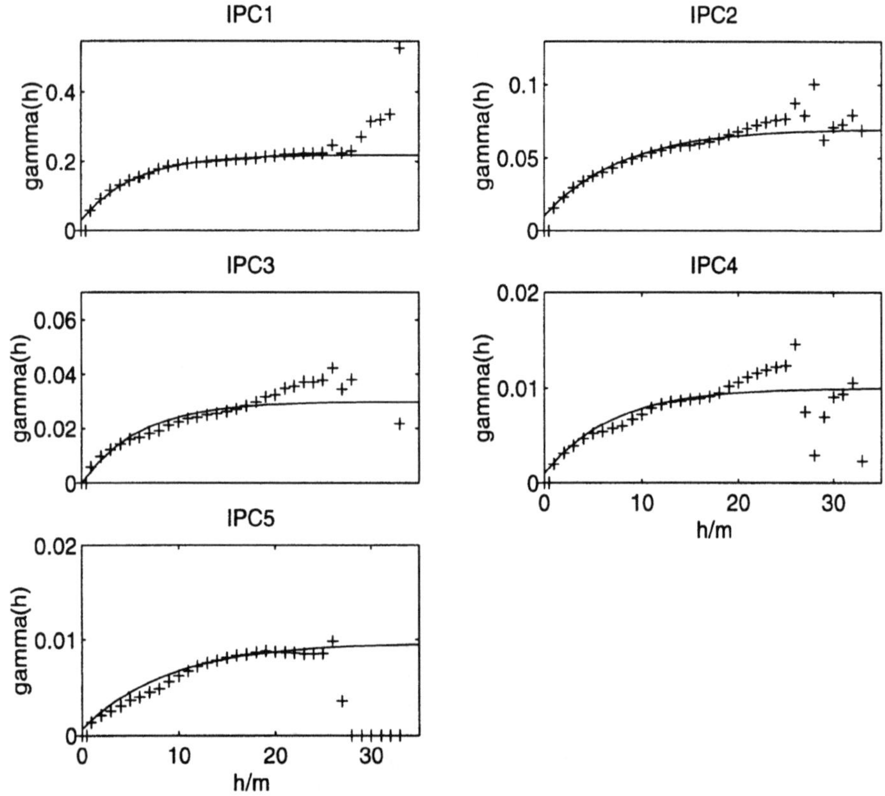

Figure 5. Experimental variograms for the indicator principal components and corresponding exponential models (solid lines).

7. Acknowledgements

This paper is based on the work conducted for Posiva Oy by the Laboratory of Engineering Geology and Geophysics at the Helsinki University of Technology. I thank Professor Heikki Niini , Associate Professor Markku Peltoniemi from the Laboratory of Engineering Geology and Geophysics and Pauli Saksa from the Fintact Oy for helpful comments during the project. I am grateful to David Warren for checking the English of the manuscript.

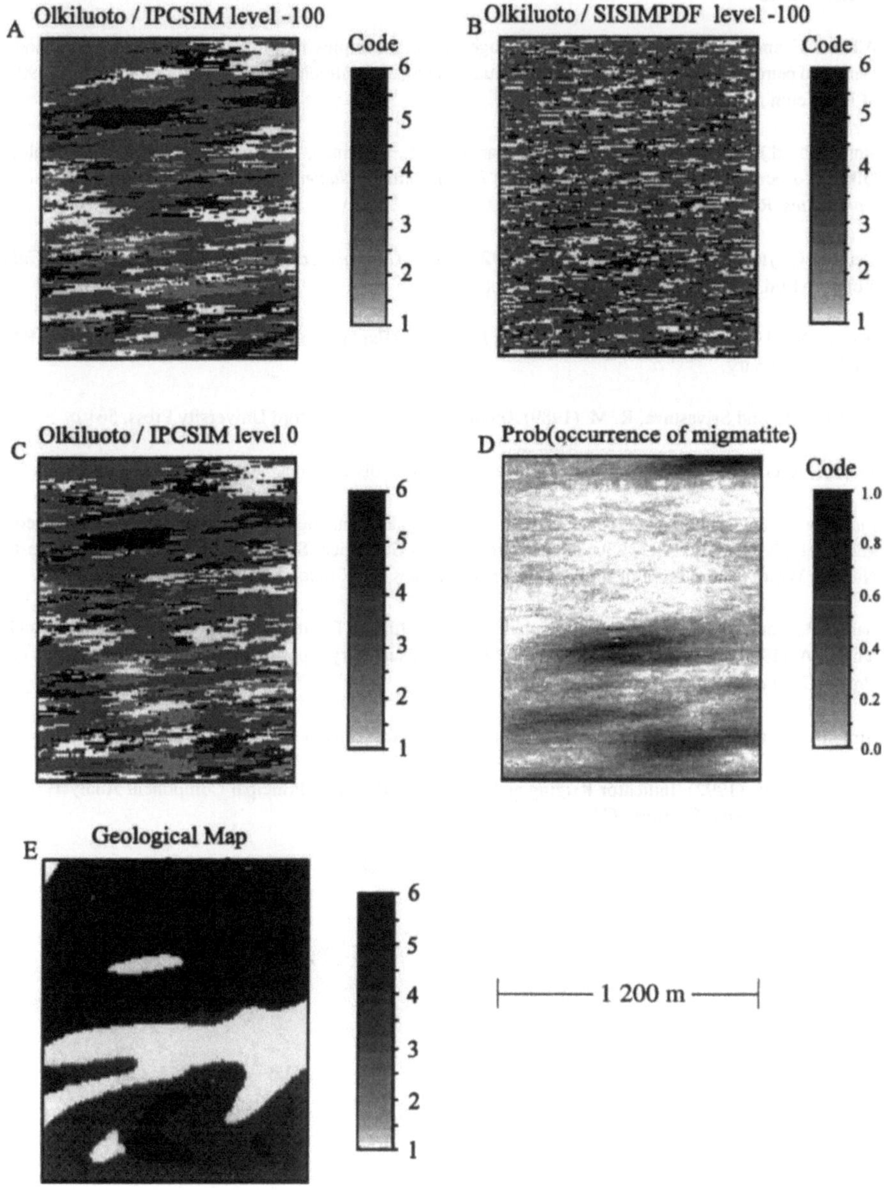

Figure 6. A) IPCSIM simulation at the level of -100 m, B) SISIMPDF simulation at the level of -100 m , C) IPCSIM simulations at a depth of 0 m , D) probability of the occurrence of the migmatite based on the 50 IPCSIM simulations at the level of -100m of the study volume and E) the geological map of the Olkiluoto area (after Anttila et al., 1992). Code 1- granite, 2- migmatite, 3- mica gneiss, 4- mafic rocks, 5 - tonalite/granodiorite and 6 - quartz feldspar gneiss.

8. References

Alabert, F. and G.J. Massonnat (1990) Heterogeneity in a complex turbiditic reservoir: stochastic modelling of facies and petrophysical variability. In 65th Annual Technical Conference and Exhibition, pages 775-790, Society of Petroleum Engineers.

Anttila, P. (ed.), Paulamäki, S., Lindberg, A., Paananen, M., Koistinen, T. and Front, K. (1992) The geology of the Olkiluoto area, summary report. *TVO/Site Investigations, Nuclear Waste Commission of Finnish Power Companies*, Report YJT-92-28.

Deutsch, Clayton V. and Journel, André G.(1992) *GSLIB, Geostaistical Software Library and User's Guide.* New York, Oxford. Oxford University Press. 340 pp.

Härme, M. (1960) Kivilajikartan selitys, lehti B1,Turku. (Explanation to the Geological Map, Sheet B1, Turku) English Summary.

Isaaks, E. H. and Srivastava, R. M. (1989) *Applied Geostatistics.* Oxford University Press, 561 p.

Journel, A. G. (1983) Non-parametric estimation of spatial distributions. *Math. Geol.*, **15**, 445-468.

Lindberg, A. and Paananen, M.(1991) Eurajoen Olkiluodon kallionäytteiden petrografia, geokemia ja geofysiikka, kairanreiät KR1, KR2, KR3, KR4 ja KR5. (In Finnish) Geological Survey of Finland.TVO/Site Investigations, Nuclear Waste Commission of Finnish Power Companies Work report 90-10. 37 p.

Saksa, P., Ahokas, H., Paananen, M., Paulamäki, S., Anttila, P., Front, K., Pitkänen, P., Hassinen, P. and Ylinen, A. (1993) Bedrock model of the Olkiluoto area, summary report. *Nuclear Waste Commission of Finnish Power Companies,* Report YJT-93-15.

Simonen, A. (1980) The Precambrian of Finland. *Geological Survey of Finland*, Bulletin **304**. 58 p.

Suro-Perez,V. (1992) Indicator Kriging and Simulation Based on Principal Component Analysis. PhD thesis, Stanford University, Stanford, CA.

SPATIAL SIMULATED ANNEALING FOR OPTIMIZING SAMPLING

Different optimization criteria compared

J.W. VAN GROENIGEN
International Institute for Aerospace Survey and Earth Sciences
P.O. Box 6
7500 AA Enschede
The Netherlands
e-mail: groenigen@itc.nl

Abstract

In this paper, a modified version of continuous simulated annealing is proposed as a tool for optimizing spatial sampling schemes at the point level. Spatial information on the sampling area, stored in a GIS, can be dealt with as constraints to the optimization process. Maps and earlier measurements can be handled as pre-information. The algorithm can distinguish different quantitative optimization criteria. In this paper, two criteria will be presented. The first ensures optimal estimation of the variogram, by reproducing a pre-specified point-pair distribution over distance- and direction classes. Compared to a previous study, the algorithm produced dramatic improvements. The second criterion aims at even spreading of the sampling points over the area. Here, the algorithm always did better than a triangular grid, especially in areas with much sampling constraints, where improvements could be up to 30%. Combination of the criteria in a phased sampling procedure is possible.

1. Introduction

As a basis for geostatistical analysis on soil and the environment, the sampling scheme plays a crucial role. Ill-designed sampling schemes can lead to much higher costs of sampling, inaccurate predictions and unnecessary health-risks in the case of environmental pollution (Groenigen and Stein, *submitted*), whereas inadequate soil information may be collected, e.g. for precision farming.

Optimizing a spatial sampling scheme requires the *optimal* spreading of a fixed number of sampling points over the research area. However, definition of optimality can differ greatly from case to case. Consider the next four cases:

A. Soares et al. (eds.), geoENV I–Geostatistics for Environmental Applications, 351–361.

i) in a study on classical sampling theory, optimality was defined as a systematic coverage of the sampling area, or as a 'perfect' random distribution of the sampling points (Gruijter and Braak, 1990).

ii) in a study on the nature of spatial variation, the optimal estimation of the anisotropic variogram was pursued. Therefore, samples would have to be taken at such locations, that maximum information on the semi-variance is gained (Warrick and Myers, 1987).

iii) in a study on spatial interpolation, optimality was defined as a sampling scheme which minimizes the kriging variance (Webster and Oliver, 1984).

iv) in a study on the sanitation of a contaminated soil, it was crucial to define the area with only a very small health risk (e.g. less than 0.025) as precisely as possible. This requires optimization of the precision of the probability map with possibly an extra focus on areas with an indistinct prediction with respect to health risks (Groenigen *et al.*, *in press*).

In all these cases, a totally different definition of optimality applies. Therefore, a tool to design optimal sampling schemes should be able to handle different optimization criteria.

In this paper, Spatial Simulated Annealing (SSA) will be applied for this purpose. This method is capable of handling several different optimization criteria, and combining them, if necessary. Furthermore, it uses a GIS to provide information on geographic features and to use earlier samplings.

First, the principles of SSA are explained. Second, two optimization criteria are introduced and optimization is demonstrated on several cases. Finally, conclusions and recommendations for further research are given.

2. Spatial Simulated Annealing

Spatial Simulated Annealing is a modified version of ordinary Simulated Annealing (SA). Therefore, first the basic principles of SA will be explained. Subsequently, the modifications made for spatial use will be presented.

2.1. SIMULATED ANNEALING

SA is a general purpose combinatorial optimization method, stemming from statistical physics (Kirkpatrick *et al.*, 1983). Other names for the same algorithm include *Monte Carlo annealing, probabilistic hill-climbing, statistical cooling, probabilistic exchange algorithm* and *stochastical relaxation.* It has been applied successfully in a wide variety of studies, where it proved to be especially insensitive to local maxima or minima (van Laarhoven and Aarts, 1987)

In geostatistics, this optimization algorithm is used as a simulation algorithm for reproducing variograms (Farmer, 1991), and for combining stochastic simulations with training images (Deutsch and Journel, 1987).

Assume a combinatorial optimization problem **P**. Let **i** and **j** represent two instances of the problem, and let the 'cost function' ϕ be defined, which has to be minimized, with costs $\phi(i)$ and $\phi(j)$, respectively. For example, instance **i** represents the temporal (and imperfect) solution to the problem, whereas instance **j** has been derived from **i** using a random perturbation in some of its variables. This perturbation is made using the so-called *generation mechanism*. Instance **j** can *replace* **i** as the new temporal solution depending on the *acceptance criterion* **P**. This criterion gives the probability that j is accepted:

$$\chi(i,j) = 1, \quad \text{if } \phi(j) \le \phi(i)$$
$$\chi(i,j) = \exp\left(\frac{\phi(i) - \phi(j)}{c(t)}\right), \quad \text{if } \phi(j) > \phi(i) \qquad (1)$$

The control parameter **c** is lowered as the optimization process evolves. The way in which the value of **c** decreases is controlled by the elapsed time **t** using a *cooling schedule*. Upon lowering of **c**, the probability of the acceptance of an inferior solution decreases. At each value of the control parameter, several transitions have to be made before the control parameter lowers any further. In this way, the (near-)optimal solution 'crystallizes' out of all the transitions (Kirkpatrick *et al.*, 1983).

2.2. MODIFICATIONS TO NORMAL SIMULATED ANNEALING

Modifications to the normal simulated annealing algorithm to obtain SSA can be divided into data-representation, the generation mechanism and the cooling schedule.

2.2.1. *Data Representation*

Let **S** represent a sampling scheme, consisting of n_S sampling points. The qth sampling point is denoted with $p_{S,q}$, with coordinates $x_{S,q}$ and $y_{S,q}$. Furthermore, the total research area is denoted by A_R, and the sub-area that can be sampled physically by $A_S \subset A_R$, thus excluding roads, houses, etc. The sampling scheme has to be optimal with respect to a quantitative function ϕ, that has to be minimized. Application of SSA reduces this problem to a combinatorial optimization problem, in which the coordinates of the sampling points are the variables that have to be optimized with respect to ϕ.

2.2.2. *Generation Mechanism*

In SSA, the random perturbation supplied by the generation mechanism is implemented as a transformation of a randomly chosen sampling point over a vector **h**, with the direction of **h** chosen at random, and |**h**| random between 0 and $\mathbf{h}_{max}(\mathbf{t})$. The variable $\mathbf{h}_{max}$ decreases with time, thus giving the sampling scheme the chance to 'freeze' in its optimal configuration by ending with very small perturbations. This is an important difference to normal SA, where major perturbations are proposed until the end of the optimization process, although they only have a small chance of being accepted. This proved to be more efficient in computing time.

2.2.3. *Cooling Schedule*

In SSA, the cooling schedule that describes **c** as a function of the progress of the optimization process, was constructed using a basic set of rules, proposed in several studies (Aarts and Korst, 1990). Using these rules, **c** decreases asymptotically:

$$\mathbf{c}_{k+1} = \alpha \cdot \mathbf{c}_k, \text{ with } \mathbf{k} = 1,2,3,\ldots \tag{2}$$

The value of $\mathbf{c}_0$ is chosen in such a way, that initially more than 95% of the proposed transitions are accepted. In this way, total randomization of the starting position is ensured.

2.3. FITNESS FUNCTIONS

Any optimization criterion has to be translated into a fitness function, in order to be used in SSA. In this paper, two of these fitness-functions have been used, both using a very different definition of optimality.

2.3.1. The Warrick/Myers criterion

This criterion, which was proposed by Warrick and Myers (1987), optimizes the estimation of the experimental variogram. The criterion demands that the point pairs are distributed over the distance- and direction-lags according to a pre-specified distribution. If the number of samples is **n**, and the number of lags is denoted by **nc**, the following function has to be minimized:
with

$$\phi_{w/m}(S) = a\sum_{i=1}^{nc} w_i(\zeta_i^* - \zeta_i)^2 + b\sum_{i=1}^{nc} \sigma_{m,i} \quad (3)$$

$$\sum_{i=1}^{nc}\zeta_i^* = \sum_{i=1}^{nc}\zeta_i = \frac{n\cdot(n-1)}{2} \quad (4)$$

where **a**, **b** and $\mathbf{w}_i$ are user-defined weights, ζ_i^* is a pre-specified number of point-pairs for the ith lag class, and ζ_i is the actual number for this lag class. In the second term of formula 3, $\mathbf{m}_i$ denotes the standard deviation from the median of the distance lag class. Following Warrick and Myers, **a** is a function of **n**:

$$a = 4[n(n-1)]^{-2}$$

The original study applied a Monte Carlo algorithm for optimization, which proved to be very susceptible to local minima, whereas complex sampling barriers could not be taken into account (Warrick and Myers, 1987).

2.3.2. The MMSD criterion

The MMSD criterion (Minimization of the Mean of the Shortest Distances) requires that all sampling points are spread evenly over the research area, thus ensuring that unvisited locations are never far from a sampling point.

An equilateral grid (fig. 1), which theoretically is almost optimal to this criterion (McBratney *et al.*, 1981), is based upon the assumption that the area is a contiguous, continuous, infinite plane. In the practice of spatial sampling, many constraints to this assumption are encountered. If we consider a soil sampling in an urban area, the following problems are met:

i) the sampling area is *never* infinite. Therefore, there will always be boundary-effects which will make the triangular grid sub-optimal. The importance of boundary effects increases with complexity of the boundaries of the area.

ii) the sampling area is almost never continuous, because of physical sampling barriers, such as houses or ponds

iii) the sampling area may not be contiguous. Often, it is subdivided into several sub-units, different in size and shape.

iv) especially in environmental studies, preliminary measurements are already available. Preferably, they should be included in the optimized sampling scheme as well.

All these constraints show that in practice the triangular grid is not an optimal solution to the MMSD- criterion. Therefore, SSA will be used to optimize the sampling schemes at the point level.

TABLE 1. A comparison of the distribution of distance classes using the Monte Carlo algorithm and SSA, case 1.

Case 1			
Distance class (m.)	ζ		ζ^*
	Monte Carlo	SSA	
0 - 20	17	30	43.5
20 - 40	18	30	43.5
40 - 60	17	30	43.5
60 - 80	19	29	43.5
80 - 100	32	32	43.5
100 - 120	51	29	43.5
120 - 140	31	29	43.5
140 - 160	46	35	43.5
160 - 180	23	37	43.5
180 - 200	26	42	43.5
> 200	155	112	43.5

The function to be minimized using the MMSD-criterion, can be described as the expectation of the distance between a randomly chosen point $\mathbf{p}_r$ from the research area $\mathbf{A}_R$, and its nearest sampling point $\mathbf{p}_{S,I}$ from sampling area $\mathbf{A}_S$:

$$\phi_{MMSD}(\mathbf{S}) = E[\min\{\delta(\mathbf{p}_r, \mathbf{p}_{S,i}), \quad i \in 1,\ldots,n_s\}]$$

where δ represents the Euclidean distance between two points. This expectation can be estimated by substituting the average value for the expectation:

$$\hat{\phi}_{MMSD}(S) = \sum_{j=1}^{n_e} \frac{\min\{\delta(p_{e,j}, p_{S,i}),\quad i \in 1,\ldots,n_s\}}{n_e}$$

where $p_{e,j} \in A_R$ represents the jth evaluation point of which there are n_e, with $n_e >> n_s$.

The evaluation points are drawn randomly from A_R, while the sampling points are spreaded over A_S by the SSA-algorithm. In this way, the algorithm ensures that sampling is optimized with respect to the whole area A_R, not just the sampling area A_S.

3. Examples

Below, several examples illustrate how SSA optimizes sampling schemes. The calculations were made using the program SIMAS (SIMulated Annealing for Spatial optimization), which is the implementation of the SSA-algorithm. The program was linked to the ARC/INFO database for spatial information on the sample areas.

3.1. TEST PROBLEMS FOR THE WARRICK/MYERS CRITERION

To test SSA on the Warrick/Myers criterion, several cases from Warrick and Myers (1987) were re-calculated (Groenigen and Stein, *submitted*). It is acknowledged that the emphasis of Warrick and Myers was rather to formulate the optimization criterion, than to apply the rather simple Monte-Carlo algorithm. Therefore, the use of SSA for these problems is a logical extension rather than a competing procedure.

Both cases assume a 400x400 meter field:

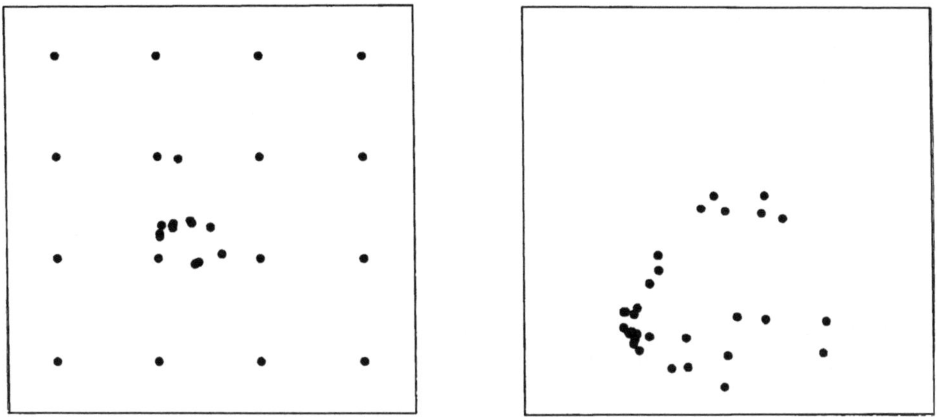

Figure 1. Optimized sampling schemes for estimation of the variogram, taking into account an earlier grid (a) and direction classes (b)

i) The first case considers 16 sampling points that were already sampled according to a rectangular grid. A set of 14 additional sampling points has to be placed optimally according to the first term of equation 3 (**b** = 0). Following Warrick and Myers, 10 distance classes of 20 m. width were distinguished, and a uniform distribution was considered optimally:

$$\zeta_i^* = \frac{\mathbf{n \cdot (n-1)}}{\mathbf{2 \cdot nc}} = \mathbf{43.5}, \quad \mathbf{i = 1, \ldots, nc}$$

The results of the optimization are given in table 1 and figure 1a. Although a complete solution is impossible because of the predefined grid, we observe a considerable improvement as compared to Monte Carlo optimization. In particular, the minimum number of sampling point pairs in a class raises from 17 to 29.

Table 2. A comparison of the distribution of distance and direction classes using the Monte Carlo, algorithm and SSA, case 2.

Case 2							
Class (m.)	ζ		ζ^*	Class (m.)	ζ		ζ^*
	Monte Carlo	SSA			Monte Carlo	SSA	
H, 0 - 20	12	22	21.8	V, 100 - 120	22	22	21.8
V, 0 - 20	8	22	21.8	H, 120 - 140	20	22	21.8
H, 20 - 40	22	21	21.8	V, 120 - 140	20	21	21.8
V, 20 - 40	20	22	21.8	H, 140 - 160	26	22	21.8
H, 40 - 60	18	22	21.8	V, 140 - 160	28	22	21.8
V, 40 - 60	24	21	21.8	H, 160 - 180	27	22	21.8
H, 60 - 80	23	22	21.8	V, 160 - 180	27	21	21.8
V, 60 - 80	27	22	21.8	H, 180 - 200	17	22	21.8
H, 80 - 100	22	22	21.8	V, 180 - 200	20	22	21.8
V, 80 - 100	22	21	21.8	> 200	5	0	0.0
H, 100 - 120	25	22	21.8				

ii) In the second case, 30 sampling locations were optimally located according to the same 10 distance classes that were distinguished as in the first case. Additionally, two direction classes were distinguished with angles along the east-west direction (**H**) and the north-south direction (**V**). Again, a uniform distribution was defined as optimal:

$$\zeta_i^* = \frac{n \cdot (n-1)}{2 \cdot nc} = 21.5, \quad i = 1,\ldots,nc$$

Table 2 and figure 1a show that with SSA no point pairs fall outside of the class range, and all combinations of distance and direction classes contain either 21 or 22 point pairs. This is a considerable improvement compared to the earlier solutions with a minimum and a maximum of 8 and 28, respectively.

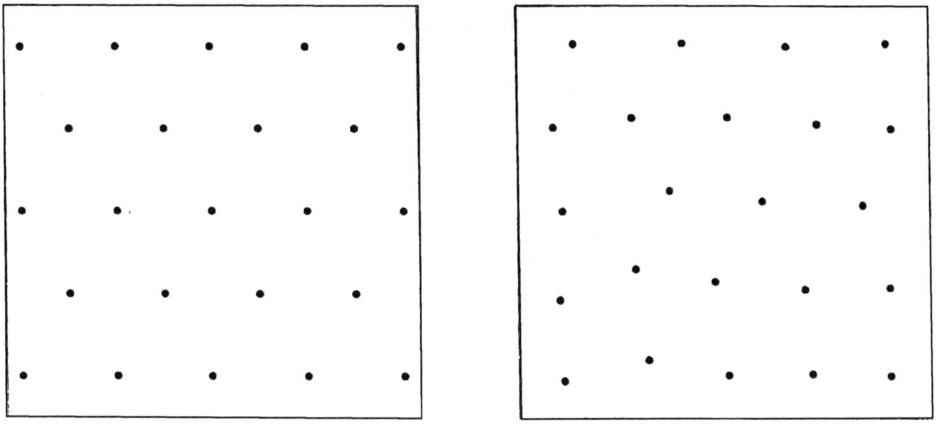

Figure 2. A triangular equilateral grid (a) and an optimized sampling scheme using SSA (b), both using 23 sampling points.

3.2. TEST PROBLEMS FOR THE MMSD CRITERION

In the following two test cases, performance of the sampling schemes is expressed in terms of the ϕ_{MMSD} -estimator, based upon 10 independent sets of test points with n_e = 20000. In this way, the standard deviation of the estimator can be calculated as a measure of reliability. Both cases assume a square area of 10x10 km:

i) Figure 2a shows an equilateral triangular grid of 23 points, which is denoted by i_1. The number of sampling points was chosen to give the equilateral grid an optimal case: it is impossible to make such a grid with 22 or 24 points. The $\phi(i_1)$ is 0.8167 km. (± 0.0013 km.). Figure 2b shows the solution i_2, which was calculated with SSA, using the same number of sampling points, and starting with a completely random scheme i_3 ($\phi(i_3)$ = 1.4344 km. ± 0.0009 km.). This optimized sampling scheme closely resembles the triangular grid, but with small deviations caused by boundary constraints. The $\phi(i_2)$ is 0.8009 km. (± 0.0009 km.), which shows that this is a small, but significant improvement.

ii) The second example illustrates the effects of research- and non-research sub-areas within the area. Figures 3a and 3b show a research area $\mathbf{A_R}$ for soil sanitation, where the upper enclosement represents an area that does not belong to $\mathbf{A_R}$ (a pond), while the lower enclosement represents an area that is part of $\mathbf{A_R}$ but not of $\mathbf{A_S}$ (a building). Figure 3a shows a sampling scheme $\mathbf{i_4}$ based upon an equilateral triangular grid, from which the points that could not be sampled were deleted, leaving 17 sampling points, and $\phi(\mathbf{i_4})$ is 0.9787 km. ± 0.0033 km. Figure 3b show the solution $\mathbf{i_5}$ using SSA, also using 17 points. This scheme has a $\phi(\mathbf{i_5})$ of 0.8875 km. ± 0.0038 km, which is an improvement of 8%. The main advantage of SSA is the consideration of differences between the upper and lower enclosements. In the lower part, the points are located at the boundary of the enclosement, to cover it as good as possible. In the upper part, it keeps distance from the (uninteresting) enclosement. In this case, prior information has a great effect upon the sampling scheme.

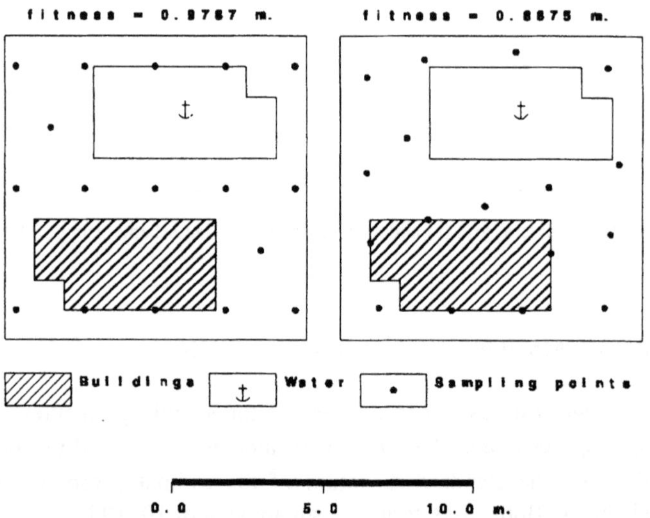

Figure 3. A sampling area with two different enclosements of a different nature, sampled by an equilateral grid (a) and optimized using SSA (b), both using 17 sampling points.

4. Conclusions and Recommendations

In this paper, it was shown that SSA can be applied to design optimal sampling schemes, taking prior information into account. Using information on earlier samplings and the spatial structure of the research area, a truly 'tailor-made' sampling scheme for the area can be composed.

It should be stressed that the given examples were deliberately kept simple, in order to make an easy insight in the benefits of SSA possible. However, SSA can be applied in all areas that can be represented in a GIS. Also, SSA is most beneficial in cases where much spatial prior information is available. In such cases, SSA can easily ensure improvements of more than 20% (Groenigen and Stein, *submitted*).

One of the main benefits of SSA is that it can handle any quantitative optimization criterion, as long as computing times are reasonable. One of the most useful will be an optimization criterion that minimizes the kriging variance, given the variogram. With such a criterion available, a truly optimal phased survey could be designed, with in the first phase a sampling scheme designed to estimate the variogram, and a second phase with additional samples for minimizing the kriging variance.

5. References

Aarts, E., and Korst, J. (1990) *Simulated Annealing and Boltzmann machines - a stochastic approach to Combinatorial Optimization and Neural Computing*, John Wiley & Sons, New York.

Deutsch, C.V., and Journel, A.G. (1987) *GS-LIB - Geostatistical software library and users guide*, Oxford University Press, Oxford.

Farmer, C. (1991). Numerical Rocks, in J. Fayers and P. King (eds), *The mathematical Generation of Reservoir Geology*, Oxford University Press, New York.

Groenigen, J.W., and Stein, A. (submitted) Spatial Simulated Annealing for designing Spatial Sampling Schemes.

Groenigen, J.W., Stein, A., and Zuurbier, R. (in press)) Optimization of environmental sampling using interactive GIS, *Soil Technology*.

Gruijter, J.J. de., and Braak, C.J.F. ter. (1990) Model free estimation from spatial samples: a reappraisal of classical sampling theory, *Mathematical Geology* **4**, 407-415.

Kirkpatrick, S., Gelatt, C.D., and Vecchi, P.H. (1983) Optimization by Simulated Annealing, *Science* **4598**, 671-680.

Laarhoven, P.J.M., and Aarts, E.H.L (1987) *Simulated Annealing: Theory and Applications*, Kluwer Academic Publishers, Dordrecht.

McBratney, A.B., Webster, R., and Burgess, T.M. (1981) The design of optimal sampling schemes for local estimation and mapping of regionalized variables, *Computers and Geosciences* **4**, 331-366.

Stein, A., Staritsky, I., Bouma, J., and Groenigen, J.W. (1995) Interactive GIS for environmental risk assessment, *International Journal of Geographical Information Systems* **5**, 509-525.

Warrick, A.W., and Myers, D.E. (1987) Optimization of sampling locations for variogram calculations, *Water Resources Research* **3**, 496-500.

Webster, R., and Burgess, T.M. (1984) Sampling and bulking strategies for estimating soil properties in small regions, *Journal of Soil Science* **31**, 127-140.

IMPROVED RAINFALL ESTIMATION BY INTEGRATION OF RADAR DATA : A GEOSTATISTICAL APPROACH

CASSIRAGA, EDUARDO F. AND GÓMEZ-HERNÁNDEZ, J. JAIME
Department of Hydraulic and Environmental Engineering
Universidad Politécnica de Valencia, 46071 Valencia, Spain

Abstract. The integration of radar data has been shown a powerful tool for the improvement of rainfall spatio-temporal estimation with respect to the estimation using only raingage data. However, current techniques are limited to standard cokriging algorithms and consider the samples as time independent. We propose the use of the co-located cokriging and kriging with an external drift algorithms, originally developed for the inclusion of geophysical data for the estimation of petrophysical attributes in the petroleum industry, to improve rainfall estimation by integration of radar data. We also propose to include time in the estimation process as a third coordinate, therefore accounting for temporal correlations in the data. These algorithms are demonstrated using part of the information gathered in the "Cévennes 86-88" experiment, consisting of radar data, exhaustively covering the area under study, and point data from 39 raingages, for a rainfall event lasting 5 hours. For this particular data set, it is concluded that accounting for the temporal correlation is not worth the extra modeling effort, in part due to the high point to point correlation between rainfall and radar data. The differences in estimation between including and not the radar data are important even for this case in which the number of rainfall gages is large.

1. Introduction

Rainfall estimation as input to real time streamflow flow models is a difficult problem. Rainfall is an intermittent phenomenon in both space and time and it displays large spatio-temporal variability. Raingage networks collect point estimates of rainfall that are interpolated in some way to provide an

A. Soares et al. (eds.), geoENV I–Geostatistics for Environmental Applications, 363–374.

estimate of the rain spatial distribution within a catchment area. In general, these interpolations provide good estimates of the total amount of rainfall but they do not model accurately the complex spatio-temporal structure of the storm. Better descriptions of the spatial variability of the rainfall can be obtained by incorporating indirect information as such obtained by metereological radar stations. Radar sensors do not measure directly rainfall but reflectivity. Reflectivity measurements are transformed into rainfall estimates through a calibration process of expressions such as:

$$Z = aR^b$$

where Z is radar reflectivity, R is rainfall, and a and b are calibration parameters which are functions of the size distribution of the rain drops. The uncertainty associated to this expression and the difficult calibration of some electronic components of the radar sensor, result in rainfall estimates derived from radar measures with relative errors as high as 100%. Therefore the radar data although abundant in time and space cannot be used directly to estimate the rainfall.

This paper discuss the use of geostatistical techniques to combine raingage data with radar measurements to improve rainfall estimation. The resulting fields honor the raingage information and at the time follow the large spatio-temporal trends present in the radar images. These estimations may be further improved if the time correlation is included in the estimation. The geostatistical techniques discussed next have been widely used in the petroleum industry in a similar context but with different attributes. In petroleum exploration is common to have a vast amount of geophysical information and only a few direct measurements of the attribute of interest. These techniques are the kriging with an external drift and the co-located cokriging (i.e., Deutsch and Journel, 1992). The results of the estimations using these methods are compared with the results obtained with ordinary kriging (not accounting for the radar data) and with full cokriging (accounting for the radar data) since they are the most common techniques used to incorporate radar data into rainfall estimation.

2. Formulation of the problem

The data used in this exercise was collected in the so-called "Cévennes 86-88" experiment (Andrieu *et al.*, 1989). The data set consists of rainfall data from 39 pluviographs and radar data covering an area of 6035 km^2 on a grid of 71 by 85 cells of 1 km^2. Both sets of data contain hourly rainfall amounts corresponding to a given storm for a total of 5 hours. The raingage data are referred to as the "hard data" and represent the primary attribute, whereas the radar derived rainfall quantities are the "soft data"

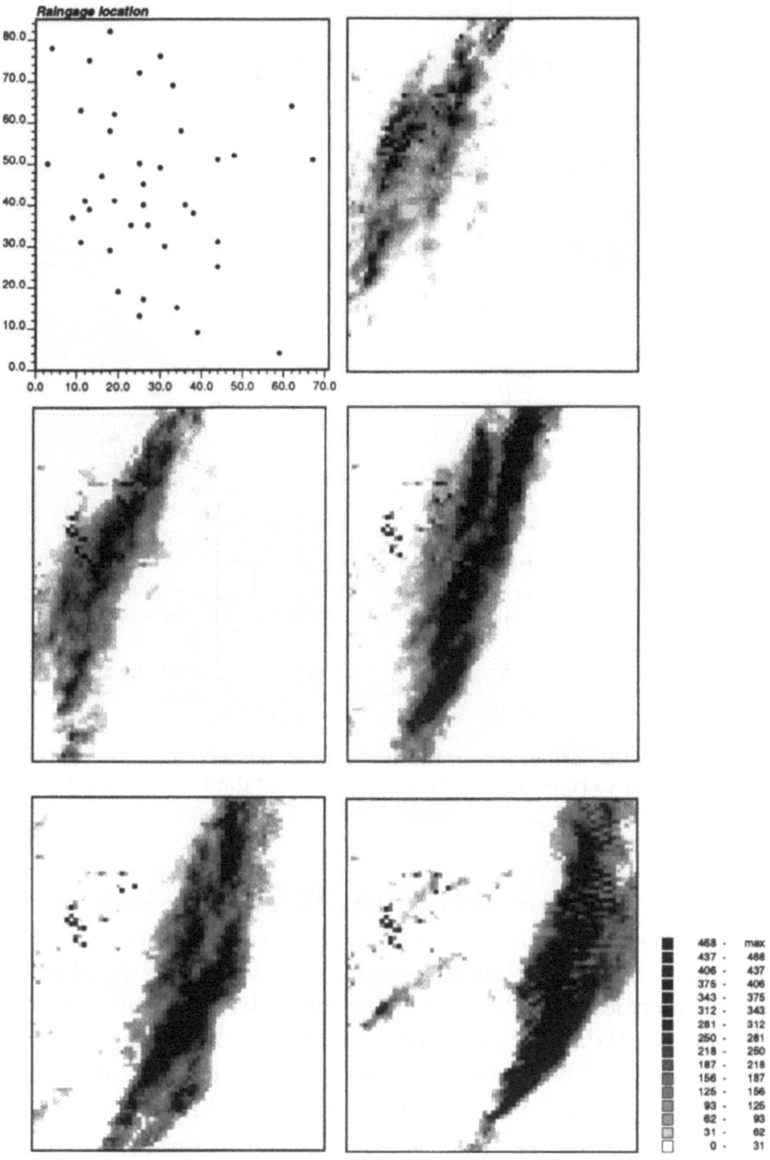

Figure 1. Raingage location and radar images. Radar images are ordered in time from top to bottom and left to right and clearly show the front movement.

and represent the secondary attribute. Figure 1 shows the location of the raingages and the radar images corresponding to the five hours recorded. In order to analyze the temporal correlation, we need to compare values

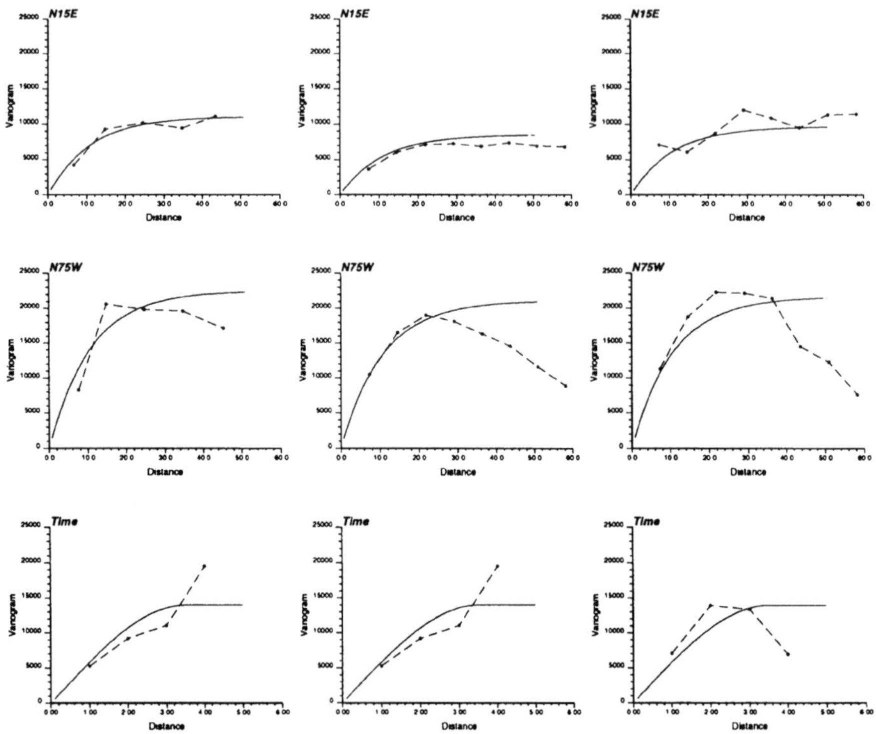

Figure 2. Experimental variograms in the three directions of maximum continuity and models fitted. From left to right: hard data variograms, soft data variograms and hard-soft data cross-variograms.

at different hours that correspond to the same relative position within the moving storm. For this purpose we have used a Lagrangian set of spatial coordinates with an origin that changes in time following the storm center of mass. The locations of the centers of mass derived from the raingage data and from the radar data are almost identical, so we have used one set of centers of mass as the position of our Lagrangian axes of coordinates. Variograms are modeled using the Lagrangian coordinates.

The spatial variograms for hard and soft data are derived from their corresponding data sets. The temporal variogram can only be derived from the radar data, since there are not enough data values to model it after the raingage data. We assimilate the temporal variogram of the hard data to that of the soft data. The experimental variograms and the models fitted are shown in Figure 2.

The first set of results correspond to the situation in which the radar data is not accounted for. A three dimensional (x, y, t) interpolation is car-

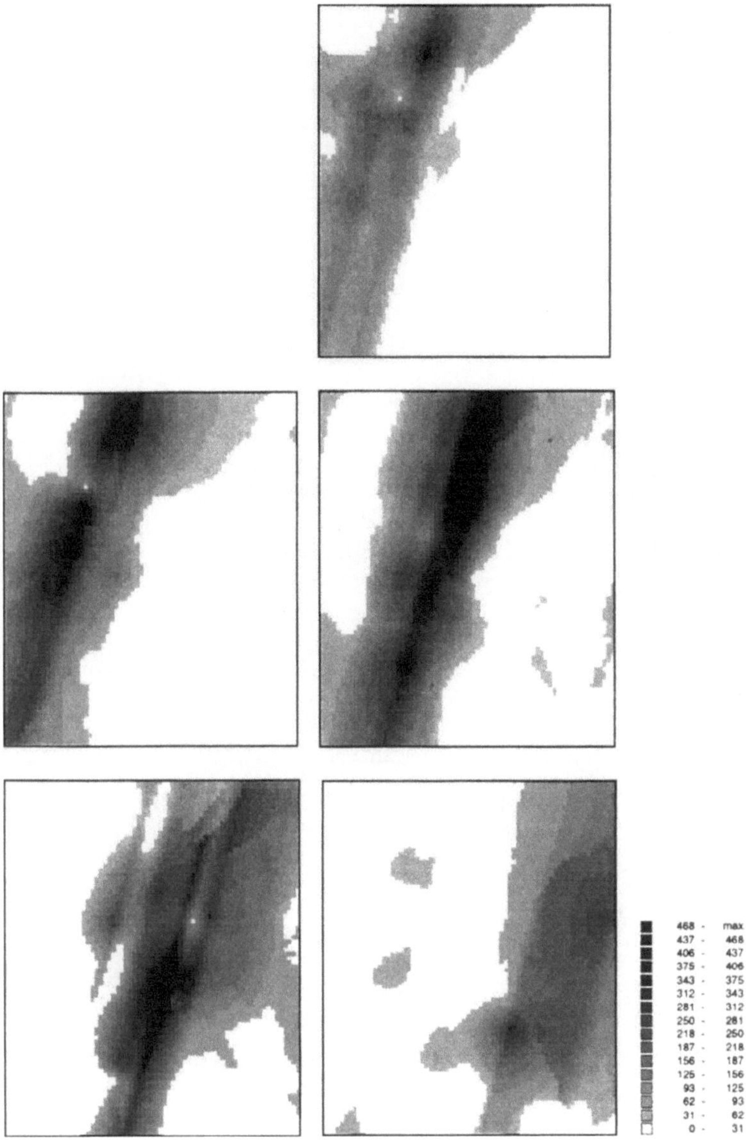

Figure 3. Ordinary kriging estimates.

ried out by ordinary kriging (OK) (Isaaks and Srivastava, 1989) using the 195 hard data (5 hourly sets of 39 rainfall measurements). Each cell over the grid of 71 by 85 cells is estimated in two ways, in the first way, the estimate is a linear combination of the 16 closest data in the same temporal plane

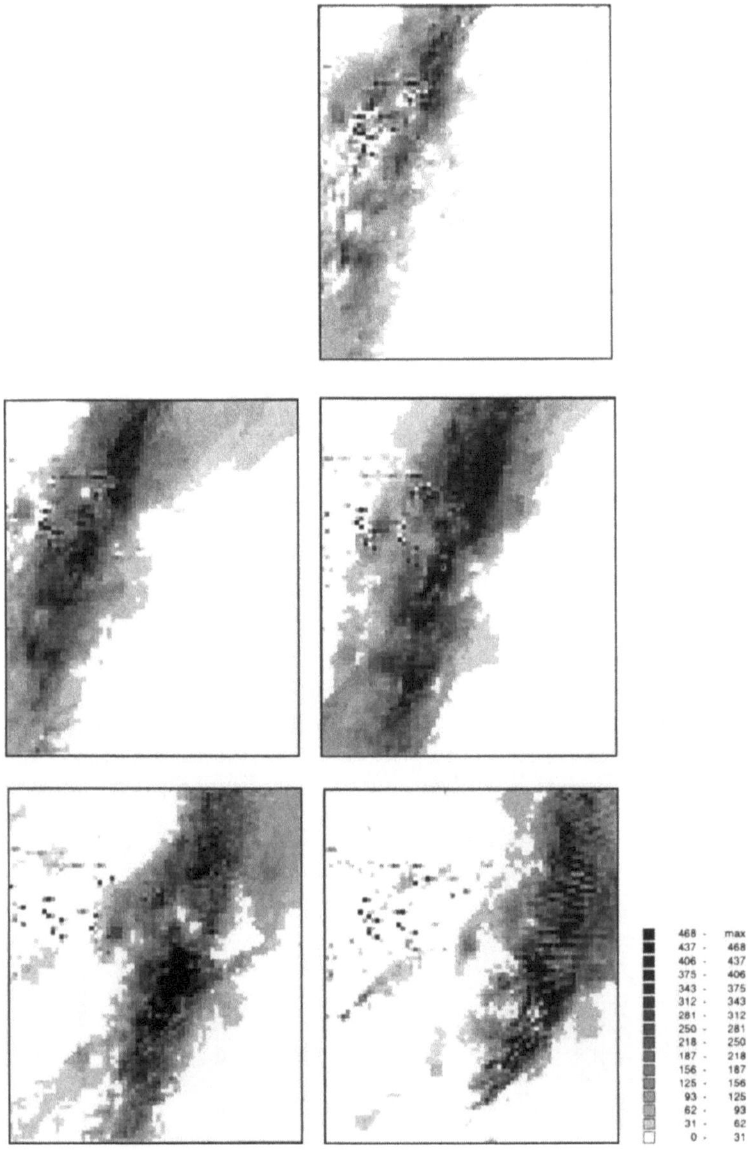

Figure 4. Ordinary cokriging estimates.

(temporal correlation is not taken into account), in the second, the estimate is a linear combination of the 16 closest data within a search neighborhood that is global in space and up to plus/minus two hours in time (temporal correlation is taken into account). Figure 3 displays the results for the case

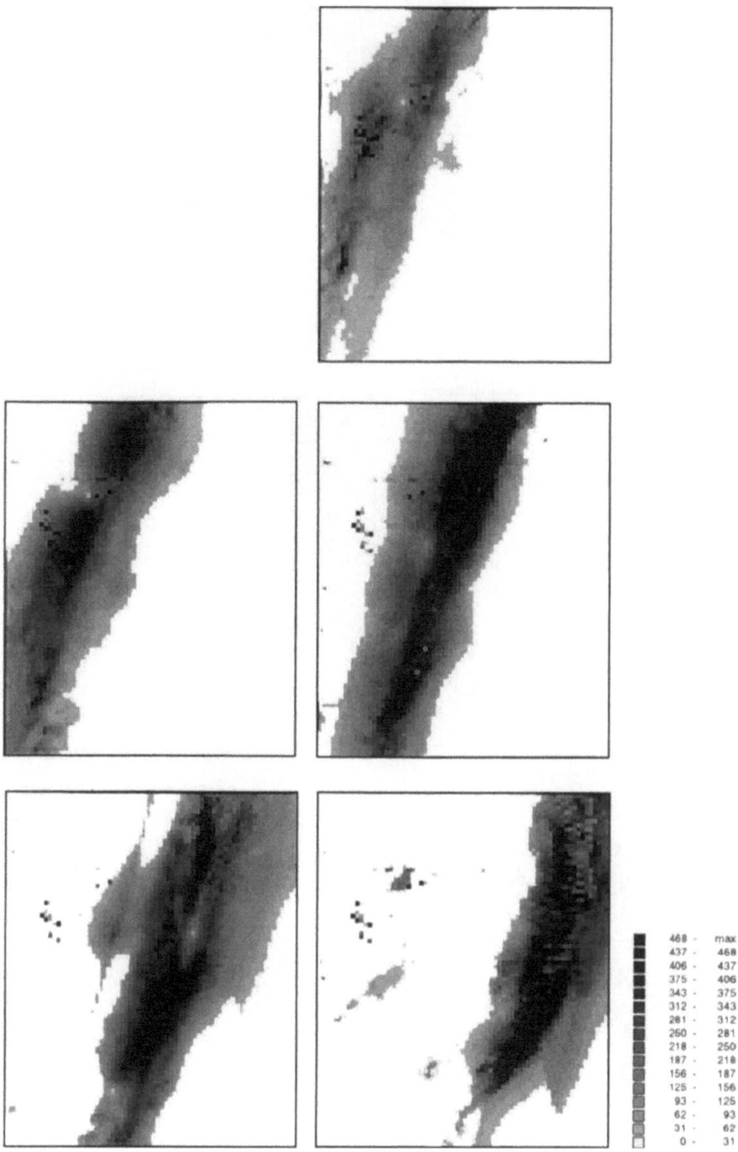

Figure 5. Co-located cokriging estimates.

in which temporal correlation is taken into account.

The second set of results uses ordinary cokriging (CK) (Isaaks and Srivastava, 1989). The models fitted to the variograms in Figure 2 correspond to a linear model of coregionalization. The estimation of each of the cells

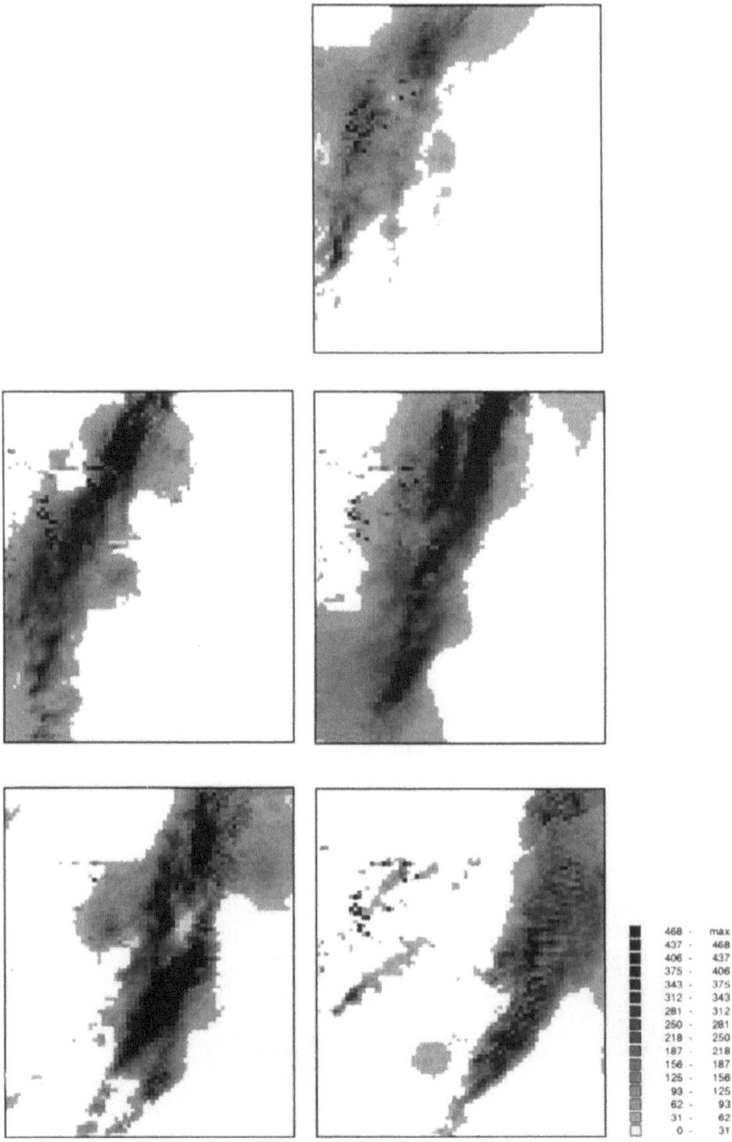

Figure 6. Kriging with external drift estimates.

in the grid is obtained as a linear combination of the 16 closest hard data and the 16 closest soft data. The estimation was repeated with and without search in the time direction in order to analyze the influence that accounting for temporal correlation has in the estimates. Figure 4 shows the results of the ordinary cokriging estimation accounting for temporal correlation.

The third set of results uses co-located cokriging (CoCK) (Xu *et al.*, 1992). This algorithm is a simplified version of ordinary cokriging in which the soft data used for the estimation is limited to the single soft datum co-located on the node being estimated. This method requires modeling of the hard data variogram plus the coefficient of correlation between hard and soft data. This coefficient of correlation is derived from the 195 pairs of hard and soft data available at raingage locations and, for this data set is 0.8. As in the previous cases, the closest 16 hard data are used in the estimation of each node of the grid. Estimation with and without temporal correlation are also performed. Figure 5 shows the results accounting for temporal correlation.

The fourth and final set of results are obtained by kriging with an external drift (KED) (Deutsch, 1991). This type of kriging assumes that the data should be modeled as a drift term plus a residual and that the drift term is an (unknown) linear function of the soft data. Application of kriging with an external drift requires modeling the variogram of the residuals. In this case, we have use an isotropic variogram equal to the variogram in the direction of maximum continuity of the hard data, under the assumption that the drift is less pronounced in the direction of maximum continuity. As in all other cases, the 16 closest data are used to carry out the estimation. The results can be seen in Figure 6.

3. Analysis of the results

3.1. LARGE SCALE VARIABILITY

A good estimation procedure should reproduce the complex spatio-temporal structure observed in the radar images. It is important to reproduce the formation and displacement of convective cells, and the abrupt transitions in rain intensity displayed by trailing fronts. Analyzing the previous figures from a purely visual perspective we can conclude that: a) Ordinary kriging of the hard data cannot reproduce the complex spatial patterns observed in the radar images, b) Ordinary cokriging, is capable of producing estimates with the same variability patterns as the radar images, as it has already been shown by others (Krajewski, 1987; Creutin *et al.*, 1988; Azimi-Zonooz *et al.*, 1989; Seo *et al.*, 1990a,b). However, the modeling of the direct and cross-variograms is time consuming and difficult to automatize, and the estimation itself is CPU-intensive. c) Co-located cokriging can be considered a good alternative to standard co-kriging. The estimated fields produced by co-located cokriging are very similar to those produced by standard co-kriging with the advantage that co-located cokriging only requires the coefficient of correlation between the two variables in addition

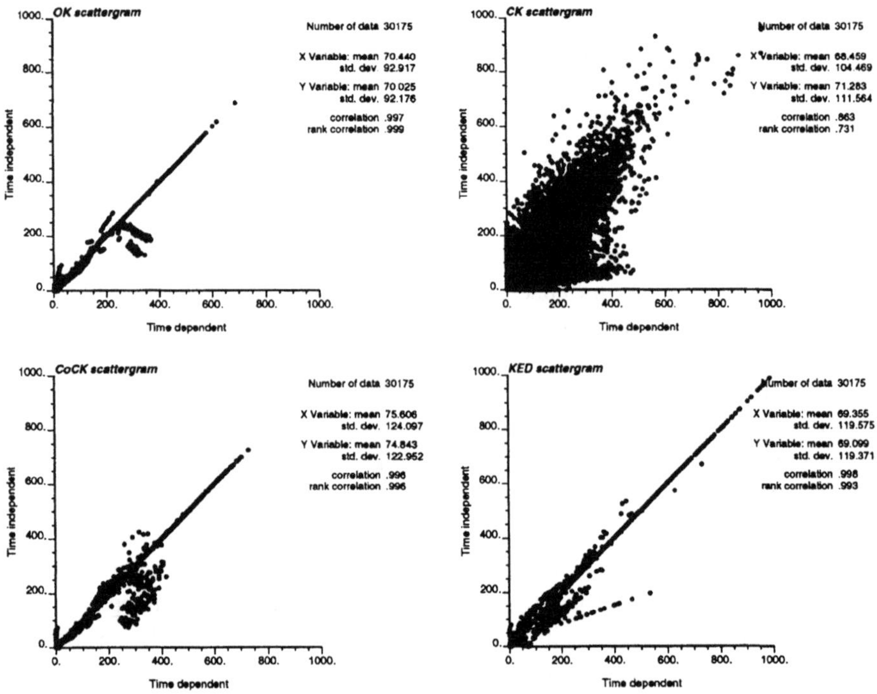

Figure 7. Scattergrams of the estimates with and without accounting for the temporal correlation.

to the variogram of the hard data and the computation of the estimates is much faster than standard cokriging (in this case 15 times faster) This may be a factor to be considered if we wish to consider these method in real-time streamflow flow forecasts. d) Kriging with external drift is the fastest method considered; it appears gives the best results, in the sense that it is the method that bests reproduces the complex patterns observed in the radar images. This method has the disadvantage requiring modeling of the variogram of residuals, and the technique should be used with care with sparse hard data since in the case of few hard data may produce artifacts in the estimates.

3.2. TEMPORAL CORRELATION

Scattergrams of the estimates with and without accounting for the temporal correlation are plotted in Figure 7. For this specific data set, the extra effort required to model the temporal variograms is not worth the effort since the differences are minimal. Part of the reason for this behavior is the

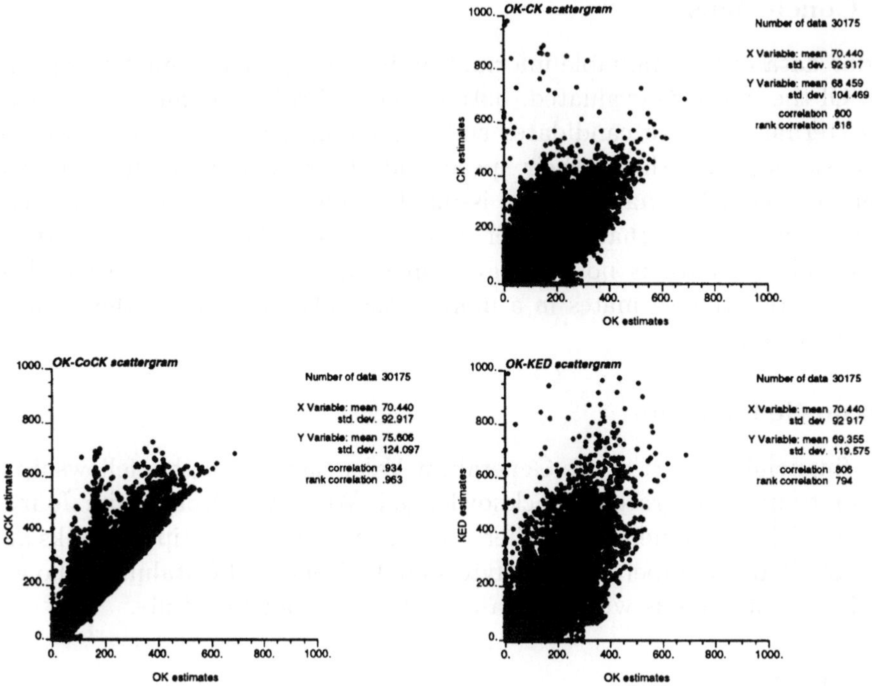

Figure 8. Scattergrams of the estimates obtained by ordinary kriging and obtained by the other methods.

large correlation coefficient between soft and hard data which enforce the Markovian assumption underlying co-located cokriging, in the sense that the single soft datum co-located at the cell being estimated carries so much information about the hard datum that it screens out much of the weights given to the other (hard and soft) data. We believe that for other data sets for which the point correlation between hard and soft data is not as high as in this case, including the temporal correlation may improve noticeably the rainfall estimates.

3.3. RADAR DATA

Figure 8 shows scattergrams of the estimations obtained by ordinary kriging and obtained by the other methods. It is evident that including the radar data produces a substantial change in the estimates.

4. Conclusions

Radar data carries valuable information for the spatial estimation of rainfall. Of the methods evaluated, both co-located cokriging and kriging with an external drift are candidates suitable for incorporation in a real-time forecast model, both for their speed and the relatively small amount of inference needed. Finally, on the issue of accounting for the temporal correlation, we believe that in other cases in which the correlation between hard and soft data is not so high, using data from adjacent time slices may improve the estimates in a more noticeably way that in the example presented here.

Acknowledgements

The financial support of the Generalitat Valenciana through a fellowship to the first author is gratefully acknowledged. We thank Prof. André Journel for his critical comments to earlier versions of this manuscript. Thanks also to Prof. Daniel Sempere from Polytechnic University of Catalunya who was so kind to provide us with the data set to carry out the study.

References

Andrieu, H., J. D. Creutin, J. Delrieu, J. Leoussof, and Y. Pointin (1989). Radar data processing for hydrology in the Cévennes region. In Remote Sensing and Large-Scale Global Processes (Proceedings of the IAHS Third Int. Assembly, Baltimore, MD, May 1989. IAHS Publ. no 186.

Azimi-Zonooz, A., W. F. Krajewski, D. S. Bowles, and D.-J. Seo, (1989). Spatial rainfall estimation by linear and non-linear co-kriging of radar- rainfall and raingage data. Stochastic Hydrology and Hydraulics., 3, 51-67.

Creutin, J. D., G. Delrieu, and T. Lebel, (1988). Rain measurement by raingage-radar combination: A geostatistical approach. J. Atmos. Oceanic Technol., 5 (1), 102-115.

Deutsch, C. V. (1991). The relationship between universal kriging, kriging with an external drift, and cokriging, *Stanford Center of Reservoir Forecasting (SCRF), Report 4*, May 1991.

Deutsch, C. V. and Journel, A. G. (1992). *GSLIB, Geostatistical Software Library and User's Guide.* Oxford University Press, New York.

Isaaks, E. and R. M. Srivastava (1989). *An Introduction to Applied Geostatistics,* Oxford University Press, New York.

Krajewski, W. F., (1987). Cokriging radar-rainfall and rain gage data. Journal of Geophysical Research, 92, (D8), 9571-9580.

Seo, D.-J., W. F. Krajewski, and D. S. Bowles, (1990a). Stochastic interpolation of rainfall data from rain gages and radar using cokriging. 1. Design of experiments. Water Resources Research, 26 (3), 469-477.

Seo, D.-J., W. F. Krajewski, A. Azimi-Zonooz and D. S. Bowles, (1990b). Stochastic interpolation of rainfall data from rain gages and radar using cokriging. 2. Results. Water Resources Research, 26 (5), 915-924.

Xu, W., Tran, T. T., Srivastava, R. M. and Journel, A. G. (1992). Integrating seismic data in reservoir modeling: The co-located cokriging alternative, SPE paper # 24742.

STOCHASTIC SIMULATION OF LICHEN BIODIVERSITY USING SOFT INFORMATION FROM REMOTE SENSING DATA

A. SOARES[a], M. J. PEREIRA[a], C. BRANQUINHO[b], F. CATARINO[b]
[a] *Instituto Superior Técnico, Av. Rovisco Pais 1096 Lisboa Codex, Portugal*
[b] *DBV, Faculdade de Ciências da Universidade de Lisboa,*
R. Ernesto Vasconcelos, Bloco C2, 4° Piso, 1700 Lisboa, Portugal

Abstract: This paper aims at modelling the air quality around a mine site, combining hard data - the field measurements of the number of epiphytic lichen species - and soft data - a remote sensing image of the region. The use of epiphytic lichens as bioindicators of air pollution is due to their identifiable reactions to different degrees of pollution. In a first step a calibration between the hard and soft data is carried out through a Probabilistic Neural Network classification algorithm. A new approach is proposed for the estimation of local conditional distribution functions with hard and soft derived data. An important tool for the air quality control due to mining activity is the stochastic simulation based on local pdf^s providing a set of equiprobable images set.

1. Introduction

This paper proposes a new method to combine remote sensing data with field measurements for the stochastic simulation of total number of epiphytic lichen species, a current air quality indicator.

The total number of epiphytic lichen species in a discrete sample grid around a mine site is related with the air quality and the vegetation healthiness due to possible damage by heavy metal particle emissions from the mine activity.

A remote sensing image, which may account for the vegetation healthiness through its spectral properties, was used in this study as a secondary variable for the air quality characterisation.

Taking advantage of a nice property of the probability field simulation (Srivastava 1992, Froidevaux, 1993), in which the estimation of local probability distribution functions (pdf) is prior to and independent of the simulation algorithm, the key idea of the proposed method is to build local pdf^s conditioned to hard data - field measurements

A. Soares et al. (eds.), geoENV I–Geostatistics for Environmental Applications, 375–387.

- and to a dense grid of soft data - remote sensing images -, before the simulation of stochastic images of the air quality indicator.

Two main steps characterise this methodology:

i) In a first step a probabilistic neural network - PNN - was trained for classifying (in classes of a cumulative histogram of lichen species) a remote sensing image of the studied area. This is achieved by using the energy levels and a vegetation index from the remote sensing data and the epiphytic lichen's biodiversity measured at sampling locations. The output of this stage is the image of cumulative probabilities of the primary variable that results from transforming the remote sensing image by the PNN classification technique.

ii) Secondly, the soft derived map of i) is conditioned to hard data in order to make it pass through the experimental control points of sample data: the final conditioned pdf^s are obtained by adding a deviation from the soft derived map to the estimated map of hard indicator data. This resulting pdf^s over whole image of studied area always makes the influence of hard data prevail over the soft data.

Finally, based on these pdf^s, the probability field simulation is performed to obtain the set of equiprobable images of the air quality indicator, which are an important tool for the environmental impact control of the mine activity.

2. Description of Proposed Methodology of Stochastic Simulation

2.1 GENERAL DESCRIPTION

The incorporation of different kinds of information (soft and hard data, related to one or more attributes) in simulation models has been performed, through corregionalization models, by many authors. Doyen (1988) used cokriging with seismic and well data to model oil reservoir properties. Nevertheless, cokriging usually implies a cumbersome practice aiming at modelling the corregionalizations between very different (in structural terms) types of data, which makes this attractive technique difficult to use in most of those situations. Journel and Zhu (1990) proposed a Markov-Bayes formalism to integrate seismic and well data to characterise internal properties of oil reservoirs. Recently, Colin *et al.* (1996) succeed to integrate seismic data and field measurements, for soil pollution characterisation, with the simple indicator kriging with a bivariate histogram. Brito and Rogado (1996) presented a stochastic simulation with very dense grid of soft data - seismic data - to characterise the weatherability of a stone quarry.

The methodology presented in this paper can be summarised by the following sequence of steps:

- A bi-variate analysis between a primary variable - lichen's biodiversity sampled in some field points - and a secondary soft variable - remote sensing image of the same region - is processed through a Probabilistic Neural Network - PNN classification technique. The PNN was trained with the control points with the objective of assigning to each pixel the probability of belonging to the different classes of the cumulative histogram of primary variable.
- In a second step, these soft PNN derived images are conditioned to hard information in order to obtain a local pdf^s for each grid cell. These local pdf^s prioritise the influence of hard data over soft data.
- Finally, the set of equiprobable images are obtained by stochastic simulation - probability field simulation.

2.2 PROBABILISTIC NEURAL NETWORKS FOR CLASSIFICATION OF SOFT DATA

The Probabilistic Neural Networks technique has two advantages when compared with other parametric Bayes classifiers: it makes no assumptions about the distribution of the variables and it can make highly non-linear decisions. Therefore, whenever the variables involved in a classification process are highly non linear, the neural networks technique has the potential of outperforming a parametric Bayes classifier (Masters, 1995).

The PNNs are based on the Bayes classification method (Specht, 1990) which objective is the minimisation of the expected cost of misclassification. Consider a collection of random samples, each represented by vector $\mathbf{x}=[x_1,\ldots,x_m]$, from K populations (k=1,...,K). Different populations may have different *prior probabilities* h_k to generate samples. If the cost associated with the misclassification of a sample which truly belongs to the population k is designated by c_k, then the unknown sample $\mathbf{x}$ is classified into population i if:

$$h_i\, c_i\, f_i\,(\mathbf{x}) > h_j\, c_j\, f_j\,(\mathbf{x}) \qquad \forall\, j \neq i \tag{1}$$

It can be proved that if the true probability density functions $f_k(\mathbf{x})$ are known for all populations, then this is a Bayes optimal decision rule.

The probability distribution function of each population is estimated by the Parzen's method (Parzen, 1962) in which the estimator is the sum of a weight function W (usually designated by kernel) for sample cases. If we consider the univariate case and a set of n samples, the Parzen estimator is given by the following expression:

$$f_k(x) = \frac{1}{n\sigma}\sum_{i=1}^{n} W\left(\frac{x - x_i}{\sigma}\right) \tag{2}$$

The σ is a scale parameter that defines the extent of influence of the sample in the $f_k(\mathbf{x})$ estimation. The Parzen's method can be extended to the multivariate case (Cacoullos, 1966).

The architecture of the PNNs(fig. 1) consists of four neuron layers (Pereira, *et al.*, 1996): the input layer, the pattern layer, the summation layer and the output layer. The input layer has as many neurons as the number of variables involved in the classification process. The pattern layer contains one neuron for each *training case*, *i.e.* a set of values for the input variables which characterise a certain class (population). The summation layer has a neuron for each class; the output class has only one neuron. The training process uses an optimisation algorithm to estimate σ values.

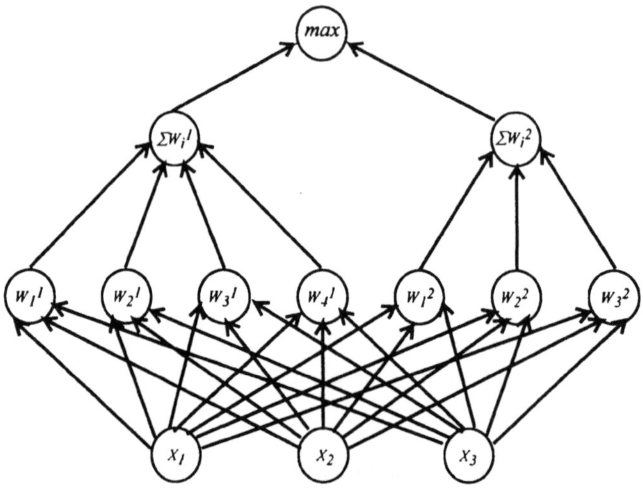

figure 1. PNN architecture.

The classification process starts by presenting simultaneously the input vector to every neuron at the pattern layer. Each neuron of this layer calculates the distance *d*, between the input and the training case represented by the neuron, which is the argument of the activation function - kernel function with a certain σ - of the neuron. Then, each neuron of the summation layer sums the values of the neurons of the pattern layer that belongs to its class. Finally, the neuron of the output layer selects the class corresponding to the maximum value between the summation layer neurons.

When the classes are mutually exclusive and cover all the populations, then the Bayes theorem can be applied, *i.e.* it is possible to calculate the probability that population A generate the $\boldsymbol{x}$ observation:

$$P[A|x] = \frac{f_A(x)}{\sum_k f_k(x)} \qquad [3]$$

This is one of the advantages of using the PNNs for classification purposes because, instead of allocating a unknown sample to a unique class, it gives the probability of that sample of belonging to each class.

2.3 ESTIMATION OF LOCAL PDF^S CONDITIONED TO SOFT AND HARD DATA

Consider in area A a primary variable Z(x) related with one characteristic - the number of lichen species whose spatial dispersion is to be modelled. $Z(x_\alpha)$ is known in a few sample locations x_α, $\alpha=1, N$.

A very dense regular grid of a secondary variable Y(x) - a remote sensing image, in this study - somehow related with Z(x), is also known for the entire area A. The objective is to calculate for each grid cell in A a local pdf of Z(x) conditioned to $Z(x_\alpha)$ and the soft data image Y(x), $x \in A$.

If only hard data were taken into account, the local pdf^s could be accessed by the estimated map over the entire area A (Journel, 1989):

$$F_z(x)^* = prob\{Z(x) < z | Z(x_\alpha), \alpha = 1, N\}^* = [I(x)]^* = \sum_\alpha \lambda_\alpha I(x_\alpha) \qquad x \in A \qquad [4]$$

given the indicator variable I(x) defined by a cutoff z.

Based on soft data Y(x) and hard data Z(x) the following *a priori* pdf^s can be calculated by regression:

$$F_z^o(x) = prob\{Z(x) < z | Y(x)\} = P(x) \qquad x \in A \qquad [5]$$

This calibration procedure between hard data Z(x) and soft data Y(x) is usually made through a regression technique using both set of values in hard data location (see 2.2). Afterwards the conditional pdf^s are extrapolated for the entire area A, under some spatial homogeneity and a sample representativity assumptions.

Once P(x) and I(x) have the same meaning (not the same statistics), the idea of this proposed algorithm is to make the map of soft derived data, which covers the entire area A, match the experimental data values $I(x_\alpha)$.

One way of doing this conditioning to hard data (which is to make the soft image honour the control points) is to add to estimated map of $F_z(x)^*$ [4] a deviation $e_p(x)$ between soft derived $F_z^o(x)$ [5] and an estimated $F_z^o{}^*(x)$:

$$e_p(x) = \left[F_z^o(x) - F_z^o{}^*(x)\right] \qquad [6]$$

The estimated map of P(x) is:

$$F_z^o{}^*(x) = \sum_\alpha \beta_\alpha P(x_\alpha) \qquad [7]$$

where $P(x_\alpha)$ are the soft derived values at hard data locations x_α, $\alpha=1, N$.

Thus, the final soft derived map conditioned to hard data is:

$$F_z^{'}(x) = F_z^*(x) + e_p(x) \qquad [8]$$

where $e_p(x)$ is an error measure between the estimated map using soft data and the soft derived map. $e_p(x)$ decreases with the proximity of sample locations - $e_p(x_\alpha) = 0$ in any sample location x_α, $\alpha=1$, N - hence the hard data influence prevails over the soft data:

$$F_z^{'}(x_\alpha) = F_z^{*}(x_\alpha) \quad [9]$$

On the other hand, for any location x_o far from the influence of experimental samples (given by the ranges of variograms of I(x) and P(x)), the resulting pdf is given by the soft image corrected for the mean of hard data:

$$F_z^{'}(x) = F_z^{o}(x_o) + [m_I - m_P] \quad [10]$$

If the soft derived map and hard data have the same mean, the final map is equal to soft derived map in areas of no influence of hard data.

$$F_z^{'}(x) = F_z^{o}(x_o) \quad [11]$$

Note that the final local pdf' are dependent on hard data location, soft data spatial pattern and the continuity structures of hard and soft data as revealed by the respective variograms.

Remark: $F'_z(x)$ given by [8] may not be in the [0,1] interval. Thus a single transformation to [0,1] range is needed before considering it as a probability distribution function.

3. Case Study: Stochastic Simulation of Epiphytic Lichens Biodiversity

This study was conducted in a mine site - a copper mine located at south of Portugal - in which the heavy metal particulate emissions from the copper concentrate stockpiles are an important source of air pollution.

The objective of this study is twofold: i) spatial characterisation of the effects of the air pollution through a stochastic imaging technique using hard and soft data and ii) using remote sensing data and just a few control points (field measurements) as a tool to control the dynamic evolution of the effects of mine activity on air quality.

3.1 USE OF EPIPHYTIC LICHENS AS BIOINDICATORS

The use of bioindicators of air pollution is based on the assessment of environmental changes due to air pollution on living organisms. An organism is considered a bioindicator when it presents identifiable reactions to different degrees of pollution (Nimis *et al.*, 1991). The use of organisms as indicators of air quality is based on the assumption that the reaction of living organisms to pollution phenomena is a better estimate of damage to man or other animals than quantitative data on the concentrations of single pollutants from physical sample devices (Nimis *et al.*, 1991). In the case of bioindicators, the exact nature of pollution, the concentration of substances involved and the duration of pollution incidents are not determined, but real potentialities for damage will be identified. Epiphytic lichens have been extensively used as good indicators of air pollution and/or air quality, and they have been used throughout the world (Nimis *et al.*, 1991, Catarino *et al.*, 1991) These organisms show differential sensibilities to air pollution - *i.e.*, the most sensitive lichens tend to disappear from polluted areas whereas the most tolerant species can be seen in areas with moderate pollution emissions - allowing us to use them as bioindicators. In this work we have used the number of epiphytic lichen species, a direct measure of biodiversity, to evaluate the air quality in the surroundings of a copper mine.

3.2 DATA AVAILABLE AT MINE SITE

Two types of data are available at the mine site:

- **Hard Data**: they resulted from a lichen biodiversity study - at each sampling point the number of epiphytic lichen species was recorded from 6-10 *Quercus ilex* trees and exceptionally from *Olea europeae* in an average area of 15 x 15 m. The sampling was conducted according to the pattern shown in fig. 2, between May and December 93. The number of lichens species varied from 5 at the centre of the mine to 20-30 species at 1 km from the mine (see Branquinho *et al.,* 1994, for a more detailed description).

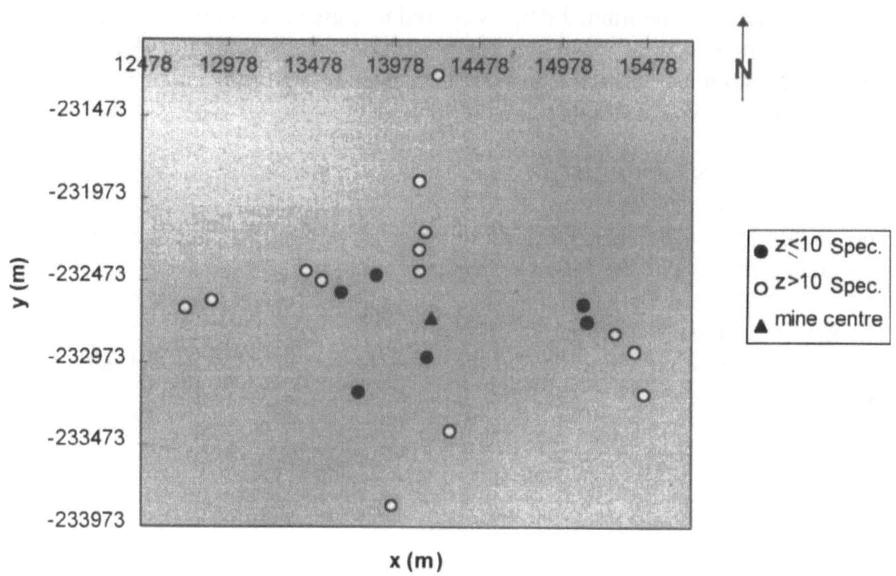

figure 2. Studied area and sample locations.

- **Soft Data:** consisted on a remote sensing image of Landsat TM (resolution of 30 x 30 m^2) referring to the hard data sampling period. The image covers an area of about 15 x 9km^2, surrounding the mine site. One vegetation index was calculated and the image was rectified using control points for geopositioning. Thereafter, the image was cut to a smaller area of 3.3 x 3 km^2, limited by the sampling area (fig. 2), which was the grid used in the following steps of the case study.

3.3 CLASSIFICATION OF REMOTE SENSING IMAGE BY PNNS

The sampling locations were identified on the remote sensing image. At these points, the energy levels of each band were recorded for the central and the 8 surrounding pixels.

For training the PNN, three variables were selected from the remote sensing image: band 1 and 2, and a vegetation index (calculated using band 3 and 4). The training set consisting of 180 samples, was divided in two classes: class 1 with less or equal than 10 species representing 25% of the total number of samples and class 2 with more than 10 species. For the surrounding area of the mine site, the average number of epiphytic lichen species is around 20-30 and 10 species is considered a low value caused by the mine activity.

The PNN was validated performing a classification with an independent data set which was not used for training the PNN. The global success of classification was of about 80.6%. Afterwards, the trained PNN was used to classify the entire image, *i.e.* for each pixel the probability of belonging to each class was calculated. Finally, as each sampling location was considered to be represented by 9 pixels, a moving window filter averaging the 9 pixels was applied to the whole image.

3.4 ESTIMATION OF LOCAL PDF^S

For the threshold - 10 lichen species - an indicator variable was defined at experimental data points (fig. 2) with the following statistics:

$$m_I = 0.30 \qquad \sigma_I^2 = 0.21$$

The omnidirectional variogram calculated (fig. 3a) reflects basically the shape of the affected vegetation coverage around the mine site. A spherical model with 350 m of range and null nugget effect was fitted to the experimental variogram.

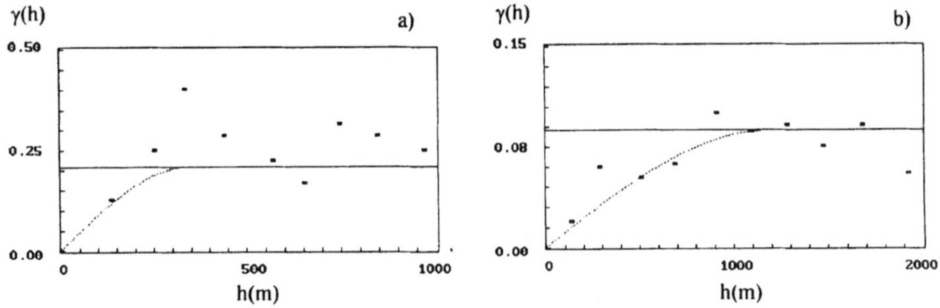

figure 3. a) Indicator variogram of hard data; b) Variogram of the uniform variable.

The omnidirectional variogram of the uniform transform of the number of lichen species variable (fig. 3b) was also calculated. A spherical model with 1200 m of range and null nugget effect was fitted to the experimental variogram. This variogram was used to generate the probability field in the stochastic simulation.

The soft derived images of PNN (of 3.3) of P(x) variable give rise to the variograms of fig. 4.

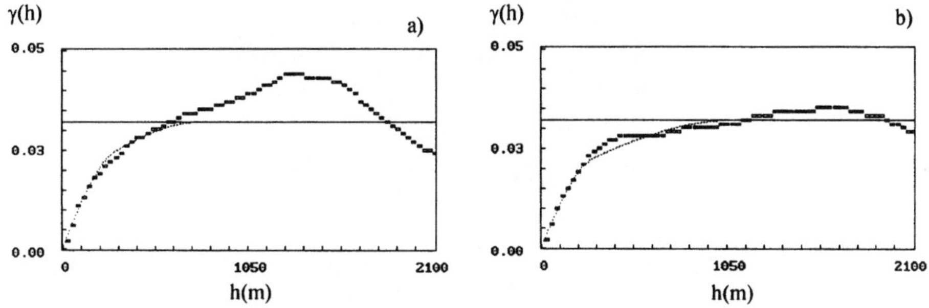

figure 4. Indicator variograms of soft derived data: a) W-E direction; b) N-S direction.

The two estimated maps [4] and [7] using hard and soft data, respectively, in sample locations were obtained by indicator kriging. Fig. 5 resumes the calculation of the final map of pdf^s [eq. 8] through the 3 maps of $F_z{}^*(x)$, $F_z{}^o(x)$ and $F_z{}^{o*}(x)$:

$$F_z\left(x\middle|Z(x_\alpha),Y(x)\right)=F_z^*(x)+\left[F_z^o(x)-F_z^{o*}(x)\right]$$

$F_z{}^*(x)$ is the estimated map of hard data (fig. 5a); $F_z{}^o(x)$ is the soft derived map (fig. 5b); $F_z{}^{o*}(x)$ is the estimated map of soft data by using the soft information corresponding to experimental sample locations (fig. 5c).

3.5 STOCHASTIC SIMULATION OF LICHEN BIODIVERSITY

A sequential gaussian simulation was used to obtain the probability field realisations p(u) corresponding to 30 images of binary clouds. p(u) is a uniform distributed variable between 0 and 1. In any location x an indicator value was simulated for each realisation p(u), using the local estimated pdf of 3.4:

$$I(x)=\begin{cases}1 & if \quad p_s(u)<F_z\left(x\middle|Z(x_\alpha),Y(x)\right)\\ 0 & otherwise\end{cases}$$

In fig. 6 it is shown 11 simulated images and an average image (bottom right corner) which is, as expected, quite similar to local pdf^s image.

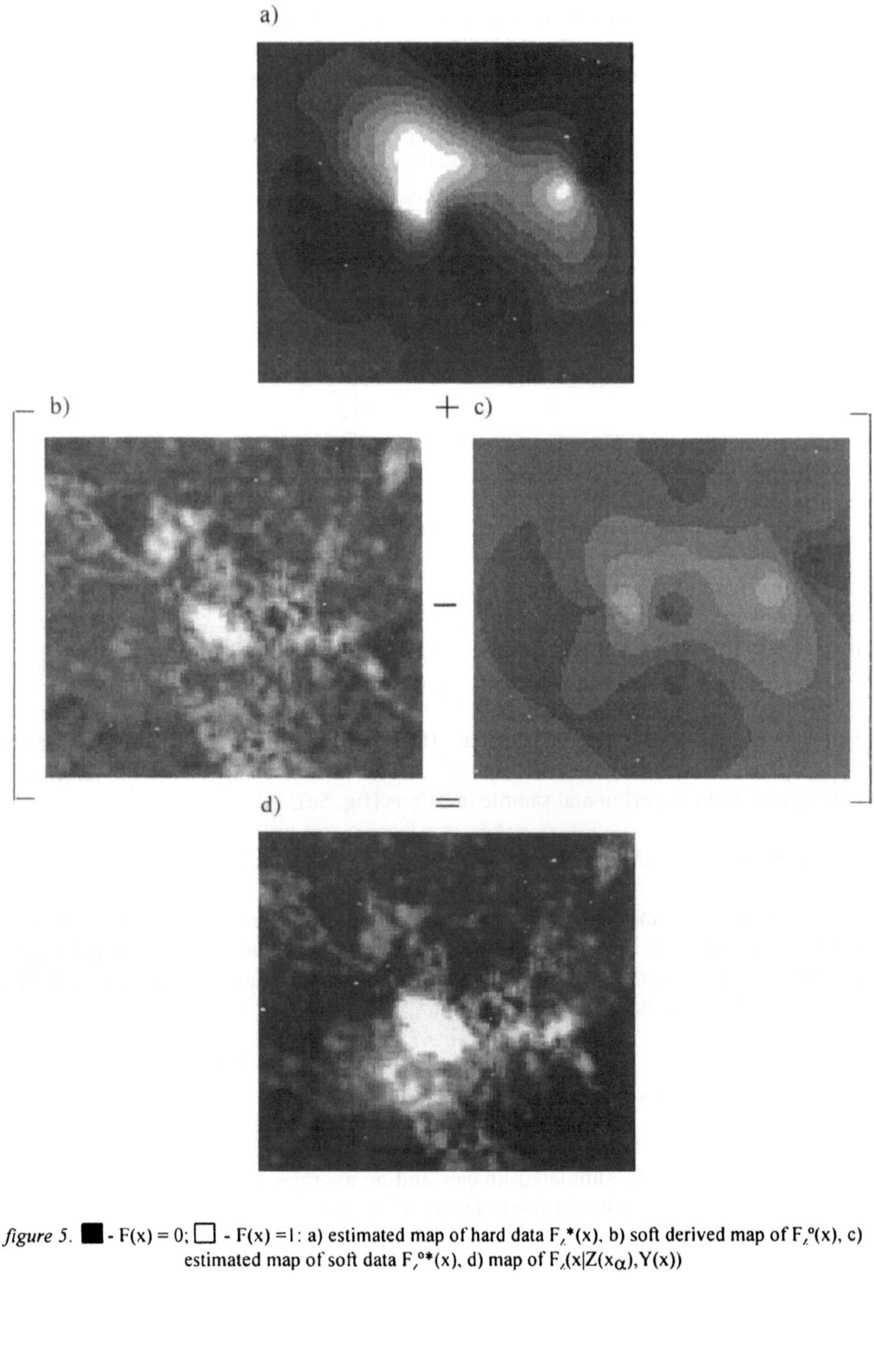

figure 5. ■ - F(x) = 0; □ - F(x) = 1: a) estimated map of hard data $F_z^*(x)$, b) soft derived map of $F_z^o(x)$, c) estimated map of soft data $F_z^{o*}(x)$, d) map of $F_z(x|Z(x_\alpha),Y(x))$

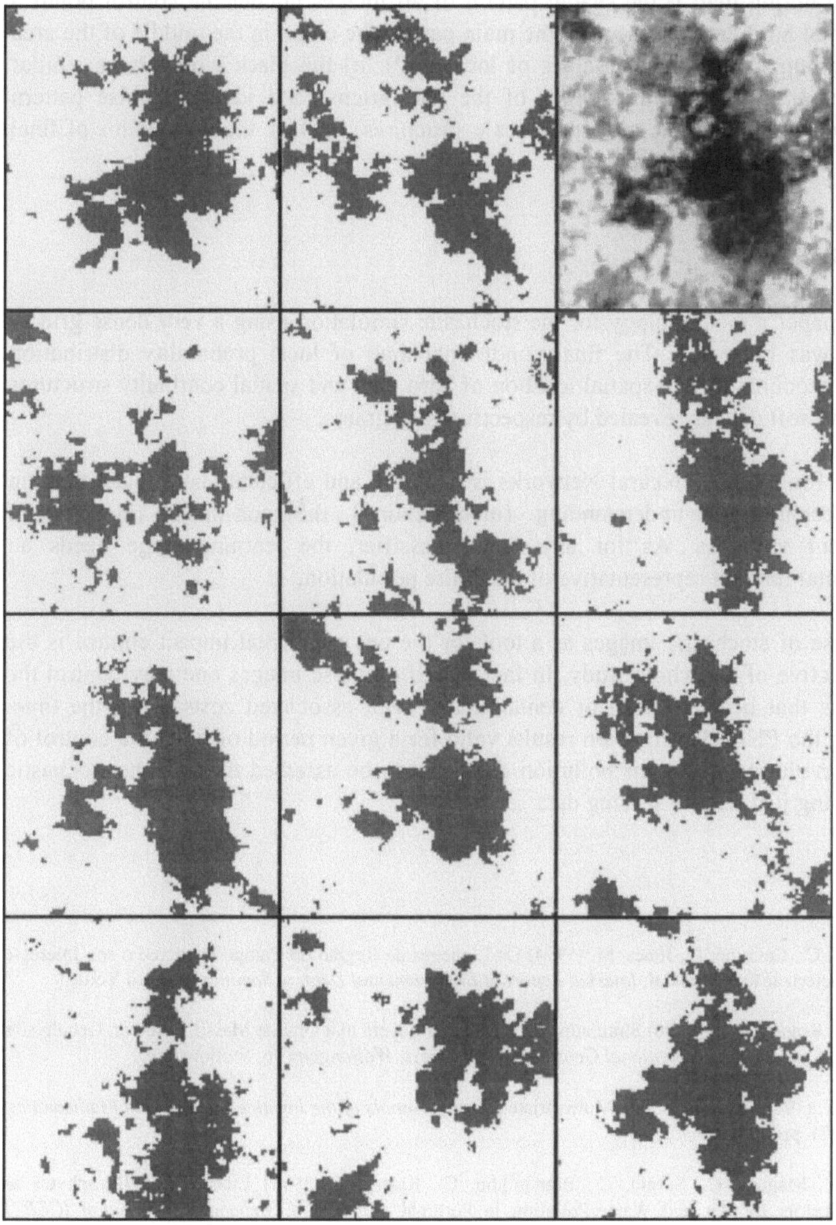

figure 6. Simulated images: ■ - z < 10 species; □ - z > 10 species.

It is worth noting that: i) the spatial pattern of all images honour the control points - experimental hard data - presenting the main part of the cloud in the middle of the area (corresponding to the highest values of local pdf'); ii) the black clouds have similar dimensions (related with the ranges of the variograms) and identical noise pattern (related with the randomness of small scale structures). In fact, the variograms of final simulated images are quite similar.

4. Final Remarks

i) In this paper a methodology for the stochastic simulation using a very dense grid of soft data was presented. The final conditional map of local probability distribution functions accounts for the spatial location of hard data and spatial continuity structures of hard and soft data as revealed by respective variograms.

ii) PNN - Probabilistic Neural Networks is a simple and efficient Bayes classification method capable of understanding (incorporating) the non-linear relations of independent variables. As for any other classifier, the learning stage needs an experimental data set representative of the entire population.

iii) The use of stochastic images as a tool for the environmental impact control is the main objective of the whole study. In fact, based on these images one may control the local areas that become clean or contaminated with associated costs along the time. Assuming the PNN classification results valid for a given period of time, the control of dynamic evolution of the air pollution damage can be assessed through the stochastic images using just remote sensing data.

References

Branquinho, C., Catarino, F., Jones, M. (1994) Os Líquenes da Região do Campo Branco e o seu Interesse na Caracterização Ambiental, *Internal Report of Environmental Dept. of Somincor*, Castro Verde.

Brito, M.G., Rogado, J.Q. (1996) Simulation of Weathered Layers in a Granite Massif Based on Geophysics Survey, *Proc. of 5th International Geostatistics Congress, Wollongong'96*, Wollongong.

Cacoullus, T. (1966) Estimation of a Multivariate Density, *Annals of the Institute of Statistical Mathematics*, 18(2), pp. 179-189.

Catarino, F., Máguas, C., Sérgio, C., Branquinho, C., Kratz, W. (1991) Lichens and Bryophytes as Bioindicators for Air and Water Pollution in Portugal, *B. Nath. Environmental Pollution ICEP 1* (European Centre for Pollution Research), 1:170-178.

Colin, P., Froidevaux, R., Garcia. M., Nicoletis, S. (1996) Integrating Geophysical Data for Mapping the Contamination of Industrial Sites by Polycyclic Aromatic Hydrocarbons: a Geostatistical Approach, *Geostatistics for Environmental and Geotechnical Applications, ASTM STP 1238*, R. Mohan Srivastava, Shahrokh Rouhani, Marc V. Cromer, A. Ivan Johnson, Ed., American Society for Testing Materials, Philadelphia.

Doyen, P.M., (1988) Porosity from Seismic Data: A Geostatistical Approach, *Geophysics*, **53**:1263-1275.

Froidevaux, R. (1993) Probability Field Simulation, in A. Soares (ed), *Geostatistics Troia'92*, Kluwer Academic Pub., Dordrecht, **1**, pp 73-84.

Journel, A. G. (1989), Fundamentals of Geostatistics in Five Lessons, in *Short Course in Geology*, **8**:1-33

Journel, A.G., Zhu, H. (1990) Integrating Soft Seismic Data: Markov-Bayes Updating, An Alternative to Cokriging and Traditional Regression, *Stanford Center for Reservoir Forecasting Annual Report.*

Masters, T. (1995) *Advanced Algorithms for Neural Networks, A C++ Sourcebook*, John Wiley & Sons, Inc.

Nimis, P. Lazzarin, G., Gasparo, D. (1991) Lichens as Bioindicators of Air Pollution by SO_2 in the Veneto Region (NE Italy), *Studia Geobotanica*, **11**:3-76.

Parzen, E. (1962) On Estimation of a Probability Density Function and Mode, *Annals of Mathematical Statistics*, **33**, pp. 1065-1076.

Pereira, M.J., Soares, A., Branquinho, C., Catarino, F. (1996) Stochastic Imaging of Air Quality by Using Remote Sensing Data, *Remote Sensing for Environment* (submitted).

Specht, D. (1990) Probabilistic Neural Networks, *Neural Networks*, **3**, pp. 109-118.

Srivastava, M. (1992) Reservoir Characterization with Probability Field Simulation, *SPE paper*, N. 24753.

CHERNOBYL FALLOUT:

REVIEW OF ADVANCED SPATIAL DATA ANALYSIS

M. KANEVSKY (1), R. ARUTYUNYAN (1), L. BOLSHOV (1), S. CHERNOV (1), V. DEMYANOV (1), N. KOPTELOVA (1), I. LINGE (1), E. SAVELIEVA (1), T. HAAS (2), M. MAIGNAN (3).
(1) Institute of Nuclear Safety, B.Tulskaya 52, Moscow, Russia;
(2) University of Wisconsin-Milwaukee, USA;
(3) Lausanne University, Switzerland

Abstract. The paper is based on the long experience in applying various spatial data analysis methods and approaches to the radioactive pollution data after the Chernobyl accident. Application of three basic approaches are described in the work: geostatistics (both parametric and non-parametric models), artificial neural networks (ANN) and monitoring network analysis, including fractals and cluster analysis. There have been developed hybrid geostatistical-ANN models, which show better performance then one approach models in many complex non-stationary cases.

1. Introduction

The present paper is dedicated to one of the outstanding case studies, dealing with spatial data analysis of environmental data. Chernobyl accident resulted in complex radioactive contamination patterns in different European regions. The work contains a review of advanced geostatistical methods, applied to the radioecological data from a different regions affected by the Chernobyl fallout.

An accident occurred at the Chernobyl nuclear power plant in Ukraine on April 26, 1986 appeared to be one of the most serious catastrophe in nuclear power. A test, designed to assess the reactor's safety margins at the fourth unit of the power plant, lead under a combination of circumstances to a steam explosion. The explosion destroyed the reactor's integrity and caused a release of radioactive materials in to the atmosphere. Vast fires and heating of the fuel due to radioactive decay resulted in continuous releases of burning graphite products, radioactive gases, aerosols, and finely dispersed fuel particles with dimensions ranging from fractions of a micron to hundreds of microns (Buzulukov et al., 1993; Israel et. al. 1990). The release lasted for 10 days and a significant part of the core content went out. Since the reactor (Unit 4) had been in use over two years, the core content was rich in a wide range of radioactive fission products. There were two major components of the release: radionuclides that

A. Soares et al. (eds.), geoENV I–Geostatistics for Environmental Applications, 389–400.

were part of the matrix of the dispersed fuel and released in the form of radioactive dust particles (contamination of the near zone), and volatile radionuclides in the form of gases or aerosols, which evaporated from the fuel (contamination of the middle and far zones).

2. Contamination of the Territories

The radioactive materials released from the destroyed reactor were transported over large distances and contaminated vast territories in Europe. There have been collected a lot of data on soil contamination and compiled in to different data bases and data banks. The present study is based on the data from BANK (Environmental Institute, Ispra Italy) and Central Data Bank of Aggregated Data (Institute of Nuclear Safety, Moscow). The contamination on caesium in Europe is described and analysed in the atlas designed by an international scientific project (De Cort and Tsaturov, 1996). The atlas contains about a 100 maps displaying isolines of CS137 contamination. Inverse distance weight method was generally used.

The goal of the present work is to review advanced methods of spatial data analysis applied to CS137 contamination. The structure of the fallout displays an extremely complex spatial phenomena due to complicated processes affected the pollution transport: meteorological conditions, sedimentation (wet and dry) orographical effects, variability of soil types. The collected data samples describe the phenomena at different scales ranging from meters to hundreds of kilometres. Contaminated areas were chosen to demonstrate high variability of contamination pattern at different scales: contamination at a populated site (Figure 1.) and contamination at a regional scales (Figure 2).

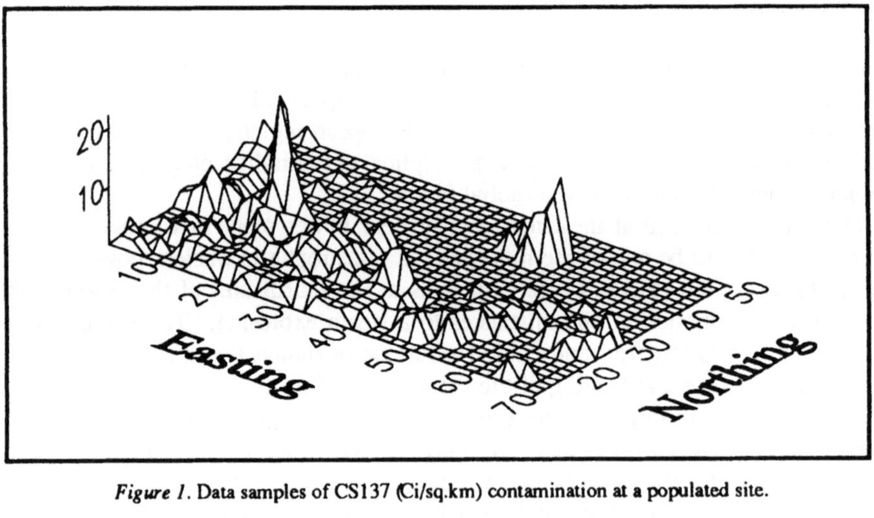

Figure 1. Data samples of CS137 (Ci/sq.km) contamination at a populated site.

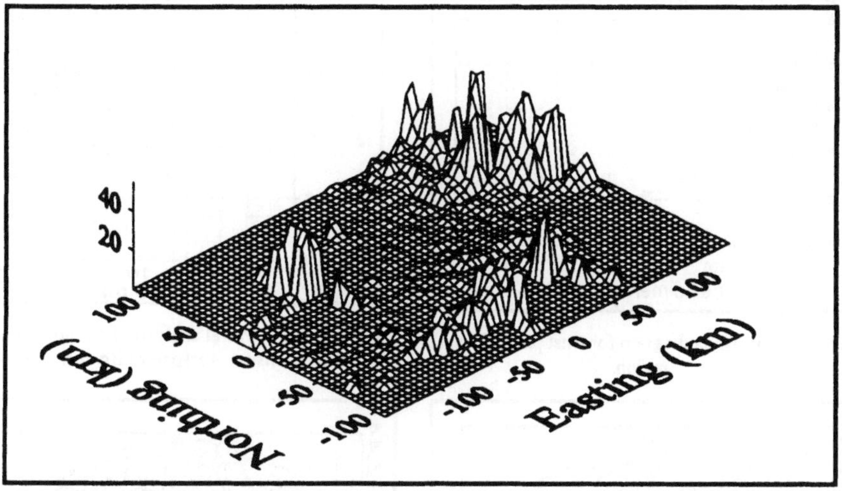

Figure 2. Data samples of CS137 (Ci/sq.km) contamination inGomel region, Belarus.

3. Monitoring Network Analysis

The analysis of spatially distributed data have to be preceded by the analysis of their monitoring networks. The ignorance of this step can strongly complicate the following analysis (statistical analysis, variography, interpolations etc.). While dealing with the monitoring network we can analyze two different but connected aspects which are the spatial (geometrical distribution of sampling points) and the dimensional network resolution, describing homogeneity. Simple geometrical analysis can be performed with the help of the histogram of squares of Voronoi polygons. Another characteristic for the quantitative description of the network is a Morishita diagram (Ouchi, Uekawa, 1986). Morishita index is estimated as follows: given area is divided into small rectangular cells of equal size (Δ). The Morishita index is

$$I_\Delta = Q\frac{\sum_{i=1}^{Q} n_i(n_i - 1)}{N(N-1)} \qquad (1)$$

where n_i (i=1,2...,Q) is the number of points in the i-th cell, and N - the total number of points. Morishita diagram is the dependence of the Morishita index on cell size. The behaviour of the diagram is connected with the clustering of the network under study. The detailed discussion on Morishita index and its connection with fractal dimension is presented in (Korvin, 1992). The Morishita diagram was used for analysis of monitoring network of the west part of Briansk region (see Figure 3). It is clearly seen that the Morishita index is decreasing with the growth of the cell size, slowly at first and much faster after the cell of dimension 20 km. It means that the network of this region is clustered and the characteristic cluster size is nearly 20 km.

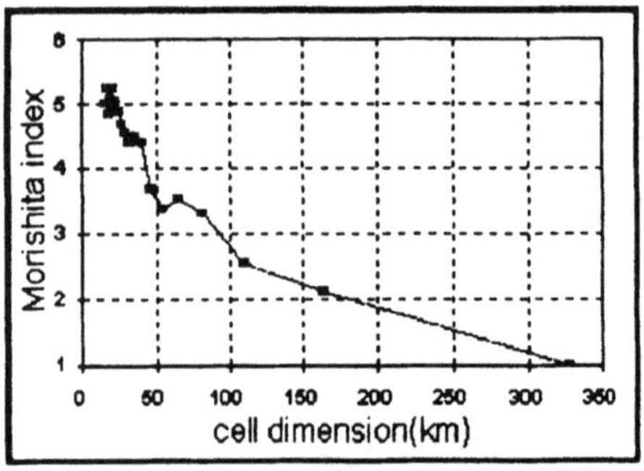

Figure 3. Morishita diagram for west part ofBriansk region.

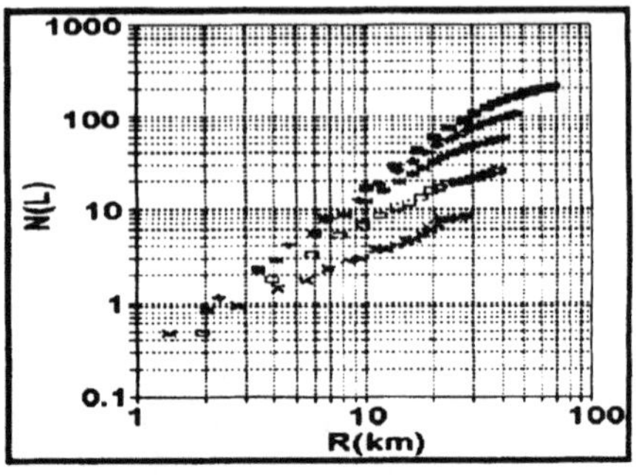

Figure 4. Multi-fractal analysis of monitoring network resolutions for westpart of Bryansk region.

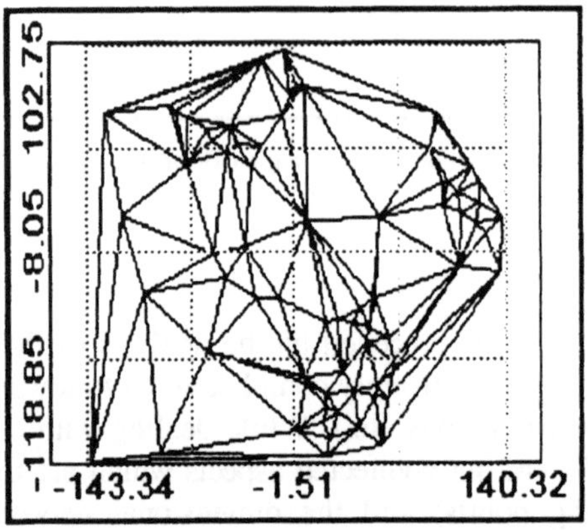

Figure 5. Austria, original monitoring network.

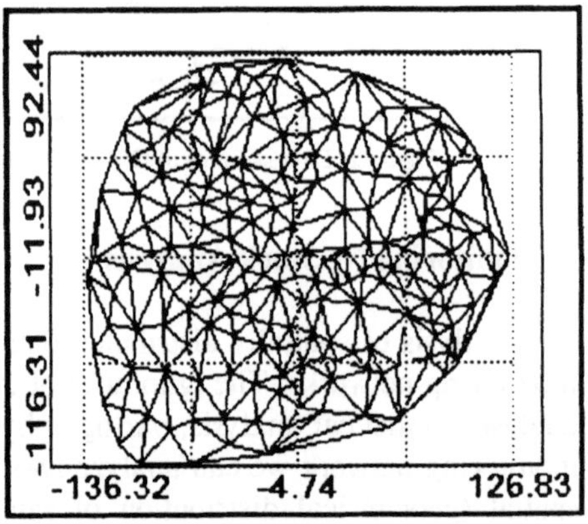

Figure 6. Austria: network after points' generation.

The second problem deals with monitoring network and its dimensional resolution. If both the network and the phenomena can be considered as fractal objects in 2D, the resolution of the network is D_n, where D_n is fractal dimension of the network (Lovejoy, et. al., 1986) . There are a lot of methods for the estimation of fractal dimension (Korvin, Boyd, O'Dowd, 1990). Here the "sand-box counting" method was used. For the network mentioned above the fractal dimension is 1.59. The phenomena with the fractal dimension less or equal $2\text{-}D_n$ cannot be detected by this network. Multifractal analysis have been performed by using data on CS137 contamination of west part of Bryansk region. We used 5 values of thresholds: 10, 15, 20, 25, 30 Ci/sq.km, and the corresponding fractal dimensions are: 1.5, 1.5, 1.39, 1.25, 1.01. The fractal analysis for networks with corresponding levels are presented in Figure 4. It is obvious that fractal dimension is decreasing with the growth of threshold value after the value of the threshold becomes rather large (approximately 20 Ci/sq.km).

Monitoring network analysis is closely connected with the presentation of the data after interpolation or simulation. Here is proposed a way, to complete the initial network by adding new points step-by-step to make it more and more homogeneous. In-

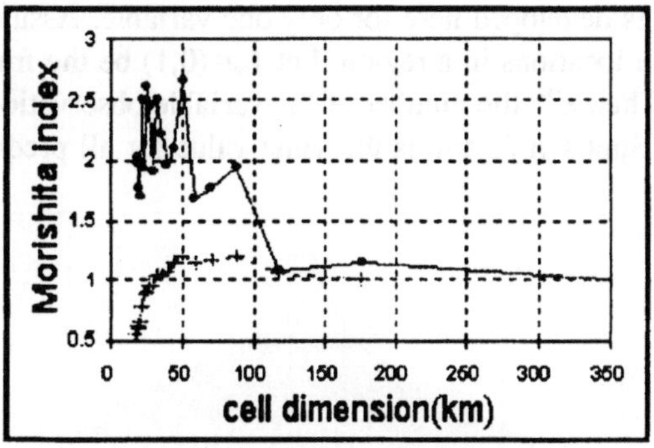

Figure 7. Morishita diagram for Austria before (•) and after (+) points' generation

creasing network homogeneity leads to more representative results. The method is based on the idea of computational geometry. Objects of triangulation - triangles generate points for the new network. To improve the homogeneity of the initial network triangles, that are situated within a cluster and having acceptable spatial and dimensional resolution, do not generate new points. Only sparse regions generates new points. In general, the method belongs to algorithms of pattern completion. This method was used for Austria (Figure 5). After 2 generation steps the fractal dimension of the network increased from 1.37 to 1.82. The Morishita diagram (see Figure 7) shows that the cluster structures nearly disappeared. The network after 2 steps is also presented in the picture (Figure 6). This algorithm is not connected with a specific interpolation method and can be used with different interpolation and stochastic simulation models (Kanevsky, Savelieva, 1995).

4. Prediction Algorithms

The main objectives of the spatial data analysis of Chernobyl fallout deal with the use and developing of models for the analysis, processing and presentations of spatially distributed information for the qualified decision making. It is well known that there are a lot of interpolation methods which can be used in spatial data analysis. Most of them fall into two major classes - deterministic and geostatistical. The first assume a deterministic formula, modelling spatial distribution of the phenomena. Most frequently used deterministic methods are inverse distance power, multiquadric equations, splines. Geostatistical approach considers spatially distributed data as realisations of a random function. This allows to use powerful statistical tools to model spatial correlation between the samples and then use this information in regression algorithms. Different types of kriging - simple, ordinary, co-kriging, kriging with trend, lognormal kriging - have been widely used to make spatial predictions. An important practical objective of the study was in introducing probabilistic aspects for the analysis and mapping of radioecological data.

4.1. MOVING WINDOW REGRESSION RESIDUAL KRIGING

The Moving Window Regression Residual Kriging (MWRRK) is an advanced nonlinear predictor, which helps to assess problems of non-stationarity in the data and global trend. The MWRRK algorithm is developed here for only one variable. Assume the considered variable is observed at n locations in a region. Let $f_w \in (0,1)$ be the fraction of n that is used for a prediction. Then $n^{(w)}$ the number of the variable observations used to calculate the prediction at $\mathbf{x}_0$, equals $n\,f_w$ and is the same value for all predictions. Call the subregion that contains these $n^{(w)}$ observations *the prediction window.*

Let **Y** be the partitioned vector containing the within - window observed random variable of the processes located at $\mathbf{x}, K, \mathbf{x}_{n^{(w)}}$. Within this window, the model for $Y(\mathbf{x}_0)$ consists of a spatial trend component plus an error or stochastic component:

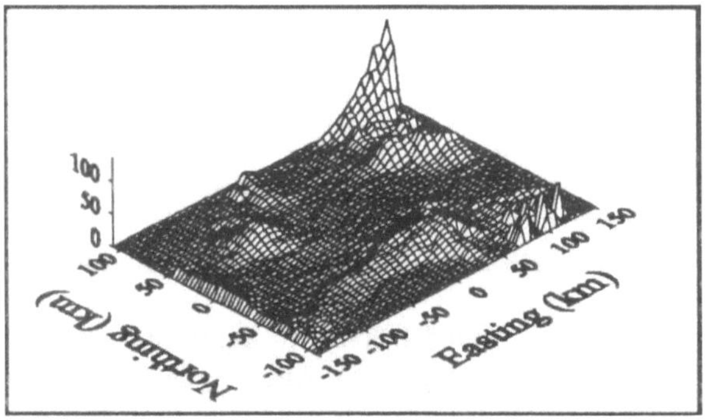

Figure 8. MWRRK estimates of CS137 (Ci/sq.km) in Gomel region, Belarus.

$$Y(\mathbf{x}_0) = \mu\left(\mathbf{x}_0, \beta, Y^*(\mathbf{x}_0)\right) + R(\mathbf{x}_0) \tag{2}$$

where $Y^*(\mathbf{x}_0)$ is the prediction of the variable at location $\mathbf{x}_0$. The vectors $\mu(\cdot)$ and R are variable's trend model values at the locations $\mathbf{x}, K, \mathbf{x}_{n^{(w)}}$, and the associated vector of variable's residual process random variables, respectively. The vector β contains the linear parameters defining the mean function μ. Then, the model can be written as: $\mathbf{Y} = \mathbf{X}\beta + \mathbf{R}$, $\mathbf{X}$ is associated design matrices.

The MWRRK predication procedure includes building a prediction window, ols linear trend model, which entails fitting cross-variogram model for residuals; then either linear or non-linear trend is fitted by gls; the remaining residuals are brought through the ordinary co-kriging procedure; and finally, the prediction is made by summing the trend and residual estimates. For the mathematical details of the MWRRK predictor and how a positive definite cross-covariance function is estimated, see Haas (1996).

MWRRCK prediction was carried out for Bryansk region on strontium 90 (SR90) and caesium 137 (CS137) as a secondary variable. Unlike in ordinary co-kriging, prediction error for MWRRCK is distributed according to spatial variability and data values.

MWRRK prediction algorithm includes the following steps:

Step 1. Centre a circular window at the prediction location, $\mathbf{x}_0$ and find the spatially closest n^w observations to $\mathbf{x}_0$. The window's radius, r_w is the distance of the furthest of these sites to $\mathbf{x}_0$.

Step 2. By assuming $V=\sigma^2 I$, fit the local trend model with ordinary least squares:

$$\beta^{(1)} = (X'X)^{(-1)} X'Y. \tag{3}$$

Using weighted least squares, fit a semivariogram model to the predicted residuals, $R^{(1)}$. Use the variogram model to compute $V^{(1)}$, an estimate of V.

Step 3. By assuming $V=V^{(1)}$, perform a generalised least squares (gls) estimate of the trend: $\beta^{(2)}=(X'(V^{(1)})^{-1}X)^{-1}X'(V^{(1)})^{1}Y$. Use the residuals, $\mathbf{R}^{(2)}$ from this gls-estimated spatial trend model to estimate a second semivariogram model. Use the corresponding variogram to form $V^{(2)}$ - a second estimate of V.

Step 4. Use the variogram model derived in step 3 to predict the residual process at $\mathbf{x_0}$ via residual kriging. This kriging system is:

$$\begin{bmatrix} V^{(2)} & \mathbf{1} \\ \mathbf{1}^T & 0 \end{bmatrix} \cdot \begin{bmatrix} w \\ -l \end{bmatrix} = \begin{bmatrix} \mathbf{U}^{(2)} \\ 1 \end{bmatrix} \tag{4}$$

The weight vector **w** and Lagrange parameter are defined as in the case of OK. The notation $\mathbf{U}^{(2)}$ indicates that **U** has been estimated with step 3's derived variogram.

The kriging prediction of the residual and the kriging variance at $\mathbf{x_0}$ are:

$$R^*(\mathbf{x_0}) \equiv \mathbf{w}' \mathbf{R}^{(2)}, \; \sigma_K^2(x) \equiv Var[R^{(2)}()] - \mathbf{w}'\mathbf{U}^{(2)} + l$$

Step 5. The final MWRRK prediction is:

$$Y^*(\mathbf{x_0}) \equiv \mu^{(2)}(\mathbf{x_0}, \beta^{(2)}) + R^*(\mathbf{x_0}), \tag{5}$$

MWRRK estimates and estimation errors are presented in Figures 8,9. MWRRK kriging errors reflect not only the spatial distribution of sampling points, but also variability of the contamination pattern. Note, that in effect the MWRRK predictor is using a non-linear model of the spatial trend with variogram estimation based only on local data. More detailed analysis is presented in (Kanevsky, et al, 1995).

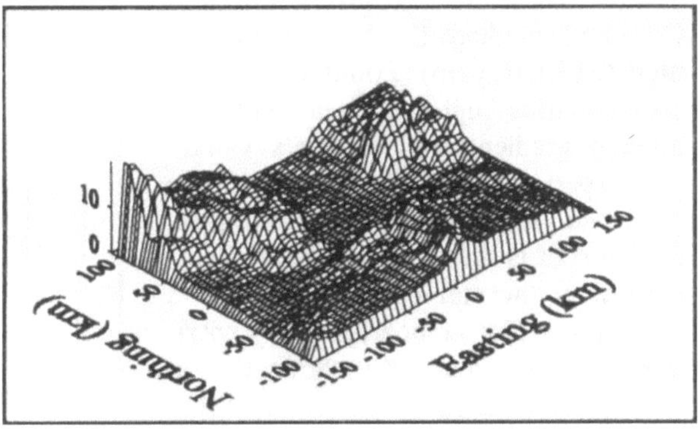

Figure 9. MWRRK estimation errors of CS137 (Ci/sq.km) in Gomel region, Belarus.

5. Local Distribution Function Estimation

Very often it is demanding to have the quantitative description of the local distribution at the estimation point. The local distribution function should lie around the unknown "true" value at the chosen location. In the case of Chernobyl fallout usually parametric models (normal, log-normal) have been used to analyse data and for the decision making. Sometimes it was mentioned that such simple models could not explain available data. In order to clarify this question indicator kriging (IK) have been applied. It allows to estimate a conditional cumulative distribution function (ccdf) at a certain location by using available hard and soft data. Indicator kriging is a non-parametric method which deals with indicator variables for a set of selected thresholds z_k. The computed ccdf is compared with a normal one.

Indicator kriging was applied for CS137 fallout in Gomel region, where the spatial pattern contains several highly contaminated spots. There were chosen some check points in areas with different contamination levels: low (0-5 Ci/sq.km), medium (5-15 Ci/sq.km) and high (>15 Ci/sq.km). Points of medium contamination represent areas with significant gradients of variable's values. The experimental and normal ccdfs are presented in Figures 10-12. Experimental ccdf is drawn by a thin line, corresponding normal cdf is drawn by a thick line. The comparison of experimental and normal cdf's shows that at low contamination levels local distribution is close to normal. Influence of outliers, which is also difficult to account by variogram model, brings the deviation from normality at higher contamination levels.

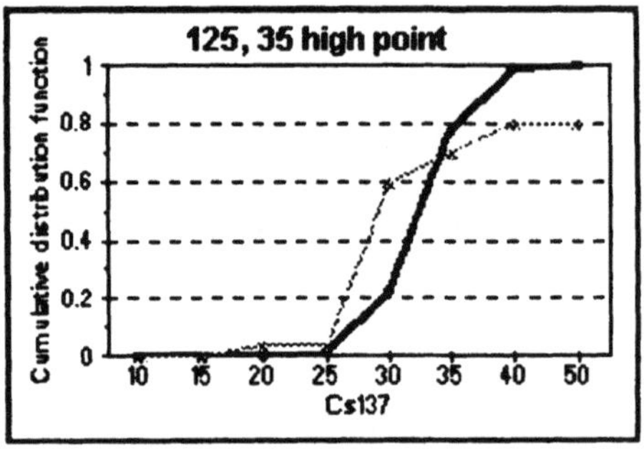

Figure 10. Local cdf at a high contamination level.

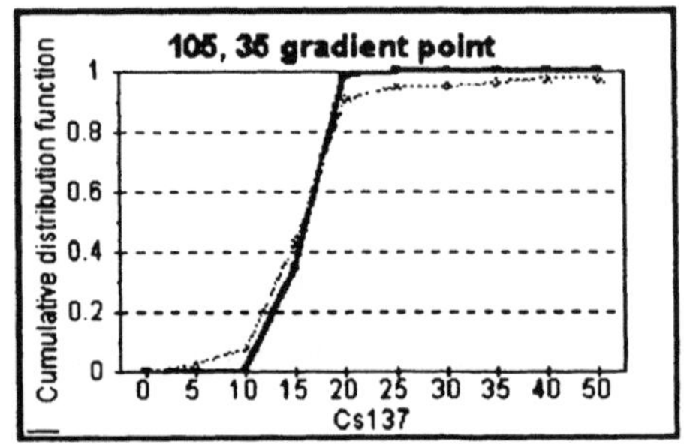

Figure 11. Local cdf at a medium contamination level.

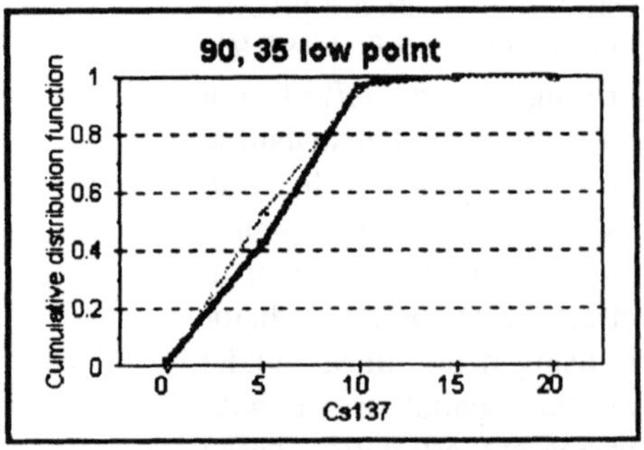

Figure 12. Local cdf at a low contamination level.

6. Artificial Neural Networks

Artificial neural networks (ANN) are analytical systems that address problems whose solutions have not been explicitly formulated. Artificial neural networks consist of numerous, simple processing units (neurones) that can be globally programmed for computation. Neural networks can store, recognise, and associatively retrieve patterns; to filter noise from measurement data; to control ill-defined problems - in summary: to estimate sampled functions when the underlying process can not be explicitly modelled. Unlike statistical estimators, they estimate a function without a mathematical model of how outputs depend on inputs. Neural networks are model-free estimators - they learn from experience with numerical and, sometimes, linguistic sample data.

ANN consists of neurones - processing elements, interlinked by a system of connections. The number of neurones and connections form network topology. Feedforward neural networks, in which information flows only in one direction from the input to the output and there is no feedback. Multilayer neural networks nave hidden (intermediate) layers between input and output neurones.

Artificial neural networks perform as a super regression model, capturing the spatial structure of the contamination according to the number of layers and neurones used. Too few neurones lead to rough estimate, whereas too complex neural network tries to model local peculiarities which usually have random origin. Neural network applied to radioactive contamination gave good results according to accuracy, validation and cross-validation tests (Kanevsky, 1995). Accuracy test shows what was learned - how well neural network estimates values at the point from the training set. In the present study data on soil contamination by Chernobyl radionuclides in the South part of FRG are used. Accuracy test was carried out by trained feedforward neural network [2-5-0-1]: 2 input neurones (corresponding to spatial co-ordinates), 5 neurones in one hidden layer, and 1 output neurone (corresponding to contamination). Correlation between measured and estimated values is high (0.834). The network is considered to have learned a task adequately if, for the inputs of the training set, the net error (determined by the difference of the network outputs from the desired outputs) is smaller than some pre-defined level. Residual analysis is made to check if the residuals have spatial correlation. Considering complex highly variable data, it is often happens that ANN can model only major large scale spots and small scale structures are not grasped. In this case ANN works as a detrending model and a kriging model applied to the residuals add peculiarities due to local structures. Raw variogram of CS137 in South part of FRG features two different structures: short range (40 km) and long range trend (300 km). ANN models the large scale structure, whereas residuals, left after applying ANN appear easy to fit variogram, representing the small scale structure of the spots. The variogram model of the residuals are further used in ordinary kriging algorithm for ANN residuals. Final estimates are computed as a sum of ANN estimates and OK estimates of residuals.

Figure 13 shows estimated contamination surface for CS137 in the South part of FRG (1 Ci/sq.km = 37 kBq/sq.m). The described method, called Neural Network Residual Kriging (NNRK) significantly improves ANN estimates, being the exact estimator. The more detailed description along with discussions are presented in (Kanevsky, et. al. 1996).

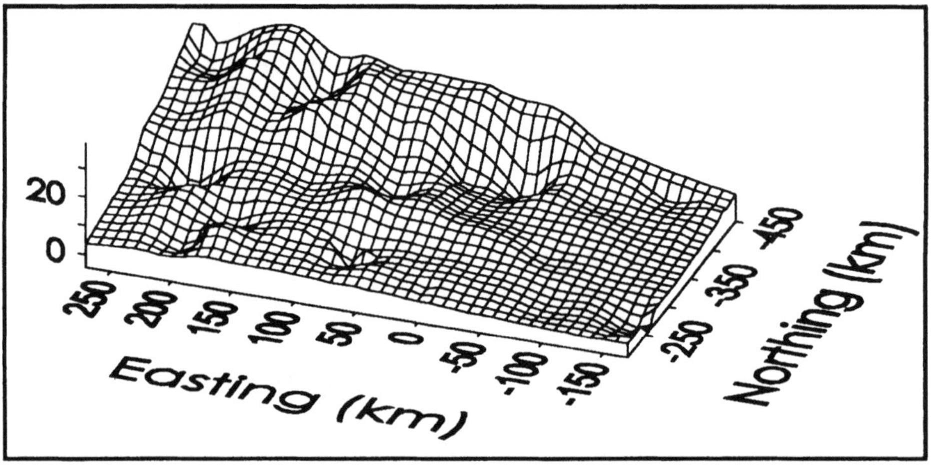

Figure 13. Neural Network Residual Kriging estimates of CS137 (kBq/sq.m), South part of FRG.

7. Spatial Data Analysis and Software

Spatial data analysis methodology that was followed in exploring Chernobyl fallout is a sequence of steps including statistical, geostatistical, fractal, stochastic and artificial intelligence methods. Applying different methods allows to get information of different type about the spatial phenomenon. The methods used in the study have been described in the relevant literature. There are many computer realisations of the techniques described above. While carrying out the present work, there have been developed a number of computer programs to aid in the research and put together the existing spatial data analysis techniques on the contemporary software level.

UpFile program, that was designed initially as a MS Windows post processor for the GSLIB library (Deutsch and Journel, 1992), then was developed into 3Plot software. 3Plot provides different types of data post plots including triangulation and Voronoi polygons and handles data formats commonly used in contemporary spatial data analysis packages. 3Plot also includes summary statistics and moving window statistics calculation modules and has a link to geographical information systems in order to present the predictions on geographical maps. For user's convenience there have been developed a number of image processing tools including co-ordinate and value tracing, zoom, etc. Sample 3Plot interface is shown in Figure 14.

Data preparation stage, which is essential when dealing with any kind of information requires a number of operations with data bases. Usually there is a lot of processing to be done using expert opinion to prepare data base for the spatial prediction. For this purpose DataMan and DBG Convertor programs for MS Windows have been designed. They handle dBase format as most widespread in contemporary data base management systems and Geo-EAS format as conventional in geostatistical software packages. DBG Convertor is a simple dBase↔Geo-EAS converter that works reliable with huge files and is easy to use. Data management system DataMan includes most common file operations and operations with data as well as specific tools selected according to the experience with environmental data analysis.

Predictions themselves were obtained from KRIG software (MWRRK) (Haas, 1996) and GSLIB (Deutsch, Journel, 1992). The last one was compiled on PC and contributed by a WinGSLIB pre-processor, which allows to create, edit and run GSLIB parameter files in a user-friendly dialogue way under MS Windows.

The software described above appeared to be extremely useful and efficient for dealing with spatially distributed data. The programs are designed in a helpful way to meet the needs of a user.

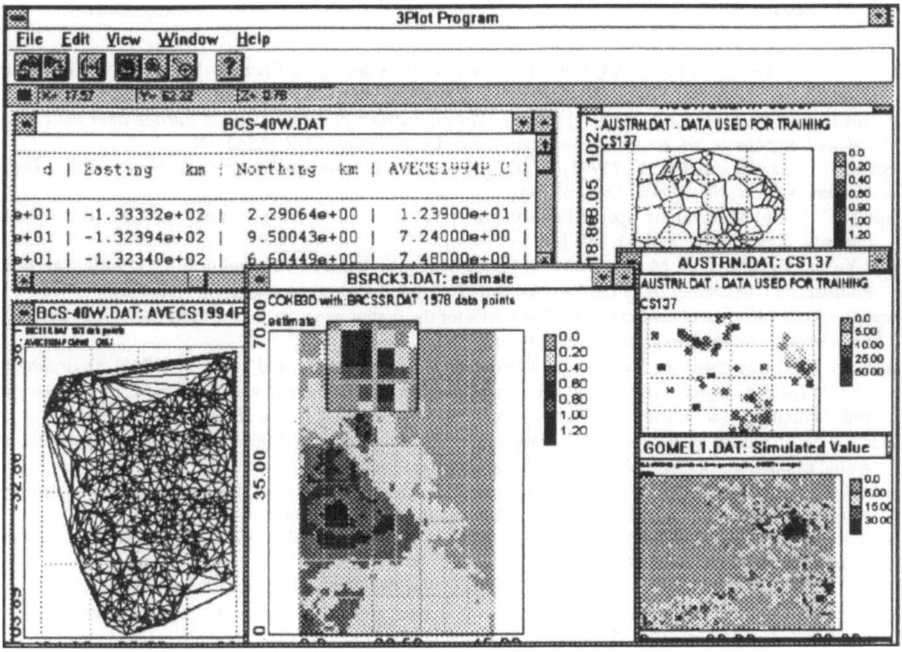

Figure 14. Examples of 3Plot outputs: data viewer, post plot, triangulation\Voronoi polygons, mosaic maps of estimates and simulations.

8. Conclusions

The present work presents general overview of the several advanced methods applied to spatial data analysis of Chernobyl fallout. There complexity of the phenomena requires various approaches to be used for deep and successful analysis. Statistical, geostatistical and artificial intelligence methods helped to describe and to model the nature of soil contamination, which is extremely variable at different scales and complicated. The presented review includes application of indicator kriging, Moving Window Regression Residual Kriging, artificial neural networks and Neural Network Residual

Kriging. While carrying out the research a number of software have been developed for data processing, analysis, and presentation according to contemporary spatial data analysis methodology. The latest developments deal with 3D mapping by modelling of radionuclides migration in environment and spatio-temporal modelling.

Acknowledgement. The work was supported in part by the grant INTAS 94-2361. The authors thank to Marc de Cort for the access to the early data of REM BANK.

9. References

Buzulukov Yu., Dobrynin Yu. (1993) Releases of radionuclides during the Chernobyl accident, in S.E. Merwin, and M.I. Balonov (eds.),*The Chernobyl Papers*, Research Enterprises,Richland, 3-21.

De Cort M., and Tsaturov Yu.S. (editors) (1996) *Atlas on caesium contamination of Europe after the Chernobyl nuclear plant accident,* European Commission, Luxembourg, EUR 16542 EN.

Deutsch C. V., and Journel A.G. (1992) *GSLIB Geostatistical Software Library and User's Guide*. Oxford University Press, New York, Oxford, 340 p.

Haas T. C. (1996) Multivariate Spatial Prediction in the Presence of Nonlinear Trend and Covariance Nonstationarity. *Environmetrics* 7(1), 145-166.

Israel Yu. A., Vakulovslii S. M., Vetrov V. A., Petrov V. N., Rovinskii F. Ya., Stukin E. D. (1990). *Chernobyl: Radioactive Contamination of the Environment*. Leningrad: Gidrometeoizdat, (in Russian.).

Kanevsky M. (1995) Use of artificial neural networks for the spatial interpolations of radioecological data, *Izvestija Akademii Nauk. Energetika* 3, 26-33, (in Russian).

Kanevsky M., Arutyunyan R., Bolshov L., Demyanov V., Savelieva E., and Haas T. (1995) *Environmental Spatial Data Analysis. Case Study: Chernobyl Fallout.* Tenth International Conference on Mathematical and Computer Modelling and Scientific Computing. Boston, USA. Books of Abstracts, p. 179.

Kanevsky M., Arutyunyan R., Bolshov L. Demyanov V., Maignan M.. (1996) Artificial neural networks and spatial estimations of Chernobyl fallout*Geoinformatics* 7, nos. 1-2, 5-12.

Kanevsky M., and Savelieva E. (1995) *Environmental monitoring networks and quantitative description of clustering*. IAMG'95 Annual Conference. Abstracts for Technical Programs, 31-32.

Karaoglou A., Desmet G., Kelly G.N., Menzel H.G. (eds.) (1996) *The radiological consequences of the Chernobyl accident*. EUR N16544 EN. 1192 p.

Kovin G. (1992). *Fractal Models in the Earth Sciences*. Elsevier Science PublishersB.V. Amsterdam.

Korvin G., Boyd D.M., O'Dowd R., (1990). Fractal characterisation of the South Australian gravity station network. *Geophys J Int*. **100**, 535-539.

Lovejoy S., Schertzer D, Ladoy P. (1986) Fractal characterisation of inhomogeneous geophysical measuring networks.*Nature* **319**, 43-44.

Ouchi T., Uekawa T., (1986) Statistical analysis of the spatial distribution of earthquakes - variation of the spatial distribution of earthquakes before and after large earthquakes. *Physics of the Earth and Planetary Interiors* **44**, 211-225.

Raes F., De Cort M., Graziani G., (1991) Multi-fractal nature of radioactivity deposition on soil after the Chernobyl accident.*Health Physics* **61**, 271-274.

SPACE-TIME ANALYSIS OF AIR POLLUTION IN LISBON

JÚLIA SEIXAS, FRANCISCO FERREIRA, CARLA NUNES,
JOÃO PEDRO SILVA

Faculdade de Ciências e Tecnologia
Universidade Nova de Lisboa
Quinta da Torre
2825 Monte de Caparica, Portugal
[mjs, ff, crn]@mail.fct.unl.pt; jps@uninova.pt

Abstract

The environmental decision-making process is preceded by data analysis, which is conditioned by a specific temporal and spatial scale. Air quality management decisions are different from a regional to local and very local spatial frame, and from hourly or daily patterns, according to the different types of air pollution sources, different meteorological conditions and different pollutant characteristics. This paper aims to present a methodology that integrates the time-space framework of air quality data to infer the temporal pattern and spatial variability that could be interpreted for environmental decision purposes. Variograms that accommodate time and space lags were used for the analysis and proved to be effective. Temporal and spatial trends were found for data collected on an hourly and daily basis and its environmental meaning is discussed. Visualization of spatial patches of air pollution in Lisbon during a working day is performed through the use of an image processing technique, named morphing. Scientific visualization has becoming a very powerful approach to explore and understand data, mainly, spatio-temporal data.

1. Introduction

Environmental management must include three sequential tasks: data capture, data analysis and decision making. Since any environmental issue is dynamic by nature and manifests itself on a certain spatial scale, the process of environmental data capture and analysis has to accommodate its twofold nature, the spatial and the temporal dimension. Only within this framework, the environmental problems are properly understood and correct decisions can be taken. Ideally, environmental monitoring should be proceeded on a continuous-space and continuous-time frame, considering the appropriate spatial and temporal scale for the phenomena under study. The process of environmental data

A. Soares et al. (eds.), geoENV I–Geostatistics for Environmental Applications, 401–413.

capture depends on the type of the environmental variable, on the available monitoring technology, as well as on the available economic resources. However, most sampling strategies are generally sparse, point-source in nature, and irregularly distributed in space and/or time. Typically, only a limited number of samples in space and/or time are available to assess a continuous, dynamic process whose state variables change with respect to both space and time, as are almost of the environmental problems and processes. Point source observations provide only small, sometimes inadequate clues to a much larger problem.

Appropriate exploratory data analysis must answer questions about patches, defined as some spatial extension of significant data, besides point data, as well as questions regarding meaningful periods of time. For example, let us consider the quartiles regions of a pollutant load surface, by classifying the whole data set within the four interquartile regions. The following assertions about statistical patches can be stated: (a) what is the movement of the P_{75} cloud of air pollution during a working day? (b) the patches occur randomly over the region, or they move systematically for the north region? (c) Its shape increases slightly from the center to all directions, or they disappear gradually for the considered time range?

Air pollution data is usually obtained from a spatial irregularly network of monitoring stations, at high temporal resolutions (1- or 15- minutes), which represents a continuous-time and discrete-space sampling network. From an environmental management perspective, several air quality assertions can be stated from that monitoring network: (a) at the regional level one is interested to infer the average behavior of air pollution loads, and relating them with the different classes of air pollution sources, and with seasonal meteorological conditions (like winter and summer); (b) at the city level one wants to infer the air pollution pattern during a working day, associating the air pollution rush hours with traffic corridors; and (c) at the street-crossing level, one aims to understand the direct cause of the traffic structure on air quality levels. Different highlights and decisions can be taken from that analysis, from technological investments in factories, to changes on the corridors in the entrance city and to limit the circulation of bus and trucks at specific streets, for example.

The goal of this paper is to present a methodology to assess simultaneously the spatio-temporal patterns of air quality at the city of Lisboa, in order to assist the environmental management decision-making. The emphasis is on exploratory data techniques, specifically on variogram methods, whose parameters reveal air quality trends that can be interpreted for environmental management purposes. Since environmental management is better issued on a spatial basis, one also aims to generate air quality spatial grids, being applied the kriging technique. Scientific visualization methods was used to track the dynamics of those grids. The use of visual stimulus has proved to be very effective in spatial analysis tasks and the time domain can be added using animation procedure. A morphing process between consecutive images was implemented to follow air pollution hourly patterns during a working day, and daily patterns during a week.

2. Environmental Data Analysis

2.1. SPATIAL ANALYSIS

Spatial analysis methods are conditioned by the underlying data model, i.e., the spatial "discretization" of state variables, which derives from the spatial sampling frame as well as the monitoring data source. For this work, the goal concerns more the pattern of where and how much the values change that the values itself, which leads for the approach of exploratory data analysis.

Quite different approaches are being used to deal with the spatial nature of the data, either from a geostatistical, or from a modeling perspective, or simply from spatial measures, either global or local. The underlying rational in each of these methods is that the computed parameters express a kind of "natural characteristic" of a spatially contiguous set of pixels (De Jong, 1995). Although the individual point values may vary, the pattern can be distinctive.

Exploratory spatial data analysis (ESDA) include traditional techniques for visualizing spatial data such as the display of box plots by distance bands or selected regions, or maps of locations of data statistical properties, such as post plots of the quartiles of data values. More sophisticated visual tools include, among others, the spatial lag pies, the spatial lag scatterplots, and pocket plots, being these tools inappropriate for large data sets and for directional analysis. Besides graphics, traditional indicators of spatial associations tend to summarize a complete spatial distribution into a single number, such as the coefficients of spatial autocorrelation, (Moran´s I coefficient). However, appropriate ESDA techniques should measure and display local patterns of spatial association, on indicating local non-stationary, on discovering "islands" of spatial heterogeneity. Recent spatial indicators accommodate these features, being some examples, the Gi statistics, the local indicators of spatial association, and the box plot of variogram clouds suggested by Getis and Ord, (1992), and Anselin, (1993, 1994).

Spatial patterns can be described quantitatively in terms of the semivariance function, which is based on the idea that statistical variation of data is a function of distance. Although the variogram modeling has been used as a basis for spatial interpolation, it can also be adopted as an exploratory analysis tool, being its parameters reliable for the interpretation of environmental issues. For a depth revision of these techniques, see Isaacs and Srivastava (1989), Cressie (1991).

2.2. INTEGRATING TIME IN ENVIRONMENTAL DATA ANALYSIS

The issue of spatial data analysis has been considered from a static perspective, as illustrated from the above revision. The inclusion of time in spatial data analysis has been proposed rather on a statistical modeling approach, than as an exploratory tool. In fact there is an ever-increasing number of papers about space-time models and analysis, as referred by Cressie (1991), being the generic problem of space-time geostatistical data related with estimation. In these models, there is usually an assumption that the large-scale spatial and temporal components do not interact or that the small-scale components are homogeneous, which allows more precise estimates of model parameters.

The consideration of time within the spatial data analysis framework has been proposed by several authors, either within the variogram modeling (Soares, 1993, Miller, 1996), or within the spatial adaptive filtering based on a moving-window scheme (Seixas, 1996). Considering the data time range, one can state that the probability of the data to be more alike at successive time slices is higher than for distance-lag time slices. This statement is different according to the temporal scale of the data (hour, day, annual, minute) and to the randomness nature of the variable (air quality data is at the opposite side of soil variables, considering the same time scale).

Experiments of visualization of spatial data over time have been proved to be very useful for environmental problem analysis and exploration, for decision support, as well as for public presentation. The thesis for this paper is that animated visualization techniques provides the basis for integrating time in spatial data analysis in a well-understandable basis.

2.3. AIR QUALITY VISUALIZATION

The visualization process is a sequence of transformations that convert a data set into a displayable image (Senay, 1994), namely: (a) data manipulation, which converts a given data set into a form suitable for subsequent visualization, using methods of griding and interpolation; (b) visualization mapping that defines an abstract visualization technique by establishing a set of bindings between the manipulated data and visualization primitives such as positional parameters, color, texture and animation, which must effectively convey the informational content of the data; and (c) rendering the image according to the visualization mapping, using operations like shading, projection, blending, and hidden surfaces removal.

Environmental visualization attempts to represent abstract environmental data using visual metaphors for goals like data interpretation, problem identification, model validation and decision making. Frequently, earth sciences data is geographically registered, which allows for the statement of topological relations. Topology is very useful, even essential, for the comprehension of environmental data visualization, and the spatial context has to be assured by some geographic frame, like remote sensing images or river and roads network maps.

According to Rhyne (1994), there are three classes of visualization tasks usually associated with environmental data analysis: (a) analysis and exploration, being visualization used as an exploratory tool for examining data integrity and gaining insight into the data set as well as data validity issues; (b) decision support, being visualization used to develop environmental quality standards, and to examine the multi-dimensional environmental impacts; and (c) presentation, being visualization used to educate and inform the general public and high level decision makers. Usually animation sequences and sound are required for a good final product.

The integration of time in scientific visualization schemes is always performed by an animated sequence of frames. Animation is a useful technique to scan data from many time periods and is crucial for analyzing time-varying data. If the data from each time period is correlated, the resulting animation show smooth evolution. Unusual changes are then jarring and then readily apparent (Becker, 1995). Therefore, and within the context of this work, once environmental data has been stated on a spatial framework, the temporal tracking of significant patches can be assessed using animation techniques. However, a comprehensible pattern only appears if there are a sufficient number of frames, and sometimes the pattern is somehow diffuse.

Since, one can assume that observations close together in time are more alike that those further apart, a more realistic temporal tracking of spatial data can be performed by using a morphing algorithm. Morphing algorithms rely on mesh points to define the correspondence points between two images. The morphing process then warps and fades from one image to the other, using the mesh of correspondence points to define the warping function. With well-placed mesh points, this process generates a natural sequence of intermediate images and "natural" transitions according to the criterion of human perception can be produced. An in-depth review of image warping methods can be made in Wolberg, (1990).

3. Space-Time Air Quality Data Analysis: Proposed Methodology

The sampling scheme of air quality data usually includes a set of monitoring stations at specific points, measuring pollutant concentration on a quasi-continuous fashion. Therefore, air quality monitoring is extremely exhaustive in the time domain but sparse in the space one, which influences the type of data analysis and estimation methods of air quality patterns.

Some experiments on air quality analysis have been unsuccessful if only the spatial dimension of the data is considered. The lack of data spatial structure can be due to the very dynamic behavior of the air quality variables at each monitoring station, which is influenced by environmental factors such us wind patterns, building architecture, and air pollutant sources characteristics. However, for each station one can expect a temporal pattern, and the assessment of the similarities and/or differences among them can highlight some smoothing of the spatial dimension of the data.

Let us denote by $Z(x,t)$ the air quality concentration at a given time t at the monitoring station located at x, usually defined by a two-coordinate pair on a plane reference system. Adopting the rational beyond spatial analysis, one can state that observations close in time tend to be more similar than those further apart, at least when considered the quasi-continuous sampling time domain. Thus, for a specific location (x_0) in the space domain, the semivariance of z (x_0,t) can be computed, considering pairs of data at successive increasing temporal lags N(d), and using the following expression:

$$\gamma(d) = \frac{1}{2N(d)} \sum_{t=1}^{N(d)} \left[z(x,t) - z(x,t+d)\right]^2 \qquad (1)$$

If there is a semivariance function, the variogram parameters can reveal some meaningful temporal patterns of air quality, mainly the time range. This analysis has to be conducted for all the spatial locations available for the study. The temporal range (d_t) concluded from the above model, and for the condition of its similarity for all the spatial locations, can now be incorporated for the spatial variability assessment, by the expression:

$$\gamma(h) = \frac{1}{2d_t N(h)} \sum_{x=1}^{N(h)} \sum_{t=1}^{d_t} \left[z(x,t) - z(x+h,t)\right]^2 \qquad (2)$$

Using that approach, time stationarity is assured for the spatial analysis of air quality data which gives a major probability of success on the use of the variogram for the spatial analysis of air quality data, as in the case of air pollutant concentrations. The integration of the time domain in the computation of the traditional semivariance has been suggested by some authors, namely Soares (1992) and Miller (1996). Since a major goal of this work is to visualize the spatial patterns of air quality data along a time range (e.g. working day or week), one needs to generate the correspondent grids, where ordinary kriging was used.

4. Application

The proposed methodology was applied for the city of Lisboa, and for the NO_2 pollutant, which derives mainly from the traffic activity. In Lisbon, the concentrations of air quality variables are measured at eight monitoring stations irregularly located, but concentrated at down town. Figure 1 shows the spatial distribution of the air quality monitoring stations referenced on a 10 m panchromatic SPOT image, as well as the main traffic corridors. As shown, monitoring stations are concentrated at down town with high pollutant concentration levels, and at the southeast border of the city, where there is not significant pollution loads, due to the good dispersion conditions provided by the Tejo river. Two different NO_2 data sets were analyzed, namely the hourly average from April 18, 1995, to assess the air quality behavior during a working day, and the daily averages from January to October, 1995, to identify weekly patterns of air pollution. The results will be presented in the next two sections.

Figure 1: Air quality monitoring stations and main traffic paths referenced on the SPOT image (10 m, panchromatic mode).

4.1. HOURLY-BASED AIR POLLUTION DATA

Urban air pollution becomes a serious environmental problem being data analysts and policy makers interested in identifying the hourly evolution of major spatial patterns during a working day. Semivariance values for the time domain were computed, according to expression 3. Due to the wide range of air pollutants concentrations for the different monitoring stations, the semivariance values are very different, which make difficult to find the existence of similar patterns among those stations. A better comparison is obtained if the semivariance values are divided by the data variance, as shown in Figure 2(a), which presents the temporal variogram for the NO_2 hourly data, for all the monitoring stations. The experimental temporal variogram is used only for time patterns interpretation, and not for modeling purposes. A time-range of 6 hours can be identified. This trend is similar for all the monitoring stations as a consequence of the rush hour period during a working day.

The time range found in the experimental temporal variogram was used to compute the spatial variogram. According to formula (2), the experimental semivariance values were calculated considering the square differences at each lag distance station pair, summed for 6 consecutive hours. The choice of the data series of 6 hours was random, and the same time range was considered for all monitoring stations, due to the similar temporal pattern. The experimental variogram presents a spatial range around 3.5 Km, where data are spatially correlated following a spherical model, presented in Figure 2(b). This might be related with the similarities between traffic patterns among the different air quality monitoring stations in Lisboa during the day. For kriging interpolation the experimental variogram was fitted to the spherical model, according to expression (3). The kriged

NO_2 concentrations as well as its standard deviation were estimated for a 10 km grid. The spatial resolution was set at 50m in both the northings and the eastings from the sampling grid.

$$\gamma(h) = 4500 + 4241.2\left(1.5\frac{h}{4000} - 0.5\frac{h^3}{4000^3}\right), h < 4000 \qquad (3)$$

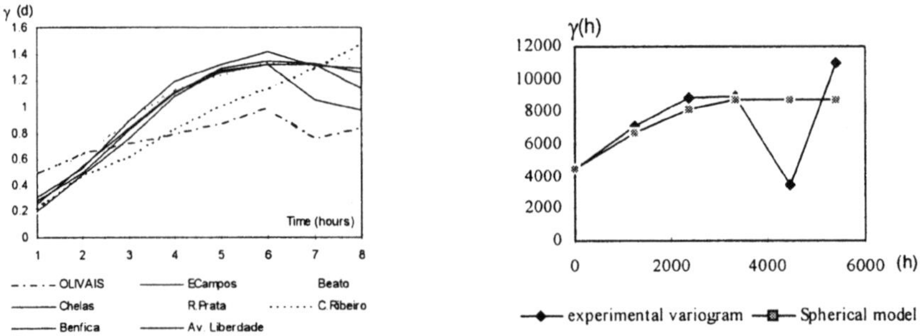

(a) (b)

Figure 2: NO_2 hourly data from Lisboa: (a) temporal, and (b) spatial variograms.

Figure 3 displays examples of air quality isosurfaces for the city of Lisbon, referring to the morning rush hour (7:00 (a) and 8:00am(b)), to lunch-time (2:00pm, (c)), and to steady-traffic conditions (9:00pm, (d)). It is quite visible the movement of higher concentrations towards south along the day, which is related with higher traffic from Lisbon entrances to downtown. The presence of more extreme air pollution patches during the morning rush hour when compared with the afternoon, is due to the wind frequency at 18 April. In fact, during the morning, the hourly average of wind frequency was about 8 km/h, and during the afternoon and evening about 27 km/h, which provide better dispersion conditions. Figure 3 (e) shows the kriged standard deviations for NO_2.

4.2. DAILY-BASED AIR POLLUTION DATA

The temporal pattern of the daily NO2 data was assessed using the daily averages between January 1, 1995 and October 31, 1995. Figure 4 (a) shows the experimental temporal variogram for five monitoring stations. As before, the semivariance values were divided by the data variance, in order to get better comparisons. A very similar trend among those stations was found, and two plateaus (2 and 7 days) can be identified, probably corresponding to two different temporal patterns. In fact, considering the daily air pollution behavior at the city, the time range of 2 days should be associated with weekends, while the time range of 7 days should be associated with the whole week. If we assume the existence of the two structures, the second one is related with the working-week.

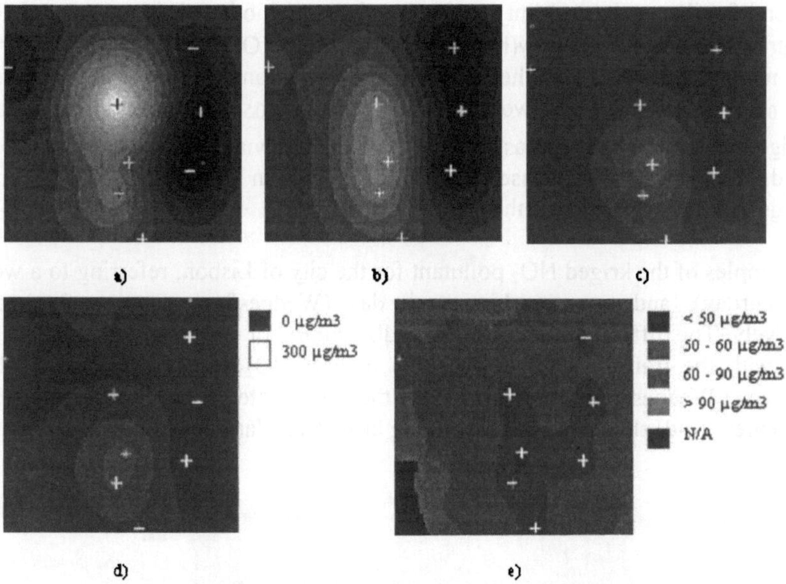

Figure 3: Map of the kriged NO2 values (a-d) and correspondent standard deviations(e).

Therefore, two spatial semivariance functions were computed in order to identify the corresponding experimental variograms. As for the previous case of the hourly data, a random sample of two consecutive (week-ends) and five consecutive (working-day) daily values, were taken into account to sum up the square differences among each lag-pair station. Figure 4 (b) displays the spatial variogram for the weekends.

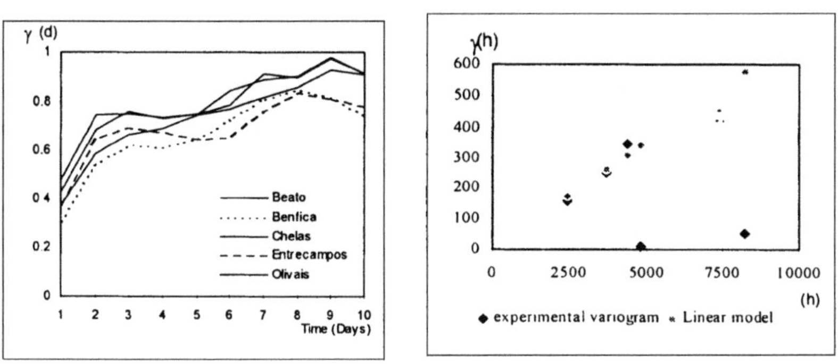

Figure 4: NO_2 daily data of Lisboa, (a) temporal variogram and (b) spatial variogram for the week-ends.

For both cases, the experimental variograms show the first three points aligned until a spatial distance of about 4.5 km. In terms of environmental science, this means that the dispersion factors influence similarly the NO_2 daily values, either week-end days, or working days, resulting in a similar spatial variability structure. The two extremely low values presented at the week-end variogram, is associated with the monitoring stations located at the suburb areas, which at the week-end days, have similar air pollutant levels (week-end familiar activities) but are far way from each other. It is the case of Benfica and Entre Campos (h=4800 m, γ(h)=11.5) and Benfica e Olivais (h=7800 m, γ(h)=25.6). Linear models were fitted to the experimental variograms, namely for the weekends ($\gamma(h) = 0.075 * h, h < 4000$) and working-week ($\gamma(h) = 0.051 * h + 90, h < 4500$) patterns. The kriged NO_2 concentrations as well as its standard deviations were estimated for a 10 km grid. As for the hourly case, the spatial resolution was set at 50m in both the northings and the eastings from the sampling grid.

Two examples of the kriged NO_2 pollutant for the city of Lisbon, referring to a weekend day (Saturday), and to a working-week day (Wednesday), Figures 5 a) and b) respectively. The differences on the spatial distribution of NO_2 concentrations is clearly related with the traffic patterns, since the highest NO_2 values are located close to downtown at Wednesday, while at Saturday they are located at the suburbs, like Benfica area. Figure 5 (c-d) shows the corresponding kriged standard deviations.

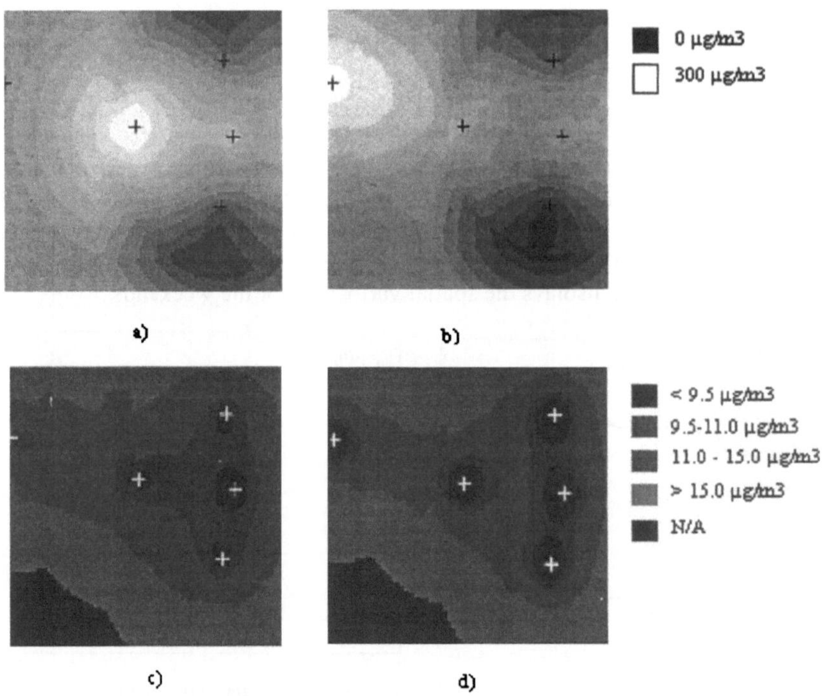

Figure 5: Map of the kriged NO_2 values for Wednesday (a) and Saturday , (b) and respective standard deviations(c-d).

4.3. VISUALIZATION OF AIR POLLUTION DATA

As stated before, temporal-spatial analysis could be enriched by means of visualization. In fact, animation schemes, such as morphing (Wolberg, 1990), are being used successfully to accommodate and follow the temporal variations of spatial data. For clearness of the visualization of NO_2 pollution patches, a threshold value was calculated, considering the whole time period. Then, a reclassification of the kriged images were performed in order to get the regions above that threshold, which represent polluted areas, with interest for air quality analysis. The visualization of the polluted patches over Lisbon was performed for a working day, based on a subset of the hourly kriged maps.

Morphing algorithms are applied to objects with well-defined edges. Usually, the mesh point selection is a time-consuming but conceptually easy task. When applied to air quality isocontours, the mesh point selection appears more difficult, since one has to recall for exogenous variables such as wind direction, to assist on the correspondence points between the two images. In spite of generating some noise in the interpolated images, morphing techniques proved to be efficient in showing the time pattern of the data. In fact, morphing methods provide a better sense of time continuity because of the smoothing behavior obtained between two consecutive time frames, and because it can generate any number of frames, which increases the time resolution.

The interpretation of the Lisbon-air-quality movies referring to a working day reveals that the paths of the NO_2 patches are roughly associated with the traffic entries in Lisbon, for example from Entrecampos at 7:00 am to the inner city at 9:00 am, becoming less important as the later morning arrive. The presence of more extreme air pollution patches during the morning rush hour when compared with the evening, is due to the wind frequency at 18 April. In fact, during the morning, the hourly average of wind frequency was about 8 km/h, and during the afternoon and evening about 27 km/h, which provide better dispersion conditions.

In terms of implementation issues, image analysis was performed on a 486DX using the IDRISI 4.1 software, which is a Geographic Information System software. Variogram analysis as well as kriging interpolation was run on Visual Basic, within the Microsoft Excel software. The morphing sequences were produced on an Apple® Macintosh™ Quadra™ 900 computer with 20Mbytes of RAM and using Elastic Reality™ 1.0 as the morphing software. Elastic Reality™ is an image warping and morphing software that enables the transformation of an initial image into a final one following user defined meshes. This is done by defining shapes on the starting and final images and then defining correspondences between them. Two sequences were produced with two different image sizes: 250x250 pixels Quicktime™ clip and a sequence of fourteen 1000x1000 pixels images for comparison purposes.

5. Conclusions

Interpreting and understanding environmental problems requires to consider the appropriate spatial and temporal context frame. Air pollution data is usually obtained from a spatial irregular network of monitoring stations, at continuous high temporal resolutions (1- or 15- minutes). However, this study showed that the use of different time-scale based data (hourly and daily) drives to different time patterns, which would lead to different interpretation within the environmental management framework.

A methodology to identify and integrate the temporal trends of air pollution data, in the assessment of its spatial variability patterns, based on variograms, is presented. Hourly NO_2 data leads to a clear temporal trend relating with the urban traffic behavior during a working-day. Similar trends were found for the 7 monitoring stations, which reveals common temporal traffic patterns to the several stations in a city. Temporal patterns similarity was also identified for the NO_2 daily data among the monitoring stations. However, within this time frame, two variogram structures were found, referring to the week-ends and to the working-week.

Integrating the significant temporal trends over space, temporal-spatial variograms show that in some situations, such as the case-study from Lisboa, a linear trend can be found. However, it is also demonstrated that each time and spatial scale selected may lead to different interpretations of the data, and consequently to different air quality related decisions. It must be noticed that almost the same spatial range (4 km) of spatial correlation was assessed for the hourly and daily NO_2 data, the associated model was set different, the daily NO_2 following a linear model and the hourly a spherical one.

Kriged NO_2 were estimated for a (10*10) km grid, with a resolution of 50 m. Considering the hourly data, maps of air pollution concentrations were hourly estimated for a period from 6:00 am to 9:00 pm, while for the daily data, those maps were estimated for all the week days, from monday to next sunday. The hourly NO_2 patches were tracked over time, through an animated visualization scheme, called morphing. The morphing approach to follow through time air pollution patches proved to be efficient in identifying the main causes of air pollution in Lisbon.

6. References

Anselin, L. (1993) Exploratory Spatial Data Analysis and Geographic Information Systems, *DOSES Workshop on Spatial Analysis and GIS*, Lisboa.

Anselin, L. (1994) Local indicators of spatial association-LISA, *Research Paper 9331*, Regional Research Institute, West Virginia University.

Becker, R., A., Eick, S. G. , Wilks, A. R. (1995) Visualizing Network Data, *IEEE Transactions on Visualization and Computer Graphics* **1**, 1, pp. 16-28.

Cressie, N., (1991) *Statistics for Spatial Data*, John Wiley & Sons, Inc., New York.

DeJong, S. M., and Burrough, P. (1995). A Fractal approach to the classification of mediterranean vegetation types in remotely sensed images, *Photogrametric Engineering & Remote Sensing*, **61**, 8, pp. 1041-1053.

Getis, A. and J. K. Ord, (1992) The Analysis of Spatial Association by Use of Distance Statistics, *Geographycal Analysis*, **24**, 3, pp. 189-206.

Isaacs, E. and Srivastava., R. (1989) *An Introduction to Applied Geostatistics*,. Oxford University Press, New York.

Miller, E. (1996), Towards a Four-Dimensional GIS: Four-Dimensional Interpolation Utilizing Kriging, GIS Research UK, University of Canterbury.

Rhyne, T., Bolstad, M., Rheingans, P. (1993) Visualizing Environmental Data at the EPA, *IEEE Transactions on Visualization and Computer Graphics* **13**, 2, pp. 34-38.

Seixas, J. (1996) Exploratory spatial-time data analysis: the case of desertification, accepted for the 2nd ESF-NSF GISDATA Summer Institute, Berlin, 24 July-1 August.

Senay, H., E. Ignatius, (1994) A knowledge-based System for Visualization Design, *IEEE Transactions on Visualization and Computer Graphics* **14**, 6, pp. 36-47.

Soares, A., Távora, J., Pinheiro, L., and Almeida, J. (1993) Predicting Probability Maps of Air Pollution Concentration: A Case Study on Barreiro/Seixal Industrial Area, *Geostatistics Troia'92*, Ed. Amilcar Soares, Klywer Academic Publishers, Holland, pp.625-636.

Wolberg, G.(1988) *Digital Image Warping*, IEEE Computer Society Press Monograph.

Ackowledgments

We would like to thank people from Comissão de Gestão do Ar de Lisboa who provide us with real air pollution data. We also thank CNIG for the SPOT image of Lisboa. This work is developed under the JNICT project Peam/P/GAI/192/93. PRAXISXXI also supports partially this work, under research contract Praxis/3/3.2/AMB/04/94.

HABITAT EVALUATION USING LOGISTIC REGRESSION

H. MARTINS[1,2], T. DOMINGOS[1,3], F. REGO[4], R. BORRALHO[5]
and J. BUGALHO[5]
[1]*Junitec - Júnior Empresas do Instituto Superior Técnico*
Av. Manuel da Maia, 36 c/v Dtª, 1000 Lisboa
Portugal
Tel.: 351-1-8461070 Fax: 351-1-8461071
[2]*Dep. Eng. Florestal - Instituto Superior de Agronomia*
Tapada da Ajuda - 1399 Lisboa Codex
[3]*Dep. Física - Instituto Superior Técnico*
Av. Rovisco Pais, 1 - 1096 Lisboa Codex
Portugal
[4]*Dep. Matemática - Instituto Superior de Agronomia*
Tapada da Ajuda - 1399 Lisboa Codex
Portugal
[5]*Erena - Ordenamento e Gestão de Recursos Naturais*
Av. Visconde Valmor, 11-4º - 1000 Lisboa
Portugal

1. Abstract

A method for wildlife habitat evaluation, based on statistical methods and Decision Theory is developed. The aim is to provide tools for an adequate management of wildlife resources, replacing empirical appreciation of habitats with scientific evaluation, increasing predictability and reducing dependence on the empirical knowledge of experienced practitioners. The method uses multiple logistic regression to predict the probability of occurrence of the studied species, based on a set of environmental variables. The transformation from probability values to occurrence predictions is done using Decision Theory, which also establishes the conditions of applicability of the model. The method is easily integrable with Geographic Information Systems, allowing the efficient use of large sets of environmental data and the application of different decision criteria for each management unit. The method is illustrated by the application to a case study: the distribution in a private game reserve of wild rabbit (*Oryctolagus cuniculus* L. 1758), one of the most important Portuguese game species.

A. Soares et al. (eds.), geoENV I–Geostatistics for Environmental Applications, 415–426.

2. Introduction

Portuguese law recognises the importance of game exploitation for agriculture, regional development and the national economy. This implies the need to guarantee the sustainability of this natural resource, by the application of adequate management and exploitation norms.

To achieve these goals it is important to develop management models with a scientific basis, decreasing the dependence on experienced professionals (Williams *et al.*, 1978; Carmo *et al.*, 1986). Based on ecological knowledge, these models relate the existence of animal species to habitat characteristics, allowing the prediction of animal responses to environmental changes and indicating the most favourable conditions for the occurrence and maintenance of certain species and/or certain population levels (Berry, 1984; Schamberger and O'Neil, 1984). Game managers may use these indications to define intervention strategies and to predict their consequences.

Thus, it makes sense to develop habitat models using a decision and risk analysis, identical to the one applied by economists in decision making under uncertainty and multiple criteria (Winterfeldt and Edwards, 1986). Within this perspective, habitat models are used to produce outputs for decision-making. However, in doing this, particular consideration must be given to the degree of accuracy and precision and the level of risk that will be accepted under different circumstances and management scenarios (Norton and Williams, 1992). This is even more important because the cost-effective application of models is essential since the intellectual, technological and other resources involved in wildlife modelling are not insignificant (Norton and Possingham, 1993).

Statistical habitat models are often preferred for this application because they are inductive models that minimise validation problems when tested appropriately against real habitat use (Schamberger and O'Neil, 1984). Besides, they are easily integrated with Geographical Information Systems (GIS). These systems have been applied in the most recent developments in habitat evaluation, due to their capacity for storing and analysing spatial multivariate data (Eby and Bright, 1985; Palmeirim, 1985; Donovan *et al.*, 1987; Pereira and Duckstein, 1993).

Among the several statistical techniques that have been used in habitat evaluation, we stress the application and interpretation of linear regression (Gutzwiller and Anderson, 1984; Marzluff, 1984), Discriminant Analysis (Fabricius and Mentis, 1990 *cit. in* Fabricius and Coetzee, 1992) and logistic regression (Pereira and Itami, 1991; Mladenoff *et al.*, 1995). One of the most common approaches in statistical habitat models is to infer habitat quality for an animal species from the probability of its presence (Pereira, 1989; Pereira and Itami, 1991; Mladenoff *et al.*, 1995).

According to this perspective of animal resource management we developed a method integrating a multivariate statistical prediction model and decision theory. This method predicts habitat quality based on the probability of occurrence of a given species, according to statistical, ecological and economic decision criteria. It was applied to wild rabbit (*Oryctolagus cuniculus* L. 1758) populations in an agricultural area of the Centre of Portugal.

3. Methods

3.1. MULTIVARIATE MODELS

The selection of variables begins with the identification of the parameters that may influence the presence of a species in a given area. This demands prior knowledge of its ecology (Williams *et al.*, 1978). Variables should also be rapidly measurable, their value should be predictable *vis à vis* future conditions, and they should be sensitive and susceptible to change (Schamberger and O'Neil, 1984).

Multivariate statistical models were chosen as a conceptual frame of relations between variables because (Shugart, 1981): (i) they are intrinsically adapted to the problems and ecological data associated with habitat selection; (ii) most seem to be robust with respect to slight deviations in their assumptions; (iii) they agree with the hypergeometric interpretation of the ecological niche, which can be seen as a multivariate sampling space.

Among multivariate statistical models logistic regression was chosen, since it predicts the probability of presence of a species at a site. This may be thresholded to obtain dichotomous data (Pereira and Itami, 1991), directly comparable to the occurrence (presence or absence) field data. Thresholding was done by varying the cut-off point in the interval [0, 1]. For each cut-off point, sites were classified as active (1) when the logistic value was above its value and as inactive (0) when the logistic value was below its value. The aim was to identify the cut-off point that maximises the accuracy of the model, according to a simple statistical criterion and Decision Theory criteria.

The statistical criterion was the maximisation of the *point correlation coefficient*, V (Poole, 1974) or ϕ_2 (Zar, 1996). This coefficient varies between -1 and 1. The value -1 corresponds to a completely erroneous classification, 0 to a random classification, and 1 to an optimal classification. It is analogous to the Pearson correlation coefficient, to which it may be compared.

3.2. DECISION THEORY

The fundamental concepts for the application of decision theory are given in TABLE 1. For each site, we have the observed (lower case symbols) and predicted values (upper case symbols), which may be active (a or A) or inactive (i or I). There are four possible outcomes: a Hit (H) when the model correctly detects an active site; a Correct Rejection (CR) when the model correctly predicts an inactive site; a False Alarm (FA) when the model mistakenly predicts an active site and a Miss (M) when the model fails to detect an active site. The two last outcomes are analogous to Type I and Type II errors, respectively, in statistical inference.

Changing the cut-off point ("bias") will influence the types of errors that occur. A higher cut-off level increases P(M), whilst a lower cut-off level increases P(FA). In the first case we have a conservative strategy, in the second we have a risky strategy.

Choosing the best cut-off point along the conservative-risky continuum depends on two factors (Saveland and Neuenschwander, 1990):

1. The prior probabilities, i.e., p(a) and p(i). If p(a) is low, a conservative criterion should be chosen. If p(a) is high, a risky criterion should be chosen.

2. The payoffs ("benefits minus costs") associated to each outcome.

Both these criteria are integrated by choosing the cut-off point which maximises the expected value (Egan, 1975). We define BC(H), BC(M), BC(CR), BC(FA) as the "benefits minus the costs" associated to each outcome, taking them as clearly definable and measurable economic costs and benefits. The expected value is given by (Saveland and Neuenschwander, 1990):

$$E(\text{cut-off}) = p(A|a)\, p(a)\, BC(H) + p(I|a)\, p(a)\, BC(M) + p(I|i)\, p(i)\, BC(CR) + p(A|i)\, p(i)\, BC(FA) \quad (1)$$

where only the conditional probabilities depend on the cut-off value.

TABLE 1. Possible relations between observed and predicted values (adapted from Saveland and Neuenschwander, 1990).

		Observed	
		a (Active)	i (Inactive)
Predicted	A (Active)	Hit p(A| a)	False Alarm p(A| i)
	I (Inactive)	Miss p(I| a)	Correct Rejection p(I| i)

3.3. MANAGEMENT STRATEGIES

We consider as the basic management decision the choice of the sites where to hunt. There are two possible strategies: hunt everywhere (S1) or hunt just in the predicted active sites (S2). However, an alternative use may also be considered, making possible two additional strategies: apply an alternative use in the predicted inactive sites and hunt in the predicted active sites (S3) or apply an alternative use everywhere (S4). We assume that this alternative use is not compatible with simultaneous hunting. The payoffs for each strategy are presented in TABLE 2 and the expected values are given by the following expressions:

$$E(S1) = p(a)\, BC(\text{hunt}) - p(i)\, C(\text{hunt}) \quad (2)$$

$$E(\text{cut-off; } S2) = p(A|a)\, p(a)\, BC(\text{hunt}) - p(A|i)\, p(i)\, C(\text{hunt}) \quad (3)$$

$$E(\text{cut-off; } S3) = p(A|a)\, p(a)\, BC(\text{hunt}) + p(I|a)\, p(a)\, BC(\text{other}) + p(I|i)\, p(i)\, BC(\text{other}) - p(A|i)\, p(i) C(\text{hunt}) \quad (4)$$

$$E(S4) = BC(\text{other}) \quad (5)$$

We assume that BC(hunt), C(hunt) and BC(other) are nonnegative and that BC(hunt) > BC(other). Otherwise, the management solution would be trivial: always use S4. The first and fourth terms of E(S3) are equal to E(S2). Since the second and third terms of E(S3) are nonnegative, we have E(S3)≥E(S2), for all cut-off points. Choosing the cut-off point equal to 0, p(A| a)=1 and p(A| i) ≤1. Thus

$$p(A|\,i)\,p(i)\,C(hunt) \leq p(i)\,C(hunt) \Rightarrow E(0;\,S2) = p(a)\,BC(hunt) - p(A|\,i)\,p(i)\,C(hunt) \geq p(a)\,BC(hunt) - p(i)\,C(hunt) = E(S1) \qquad (6)$$

We define $cutoff_{max}$ as the cut-off point which maximises E(cut-off; S2), i.e. E($cutoff_{max}$; S2)≥E(0; S2). Thus, E(S2)≥E(S1), independently of the cut-off point.

TABLE 2. Payoffs for different strategies. BC(hunting) - payoff from hunting. C(hunting) - cost from hunting. BC(other) - payoff from an alternative use.

Strategy 1 (S1)	a	i
A	BC(hunt)	-C(hunt)
I	BC(hunt)	-C(hunt)

Strategy 2 (S2)	a	i
A	BC(hunt)	-C(hunt)
I	0	0

Strategy 3 (S3)	a	i
A	BC(hunt)	-C(hunt)
I	BC(other)	BC(other)

Strategy 4 (S4)	a	i
A	BC(other)	BC(other)
I	BC(other)	BC(other)

S1 and S4 are null strategies where the output of the model is not used. For these strategies the expected value is independent of the cut-off point. S2 and S3 use the output from the model. The difference between them is that S3 has a positive payoff from predicted inactive sites, whilst S2 has zero payoff from these sites. Thus, for equal prior probabilities and payoffs, S3 will bias its choice in favour of inactive sites, typically having a higher cut-off point.

Since we are interested in the relative expected values, we normalise one of the payoffs, reducing the degrees of freedom to three. We choose BC(other)=1, and the degrees of freedom are thus p(a), p(a)BC(hunt), p(i)C(hunt). With given prior probabilities, there are just two degrees of freedom: BC(hunt) and C(hunt).

We now analyse the benefits from using the predictive model. Consider a situation where there is no possible alternative use. If BC(hunt) p(a) is very high, S1 will be the most worthwhile, since developing and using predictive models will not be cost-worthy. Otherwise, the predictive model will always be useful, since it avoids the costs of hunting in a great number of sites where there will be no profit.

Consider now the situation where the alternative use is available. If BC(hunt) p(a) is small, S4 will be best. Even for medium values of BC(hunt) p(a), S4 will still be best, since it avoids the costs of using the model. For high values of BC(hunt) p(a), S3 will be best, since it takes profit from using some sites for hunting.

Thus we have the interesting situation depicted in TABLE 3; the availability of an alternative use changes the conditions for choosing the use of the model. If the predictive capacity of the model increases, its domain of applicability will widen. In the limit of perfect predictive capacity, it is always useful to apply the model. Conversely, in the limit of null predictive capacity (random classification) it is never useful to apply the model.

TABLE 3. - When to use the predictive model.

	BC(hunt) p(a) is small	BC(hunt) p(a) is large
Alternative use is unavailable	Use model (S2)	Do not use model (S1)
Alternative use is available	Do not use model (S4)	Use model (S3)

4. Case Study

4.1. STUDY AREA

We applied the method described above to a case study: wild rabbit space use in Quinta da França. The latter is a farm in central Portugal (40°16' N, 7°25' W), with an area of 500 ha. About 40% of this area is allocated to agriculture and livestock, with cereal production (maize, oats, barley and wheat), tobacco, and pastures for sheep, goats and horses; 50% of the area is forest, with oak, pine and eucalyptus stands; the remaining 10% is fallow land. Quinta da França is a game reserve, althoug has not yet been subject to any type of game management yet. The study area was divided in a lattice of 50 m x 50 m (0,25 ha) cells, chosen for ease in measuring environmental data. We studied the rabbit, since it is the major game resource of this reserve.

4.2. DATA AND LOGISTIC REGRESSION

We carried out fifteen parallel linear transects, crossing the area in a W/E direction and separated 250 m from each other, comprising a total length of 19.617 km. Each transect was done twice, in opposite directions, surveying just one side, to detect signs of presence of rabbits (e.g., burrows or warrens). For each detected sign, the following data were gathered:

a) The distance to the beginning of the transect (using a pedometer calibrated to the average step of 0,55 and rounding to an integer number of steps);

b) The perpendicular distance of the sign to the line of the transect (estimated visually, within the distance classes: 0-0.5m, 0.5-1m, 1-2m, 2-3m, 3-5m).

c) The habitat where the sign was found.

The randomness of the sequential distribution of detections was tested for each transect in the 0-0,5 m class and the aggregated classes 0-1 m and 0-2 m (which comprise 83% of the signs). For each transect the series of 1's (detections) and 0's (no detections) corresponding to each integer number of steps was used. The bilateral one sample runs test (Daniel, 1978) was used, with a significance level of 5%, values of n_1 (number of 0's) or n_2 (number of 1's) in excess of 20 and the r distribution approximated by the reduced normal distribution (z). The null hypothesis of independence of detections was not disproved for any transect and perpendicular distance class ($z \leq 0.797$, $p \leq 0.05$).

Inspection of histograms of detections according to distance class indicates that there is loss of visibility with increasing distance from the transect. If loss of visibility is equal for all habitats, all the data may be used in the regression. Otherwise, using all the data would bias the regression, favouring the habitats with higher visibility. To test for this, a coefficient of visibility loss was calculated, given by:

$$cvl_i = \sum_{n=1}^{N} x_{i,n} \Big/ \sum_{n=1}^{N} x_{1,n} \qquad (7)$$

where cvl_i is the loss of visibility from the first distance class to the *i*th-class, *i* is the distance class, *n* is the habitat, *N* is the number of habitats, $x_{i,n}$ is the number of detections in the *i*th-distance class and the *n*th habitat. The null hypothesis of equal visibility loss in all habitats was considered. A χ^2 test for each distance class was carried out, with N-2 degrees of freedom and the expected frequencies, $x'_{i,n}$, given by $x'_{i,n} = cvl_i \, x_{i,n}$. Habitats not respecting Cochran's (1954) rule of minimum frequencies were not considered in this test. The null hypothesis was not rejected for all distance classes (χ^2 = 5.54, 2.21, 0.37, 5.06, df=2, α=0.05). Thus the data from all distance classes were aggregated, providing the data for the dependent variable in the logistic regression.

We considered an extensive list of independent variables contributing to spatial selection by the rabbit, i.e. variables influencing the satisfaction of nutritional needs, reproduction and shelter (Table A.1.; Table A.2.). Variables were given their values at the geometrical centre of the cells. Categorical variables were transformed to dummy variables (Table A.3.; Table A.4.).

The statistics package SPSS for MS WINDOWS 6.1 was used to fit the logistic regression model. The independent variables were chosen using a stepwise process with a value of "α-for-entry" of 0.05 and "α-for-exit" of 0.1, as suggested by Draper and Smith (1981). A total of 393 sampled cells were considered. The bias in the constant, b_0, due to a larger number of inactive cells (n_i), compared to active cells (n_a), was corrected using the expression (Maynard, 1981 *cit. in* Pereira, 1989):

$$b_{0(cor)} = b_0 + \ln(n_i/n_a). \qquad (8)$$

4.3. RESULTS AND DISCUSSION

We obtained the logistic regression model given in TABLE 4. The maximum value of V was obtained for a cut-off point of 0.6, with V=0.38, which is low when compared to r values in similar studies (Pereira 1989, ρ=0.55; Rice *et al.* 1993, r=0.58; Borralho *et al.* 1996, r=0.95, r=0.74).

TABLE 4. Logistic models for prediction of occurrence probability. The variables are presented according to inclusion order in stepwise regression, with the regression coefficients (b) and significance level (p).

Model	Variables	b	p	n_i	n_a	$\ln(n_i/n_a)$	b_0
	DIM	0,035	0,000	311	82	1,333	-2,072
Rabbit	IPG	0,397	0,016				
p<0,0000	LENAT	-0,640	0,005				

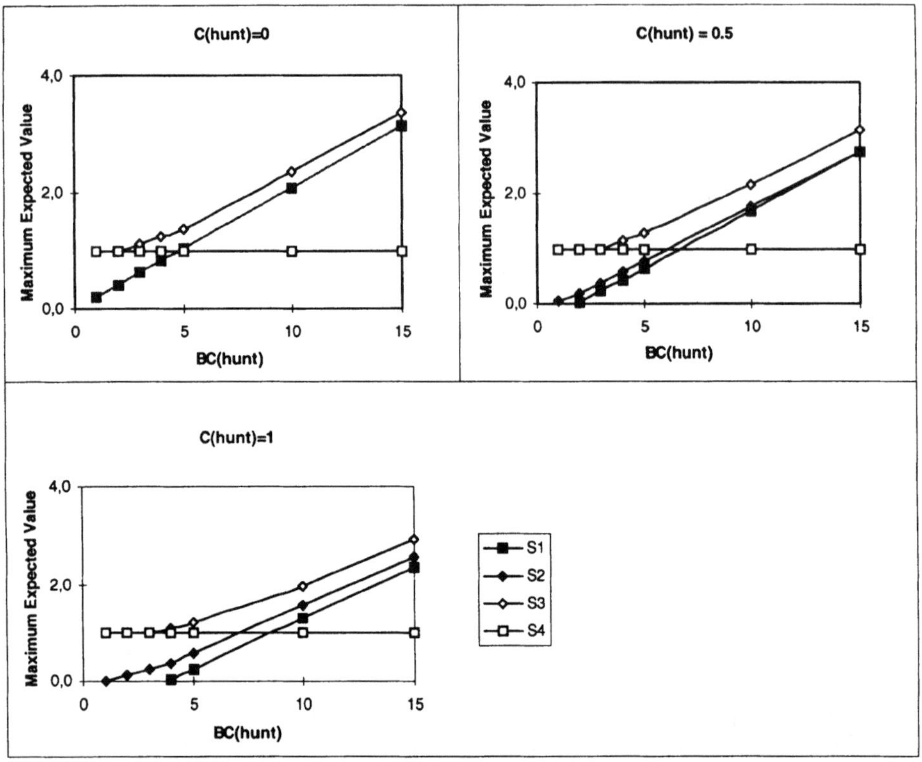

Figure 1 - Expected values as a function of BC(hunt) for the four strategies and three different values of C(hunt). The normalisation BC(other) = 1 is used. For C(hunt)=0, S1 and S2 are superimposed.

We now consider the expected values for each strategy, S1 to S4, presented above. Since the prior probabilities are given, p(a)=0.21 and p(i)= 0.79, there are only two degrees of freedom, BC(hunt) and C(hunt). The behaviour of the maximum expected values as a function of BC(hunt) and for different values of C(hunt) is presented in Figure 1 (for each value of BC(hunt) the cut-off value maximising the expected value; the corresponding expected value is plotted). The charts illustrate the relations between strategies discussed above: E(S3) ≥ E(S2) ≥ E(S1). We see that, for low values of BC(hunt), the best strategies are S3 and S4, with the same payoff. In this situation, is better to strategy S3 to have a cut-off value of 1.0, thus allocating every cell to an alternative use. As BC(hunt) increases, S3 becomes better than S4. At this point, it becomes worthwhile to use the model (this would happen for lower values of BC(hunt) if the p(a) were higher for this study). For higher values of BC(hunt), S1 and S2 also overtake S4. For C(hunt)=0, E(S1) = E(S2). As C(hunt) increases, the difference between E(S1) and E(S2) increases. However, this difference becomes insignificant for large values of BC(hunt). For both S2 and S3, as BC(hunt) increases, more and more cells are allocated for hunting, due to a decrease in the cut-off value.

5. Conclusions

A set of four different management strategies was identified, two of them using the predictive model (S2 and S3) and two of them not using the predictive model (S1 and S4). S3 is always better than S2. We showed that the availability of an alternative use changed the conditions for application of the model. If no alternative use is available, the predictive model should be used for low values of BC(hunt). If an alternative use is available, the predictive model should be used for high values of BC(hunt).

Despite the low fit of the logistic model to the data, we showed that, for certain payoffs, the model is a useful management tool, i.e., it allows a cost-effective increase in the expected payoff. Namely, if there is no alternative use and the cost of hunting is nonnegative, it is always useful to apply the model (as long as its predictive capacity is better than a random model). This is important, since it lends support to creating statistical models for management even when the available data set is not too large.

After calibrating the model for a given situation, as we did with the case study, it is applicable to other situations. In these applications, no fieldwork is necessary. In the spirit of Bayesian inference, the prior probabilities are given by subjective estimates, and the conditional probabilities are given by the model predictions based on the environmental data of the application.

This approach can be useful in large scale conservation decisions. Multivariate models can predict the best conservation areas, using easily obtainable geographic data, stored and manipulated in a GIS. After establishing conservation priorities and formalising them as payoffs, this method will be able to identify areas of conservation interest.

6. Acknowledgements

This work is integrated in Project MISART - Integrated Modelling of an Environmental, Rural and Tourist System, supported by the European Commission, under contract LIFE94/P/A151/P/01307/CEN with Junitec - Junior Empresas do Instituto Superior Técnico, and by Programa Praxis XXI, through grants BD/3672/94 (T.D.) and BD/3847/94 (R.B.).

7. References

Berry, K.H. (1984) Introduction: development, testing and application of Wildlife-Habitat Models, in J. Verner, M. Morrison and C.J. Ralph (eds.), *Wildlife 2000. Modelling Habitat Relationships of Terrestrial Vertebrates*, The University of Wisconsin Press, pp. 3-4.

Borralho, R., Rego, F., Palomares, F., and Hora, A. (1996) The distribution of the Egyptian mongoose *Herpestes ichneumon* (L.) in Portugal, *Mammal Rev.*. **26**(1), 1-8.

Cardoso, J.C. (1965) *Os solos de Portugal. Sua classificação, caracterização e génese. 1- A sul do rio Tejo*, Secretaria de Estado da Agricultura, Direcção-Geral dos Serviços Agrícolas, Lisboa.

Cardoso, J.C. (1974) A classificação dos solos de Portugal. *Boletim de Solos do Serviço de Reconhecimento e Ordenamento Agrário* **17**, 14-46.

Carmo, P.L., Romão C.O. and Santos, A.M. (1986) Avaliação de potencialidades cinegéticas - método HEP, *Comunicações do I Congresso Florestal Nacional*, Lisboa.

Cochran, W. G. (1954) Some methods for strengthening the common χ^2 Tests, *Biometrics* **10**, 417-451.

Daniel, W.W. (1978) *Applied Nonparametric Statistics*, Houghton Mifflin Company.

Donovan, M.L., Rabe, D.L., and Olson, C.E. (1987) Use of Geographic Systems to develop Habitat Suitability Models, *Wildl. Soc. Bull.* **15**, 574-579.

Draper, N.R. and Smith, H. (1981) *Applied regression analysis*, Second Edition, Wiley Series in Probability and Mathematical Statistics, New York.

Egan, J.P. (1975) *Signal Detection Theory and ROC analysis*, Academic Press. New York.

ERENA (1994) *Plano de Ordenamento e de Exploração Cinegéticos e Plano de Aproveitamento Turístico*, TERRAPRIMA - Sociedade Agrícola, Lda. Lisboa.

Eby, J.R. and Bright, L.R. (1985), A digital GIS based on LANDSAT and other data for elk habitat efectiveness analysis, *Proceedings of the 19th International Symposium of Remote Sensing and Environment* pp. 855-864.

Fabricius, C. and Mentis, M.T. (1990) Seasonal habitat selection by eland in arid savanna in South Africa, *S. Afr. J. Zool.* **25**, 238-244 (*cit. in* Fabricius e Coetzee 1992).

Fabricius, C. and Coetzee, K. (1992), Geographic information systems and artificial intelligence for predicting the presence or absence of mountain reedbuck, *S. - Afr. Tydskr. Natuurnav* **22**(3) 80-86.

Gutzwiller, K.J. and Anderson, S.H. (1984), Improving Vertebrate-Habitat Regression Models in Verner, J., Morrison, M., and Ralph, C.J. (eds.), *Wildlife 2000. Modeling habitat relationships of terrestrial vertebrates*, The University of Wisconsin Press, pp. 161-164.

Marzluff, J.M. (1984) Assumptions and design of regression experiments: the importance of lack-of-fit testing, in J. Verner, M. Morrison and C.J. Ralph (eds.), *Wildlife 2000. Modeling Habitat Relationships of Terrestrial Vertebrates*, The University of Wisconsin Press, pp. 165-170.

Maynard, P.F. (1981) *The logit classifier: a general maximum likelihood discriminant for remote sensing applications*, Unpublished M. A. Thesis, Department of Geography, University of California, Santa Barbara, California (*cit. in* Pereira 1989).

Mladenoff, D. J., Sickley, T.A., Haight, R.G., and Wydeven, A.P. (1995), A regional landscape analysis and prediction of favorable gray wolf habitat in the northern Great Lakes region, *Conservation Biology* **9**(2), 279-294.

Norton, T.W. and Williams, J.E. (1992) Habitat modelling and simulation for nature conservation., *Mathematics and Computers in Simulation* **33**, 379-384.

Norton, T.W. and Possingham, H.P. (1993) Wildlife modelling for biodiversity conservartion. *Modelling Change in Environmental Systems*. John Wiley and Sons.

Palmeirim, J.M. (1985), Using LANDSAT TM imagery and spatial modelling in automatic habitat evaluation and release site selection for the Ruffed Grouse (*Galliformes: Tetraonidae*), *Proceedings 19th International Symposium of Remote Sensing and Environment* pp. 729-738.

Pereira, J.M.C. (1989) *A spatial approach to statistical habitat suitability modeling: The Mt. Graham Red Squirrel Case Study*, Unpublished Ph. D. Dissertation, School of Renewable Natural Resources, University of Arizona, Tucson, Arizona.

Pereira, J.M.C. and Itami, R.M. (1991) GIS-Based habitat modeling using logistic multiple regression: A study of the Mt. Graham red squirrel, *Photogrammetric Engineering & Remote Sensing* **57**(11), 1475-1486.

Pereira, J.M.C. and Duckstein, L. (1993). A multiple criteria decision-making approach to GIS-based land suitability evaluation. *Int. J. Geographical Information Systems* **5**(7), 407-424.

Poole, R.W. (1974) *An Introduction to Quantitative Ecology*, McGraw-Hill Book Company, New York.

Rice, S.M. , Guthery, F.S., Spears, G.S., DeMaso, S.J. and Koerth, B.H. (1993) A precipitation-habitat model for northern Bobwhites on semiarid rangeland, *J. Wildl. Manage.* **57**(1), 92-192.

Saveland, J.M. and Neuenschwander, L. F. (1990) A signal detection fremework to evaluate models of tree mortality following fire damage, *Forest Science*, **36**(1), 66-76.

Schamberger, M.L. and O'Neil, L.J. (1984) Concepts and constraints of habitat-model testing, in J. Verner, M. Morrison and C.J. Ralph (eds.), *Wildlife 2000. Modeling Habitat Relationships of Terrestrial Vertebrates*, The University of Wisconsin Press, pp. 5-10.

Shugart (Jr.), H.H. (1981) An overview of multivariate methods and their application to studies of wildlife habitat, in D.E. Caper (ed.), *The use of multivariate statistics in studies of wildlife habitat*. Proceedings of a workshop, Burlington, pp. 4-10.

Williams, G.L., Russel, K.R. and Seitz, W.K. (1978) Pattern recognition as a tool in the ecological analysis of habitat, in A. Marwelstein (ed.), *Classification, inventory and analysis of fish and wildlife habitat - The Proceedings of a National Symposium*, Phoenix, Ariz. U. S. Dep. Inter., Fish and Wildlife Serv., Off. of Biol. Serv., FWS/OBS-78/76 USD, FWS, OBS, Washington, D.C., pp. 521-531.

Winterfeldt (von), D. and Edwards, W., (1986), *Decision Analysis and Behavioral Research*, Cambridge University Press, New York, 604 pp.

Zar, J.H. (1996) *Biostatistical Analysis*, Prentice-Hall, Third edition, London.

8. Appendix

TABLE A. 1 Environmental Variables.

Variable	Code	Type	Unit	Range	Categories
		for individual cells			
altitude	ALT	ratio	m	417-486	
density of the herbaceous layer	COBHER	ratio	%	15-100	
density of the shrub layer	COBARU	ratio	%	0-80	
density of the arboreous layer	COBARB	ratio	%	0-50	
height of the herbaceous layer	ALTHER	ratio	m	0,11-1,00	
height of the shrub layer	ALTARU	ratio	m	0,42-3,00	
height of the arboreous layer	ALTARB	ratio	m	2,00-15,00	
soil	*	nominal			soil types (TABLE A. 2)
soil use	*	nominal			habitats (TABLE A. 2)
distance to waterlines	DTLA	ratio	m	0-225	
distance to permanent water points	DTPAP	ratio	m	0-1220	
distance to habitat edge	DTORL	ratio	m	0-720	
distance to railway	DTCAMF	ratio	m	0-1588	
distance to the main road	DTCAM	ratio	m	0-1446	
grazing intensity	IPG	nominal			intense
	IPM				moderate
	IPF				weak
		for groups of 3, 5 and 7 cells			
diversity	ISW3	ratio		0,00-1,10	
(Shannon-Wiener Index)	ISW5			0,00-1,19	
	ISW7			0,00-1,32	
fragmentation	FRG3	ratio	no. of	1-3	
	FRG5		different	1-4	
	FRG7		classes	1-4	
		for groups of cells in the same contiguous area of habitat			
dimension	DIM	ratio	ha	1-121	
perimeter	PER	ratio	m	432-5113	
edge availability	DIORL	ratio	m/ ha	42-518	
dominance	DOM	ratio	%	3-25	

TABLE A. 2 Coding for the attributes of the nominal variables "soil use" and "soil".

Variable	Attributes	Code
soil use (ERENA, 1994)	Fallow	P
	Pasture	Pas
	Maize	Mi
	Tobacco	Tb
	Pasture with olive orchard	OlPas
	Remains from a cereal dryland culture	Car
	Shrub	Ma
	Oak forest	Cv
	Oak forest with shrub	MaCv
	Degraded areas	Ad
	Pine	Pnb
	Eucalyptus	Eg
	Acacia	Mim
soil type[1] (Cardoso, 1965, 1974)	Solos Litólicos Não Húmicos Pouco Insaturados Normais de granitos (Pg)	SPG
	Aluviossolos Modernos Não Calcários de textura ligeira (Al)	SAL
	Litossolos dos Climas de Regime Xérico de granitos ou quartzodioritos (Eg)	SEG
	Aluviossolos Modernos Não Calcários de textura ligeira, com fases inundáveis e pedregosas (Al(i,p))	SALIP
	Solos de Baixas (Coluviossolos) Não Calcários de textura ligeira, fases gropédicas (Sbl(a))	SSBL

TABLE A. 3 Transformation of the dummy variables "soil" and "grazing intensity" to dummy variables.

Attributes	variable 1	variable 2	variable 3	...	variable n-1
attribute 1	n-1	0	0	0	0
attribute 2	-1	n-2	0	0	0
attribute 3	-1	-1	n-3	0	0
...	-1	-1	-1	....	1
attribute n	-1	-1	-1	-1	-1

TABLE A. 4 Transformation of the dummy variable "soil use" to dummy variables.

Attributes	Herbaceous vs. Woody (HER)	Natural Woody vs. Artificial Woody (LENAT)	Oak vs. Shrub (CV)	Ma/Ad vs. MaCv (MA/AD)	Pasture vs. Cultivated Land (PAST)	Mi vs. Car (MI)
P/Pas/OlPas	4	0	0	0	2	0
Ma/Ad	-3	1	-1	1	0	0
Cv	-3	1	2	0	0	0
MaCv	-3	1	-1	-1	0	0
Mi	4	0	0	0	-1	1
Car	4	0	0	0	-1	-1
Pnb/Eg	-3	-3	0	0	0	0

[1] This classification is not translatable, since it specific for Portugal.

GEOSTATISTICAL MODELING OF SPECIES DISTRIBUTIONS

Implications for Ecological and Biogeographical Studies

LAWRENCE E. HUNT
Museum of Systematics and Ecology
Department of Biological Sciences
University of California
Santa Barbara, California 93106

Abstract. The geographic distribution of an organism forms a nested hierarchy of patterns across a range of spatial and temporal scales. Generalizing distribution patterns from locality data is difficult because of the inability to maintain a given level of sampling intensity with increasing spatial scale. Geostatistics, by capitalizing on the autocorrelative character of natural systems, can be an effective tool in dealing with the logistical difficulties of multi-scale geographic pattern analysis. Geostatistical models of three soil properties were successfully used to predict the distribution of a fossorial lizard, *Anniella pulchra* Gray (Sauria:Anguidae), over an 11-hectare area based on point data collected in the field. A nested soil sampling strategy was used to identify the scale of spatial variation of these soil properties. These analyses determined that a large proportion of the spatial variation inherent in soil texture, penetrability, and bulk density occurs at distances less than 2 meters, i.e., at scales within the home range of individual lizards. The magnitude of this edaphic variation may strongly influence microhabitat selection and consequently, the local and meso-scale distribution of this species. The "bottom-up" approach used here couples detailed microhabitat data with a geostatistical analysis of soil properties to interpolate edaphic variation and generate representations of lizard distribution at larger spatial scales. These models are amenable to field-testing and refinement.

1. Introduction

Ecology and biogeography share a common purpose, the study of distribution and abundance, yet these disciplines operate at opposite ends of spatial and temporal continua. Ecological studies frequently lack the ability to extrapolate results beyond the immediate study area. Compounding this, apparent stochastic variation in local systems may actually be strongly influenced by deterministic forces operating at regional spatial and temporal scales. Linkages between scales are poorly understood, but are currently the subject of much research. Biogeographical studies rarely incorporate ecological information that governs distribution and abundance at smaller spatial scales. The origin and maintenance of the discontinuity between these divergent approaches lies in the inability to maintain a given level of sampling intensity with increasing spatial scale.

The fundamental premise of this paper is: geographic distribution forms a nested hierarchy of spatial patterns across a range of spatial scales extending from an assemblage of individuals at point locations autocorrelated in space and time up to the biogeographic limits of the species. The basic challenge in mapping geographic distribution is to generalize species data originally collected at local spatial and temporal scales, to larger realms. The most subjective representation of distribution is

A. Soares et al. (eds.), geoENV I–Geostatistics for Environmental Applications, 427–438.

an area-covering surface (e.g., Stebbins, 1985). This method, although based on point data, sacrifices accuracy for generality, and overestimates the actual distribution of the species within the range limits. A related and more informative method is to plot the marginal locality records for the species then use these locations as guide points to form a minimum area polygon (e.g., Hall, 1981). Simple plots of locality data on maps or grids, "dot" maps, provide the most objective but least general distribution maps because they represent presence only at specific locations. A much more useful representation is one where locality data is displayed upon a map of species occurrence associated with ecological requirements but this approach has been hampered by the lack of knowledge of the ecological requirement of most species and the problem of maintaining sampling intensity with increasing spatial scale.

Geostatistics capitalizes on an intrinsic property of nature, i.e., spatial and temporal autocorrelation, or rather its inverse, semivariance, and can provide an effective tool in dealing with the logistical difficulties associated with analyzing multi-scale geographic patterns. The present study employs microhabitat-based, geostatistical models of soil properties to predict the distribution of the California legless lizard, *Anniella pulchra* Gray (Sauria:Anguidae). This "bottom-up" approach creates distribution maps at ecological and regional spatial scales that can be field-tested and refined.

2. Natural History

The genus *Anniella* includes two species of lizards that are highly specialized for a fossorial existence. The California legless lizard, *A. pulchra*, ranges from San Francisco Bay in California southward to northwestern Baja California Norte, Mexico. Its distribution is roughly coincident with the California Floristic Province, a region with a Mediterranean climate characterized by a warm, dry season (May to October), and a cool, rainy season (November to April).

California legless lizards are highly fossorial and possess a suite of morphological and behavioral specializations for a subterranean lifestyle. These lizards are habitat generalists, occurring in coastal aeolian dune scrub, coastal sage scrub, chaparral, oak woodland, and montane pine woodland from sea level to elevations near 1,850 m. Despite these broad vegetation associations, this species is a microhabitat specialist. The under-canopy leaf litter environment of shrubs and trees are foci of *Anniella* activity, they are rarely found in exposed soils between plants or on the surface of the soil. Edaphic properties, especially soil texture, appear to control legless lizard presence or absence and its highly fragmented geographic distribution. Soil temperature and soil moisture control seasonal and diurnal activity patterns and spatial variation in these microclimatic properties is heavily influenced by the presence and physiognomy of plant cover. Exposed and covered soils exhibit very different thermal regimes. The extreme spatial variability of near-surface soil temperatures allows these lizards to precisely maintain a preferred body temperature by small-scale horizontal and vertical movements through the soil (Hunt, 1994).

This species was selected for study because: it is geographically widespread; head and body movements are used to burrow through the soil so its local and regional distribution is likely strongly dependent upon measurable edaphic variables; it is an ectotherm so activity patterns are closely tied to measurable microclimatic parameters, and; it has limited dispersal abilities (Miller, 1944), which allows microsites to be characterized with a precision not possible in more mobile vertebrates.

3. Methods

Study sites included coastal and inland locations scattered throughout San Luis Obispo, Santa Barbara, and Ventura Counties in central and southern California. This paper deals with one inland study site, Ozena, located in north-central Ventura County, in the upper watershed of the Cuyama River. The study site lies on the northern edge of the upper Cuyama River floodplain, at an elevation of 1,135 m, and includes a succession of three stream terraces. Soil texture on the terraces appeared to be a function of distance from the active channel of the river, i.e., terraces close to the river possessed coarse-textured soils and were presumably younger than terraces further from the active channel, which had fine-grained soils. The study site encompassed approximately 312,500 m^2. Site vegetation is open piñon-juniper woodland, with a diverse shrub understory (Vasek and Thorne, 1977). Dominant and sub-dominant tree and shrub species beneath which lizards are found include: piñon pine (*Pinus monophylla*), California juniper (*Juniperus californica*), scrub oak (*Quercus dumosa*), manzanita (*Arctostaphylos glauca* and *A. parryana*), ceanothus (*Ceanothus* spp.), deerbrush (*Purshia glandulosa*), ephedra (*Ephedra californica*), rabbitbrush (*Chrysothamnus nauseosus*), big sagebrush (*Artemisia tridentata*), buckwheat (*Eriogonum fasciculatum*), and yucca (*Yucca schidigera*).

A grid-based approach was used to quantify microhabitat selection by legless lizards, by objectively and systematically sampling the environment around individual lizards. Microsites likely to contain legless lizards were searched by carefully raking the leaf litter beneath shrubs or trees. When lizards were found, a thermometer was placed in the soil at the depth and location of the mid-body and allowed to equilibrate for 60 seconds. Because of the low thermal capacity of these small, elongate lizards, this surrogate temperature measurement allowed a very close approximation of actual field body temperature without having to handle the lizards. Immediately following measurement of body temperature a 2 m x 2 m grid, with a 2,500 cm^2 cell size was centered on the lizard collection point and oriented north with the aid of a compass. This grid size represented a compromise between microhabitat grain, apparent home range size, and sampling intensity. Soil temperature was measured at the location and depth of the lizard (grid center), and at 24 other grid node points. Soil penetrability (measured horizontally in cm at 2 cm and 8 cm depths), was recorded in the field at the lizard collection point (grid center) and four other grid node points. Soil moisture (% by mass), bulk density (gms/cc), and soil texture (% sand/silt/clay), were measured at the lizard collection point and at four other node points within the grid. These three soil properties were measured from a 250-300 g soil sample collected from the face of a small pit at each grid sample point. The sampler collected soil from the leaf litter/soil interface to a depth of 8 cm. The samples were bulked in the field to get a composite sample from this depth interval. Soil moisture, bulk density, and soil texture were determined in the lab using standard analytical procedures (Klute, 1986). This sampling strategy quantified microclimatic and microedaphic conditions within a 4 m^2 area surrounding individual lizards in the field. Forty-three grids, including 26 lizards, were sampled in this manner at the Ozena study site. These data formed the quantitative basis for predicting lizard presence or absence at larger spatial scales.

The edaphic and biophysical variables measured in these small grids exhibit widely divergent magnitudes of spatial and temporal variation, e.g., soil temperature versus soil texture. Consequently, although biophysical variables like soil temperature and soil moisture are good predictors of diurnal or seasonal activity, actual presence or absence at a particular location depends on relatively more stable edaphic features such as soil texture. Stepwise discriminant analysis of the edaphic and biophysical data collected in the grids selected a subset of three variables that significantly distinguished lizard presence or absence: sand content, soil penetrability at 2 cm deep, and bulk density of the soil.

A nested sampling strategy was used to determine the scale of spatial variation, obtain estimates of the sample semivariograms, and direct subsequent sampling strategies for these soil properties. Nested sampling is a modification of multistage hierarchical sampling in which the stages of the design represent particular sampling intervals. The technique is detailed in Oliver and Webster (1987) and Webster and Oliver (1990). The design for the Ozena study site consisted of five stages covering a range of 2 m to 200 m. Sampling intervals were increased in geometric progression from the smallest interval expected to be of consequence to the largest, so that the components of variance could be regarded as independent. Six primary "centers" were spaced 200 m apart in two rows of three points (stage 1). All other points were then chosen on random orientations from these centers as follows. From each of the six centers a second site was chosen 50 m away (stage 2). From each of these 12 sites another point was chosen 20 m away (stage 3). At the fourth stage the procedure was repeated to locate points 5 m away from those of the third stage, giving 48 points. At the fifth stage, only half the fourth stage points were replicated by sampling points 2 m away, to give a total of 72 sampling sites. This unbalanced sampling strategy gives a 25% economy in sampling effort over a fully balanced design, which doubles the sampling effort at each stage (Oliver and Webster, 1987). A nested analysis of variance (ANOVA) was performed on these data to estimate the components of variance contributed by each stage or sampling interval (Webster, 1977; 1985; Oliver and Webster, 1986). The proportion of variance explained was plotted against the distance between samples, to give a crude sample semivariogram of each soil property.

A uniform sampling strategy was implemented based on the results of the nested sampling survey. A uniform sampling interval of 36 m was chosen, based on the average distance at which 75% of the variation in each soil property was reached, as extrapolated from the nested ANOVA semivariogram plot (Fig. 1). Based on this result six, 540 m transects, spaced 36 m apart, were aligned in an east-west direction. These transects formed a grid 180 m x 540 m, which encompassed an area of 97,200 m^2 and included portions of all three stream terraces found within the study site. A transect length of 540 m allowed a minimum of 15 comparisons at the first lag interval (36 m). At each sample point the same three soil properties were recorded, as originally described for the 2 m x 2 m and nested sampling surveys.

Sample semivariograms from the uniform sampling survey were computed using the standard semivariance equation described in Oliver and Webster (1987). Geometric anisotropy in the data set was analyzed by creating directional semivariograms in four directions, 0, 45, 90, and 135 degrees, with a tolerance angle of 22.5 degrees. This effectively divided the sample pairs in the omni-directional semivariogram into four subsets. These directional semivariograms exhibited a common pattern of progression and similar maxima, indicating that significant anisotropy did not exist in these soil properties at this site. Isotropic semivariogram models were selected and fitted to the sample semivariograms using best-fit analysis. This technique relied on a combination of Marquardt non-linear procedures to fit initial model parameters, then stepwise iteration to refine these initial estimates. A particular model was considered to be best-fit when the reduced sums of squares (RSS) was not further minimized by significant changes in model parameters. RSS was calculated for the regression of actual versus model-estimated semivariance at each lag class distance (r^2).

Sand content, penetrability, and bulk density values were interpolated at 15 m intervals across the 97,200 m^2 sample area using the best-fit semivariogram model for each soil property and simple punctual kriging. The high nugget variance:sill ratio for all soil properties, indicated that a large neighborhood size and a large search radius was necessary to krig with an acceptable point-to-point estimation variance. Consequently, 8-12 neighbors and a neighborhood size equal to the maximum lag distance (569 m) were used during kriging. The kriged interpolation grids portrayed variation in each soil property across the study site. Contour intervals on the interpolated soil maps were adjusted so the maps portray ecological information relevant to lizard distribution. Map contour intervals were based on the range and 95% confidence intervals of the mean soil value at lizard collection points in the 2 m x 2 m grids. For example, intervals for sand content represent: minimum value for site to minimum value for range of lizard occurrence (< 70.1% = species absent); minimum range value to lower 95% confidence interval for mean value (70.1% - 77.1% = low probability of occurrence); lower 95% confidence interval to maximum value for site (77.1% - 93.5% = high probability of occurrence). Contour intervals for bulk density were subjectively determined because of a lack of sample and interpolated values within the range of lizard occurrence (i.e., between 0.757 and 1.235 gms/cc), based on the uniform sampling survey (Fig. 2).

These kriged maps were used to predict lizard presence or absence. A composite predicted distribution map was produced by visually superimposing these maps and noting the co-occurrence of interval boundaries. Equal weight (predictive ability) was given to each soil property. Absence was defined as areas where all three variables exceeded the 95% confidence limit boundaries. Areas predicted as supporting lizards had at least one variable within the confidence limits for lizard presence. A thorough survey of the entire study site was conducted before, during, and following the soil sampling surveys, in order to locate as many lizards as possible. The locations of these individuals were mapped in the field and superimposed on the composite map to qualitatively assess concordance of predicted versus actual lizard distribution.

4. Results

The results of the nested sampling strategy and nested ANOVA are clear (Fig. 1). At least 80% of the total spatial variation in bulk density, penetrability, and sand content occurs at distances less than 50 m, indeed, at least 30% of the variation occurs at stage 5, representing unresolved variation over all distances less than 2 m, plus any measurement error.

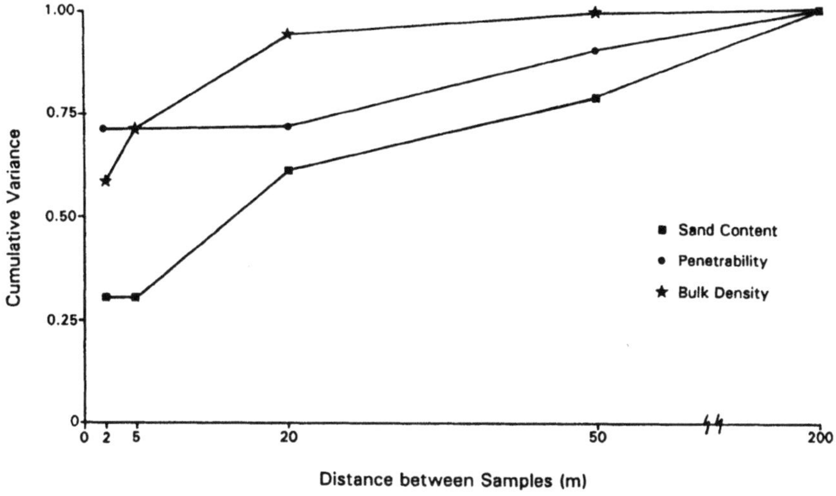

Figure 1. Cumulative components of variance versus sampling interval.

Histograms and sample semivariograms for each soil property based on the uniform sampling survey support are shown in Fig. 2. Sand content was significantly skewed towards lower values ($p < 0.05$), requiring a log-normal transformation to normalize the data prior to semivariance analysis. Bulk density and penetrability did not require transformation.

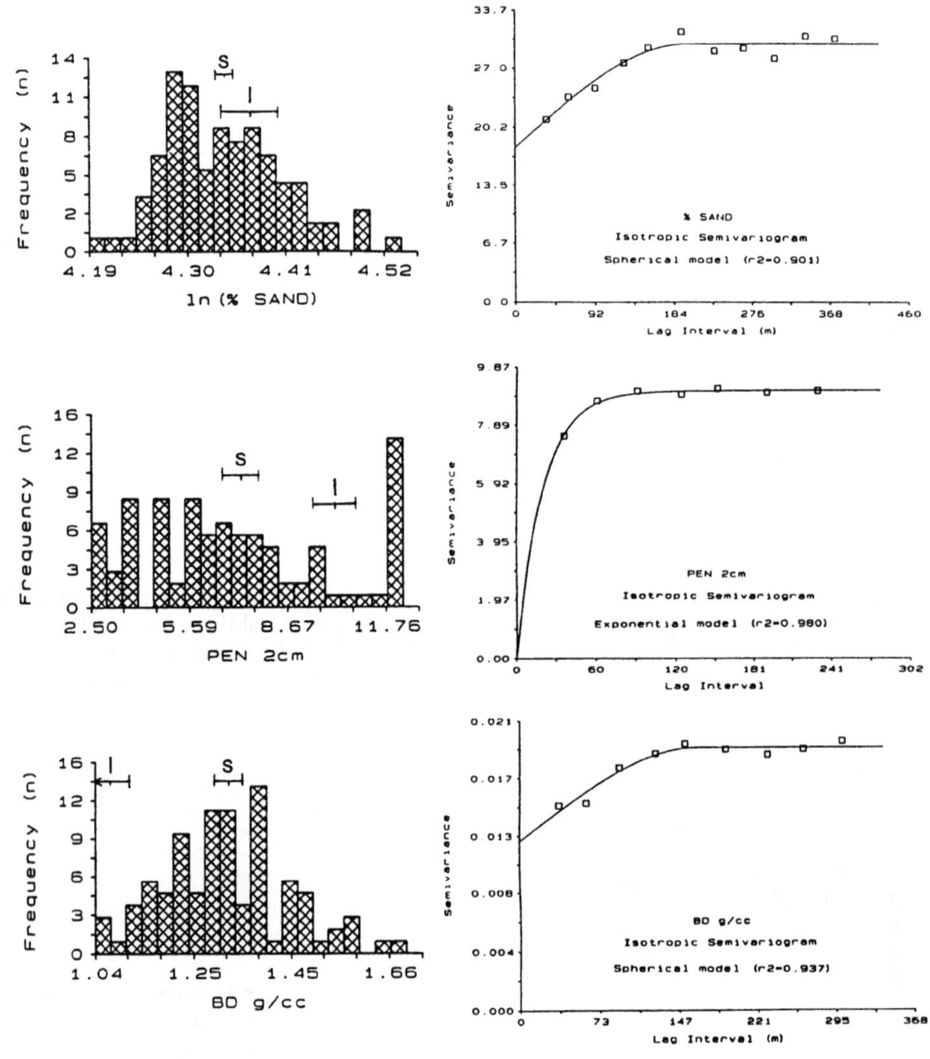

Figure 2. Frequency distributions (left) and sample semivariograms with models fitted (right) from the uniform sampling survey for sand content (top), penetrability (middle), and bulk density (bottom). Means and 95% confidence limits of site values (s) and lizard collection points (l), are shown on histogram plots.

Model semivariogram type and estimates of model parameters are provided in Table 1. The spherical model provided the best fit for sand content and bulk density sample semivariograms. This model reaches its sill at a finite range, A_0, which is the distance parameter. The gaussian model, provided the best fit for penetrability ($r^2 = 0.988$), but was not selected because of the high nugget variance present in this soil property (Fig. 1). Instead, an exponential model was fitted to the sample semivariogram. This model is similar to the gaussian model in that it reaches its sill asymptotically.

TABLE 1. Model type and parameters for the semivariograms.

Variable	Model Type (r^2)	Nugget Variance	Sill	Nugget: Sill Ratio	Effective Range, A_0 (m)
Sand Content (%)	Spherical (0.901)	17.89	29.79	0.60	192
Penetrability at 2 cm Depth (cm)	Exponential (0.980)	0.01	9.07	0.001	60*
Bulk Density (g/cc)	Spherical (0.937)	0.012	0.019	0.63	166

* Major range was multiplied by 3 to get effective range.

In the exponential model, A_0 is not the range but rather a parameter used in the model to provide range. The effective range of spatial dependence in this model is typically estimated as $3A_0$, although this tends to slightly overestimate the range parameter (Webster, 1985). The model semivariograms for sand content and bulk density show essentially the same progression; semivariances increase with increasing lag to about 170 m beyond which they fluctuate around constant values (Fig. 2). The nugget variance for these variables was large, comprising at least 60% of the sill (Table 1), which corroborated the results of the nested ANOVA. Penetrability shows spatial dependence up to about 75 m. The average range of spatial dependence defined by the models is approximately 139 m. The magnitude of small-scale variation was greater than anticipated and reinforced the fact that a much smaller sampling interval is needed to resolve small-scale variation in these soil properties. In order to ensure that unresolved variation does not exceed that shown in Fig. 1, the sampling interval likely has to be less than 5 m.

TABLE 2. Summary statistics and 95% confidence limits for mean values of soil properties at lizard collection points. The confidence limits provided the boundary criteria for lizard presence or absence.

Variable	Mean (s.d.)	Range	Sample Size	95% Confidence Limits
Sand Content (%)	79.3 (5.74)	70.1-88.8	26	77.1-81.5
Penetrability at 2 cm depth (cm)	10.2 (2.00)	5.8-12.3	26	9.5-11.0
Bulk Density (g/cc)	1.08 (0.132)	0.757-1.235	26	1.024-1.125

Kriged maps of bulk density, penetrability, and sand content are shown in Figure 3. The soil properties show unique patterns of variation overall, with some concordant areas of high values. The maps of sand content and bulk density portray particular site conditions, e.g., the dark area along the northern edge of the bulk density map corresponds to the uppermost (oldest) stream terrace, and the two extensive dark areas

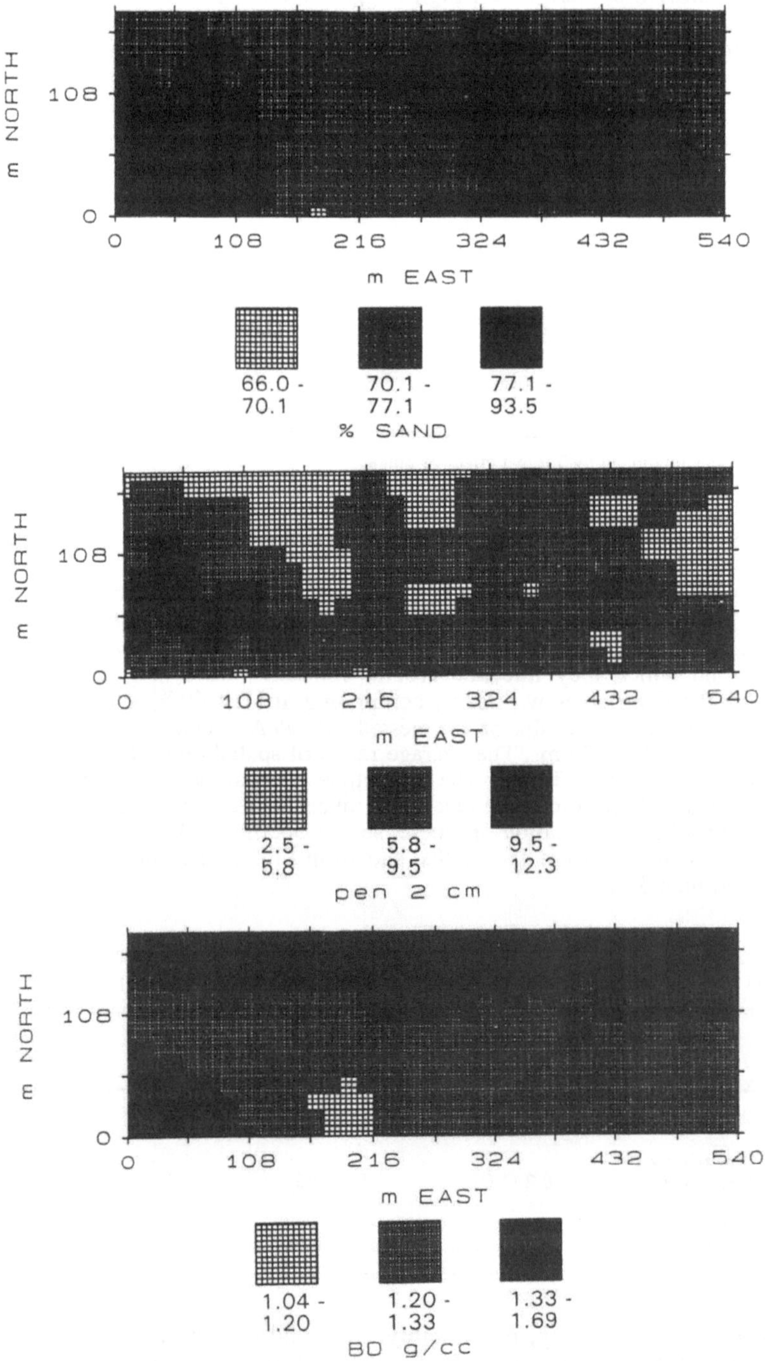

Figure 3. Kriged maps of the 97,200 m^2 sample grid for sand content (top), penetrability (middle), and bulk density (bottom), at 15 m resolution, modeled with the semivariograms in Fig. 2. Contour intervals have been adjusted to reflect range and 95% confidence limits relevant to legless lizard distribution (see Table 2 and text for further explanation).

on the map of sand content are separated by an old streambed. The penetrability map is more spatially heterogeneous, reflecting significant point-to-point variation in this soil property. Soils with more than 77% sand, horizontal penetrability greater than 9.5 cm, and bulk density less than 1.20 gms/cc, were considered suitable for lizards (Table 2). Soils outside these limits are likely too dense to permit burrowing by this species. The kriged maps of these soil properties suggest lizards should be found in discrete, isolated patches and should be absent from or have a low probability of occurring over most of the study site (Fig. 3).

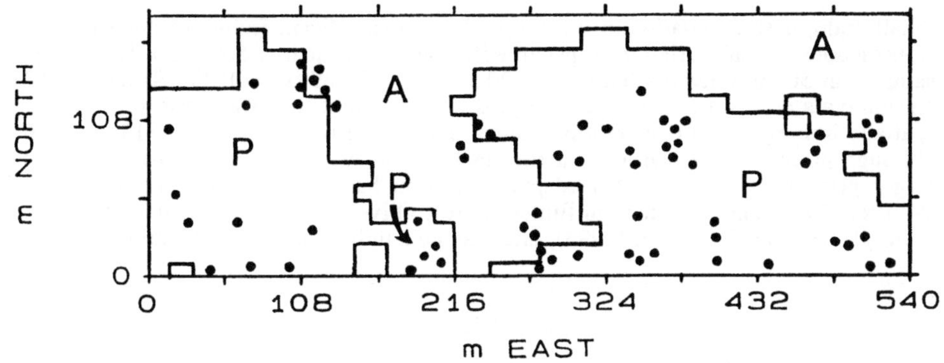

Figure 4. Predicted lizard distribution at 15 m resolution, based on composite overlay of kriged maps of soil properties (P = presence; A = absence). Dots show actual occurrence of lizards.

Species occurrence is usually not dependent upon a single environmental variable unless the value of that variable is at or near the organisms' biophysical or physiological limits. Legless lizard occurrence depends upon a combination of microclimatic, edaphic, and vegetative features, whose relative importance varies depending on the temporal and spatial scale of observation. The composite predicted distribution map produced by superimposing the kriged soil property maps showed high concordance with the actual distribution of lizards found on the study site (Fig. 4). Approximately, 19% of the lizard collection points fall outside the boundaries predicted for presence, but all are close to the boundary lines, indicating that more intensive sampling of soil properties in these areas may resolve the boundary between presence and absence more precisely. Concordance with predicted areas of absence was high, "outliers" occur in areas of predicted low probability of occurrence. Suitable edaphic conditions occur within these areas, but could not be resolved with a 36 m sampling interval.

The sampling support and subsequent large nugget variance in the model semivariograms (Table 1), constrained greater resolution of soil properties in Fig. 3. Consequently, many of the locations where lizards were found (Fig. 4), fall within areas of predicted low probability of supporting lizards on the maps of bulk density and penetrability. This discrepancy is due to the moderating effect that shrubs and trees have on soil properties immediately beneath the canopy, which was not captured with a transect sampling strategy, and the high affinity lizards show for these microsites. Lizards are typically found at or below the leaf litter/soil interface beneath shrubs, trees, or cover objects. Soils at these microsites tend to be more friable and less dense than exposed soils outside the plant canopy just a few meters away. Capturing edaphic variation experienced by individual lizards requires either more intensive sampling or non-random sampling such as that employed for the 2 m x 2 m grids. Support based on transect sampling, especially at a 36 m interval, tends to underestimate the actual distribution of lizards because of the low probability of sampling microsites likely to contain lizards. Kriging at 5 m intervals did not resolve greater spatial variation, but only further resolved the boundaries of patterns already discerned by the 15 m kriging operation. Actual lizard occurrence was highly concordant with the areas of predicted high probability of lizard occurrence on the map of sand content.

5. Discussion

Small-scale, nested variation in soil properties is a significant feature of this study site. Conspicuous spatial variation in penetrability was evident in a 500 cm x 500 cm grid sampled at 50 cm intervals (mean = 4.20 cm; s^2 = 2.789; n = 121). Sampling at 10 cm intervals across a 50 cm x 50 cm grid nested within the larger grid did not yield spatially homogeneous penetrability values (mean = 2.62 cm; s^2 = 0.7813; n = 36), although point-to-point variance in the smaller grid was significantly less than in the larger grid ($p<0.01$). Nested edaphic variation has been reported by Beckett and Webster (1971) and Webster and Butler (1976), Oliver and Webster (1987), Robertson *et al.* (1988), and Bell *et al.* (1993), and may be a universal feature of soil properties.

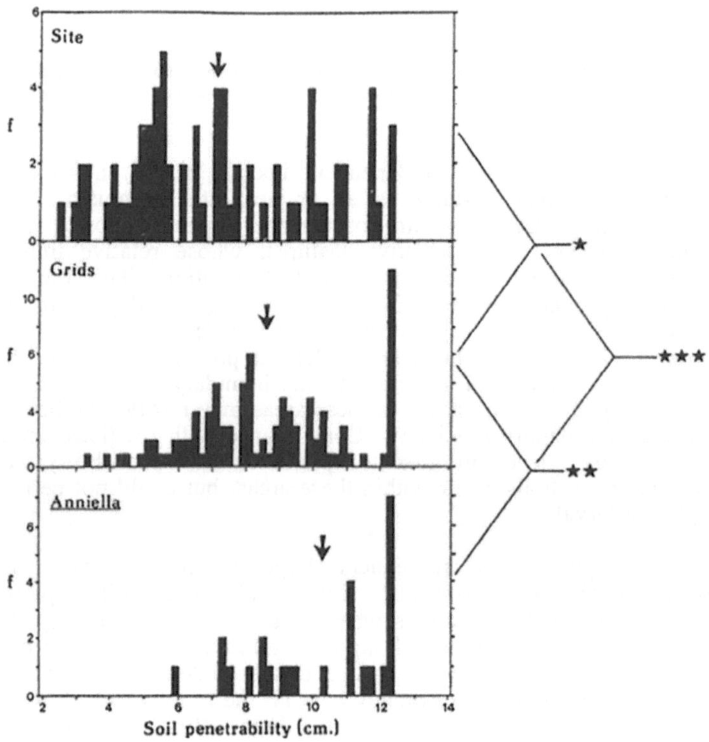

Figure 5. Nested hierarchy of spatial variation in soil penetrability as illustrated by microhabitat selection by legless lizards for study site in general (top), 2 m x 2 m grids within the study site (middle), and lizard collection points within the grids (bottom). Group means are indicated by the arrows. *p*-values for comparisons of group means are: * = $p<0.05$; ** = $p<0.01$; *** = $p<0.001$.

Legless lizards capitalize on small-scale edaphic variation, as reflected in their highly fragmented geographic distribution. Figure 5 illustrates these points. Soil penetrability ranges from 2.5 cm to 12.3 cm across the 97,200 m^2 study site. The locations of the 2 m x 2 m grids within the study site were, in a sense, "selected" by individual lizards, yet these small grids captured almost the same range of penetrability values found over the entire study site from uniform sampling. Microhabitat selection significantly increased mean grid penetrability relative to mean site penetrability. Likewise, mean point-to-point penetrability values where lizards were found was significantly greater than either mean grid or mean site penetrability.

The kriged soil property maps for sand content and bulk density variation closely match stream terrace boundaries on the study site (Fig. 3). The relative age and stability of these terraces increases in a northward direction away from the active channel of the Cuyama River. Soil and vegetation development show concomitant patterns. Together these hydrologic, edaphic, and biotic factors appear to have a strong influence on the local distribution of legless lizards at this site. Lizards are unlikely to be found on the oldest, highest floodplain terrace along the northern edge of the site because the fine-textured, relatively dense soils found there appear to limit burrowing activity. The broad area of predicted lizard absence left of center in Figure 4, is coincident with an old streambed channel. Soils in this region are mostly silts or clays, with some sand, but invariably containing a high stone content. More uniform, sand-dominated soils appear to be developing around the edges of this area as a result of fluvial transport of sand from the surrounding areas and shrub succession. Consequently, lizards occurring along the edges of this old stream channel may colonize this region only as suitable microsites evolve with vegetation development. Woodward (1987) suggested that the temporal dynamics of species are dependent upon the relation between scale of environmental variation (e.g., seconds, days, years, centuries) and the time scale of organismal response. A taxon responds to environmental variation at or above the spatial grain it senses. This represents the home range of individual animals or the dispersal range of populations. Population dispersal in *Anniella* likely depends upon the spatial and temporal dynamics of soil and vegetation development. The semivariograms and kriged maps of soil properties not only illustrate the environmental grain experienced within the home range of individual lizards but, indirectly, their response to this variation through precise microhabitat selection.

Geographic distribution is ordered by spatial extent, spatial grain (resolution), temporal extent, and temporal grain. A "top-down" approach to modeling meso- and large-scale representations of geographic distribution requires that the various sources of information be categorized so variation in data collected at disparate spatial and temporal scales can be differentiated. At present the greatest difficulty that hampers a "top-down" probabilistic representation of geographic distribution is the lack of scientific understanding of the factors that influence the distribution patterns of species, populations, and individuals; factors which differ in importance at different spatial and temporal grains (Hollander *et al.*, 1994).

Accurate analysis of species occurrence data, must distinguish between data extent (the total area analyzed), spatial grain of the data, and sampling intensity (fraction of extent sampled at a given grain) (Turner *et al.*, 1989). Field data provide the most concrete data for reconstructing the range of a species at biogeographic extent however, these data are often patchy, of indeterminate grain, and cover only a fraction of the total extent. This is where geostatistical modeling may be of use. The "bottom-up" approach presented in this paper combined detailed microhabitat data with a geostatistical analysis of surrogate distribution parameters to identify the scale of variation, interpolate environmental variation over a known extent, and generate representations of distribution at spatial scales involving the home range size of individuals and populations. These are the spatial scales of interest to ecology, ecological biogeography, and population genetics, and which are difficult or impossible to map using remote sensing technology. These "bottom-up" data are geographically well-referenced and are directly measured at the spatial grain of the organism. Iterative improvements in the models can be made with additional field-testing of predicted range boundaries and sampling intervals. Work in-progress is focusing on refining the kriged maps of environmental variables and integrating them into a GIS format which will allow production of explicit classification trees for krig point allocation and formulation of classes of distribution models. Existing maps of local and regional environmental variation (e.g., U.S. Soil Conservation Service soil survey maps) are will be incorporated into models of *Anniella* distribution at regional spatial scales.

6. References

Beckett, P.H.T. and R. Webster (1971) Soil variability: a review, *Soils and Fertility* **34**, 1-15.

Bell, G, M.J. Lechowicz, A. Appenzeller, M. Chandler, E. DeBlois, L. Jackson, B. Mackenzie, R. Preziosi, M. Schallenberg, and Tinker, N. (1993) The spatial structure of the physical environment, *Oecologia* **96**, 114-121.

Hall, E.R. (1981) *Mammals of North America, 2 vols.*, Roland Press, New York.

Hollander, A.D, F.W. Davis, and Stoms, D.M. (1994) Hierarchical representations of species distribution using maps, images and sighting data, in R.I. Miller (ed.), *Mapping the Diversity of Nature*, Chapman & Hall, London, pp. 71-89.

Hunt, L.E. (1994) Field temperature and soil moisture selection by the California legless lizard, *Anniella pulchra pulchra* Gray (Sauria:Anguidae), Dept. Biol. Sci., Univ. California-Santa Barbara, *Ph.D. diss.*, **Ch. 1**, pp. 1-32.

Klute, A. (ed.) (1986) *Methods of Soil Analysis, Part 1: Physical and Mineralogical Methods, 2nd. ed.*, Soil Science Society of America, Madison, Wisconsin.

Miller, C.M. (1944) Ecologic relations and adaptations of the limbless lizards of the genus *Anniella*, *Ecol. Monographs* **14**, 271-289.

Oliver, M.A. and Webster, R. (1986) Semi-variograms for modeling the spatial pattern of landform and soil properties, *Earth Surface Proc. and Landforms* **11**, 491-504.

Oliver, M.A. and Webster, R. (1987) The elucidation of soil pattern in the Wyre Forest of the West Midlands, England. II. Spatial distribution, *J. Soil Science* **38**, 293-307.

Robertson, G.P, M. Huston, F. Evans, and J. Tiedje (1988) Spatial variability in a successional plant community: patterns of nitrogen availability, *Ecology* **69**, 1517-1524.

Stebbins, R.C. (1985). *A Field Guide to Western Reptiles and Amphibians, 2nd. ed.*, Houghton-Mifflin Company, Boston.

Turner, M.G., Dale, V.H., and Gardner, R.H. (1989) Predicting across scales: theory development and testing, *Landscape Ecology* **3**, 245-252.

Vasek, F.C. and Thorne, R.F. (1977) Transmontane coniferous vegetation, in M.G. Barbour and J. Major (eds.), *Terrestrial Vegetation of California*, John Wiley & Sons, New York, pp. 797-832.

Webster, R. and F. Butler (1976) Soil classification and survey studies at Ginninderra. *Austr. J. Soil Research* **14**, 1-24.

Webster, R. (1977) *Quantitative and numerical methods in soil survey and classification*, Clarendon Press, Oxford.

Webster, R. (1985) Quantitative spatial analysis of soil in the field, in B.A. Stewart (ed.), *Advances in Soil Science* **3**, Springer-Verlag, New York, pp. 1-70.

Webster, R. and M.A. Oliver (1990) *Statistical Methods in Soil and Land Resource Survey*, Oxford University Press, New York.

Woodward, F.I. (1987) *Climate and Plant Distribution*, Cambridge University Press, New York.

GLOBAL WARMING AND THE WEIGHTS OF RATS: USES OF THE VARIOGRAM IN THE ANALYSIS OF LONGITUDINAL DATA

J. HASLETT
Department of Statistics, Trinity College, Dublin 2, Ireland

Abstract: Monthly data have been assembled on the temperature of the Earth for the last 140 years. Time series analysis of such data have indicated that - under the 'business as usual' scenario - the world is warming. This is however somewhat sensitive to the choice of model for the covariance structure and the estimation procedure used.

Repeated measurements are common in laboratory and other experiments. One such involves monitoring the weights of different sets of rats over a period. Exactly the same time techniques as above may be used and are now being recommended in classical statistical analyses of such data.

The variogram can provide a basis for satisfactory modelling in both cases. Two estimating procedures are considered. The first is a variant on the classic 'method-of-moments' procedures, adapted for estimating in the presence of trend. The second is the use of maximum likelihood procedures. The paper will thus illustrate methods that have been recently proposed in classical statistical literature.

1. Introduction

The variogram has been the basic tool underlying geostatistics for 40 years. For stationary processes it is directly and simply related to the (auto) covariance function. It has, nevertheless, had a very limited press in the classical statistical literature, even among those who deal daily in time series. For example, the Current Index to Statistics database lists 96089 papers from the period 1983-1994. Of these 29 have variogram in the title; only 21 have the variogram in the list of key-words; only a few of these address applications other than geostatistics (Cressie and Grondona, 1992; Robinson, 1990 and critically, Diggle, 1988). By contrast 210 have 'auto-correlation' in the title and 135 in the keywords. Nevertheless, partly through the work of Diggle, the variogram is coming out of the geostatistical straight-jacket. This paper is intended to give brief account, to a geostatistical audience, of some of these applications - broadly under the heading of longitudinal data - with some emphasis on the key differences.

A. Soares et al. (eds.), geoENV I–Geostatistics for Environmental Applications, 439–450.

Effectively classical geostatistics is built not so much on the variogram as on General Least Squares. The model $\underline{Y}=X\underline{\beta}+\underline{\varepsilon}$, $\underline{\varepsilon} \sim N(\underline{0},V)$ - in which for given V the parameters $\underline{\beta}$ are estimated by GLS - is in fact the 'grand-father' of very many statistical procedures. Special cases include:

(i) with $V = \sigma^2 I$ we have most of classical statistics, including multiple regression, analysis of variance and, in a spatial context, trend surface models; here 'cases' Y_i are independent, with a mean structure determined by $X\underline{\beta}$

(ii) with $\underline{Y}_i=\underline{\mu}$ and $Var(\underline{Y}_i) = V$ we have a model for independent, identically distributed multivariate cases; here $\underline{Y}$ is a stacked vector of blocks $\underline{Y}_i$, $\underline{\beta} = \underline{\mu}$, X has a block structure and V is block diagonal.

(iii) with $X\underline{\beta} = \underline{1}\mu$ and the elements of V being $v_{ij} = \sigma^2\alpha^{|i-j|}$ we have a model for a first order auto-regressive process with a constant mean corresponding to data at times i=1,2,.....n.; all ARMA processes involve extensions of this via richer functions of lag.

(iv) with $X\underline{\beta} = \underline{1}\mu$ and the elements of V being $v_{ij} = \sigma^2 - \gamma(|\underline{x}_i - \underline{x}_j|)$, where $\gamma(.)$ is a suitable negative definite function, we have the classic model underlying ordinary kriging of a stationary random function for data $Y_i = Y(\underline{x}_i)$ at locations $\underline{x}_i$; $\gamma(\underline{h})$ is of course the variogram.

(v) with $X\underline{\beta} = F\underline{\alpha} + Z\underline{\beta}$, where each column of F is one of the classical polynomials in the co-ordinates $\underline{x}_i$ and each column of Z is a set of 'external' variables available at each site, and V as in (iv), we have universal kriging with external variables.

The even more general situation where the process is not variance stationary, but has stationary increments, may also be accommodated as an extension to the above framework. Typical geostatistical models relax this even further and require only that the process (or its increments) be locally stationary.

Although formally the same model underlies very many apparently different procedures, there are undoubtedly considerable differences of emphasis. In model (i) effort concentrates on a highly structured modelling of the mean; by contrast the variance model is extremely simple. For example the cases in a structured experimental design may be undergoing a wide variety of combinations of 'treatment'. An example below is based on 3 groups of rats being fed a variety of diets under controlled circumstances; this structure is reflected in $X\underline{\beta}$. Many situations will involve the incorporation of covariates into the mean structure. In model (iv) the mean structure is very simple indeed but much emphasis is placed on modelling the matrix V via possibly sophisticated variogram models. In model (v) the possibility arises for quite subtle modelling of the mean. Here and in the work below we discover that there are many situations where emphasis must be placed on <u>both</u> aspects of the model.

The different modelling emphases reflect the typically different objectives. The geostatistician is mostly concerned with interpolation via kriging. The global mean μ is in fact a 'nuisance' parameter, rarely computed even when meaningful (as when kriging is applied within a 'universal neighbourhood'). When kriging is applied on the basis of moving neighbourhoods, the interest in μ is relaxed even further. All the geostatistician wishes to know is that it is 'slowly changing'. By contrast many of the users of classical statistics are primarily interested in the mean structure. Do different groups of rats respond differently to different types of diet, *on average* ? In many time series applications the interest is almost as in kriging; the objective is short term prediction; here also the mean structure is of relatively little interest. Even in time series however, the interest is sometimes in the mean structure *per se*. Intervention analysis seeks to ask if, for example, a change in policy some time ago (compulsory seat belts in cars, an advertising campaign in marketing) did induce a *measurable* effect. In the study of climate change there is an interest in formally testing the null hypothesis that 'the apparent change in global temperatures over the past 140 years is not inconsistent with a statement that the underlying mean is in fact constant.'

There is an important theoretical difference between these examples and that of classical geostatistics. The basic data in geostatistics are regarded as a single partial realisation of a random function. The concept of replication makes no sense. In the example of the rats' diets a number of rats are independently subjected to similar treatment over a period of time. In the study of global warming we are concerned with the next century; the only way we can proceed is to adopt a model stating that - in some respects at least - next century is another (not quite independent) replication of the last; the study then of some aspects of last century will tell us something about the future. This theoretical statement has important practical consequences for the (geo)statistician. The classical geostatistician shuns some of the standard tools of the classic statistician and maximum likelihood in particular. He/she therefore rests his/her methodology on method-of-moment procedures. He/she is very wary of the potential for confusion of trend and auto-correlation underlying, for example, universal kriging. The formal treatment of IRF(k) is offered to by-pass this. The classical statistician, with replication, can in principle at least, distinguish between these two. Overall then we see that situations where there is:

- direct scientific interest in the underlying mean structure;
- auto-correlation in time; and
- replication

fall into the domain where the tools of both communities can become useful.

The objective then in this paper is to illustrate this thesis with reference to two examples. Some aspects of this treatment will be new to a geostatistical community. In both cases the analyses have been performed by software OSWALD (Smith and Diggle, 1994). This software uses Restricted Maximum Likelihood Estimation (REML). It has been primarily designed for repeated measures/longitudinal data analysis as in our first example above, namely a designed experiment on the diets for rats. The latter, being the analysis of global temperatures, may however more familiar in content for a geostatistical audience and we begin with that.

2. Global Warming.

The UN's International Panel on Climate Change (IPCC) has assembled meteorological data since the 1850's on measurements of surface temperature. These data are very sparse in the early part of this period - and bearing in mind that two thirds of the globe's surface is ocean - are still very spatially sparse. The data have been interpolated onto 5° grids for each month since 1856. See Jones and Briffa (1992) for details. These 'pseudo' data are available for researchers at various levels of spatial and or temporal aggregation. Below we show two aspects of an analysis of such data. These involve aggregations for the Northern hemisphere for the months of January and July 1854 to 1994. The data are referred to as 'anomalies'. By this is meant that the January temperatures are with reference to the average in the period 1950-1970 and similarly for the July temperatures. Both time series are shown below in Figure 1; both series have, on this scale, an average of zero for the period 1950-1970.

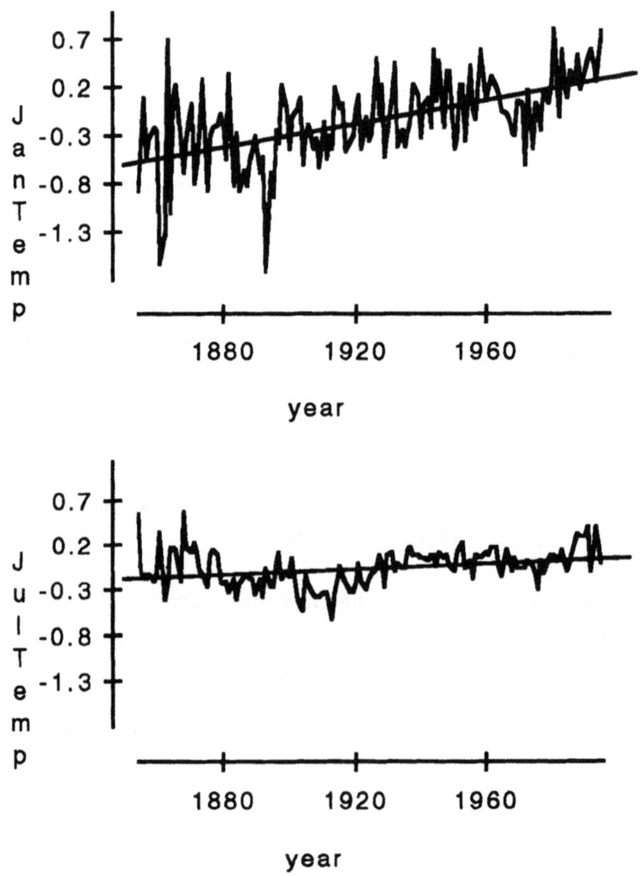

Figure 1 January and July Temperatures, 1854-1994

It is apparent that the series are quite different in character. How may we explore this more formally? We propose, for each series, the model

$$Y(t) = \alpha + \beta t + \varepsilon(t)$$

where t represents the years 1854-1994 and where $\varepsilon(t)$ is itself modelled as a potentially correlated series. We adopt for the variogram of $\varepsilon(t)$ the model

$$\gamma(h) = \gamma_1 + \gamma_2(1 - e^{-\phi|h|})$$

Figure 2 give details of the experimental and fitted variograms. It is seen that these are in close agreement. These in turn provide estimates and standard errors for the parameters which are set out in Table 1 below. It should be emphasised that the empirical variogram shown in Figure 2 is that of the classic residuals from the fitted linear model ($\hat{\underline{e}} = \underline{Y} - \mathbf{X}\hat{\underline{\beta}}$) provided via REML by OSWALD. However the variogram parameters have not been estimated by fitting the variogram model - by weighted non-linear least squares, for example - to this empirical variogram, but by OSWALD's REML procedure; see section 4 below. The good visual fit is therefore a satisfactory - and separate - confirmation of the model's fit. This is to be contrasted with the second example, below.

The implication of the analyses are that there is weak correlation in both series, that there is quite firm evidence that the slope is non-zero for January ($t = 6.4$) but very weak evidence for July ($t = .92$). Further the variograms may be decomposed into a nugget effect, representing high frequency year to year variation and a much more slowly changing effect (which is if fact auto-regressive with AR(1) parameter $e^{-\phi}$) which might be said to represent a slowly changing climate. There is therefore some evidence that this latter component is more important in July (comparison of the values of γ_1 and γ_2 for January and July) and further that it is more slowly changing in July (AR parameters .078 and .025). These details are in fact aspects of a clear seasonal pattern across all 12 months.

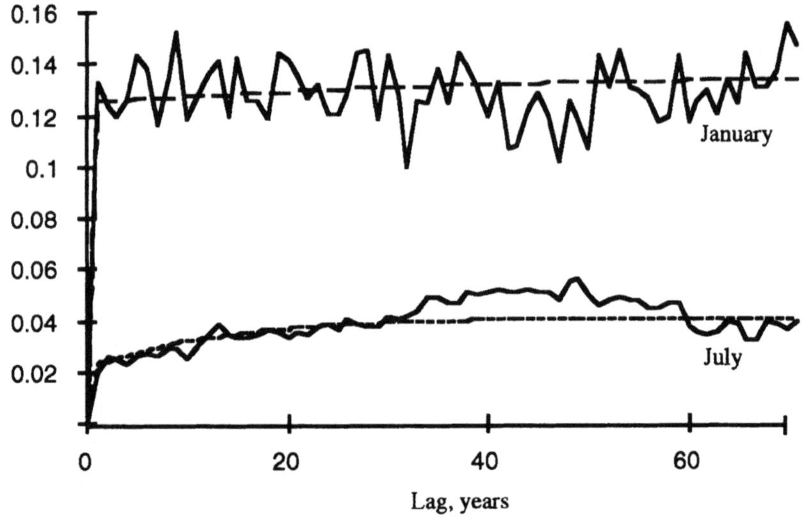

Figure 2 Fitted (REML) and experimental variograms for January and July

TABLE 1. Fitted Parameters

Mean Parameters	slope	SE	t
January	0.00597	0.000933	6.40
July	0.00114	0.001233	0.92

Variance parameters	$\gamma 1$	$\gamma 2$	ϕ
January	0.125	0.011	0.025
July	0.021	0.020	0.078

3. Rat's Weights

These data, reported in Crowder and Hand (1990) and shown in Figure 3, refer to an experiment in which the weights of three groups of rats (n = 8,4,4 respectively) were measured over a 9 week period.

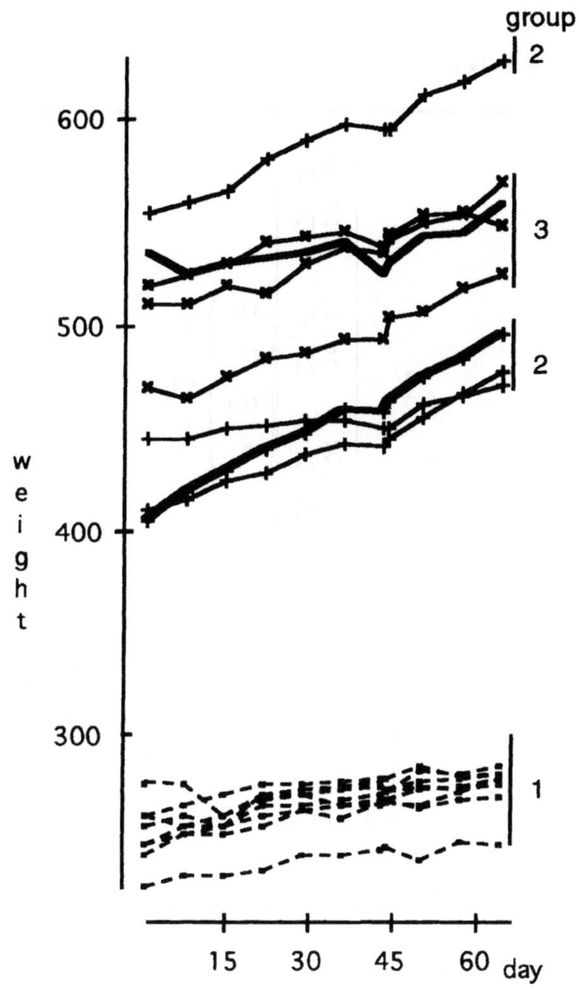

Figure 3. Weights of rats by day

The model adopted was

$$Y_{ij} = A_i + \delta t_j + \varepsilon_{ij} \qquad i=1,..,16 \text{ (rats) and } j=1,...63 \text{ (days)}$$

where the (common) variogram of the terms is $\gamma_2(1- e^{-\phi|h|})$ (again reflecting an AR(1) model) and where the A_i, denoting the intercept - or initial weight - are themselves random variables modelled as $A_i \sim N(\alpha_i, \gamma_1)$ for each of the 3 groups. This latter model reflects the fact that not only may the groups may be different by choice of type of rat (different α_i) but also that the rats within each group are random samples of rats from that group ($\gamma > 0$). Statisticians refer to this latter as a 'random effect' and to the entire model as a 'mixed model', here with the extra complication of correlated errors. Note that the fact that the A_i are treated as random induces a correlation between the weights across rats in the same group.

The full linear model is as below

$$\underline{Y} = \begin{pmatrix} X_{11} \\ X_{12} \\ \cdot \\ X_{18} \\ \hline X_{21} \\ \cdot \\ X_{24} \\ \hline X_{31} \\ \cdot \\ X_{34} \end{pmatrix} \begin{pmatrix} \alpha_1 \\ \alpha_2 \\ \alpha_3 \\ \delta \end{pmatrix} + \underline{\varepsilon}$$

where

$$X_{1j} = \begin{pmatrix} 1 & 0 & \cdot & 0 \\ 1 & 0 & \cdot & 7 \\ \cdot & \cdot & \cdot & \cdot \\ 1 & 0 & \cdot & 63 \end{pmatrix}, \quad X_{1k} = \begin{pmatrix} 0 & 1 & 0 & 0 \\ 0 & 1 & 0 & 7 \\ \cdot & \cdot & \cdot & \cdot \\ 0 & 1 & 0 & 63 \end{pmatrix} \quad X_{1m} = \begin{pmatrix} 0 & 0 & 1 & 0 \\ 0 & 0 & 1 & 7 \\ \cdot & \cdot & 1 & \cdot \\ 0 & 0 & 1 & 63 \end{pmatrix}$$

for j=1..8, k=1..4, m=1..4

The entire (stacked) vector of observations in group i thus has covariance $Var(\underline{Y}_m) = V_m = \gamma_1 J_m + \gamma_2 \Sigma$ for m=1,2,3, where J is a matrix having all elements unity, the elements of Σ are $e^{-\phi|i-j|}$ and $\underline{Y}_m$ has dimension 80,40,40 for m=1,2,3. The entire set of data therefore is modelled as

$$\underline{Y} = X^t \underline{\beta} + \underline{\varepsilon}, \ \underline{\varepsilon} \sim N(\underline{0}, V)$$

where V is block diagonal of dimension 160, $\beta = (\alpha_1, \alpha_2, \alpha_3, \delta)$ and X has the block structure shown above.

The fitted parameters are as in Table 2. The conclusion from the mean parameters is that there is firm evidence of a weight gain (δ = .603, SE =.054). Fuller analyses (for example, with different slope parameters δ for each group and consequently a vector $\underline{\beta}$ of dimension 6 and a design matrix of dimension 6x160) do not provide evidence of significantly different rates of weight gain.

TABLE 2. Fitted Parameters

	α_1	α_2	α_3	δ
Mean Parameters	243.2	465.4	507.6	.603
Standard Errors	22.4	31.2	31.2	.054

	γ_1	γ_2	ϕ
Variance Parameters	4861	105	.034

The variogram parameters are also given in Table 2. The variogram as fitted by REML is shown in Figure 4 below where it is contrasted with the empirical variogram based on the residuals from the fitted model. Note that there is, effectively, one empirical variogram per rat. Recall that there are only 10 observations for each rat and consequently only a nine pairs available at lag 7 days and only one at lag 63 days. There is, as to be expected, substantial variation between these. The averaging provides us however with the possibility of computing Standard Errors at each lag. The average of these is presented, together with ±2 Standard Errors. The 'nugget' effect (γ_1) is explicitly shown in Figure 2 but its equivalent (between rat variation) is suppressed in Figure 4 as it is very large. This has the effect of exaggerating the differences.

It is seen that there is a difference - perhaps important - between the fitted variogram and the average of the empirical variograms and some questions arise with regard to the model fit. To discuss these differences it is necessary to consider in more detail the fitting procedures used.

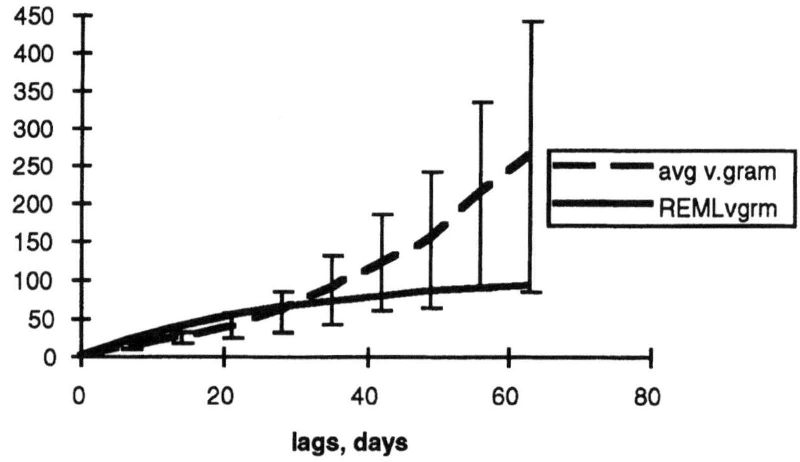

Figure 4 Variogram fitted by REML and by averaging the classical experimental variograms

4. Methodology

The OSWALD methodology for variogram fitting is not conventional in geostatistics and is therefore worthy of some discussion. The variogram itself is adopted as the central tool for analysis of this type of data as it is quite often the case that the data are irregular in time.

In geostatistics the first stage of the procedure involves computing squared paired differences $(y(\underline{x}_i) - y(\underline{x}_j))^2$ and hence an experimental variogram via averages of these for groups of pairs having $|\underline{x}_i - \underline{x}_j| \approx h$. To this is fitted, typically by some simple least squares procedure, an appropriate functional form for the variogram. This methodology fails if μ is not constant and many of the different techniques of geostatistics involve variations on this theme. Note that in the studies presented, the non-constancy of μ is the primary objective of the study.

At the second stage then the estimated variogram becomes the basis of, for example, kriging in which (at least implicitly) a General Least Squares algorithm is used to estimate the common underlying mean μ (in 'ordinary kriging') or its parameters (in for example 'universal kriging'). This methodology for estimating the variogram does not rely in any explicit way on the distributional assumptions, for example that the process is Gaussian; it is a method-of-moments procedure.

As mentioned, the fitting methodology in OSWALD is via REML. The reader is referred to Cressie(1991, p92) for a discussion. This variant of Maximum Likelihood has the effect of filtering out the unknown mean structure in the estimation of the parameters defining correlation structure R. In this sense it has similar properties to the classical geostatistical procedure with its concentration on increments. It is well recognised for example that in the case of stationary processes working with the experimental variogram avoids the need - and consequent bias - to estimate the unknown parameter μ as is necessary in computing the auto-covariance. Indeed the objective of Robinson's (1992) paper is to bring this bias, well known to geostatistics, to the attention of a purely statistical audience. REML is more general in this respect however, for not only is a constant filtered out before fitting the variance structure, but any mean with a linear structure is also filtered out. REML can be shown to have less bias than MLE.

The algorithm is iterative; the classical geostatistical reference is Zimmerman (1989). Diggle recommends that initial values are chosen by applying simple (e.g. classical geostatistical) methods to residuals obtained by applying OLS to a model with an over rich specification of the mean structure (e.g., in the context of the rat's data, the 'saturated' model with a separate parameter for each group and each time, and thus 30 parameters in this case).

The contrast between the classical and REML fits in our two examples is probably due to the very high lag 1 correlation ($\rho_1 = e^{-\phi}$) in all cases; these are respectively 0.975, 0.925 (July, January) and 0.967 (rats). The contrast is most apparent in the estimate of the sill. In REML this is estimated from $(n-p)^{-1}\hat{\underline{e}}^t R^{-1}\hat{\underline{e}}$. With an AR structure R^{-1} is a tri-diagonal and the estimate of the sill involves only lag 1 pairs $((\hat{e}_i\hat{e}_{i+1}))$ as well as terms $\hat{e}_i^2$. By contrast the geostatistical method of moments (MM) procedure will put greater weight in this estimation on well separated pairs, that is $(\hat{e}_{i+h} - \hat{e}_i)^2$ with large h. The REML procedure thus relies heavily on the AR aspect of the model to extrapolate to large lags the contrasts at lag 1. The MM estimate of the sill is therefore perhaps more robust.

It should be noted however that the variogram provides only part of the overall variance structure. In both our examples above another component of variance was introduced. The short term 'nugget' effect in the climate data is familiar in many geostatistical studies applied to time series. In the repeated measures study, other components of variance arise from dealing with 16 different rats, and additional components of covariance arise from the random effects structure of the overall design. The interpretation of the nugget effect as a 'between rats' variance is less familiar in a geostatistical study. Analysis of such components of variance is very common in classical statistics. The REML algorithm estimates these also, and at the same time. Note once more that the modelling of the mean as here is very much more structured than is typical in geostatistics.

5. Developments

The above methodology is rapidly becoming mature in the analysis of longitudinal data. It opens up possibilities in other areas such as time series in the first instance. It and REML may provide possibilities in spatial work where there is a complex underlying mean structure. There are however some theoretical problems with ML methods. Additionally there are practical problems even in time series as, if the data set is large, then the ML algorithm involves the repeated inversion of large (n^2) matrices. Our global warming work involved manipulating 140^2 matrices.

The model checking stage of any statistical procedure is an important further stage. Some suggestions arising directly from cross-validation ideas in geostatistics have recently been proposed by Haslett and Hayes (1996) for a wide variety of related models as above.

6 References

Cressie, N. (1991) *Statistics for Spatial Data*, Wiley.

Cressie, N. and Grondona, M.O.,(1992) A comparison of variogram estimation with covariogram estimation, *The Art of Statistical Science*, 191- 208.

Crowder, M.J. and Hand, D.J. (1990) *Analysis of Repeated Measures*, (1st ed) Chapman and Hall.

Diggle, P (1988) An approach to the analysis of repeated measurements, *Biometrics* **44**, 959- 971

Haslett, J . and Hayes, K. (1996) Residuals for the Linear Model with General Covariance Structure: in press

Jones P.D. and Briffa K.R. (1992) Global Surface Air Temperature Variations During the Twentieth Century: Part 1, Spatial, Temporal and Seasonal Details, *The Holocene* **2**, 165-179

Robinson, G.K. (1990) A role for variograms, *Australian Journal of Statistics* **32** , 327- 335

Smith, D.M. and Diggle, P.J. (1994) Oswald (Version 2.0) Object Oriented Software for the Analysis of Longitudinal Data in S, *University of Lancaster*, UK.

Zimmerman, D.L. (1989) Computationally efficient restricted maximum likelihood estimation of generalised covariance functions. *Mathematical Geology*, **21**, 665-672.

VARIOGRAPHIC REFINEMENT OF NORTH SEA HERRING ACOUSTIC SURVEY DATA

P. G. FERNANDES and E. J. SIMMONDS
SOAEFD Marine Laboratory,
P.O. Box 101, Victoria Road,
Aberdeen, AB11 9DB,
Scotland, U.K.

Abstract

A log backtransformed variogram was used to provide a robust estimation of the experimental variogram in the presence of highly skewed data, such as those obtained in fisheries acoustic surveys. Simulations based on North Sea herring surveys were carried out which found no evidence of greater bias in the log backtransformed variogram relative to the classical variogram. The log backtransform was then applied to six North Sea herring acoustic surveys to determine how best to utilise the available datasets. The exclusion of intertransect data and use of along-the-track data did not confer any significant additional variographic stability. When divided into eastern and western regions, values of spatial dependence (nugget as a proportion of sill) were consistently greater in the western region. The recommended dataset selection criteria is, therefore, to use the log backtransform on the whole data set (i.e. including intertransects) separated by region (east and west). Variogram modelling and the resulting inferences on herring distributions are discussed.

1. Introduction

A number of review papers have dealt with the application of geostatistics to fisheries survey data (Armstrong *et al.* 1992; Rivoirard 1992; Petitgas 1993a; Petitgas and Prampart 1993). The practise is becoming more common, however, in cases where fisheries acoustic data are considered, the analysis is often hampered by the highly skewed nature of density distributions which may lead to poor variographic structure (Murray 1995; Porteiro *et al.* 1995; Maravelias *et al.* 1996). Estimation of the variogram is widely regarded as the most critical step in a geostatistical analysis (Cressie 1991). Ultimately the goal for the use of geostatistics in fisheries is to yield abundance and variance estimates which account for spatial autocorrelation and therefore provide appropriate sampling errors. Only once confidence can be established in the generation of the correct variogram, can this goal be achieved.

A. Soares et al. (eds.), geoENV I–Geostatistics for Environmental Applications, 451–462.

There are techniques which have been reported as methods for establishing a variogram from fisheries data with highly skewed amplitude distributions (Petitgas 1993b; Simard *et al.* 1993). However, these have not been universally received because they are seen as ad hoc solutions to specific datasets. In a recent development, the application of a log backtransformation provides a robust estimation of the variogram for skewed data (Guiblin *et al.* 1995). As one might expect from a log transform, the experimental variograms are better behaved, easier to interpret and easier to model. In the present study, we use simulations to compare this backtransformation method of determining the variogram with the classical variogram and the variogram derived from the non-centred covariance. We are particularly concerned with investigating the relative bias of the backtransformation method because it does not require the data values to follow a log-normal distribution and as such it is a new technique.

Having established the merits of the log backtransformation we then employ the method to analyse data from North Sea herring acoustic surveys. The objective is to determine the most suitable selection criteria which can be applied to the data in order to maximise the stability of the variogram.

2. Acoustic survey data

Acoustic surveys for herring are carried out in July of each year in the western North Sea (e.g. Simmonds *et al.* 1995). The herring at this time are feeding and may be encountered throughout the surveyed area. Surveys consist of regular east/west parallel transects, proceeding north, up the eastern side of the Orkney and Shetland Isles, and then south, down the western side of the Isles. These surveys are carried out using a vertical echosounder and pelagic fishing gear which obtain quantitative and qualitative information respectively on herring distributions (see MacLennan and Simmonds 1992 for details). The data are collected continuously, but are collated, at intervals of approximately 2.5 nautical miles, into an acoustic index which is proportional to herring density. The data used in this study were taken from 1989-1994 (Figure 1).

3. Simulations

3.1 METHODS

The population to be surveyed was simulated by a one dimensional array of values as defined in Simmonds and Fryer (1996); facilities for the generation of full two dimensional fields were not available. For each simulation a line array of 100,000 points was generated: the "super-set". The variance of all distributions was chosen as unity, the value of the mean was ignored and the level of skew was altered from 0 to 15 (the maximum value encountered in the survey data was 12). Other statistical properties were generated to reflect those of the survey data (a nugget of approximately 50 % and a range component of between 7 and 20 nmiles). The super-set provided

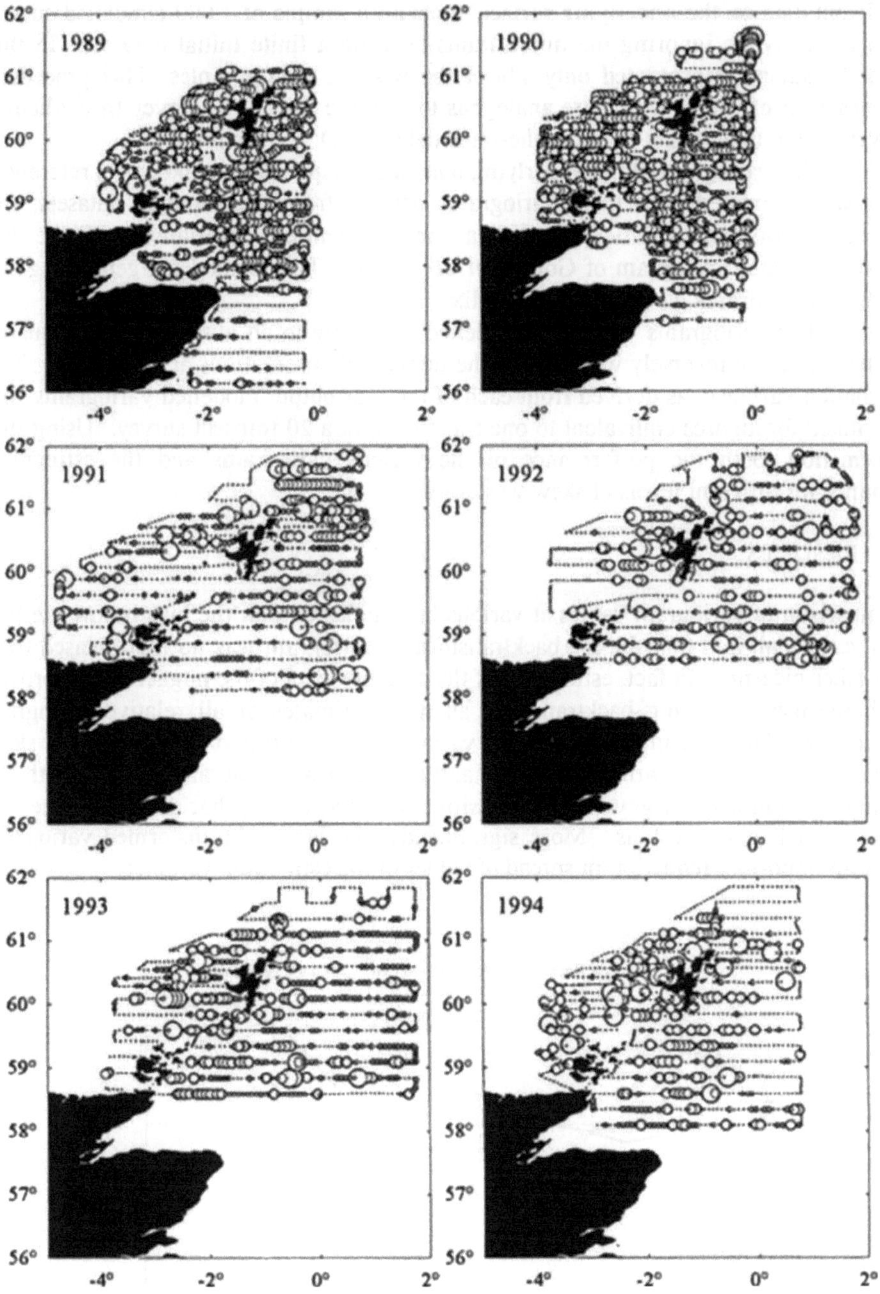

Figure 1. Post plots of an acoustic index for Orkney/Shetland herring from 1989 - 1994 (circle size proportional to log density, crosses indicate zero observations).

sufficient data on the underlying surface to obtain a sample of 1,000 simulated survey data values while ignoring the implications of using a finite initial data set. In this case the sample represented only 1% of the possible data samples. This procedure derives a set of values which are analogous to a single line (1-D) survey throughout a survey region (equivalent to along-the-track data in 2-D; see section 4.1).

The variogram of the underlying population superset was taken as a reference. This was compared with three variograms derived from the sampled datasets: the classical variogram, the variogram via the non-centred covariance and the log backtransformed variogram of Guiblin *et al.* (1995). The method for generating the log backtransform is given in the Appendix.

The variograms were then modelled according to an automatic least squares fitting procedure inversely weighted by the number of observations at each range. The estimation variance, as derived from each of the four output modelled variograms was calculated for an area equivalent to one transect from a 20 transect survey. Using this information, both the performance of the various variograms and the estimation variance for different levels of skew was evaluated.

3.2 RESULTS

Comparison of variogram values at various lags indicated that the bias in any method was very small and that the log backtransformed variogram was no more biased than any other measure. In fact, estimation of the absolute level of the nugget was improved with the use of the log backtransform as were estimates of all relative variogram parameters. The bias in the estimation variance was small and increased with skew (Figure 2). The non-centred covariance and the classical variogram (coupled with the fitting procedure used) gave a small positive bias and the log backtransform method gave a small negative bias. More significantly, the log backtransformed variogram gave an important reduction in spread of values of the estimation variance.

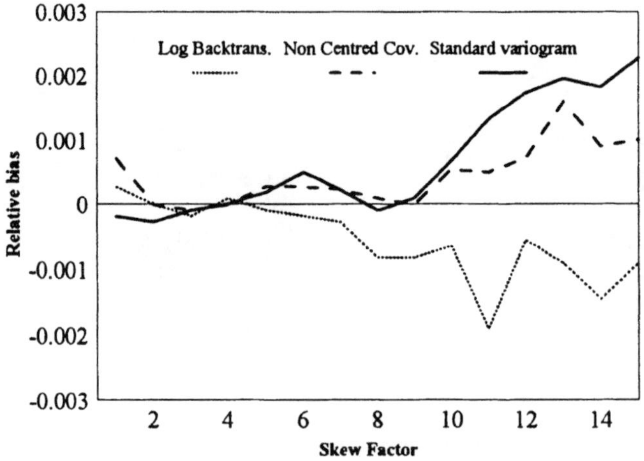

Figure 2. Relative bias in estimation variance at various skew levels

4. Variography

4.1 METHODS

North Sea herring acoustic data from each year were divided into nine subsets (Figure 3), according to location (east and west of the Orkney/Shetland Isles), to the inclusion of intertransect data, and to between-transect independence producing along-the-track data. Log backtransformed variograms were then calculated at a lag spacing of 2.5 nmiles ± 1.25 nmiles (tolerance), to a distance of 90 nmiles assuming isotropy (unidirectional variograms showed no evidence of consistent anisotropy).

A variogram model was then fitted according to a semi-automatic weighted least squares procedure. Parameters of each model were estimated to an approximate fit by eye and the final fit was determined by minimising the sum of squares weighted according to the number of pairs of points and the inverse of the lag distance squared. The value of the sum of squares provided an indication as to the goodness of fit between different models. The model parameters obtained from the best fit of either the nugget + spherical or nugget + exponential models were used to compare the 9 datasets. Values of spatial dependence were calculated after Robertson (1995) as follows:

$$\text{Spatial dependence} = \left(\frac{C}{C + C_0}\right) \times 100 \qquad \textbf{(1)}$$

4.2 RESULTS

The application of the backtransformation provided a much more robust estimation of the experimental variogram than the classical variogram in all cases (e.g. for the whole area dataset see Figure 9). This was most evident in 1991 and 1994 where the classical variograms provide little or no evidence of any spatial structure.

In many cases, either one of the nugget + spherical or nugget + exponential models may fit the data better than the other, however, in this study there were ten cases (of 54) when the fit differed by less than 5%. In every one of these cases the nugget effect was reduced for an exponential fit, and sometimes greatly reduced. In five cases the exponential model fit produced zero nugget compared to high nugget values from the equivalent spherical model fit. However, in four of these latter cases, the spherical model was actually the better fit and, therefore, it was chosen as the best model and a nugget of zero was not modelled. The practical ranges in these marginal cases were also slightly different, in most cases an exponential model fit produced slightly smaller values of practical range. The values of the sill were almost identical. It should be noted therefore, that when performing semi-automated fits such as in the procedure described here, with data where there is little information about the nugget effect, a subjective choice of model may influence the results, particularly of the nugget.

In general, the along-the-track subsets produced slightly better fits, but this is mostly due to the stability of the sill and the difference was never so great as to prevent

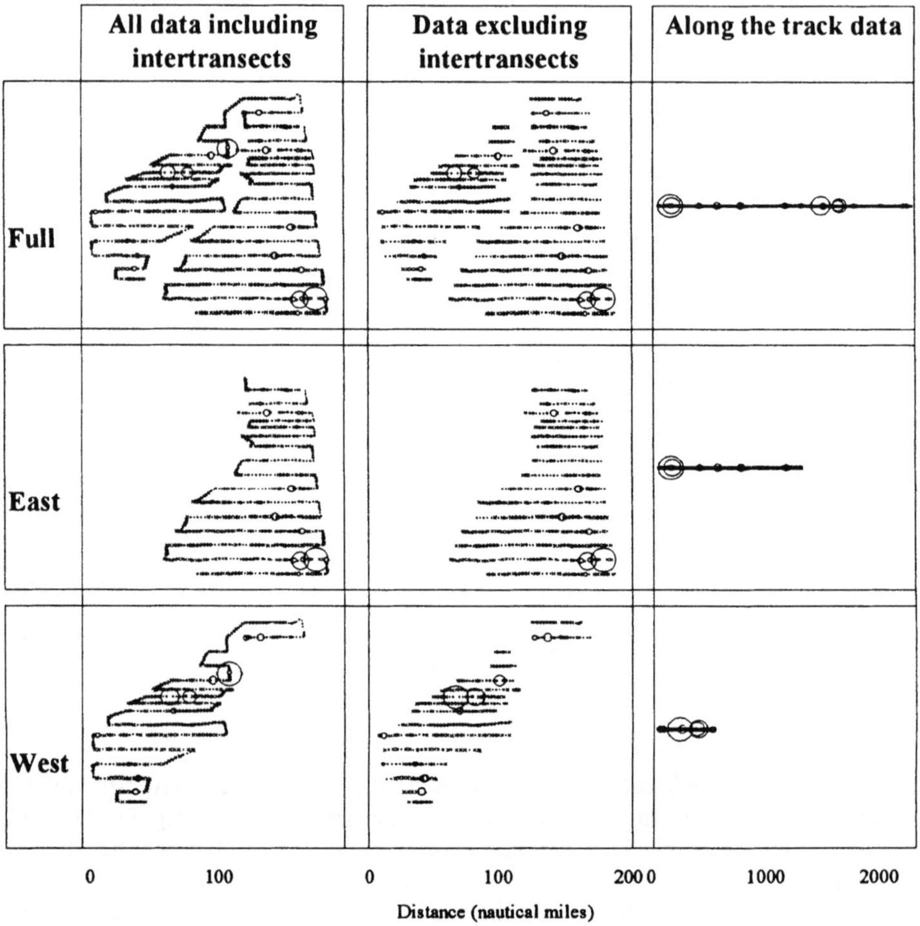

Figure 3. Graphical representation of data selection: post plots of the acoustic index from a single survey (1991).

a confident model fit to the other data subsets. The along-the-track data did not, therefore, confer any additional advantage in terms of stability or robustness. Similarly the exclusion of intertransects did not significantly improve the stability of the variogram.

In most cases, the along-the-track variograms had slightly longer practical ranges. This was quantified by fitting more complex nested hole-effect models (such as nugget + hole effect + spherical) which fit the data very closely. Any wave like pattern was almost certainly due to sampling fluctuation and so these models were not used other than to detect this subtle, but consistent, range elongation in the along the track datasets. This effect is probably due to the fact that when turning a corner, the actual distance, as would be calculated for the two dimensional datasets, is less than that calculated by the along-the-track route.

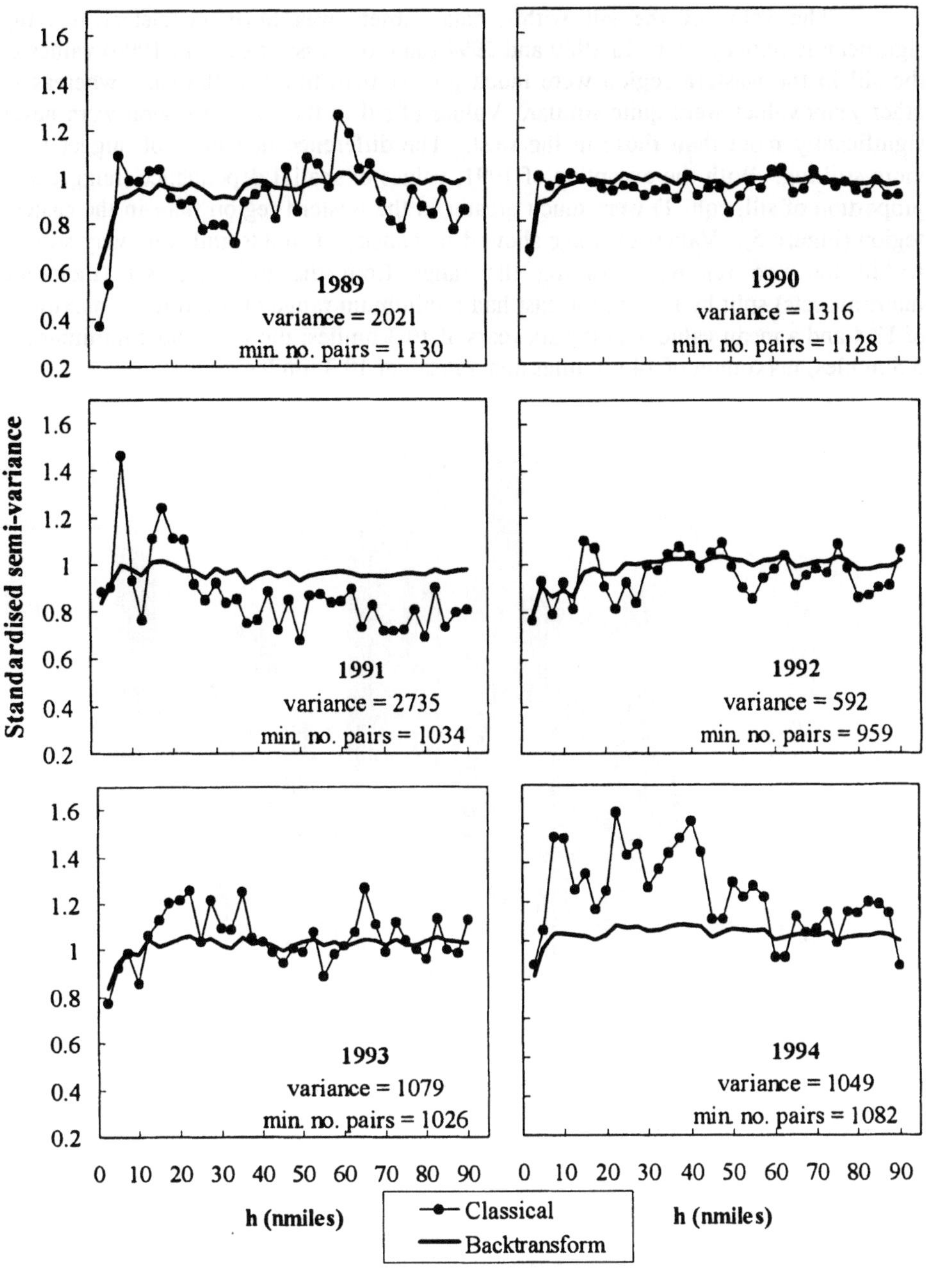

Figure 4. Omnidirectional classical and log backtransformed variograms for the abundance of Orkney Shetland herring, whole area dataset including intertransects (as in Figure 1), standardised to the sample variance. Lag = 2.5 nmiles ± 1.25 nmiles.

The value of the sill within data subsets was fairly consistent, varying significantly only by year. In 1989 and 1994 (and to a lesser extent in 1990) values of the sill in the western region were much greater than those in the east; whereas in other years values were quite similar. Values of sill in the eastern region were never significantly more than those in the west. The difference in values of nugget were more striking. With the exception of 1991, values of spatial dependence (nugget as a proportion of sill, eqn. 1) were much greater in the western region than in the eastern region (Figure 5). Values of range showed no consistent trends and were very similar within the two regions. Taking the range from the full data set (including intertransects) split by region: the east had a minimum range of 5 nmiles, a maximum of 13.4 and a mean value over the six years of 10.3 nmiles; the west had a minimum of 5.5 nmiles, maximum of 14.6 nmiles and a mean of 10.3 nmiles.

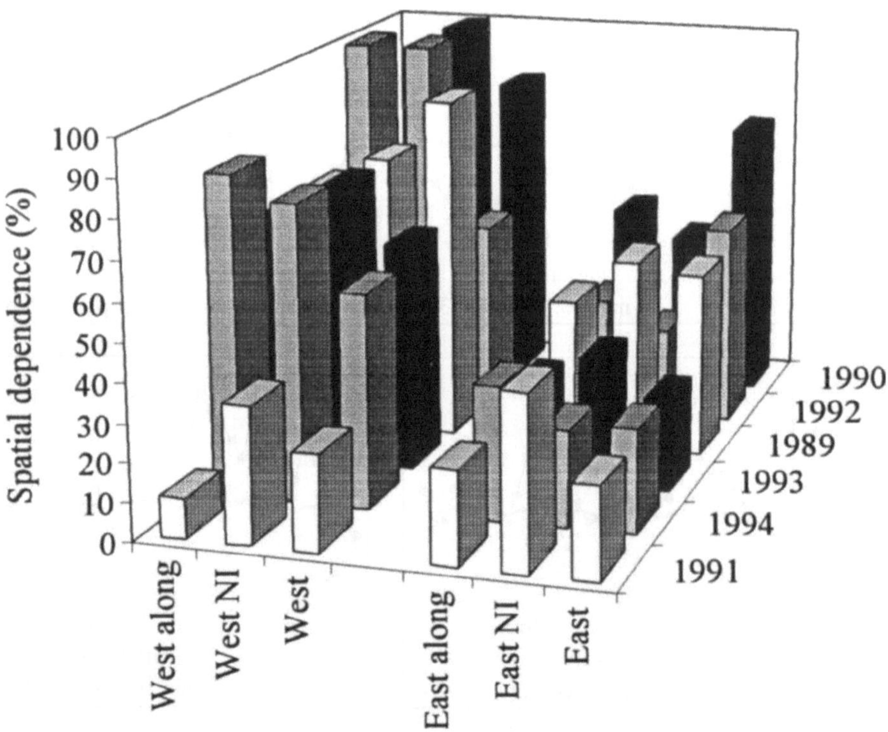

Figure 5. Comparisons of spatial dependence (nugget as a proportion of sill) as modelled by best fitting models, for east and west data subsets (along = along-the-track data; NI=No Intertransects).

5. Discussion

The general observation that it is easier to observe spatial autocorrelation in data using a log backtransform is supported by this study. While with simulation it is impossible to provide conclusive proof of the best method for estimating variogram parameters there is very strong evidence that the use of log transform and a backtransform scaled to the variance provides an improved method for estimating variogram parameters and estimation variance. No evidence of greater bias in the log backtransformed variogram relative to the classical variogram was found.

Many of the conclusions in this paper rely on the interpreted parameters which have been derived by a semi-automatic fitting procedure. These values may have been different if an alternative weighting scheme had been used or if the models had been fitted by eye. However, the methods applied were consistent and the objective was to undertake a comparative exercise looking for trends between different combinations of data. In all cases the fitted models were ones which looked reasonable enough to have been fitted by eye. The practice of fitting a variogram model to an experimental variogram by automated means is often criticised (Journel and Huijbregts 1978) and has been subject to much debate (Armstrong 1984). Any alternative relying on experience requires a skill which takes a long time to acquire and even then may be reliant on knowledge gained in the analysis of a completely different field. The use of automated fits in ecological studies is common, however, weighting procedures may vary from being absent (Jackson and Caldwell 1993; Robertson 1995; Young and Harvey 1996), to the incorporation of the number of pairs (Webster 1985) and the inverse of the variogram value squared (Cressie 1991; Pelletier and Parma 1994). There is a strong case for an efficient unbiased automated procedure which in most cases would produce similar results to that of a trained expert. The procedure presented in this paper is believed to be adequate (Rivoirard, *pers. comm.*).

There is no account of an objective choice of model within ecological studies, and equally no reasonable justification has been given for choosing one model over another. In this study either a nugget + spherical or nugget + exponential nested model was chosen purely on the basis of the best fit according to the weighted least squares procedure. There were cases when the difference between fits was small, which could have led to large differences in parameter values according to the choice of model. It should be noted that for the same fit an exponential model would produce a smaller nugget and possibly a slightly longer practical range than a spherical model. This problem might be alleviated with further knowledge of the shorter ranges, which in the case of acoustic data is available (by high resolution examination of ping based data). Until recently it has been computationally inhibitive to attempt variography on such data, however, methods to address this problem should be available soon.

The results provide enough evidence to indicate that there are differences between the east and west regions. The most important of these concern the values of nugget which again would be improved by a more detailed examination of the data. It would therefore seem pertinent, to stratify the whole area data sets into east and west regions.

Quite why such differences should exist is not clear. It seems that the herring in the eastern region are more randomly distributed than those in the west, which show greater spatial dependence. This may be related to the observation that the fish in the western region are on average larger than those in the east, or at least that the length size classes are not as variable. Larger fish may tend to congregate in larger associations than smaller fish; such behaviour has been observed in minnows and dace (Pitcher *et al.* 1983).

There seems little doubt that the acoustic data at a resolution in excess of school size exhibit spatial autocorrelation. This implies therefore, that the schools are aggregated in school groups analogous to Cram and Hampton's "shoal groups" (1976). This feature is very common amongst the clupeoids (Horwood and Cushing 1978; Blaxter and Hunter 1982) although it complicates the theory that schooling reduces predation, because school groups are more easily detected than randomly distributed schools. School groups are particularly common in herring and have been observed off the west coast of Scotland (Rusby 1977), the Moray Firth (Wood and McGee 1925) and in the southern North Sea (Tungate 1958). The diameter of the school groups in the latter studies appeared to be of the order of 10 nmiles and the single group observed by the former study was just over 4 nmiles long. These are similar values to the range components (equivalent to aggregation diameter) modelled by the single structures in the present study. A more detailed investigation of patch dynamics may come about after examination of higher resolution data.

Acknowledgements

We would like to thank Dr. J. Rivoirard and Mr. P. Guiblin of the School of Mines, Fontainebleau, for their helpful and constructive comments. We would also like to acknowledge the support of the EU AIR programme, contract C2/94/1007.

References

Anon (1993). "Report on the workshop on the applicability of spatial statistical techniques to acoustic survey data." *ICES Cooperative Research Report* **195**: 87 pp.

Armstrong, M. (1984). Improving the estimation and modelling of the variogram. In G. Verly, M. David, A. G. Journel and A. Marechal (eds.) "*Geostatistics for natural resource characterisation*". Dordrecht, Reidel. 1-19.

Armstrong, M., Renard, D., Rivoirard, J. and Petitgas, J. (1992). Geostatistics for fish survey data. Centre de Geostatistique, Ecole Nationale Superiore des Mines de Paris, Paris, France. 89 pp.

Blaxter, J.H.S. and Hunter, J.R. (1982). "The biology of the clupeoid fishes." *Advances in Marine Biology* **20**: 1-223.

Cram, D.L. and Hampton, I. (1976). "A proposed aerial/acoustic strategy for pelagic fish stock assessment." *Journal du conseil permanent international pour l'exploration de la Mer* **37**(1): 91-97.

Cressie, N. (1991). *Statistics for spatial data*. John Wiley & Sons, New York. 900 pp.

Guiblin, P., Rivoirard, J. and Simmonds, E.J. (1995). "Analyse structurale de donnees a distribution dissymetrique exemple du hareng ecossais." *Cahiers de Geostatistique* **5**: 137-160.

Horwood, J.W. and Cushing, D.H. (1978). Spatial distributions and ecology of pelagic fish. In J. H. Steele (ed.) "*Spatial pattern in plankton communities*". New York, Plenum Press Ltd. 355-383.

Jackson, R.B. and Caldwell, M.M. (1993). "Geostatistical patterns of soil heterogeneity around individual perennial plants." *Journal of Ecology* **81**: 683-692.

Journel, A.G. and Huijbregts, C.J. (1978). *Mining geostatistics*. Academic Press, London. 600 pp.

MacLennan, D.N. and Simmonds, E.J. (1992). *Fisheries acoustics*. Chapman & Hall, London. 325 pp.

Maravelias, C., Reid, D.G., Simmonds, E.J. and Haralabous, J. (1996). "Spatial analysis and mapping of acoustic survey data in the presence of high local variability. An application of geostatistics in North Sea herring" *Canadian Journal of Fisheries and Aquatic Sciences* **53**(7): 1497-1505.

Matheron, G. (1971). "The theory of regionalised variables and its applications." *Cahiers du centre de Morphologie Mathematique, Fontainebleau* **5** (Ecole Nationale Superiore des Mines de Paris, Paris, France). 211 pp.

Murray, A.W.A. (1995). "Comparison of geostatistical and random sample survey analyses of Anatarctic krill acoustic data." *ICES Journal of Marine Science* **53**: 415-421.

Pelletier, D. and Parma, A.M. (1994). "Spatial Distribution of Pacific Halibut (*Hippoglossus stenolepis*): An application of Geostatistics to Longline Survey Data." *Canadian Journal of Fisheries and Aquatic Sciences* **51**(7): 1506-1518.

Petitgas, P. (1993a). "Geostatistics for fish stock assessments: A review and an acoustic application." *ICES Journal of Marine Science* **50**(3): 285-298.

Petitgas, P. (1993b). "Use of disjunctive kriging to model areas of high pelagic fish density in acoustic fisheries surveys." *Aquat. Living Resour.* **6**(3): 201-209.

Petitgas, P. and Prampart, A. (1993)."EVA (Estimation Variance). A geostatistical software on IBM-PC for structure characterisation and variance computation." *ICES CM 1993/D:65*, 32 pp. (mimeo).

Pitcher, T.J., Magurran, A.E. and Allan, J.R. (1983). "Shifts of behaviour with shoal size in cyprinids." *Proc. Br. Freshwat. Fish. Conf.* **3**: 220-228.

Porteiro, C., Carrera, P. and Miquel, J. (1995). "Analysis of Spanish acoustic surveys for sardine, 1991-1993: abundance estimates and inter-annual variability." *ICES Journal of Marine Science* **53**: 429-433.

Rivoirard, J. (1992)."Comments on the course on geostatistics for fish survey data (Fontainebleau, 3-7 Feb 1992)." *ICES CM 1992/D:12* (mimeo).

Robertson, G.P. (1995). "The spatial distribution of nematode trophic groups across a cultivated ecosystem." *Ecology* **76**(5): 1425-1432.

Rusby, J.S.M. (1977). "Long-range survey of a herring fishery by side-scan sonar." *Rapports Process-verbaux Reunion Conseil international Exploration du Mer* **170**: 7-14.

Simard, Y., Marcotte, D. and Bourgault, G. (1993). "Exploration of geostatistical methods for mapping and estimating acoustic biomass of pelagic fish in the Gulf of St. Lawrence: Size of echo-integration unit and auxiliary environmental variables." *Aquat. Living Resour.* **6**(3): 185-199.

Simmonds, E.J., Toresen, R., Corten, A., Pedersen, J., Reid, D.G. and Fernandes, P.G. (1995)."1994 ICES coordinated acoustic survey of ICES Divisions IVa, IVb, VIa and VIIb." *ICES CM 1995/H:15*, 13 pp. (mimeo).

Simmonds E.J. and Fryer R.J. 1996. "Which are better, random or systematic acoustic surveys? A simulation using North Sea herring as an example." *ICES J. Mar. Sci.*, **53**, 39-50.

Tungate, D.S. (1958). "Echo sounder surveys in the autumn of 1956." *Fishery Investigations, London, Series II* **22(2)**: 17 pp.

Webster, R. (1985). "Quantitative Spatial Analysis of Soil in the Field." *Advances in Soil Science* **3**: 1-70.

Wood, H. and McGee, G. (1925). "Aircraft experiments for the location of herring shoals in Scottish waters." *Scientific Investigations* **1** (Sea Fishery Board for Scotland): 20 pp.

Young, B.M. and Harvey, L.E. (1996). "A spatial analysis of the relationship between mangrove (*Avicennia marina* var. *australasica*) physiognomy and sediment accretion in the Hauraki Plains, New Zealand." *Estuarine, Coastal and Shelf Science* **42**: 231-246.

Appendix

The variogram via the backtransformed log transformed data is derived by applying a logarithmic transformation to the data values, such that:

$$\mathrm{L} = \mathrm{Ln}(1 + \tfrac{\mathrm{z}}{\mathrm{b}}) \tag{2}$$

where:

z = sample value
b = constant (in this case taken as the mean of z)

The variogram of the log transformed variable **L** is then calculated according to the classical method (Matheron, 1971). The experimental variograms of the log transformed data are then backtransformed according to the formula of Guiblin *et al.*, (1995):

$$\hat{\gamma}_{\mathrm{Z}}(\mathrm{h}) = ((\mathrm{b}+\mathrm{m})^2 + \mathrm{var}(\mathrm{z}))\cdot(1 - \mathrm{e}^{-\sigma^2 \frac{\hat{\gamma}_{\mathrm{L}}(\mathrm{h})}{\mathrm{var}(\mathrm{L})}}) \tag{3}$$

where:

$\hat{\gamma}_{\mathrm{Z}}(\mathrm{h})$ = backtransformed variogram value at lag **h**
b = constant (in this case taken as the mean of z)
m = mean of **z**
var(z) = variance of **z**
$\hat{\gamma}_{\mathrm{L}}(\mathrm{h})$ = variogram value of **L** at lag **h**
var(L) = variance of **L**

and:

$$\sigma^2 = \mathrm{Ln}(1 + \frac{\mathrm{var}(\mathrm{z})}{(\mathrm{b}+\mathrm{m})^2}) \tag{4}$$

The backtransformation process requires an assumption of second order stationarity where: the expectation of the random function has a mean (m) independant of its location x (eqn. 5) and the covariance between two points is only a function of their relative position (eqn. 6). In this case the variogram is bounded (eqn 7).

$$E(Z(x)) = m \tag{5}$$

$$Cov(Z(x+h), Z(x)) = C(h) \tag{6}$$

$$V(Z(x+h) - Z(x)) = 2\gamma(h) \tag{7}$$

APPLYING GEOSTATISTICS TO EXPOSURE MONITORING DATA IN INDUSTRIAL HYGIENE

H. WACKERNAGEL AND C. LAJAUNIE
Centre de Géostatistique, Ecole des Mines de Paris
35 rue Saint Honoré, 77305 Fontainebleau, France

AND

L. THIERY, R. VINCENT AND M. GRZEBYK
Institut National de Recherche et de Sécurité (INRS)
Av. de Bourgogne, B.P. n° 27, 54501 Vandœuvre Cedex, France

Abstract. In industrial hygiene, the statistical evaluation of the exposure of a worker to a chemical agent or to noise in an industrial site during a shift is needed. The data consists typically of time series, but occasionally spatial location can also be relevant. It appears that the autocorrelation, as characterized by the variogram, is usually well structured at least within an hourly scale. For exploring and modeling the data, estimating means with corresponding confidence intervals, simulating and planning sampling, geostatistics provides tools to characterize the interplay of support (integration time), autocorrelation (time dependence), stratification (qualitative changes of working activity) and distributional transformations within a unified framework.

1. Introduction

A number of standards (e.g. [1, 2, 3, 6]) in industrial hygiene nowadays contain a statistical section. Typically the mean over a given time period (shift or weekly average) has to be estimated and usually confidence intervals based on a Student t-test have to be specified. This test is based on a hypothesis of independence which is violated if the data turns out to be autocorrelated. If the test is applied ignoring autocorrelation, the confidence intervals are optimistic in the sense that the interval is narrower than

A. Soares et al. (eds.), geoENV I–Geostatistics for Environmental Applications, 463–476.

it ought to be. In particular its upper bound is systematically underestimated. This will be shown on a series of trichloroethylene measurements (in section 3.1).

Sound is commonly expressed in decibel (dB), as the logarithm of the squared acoustic pressure, because the sensibility of the ear is roughly logarithmic. This fact has led some authors to advocate lognormal models for acoustic data. which are used in standards (e.g. [1, 3, 6]). For the geostatistician it might be tempting to use lognormal kriging [15, 17] in this context. We have experimented lognormal kriging on several noise exposure series and show (in section 3.2) both a case where it performs extremely well and another case in which lognormal kriging estimates are both biased and less efficient than the classical estimator of the mean exposure for an 8h-shift.

The paper is subdivided as follows. In section 2 we discuss the characteristics of occupational exposure data and list corresponding concepts. In section 3 we present two case studies, one on chemical data (trichloroethylene), the other on noise data. Conclusions are finally drawn in section 4.

2. Concepts

Occupational exposure to chemicals and physical agents (noise, heat,...) varies continuously in time. This variability depends on the fluctuation of the pollutant or emission sources, process cycles. mobility of the workers in the shop, seasons. Industrial hygienists have generally to conduct exposure assessments during short periods (2-3 days for example) sometimes with a low number of exposure measurements. It is in this context that they have to take the decision about compliance or non-compliance with the exposure limit values. In the following we discuss the concepts used to analyze and evaluate time series of occupational exposure data. Some of these questions have already been partially dealt with in the industrial hygiene literature (e.g. in [19, 4, 9, 8]) but geostatistics allows to approach them with a coherent, complete and efficient approach [16].

2.1. SUPPORT AND SAMPLING WINDOW

Classical statistics are usually applied in industrial hygiene using the number of samples (sample size) as the only parameter in sampling design. Actually the duration of the individual measurement, called the *support* in geostatistics, and the size of the sampling window are two other basic parameters that should be taken into account.

2.2. AUTOCORRELATION

Although there may be examples of exposure data showing no autocorrelation in time (Guichard [5] produces an unstructured experimental variogram for dust exposure data from a quarry) our experience with chemical agents (trichloroethylene, lead, styrene,...) and noise data [21, 22] is that variograms are usually well structured. In case of doubt, a simple graphical test consists in randomly permuting the sequence of the data and recomputing for comparison the variogram of the randomized sequence.

2.3. STRATIFICATION

Time series of exposure data are often not homogeneous. For example, one out of several machines in a room is turned off at a given time and turned on again later, occasioning a distinctly lower average noise level during that time interval; changes of work function may lead to variations of the average exposure to a chemical agent. If these time intervals can be properly identified, perhaps on the basis of an accurate job description for the given time period, it is preferable to stratify the series into homogeneous intervals. The statistical properties for each interval are determined separately and the results are ultimately recombined.

2.4. DISTRIBUTIONAL ASSUMPTIONS

A popular assumption in industrial hygiene is that data are lognormally distributed. This approach provides excellent estimators when lognormality is actually verified, but if the series are inhomogenous with multimodal histograms the lognormal estimators perform badly in comparison with normal estimators even if the overall distribution is left skewed.

The Larsen lognormal model [12] is mentioned in the industrial hygiene literature both for chemical [16] and for acoustic [13] data. The Larsen model rests on a hypothesis of distributional self-similarity for a given range of time scales (which is actually contradictory with lognormality). It is closely related to the de Wijsian model of geostatistics [10]. Geostatistics offers a great variety of models to describe the deformation of an arbitrary distribution when changing the support [11, 18]. However for computing dispersion variances at various supports from data at a given support, only the knowledge of the variogram is necessary [14, 16, 22].

2.5. WINDOW MEAN VERSUS LONG TERM MEAN

In classical statistics the mean to be estimated is the mean of the i.i.d. (independent identically distributed) random variables of which the data

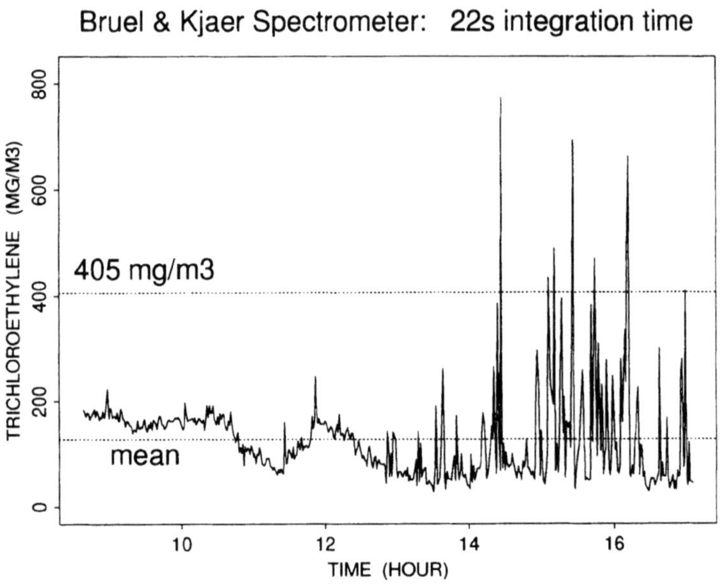

Figure 1. Time series of trichloroethylene measurements with 22s integration time.

are considered a realization.

When incorporating the regionalization in time (location, support) of the data into a probabilistic model, two different interpretations of the notion of mean estimate are possible:

- we have the (ambitious) aim to estimate the long-term mean of the process, e.g. the annual mean from data collected on a particular day;
- we wish (simply) to estimate the mean exposure for a given sampling window, e.g. an 8-hour shift on a given day from data sampled that day.

In the first situation a statement is made about the general compliance of a factory with a given regulation. In the second situation the statement is restricted to the particular day the sampling was performed and no extrapolation is attempted.

3. Two case studies

In the first case study (trichloroethylene data) we estimate the *long-term mean* and assess confidence intervals for different supports taking into account the autocorrelation of the data. We compare the confidence intervals with those obtained ignoring autocorrelation.

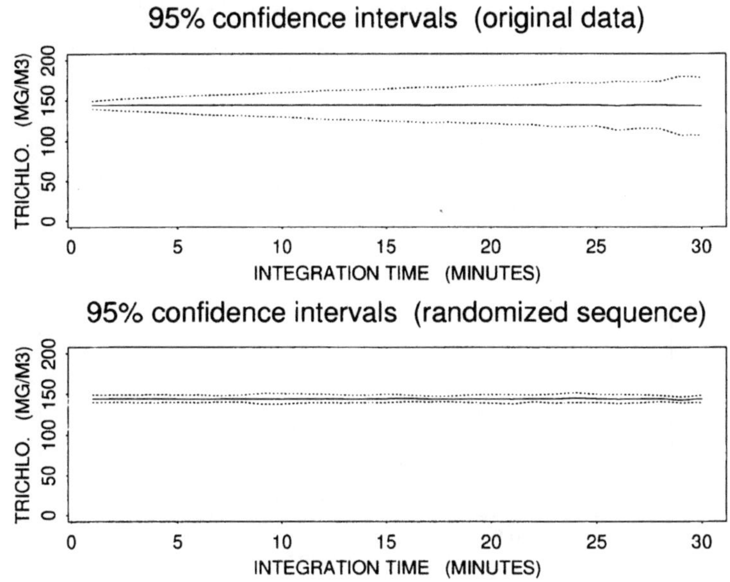

Figure 2. Confidence intervals around the mean in a classical t-test performed on the morning trichloroethylene data for integration times varying from 22s to 30mn. For the upper graph, computations have been done on the original sequence. For the lower graph, a random permutation of the samples was performed prior to integration, to generate a non-autocorrelated sequence.

In the second case study (noise data) we estimate the *mean for an 8h-shift*, comparing by resampling the performance of the classical estimator with a lognormal estimator in two situations: when the data is homogeneous and when it is multimodal due to changes in the working environment during the shift.

3.1. KRIGING THE MEAN OF TRICHLOROETHYLENE DATA

The trichloroethylene measurements were performed near a degreasing machine during an 8-hour shift. A Brüel & Kjaer (BK) photoacoustic spectrometer measured every minute a value averaged over 22s (sampling time). Charcoal sorbent tubes were also used for validation purpose and were replaced every 7-8 minutes. Superimposition of the two time series, after first averaging the BK values over 7mn, showed good compatibility between the two measurement types.

The BK spectrometer measurements are plotted against time on Figure 1. The value of the mean of the data and the French occupational limit value of 405 mg/m^3 are also shown on this graphic. It is apparent that something changed between the morning and the afternoon of this day in May:

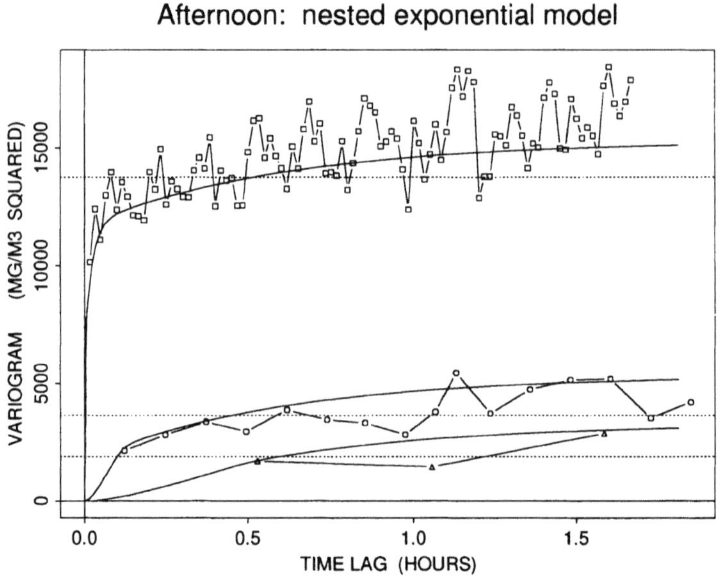

Figure 3. Three experimental variograms of the afternoon trichloroethylene data for integration times of 22s (squares), 7mn (circles) and 30mn (triangles). They were fitted with a nested exponential model (plus a nugget effect term). The fit of the regularized model is shown for these three integration times.

because of the greater heat in the afternoon, windows were opened in an adjacent workshop. The higher temperature and the air circulation explain both the greater variability and the lower mean of trichloroethylene for the second half of the shift. We have thus split the data set into two halves and used them separately to estimate means and confidence intervals.

3.1.1. *An experiment: t-test for different integration times*

As an experiment we use the morning trichloroethylene data to assess confidence limits for different supports. The complete 22s data set is partitioned into intervals corresponding to a given support in the range from 22s to 30mn. The 95% confidence intervals of a Student t-distribution are then evaluated for each integration time. The intervals should be a constant function of the integration time, because we are merely partioning with a different support the same data set (except for a leftover at the end of the series, which explains that the mean estimate is not constant). The upper graphic of Figure 2 displays the intervals obtained for each support with the morning trichloroethylene data. Surprisingly, the intervals widen steadily when the integration time is increased, instead of remaining constant.

On the lower graphic of Figure 2 we verify, by randomly permuting the sequence of the 22s values before partitioning them, that, if the indepen-

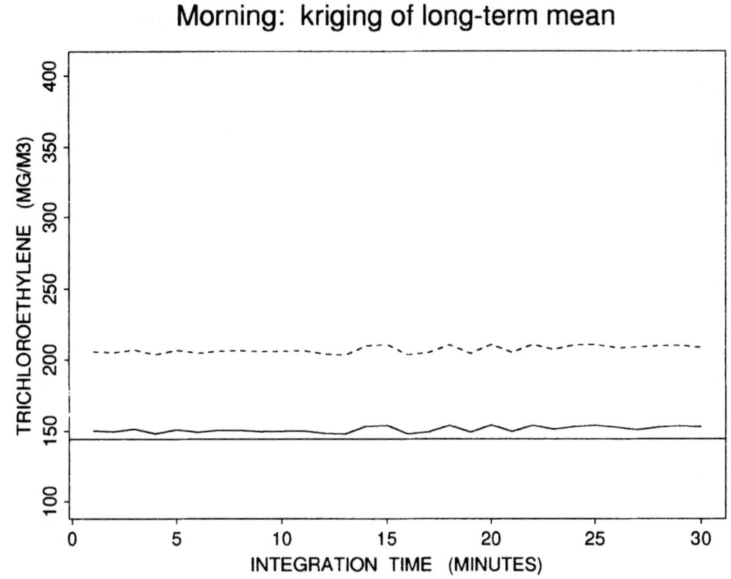

Figure 4. Kriged value of long-term mean (solid line) using the morning trichloroethylene data and upper bound of 95% confidence interval (dashes) for integration times varying from 22s to 30mn.

dence hypothesis were true, the confidence intervals would not depend on the support.

The technical explanation is the following. The Student test is based on a pivotal quantity U:

$$U = \frac{1}{\sqrt{n}} \frac{\overline{X} - \mu}{S} \qquad \text{with } S^2 = \frac{1}{n-1} \sum_{\alpha=1}^{n} \left(X_\alpha - \overline{X}\right)^2,$$

under the assumptions that (besides normality) S^2 is a consistent estimate of the population variance σ^2 and the variance of the estimated mean $\overline{X}$ behaves like σ^2/n.

Both statements are false in the case of correlated data, because:

- S^2 estimates consistently the dispersion variance within the observation window and not the variance of the process (assumed stationary); the difference between these two variances is the variance of the daily averages, not accounted when sampling during a single day.
- the decrease of the variance of $\overline{X}$ is actually slower than σ^2/n.

These two errors lead to an optimistic evaluation of the situation in the sense that confidence intervals are estimated narrower than they actually are, because of the time dependence of the data.

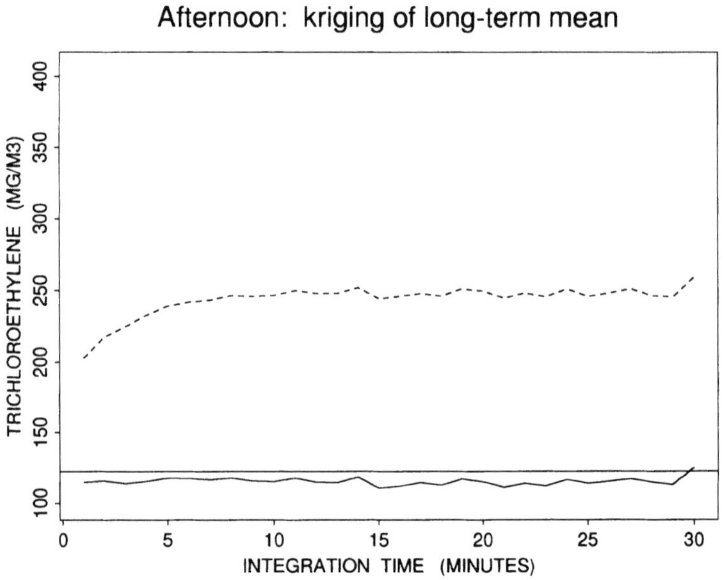

Figure 5. Kriged value of long-term mean (solid line) using the afternoon trichloroethylene data and upper bound of 95% confidence interval (dashes) for integration times varying from 22s to 30mn.

3.1.2. *Geostatistical modeling*

We now compute confidence intervals taking into account the autocorrelation of the trichloroethylene data. We shall also assume that both the variance and the autocorrelation function of the stationary process are known, deriving them from a variogram model.

We model separately the morning and the afternoon data by fitting a regularized theoretical variogram model to the variogram estimates for integration times of 22s, 7mn and 30mn. Then we estimate for integration times in the range of 22s to 30mn the long-term mean by kriging. We finally build a conventional upper 95% confidence bound by adding twice the kriging standard error to the kriged mean. (This confidence bound should actually be accurate only within a context of multinormality).

The variogram estimates of the morning data were fitted with a single regularized exponential structure with a practical range of 3 hours (plus a nugget-effect term). The Figure 3 shows the fit of afternoon data variogram estimates with a model consisting of nested exponential structures with practical ranges of 4mn and 1h47, plus a nugget effect. The regularized [7, 22] variogram model is plotted for the three integration times of 22s, 7mn and 30mn used to compute variogram estimates. Subsequently it shall be used for any support in the range of 22s to 30mn.

The results of the kriging of the long-term mean for the morning data are shown on Figure 4. The solid line represents the kriged mean for the different integration times. It is larger than the arithmetic mean of the 22s data (horizontal straight line), because the kriging estimator tends to give more weight to samples at both ends of the sampling window than to those in the middle (which are lower, especially in the interval 11h–11h30) for the morning data. The fluctuation of the mean is also partly explained by the partitioning of the series: we discarded a leftover at the end of the sampling window which was of different size, depending on the support considered. The dashes on Figure 4 represent the upper bound of the conventional confidence interval which is nearly constant over the range of integration times considered.

If we had assumed no autocorrelation, i.e. a "pure" nugget-effect, the kriging of the mean would be equivalent to the arithmetic mean and the associated kriging standard error would reduce to the classical expression $\sigma/\sqrt{n}$ (see [22] for details). The no-autocorrelation model on the basis of the 22s morning data yields a conventional upper limit of 145 mg/m^3 against a limit above 200 mg/m^3 for the 3h-range autocorrelation model!

The results of the kriging of the mean for the afternoon data, with a nested exponential model, are shown on Figure 5. The conventional upper bound now is not constant, apparently a consequence of the nesting of two exponential structures with a different practical range. (The same type of computation using nested triangular structures, instead of exponential structures, led to the same effect, i.e. constant intervals for a single structure and non-constant for the nested model).

With a no-autocorrelation model for the 22s afternoon data we would have an upper bound of 124 mg/m^3 instead of 200 mg/m^3 for 22s data (or 250 mg/m^3 for 30mn integrated data) with the nested exponential autocorrelation model!

The conclusion from these computations is that neglecting the autocorrelation in the data leads to overly optimistic confidence intervals. In particular, the upper limit of the confidence interval tends to be severely underestimated. An additional source of error can result from an underestimation of the time dependence (and corresponding variance) at a larger scale than the observation window.

3.2. LOGNORMAL KRIGING OF INDUSTRIAL NOISE

Industrial noise data have been collected using personal sound exposure meters carried by two operators in different factories. The noise exposure series are shown on Figure 6. The upper series, **regleur2**, is from a worker taking care of different machines and walking in the workshop to check

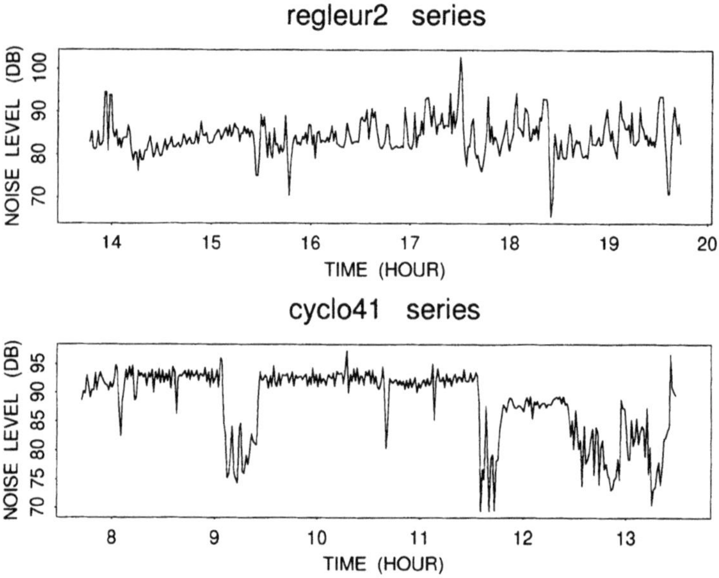

Figure 6. Times series regleur2 and cyclo41 of 1mn noise level measurements (in decibel).

the tuning of the machines. The other series, **cyclo41**, is from an operator of several machines in the same room; breaks took place from 9h10 to 9h30 and from 11h40 to 11h50. Whereas both variograms (not shown, see [21, 22] for examples) suggest ranges extending over several hours, it has to be mentioned that the **regleur2** variogram exhibits also a marked structure at the 10mn-scale.

The histograms of the two series are represented on Figure 7. For **regleur2** the assumption of lognormality seems satisfactory with a fairly symmetric histogram in the decibel scale. The same cannot be said about the histograms of the **cyclo41** series.

The classical estimator of the average noise level, termed L_{eq}, consists in computing the arithmetic mean of the sound exposure (which is the additive quantity) and converting the result into the decibel scale [1, 3, 6]. The lognormal kriging is applied in the same way: the 8h-shift mean of sound exposure is computed by lognormal block kriging [15, 17] and the result is then expressed in decibel.

For each of the two series, 1000 sets of 36 samples were taken at random. Both estimators were applied to each set of samples and the histograms of the estimated values converted into dB are shown on Figure 8.

For **regleur2** (upper graphics on Figure 8) the standard error of lognormal kriging is downsized almost by half in comparison with the classical L_{eq}. The second mode at 90 dB on the classical L_{eq} histogram, due to an ex-

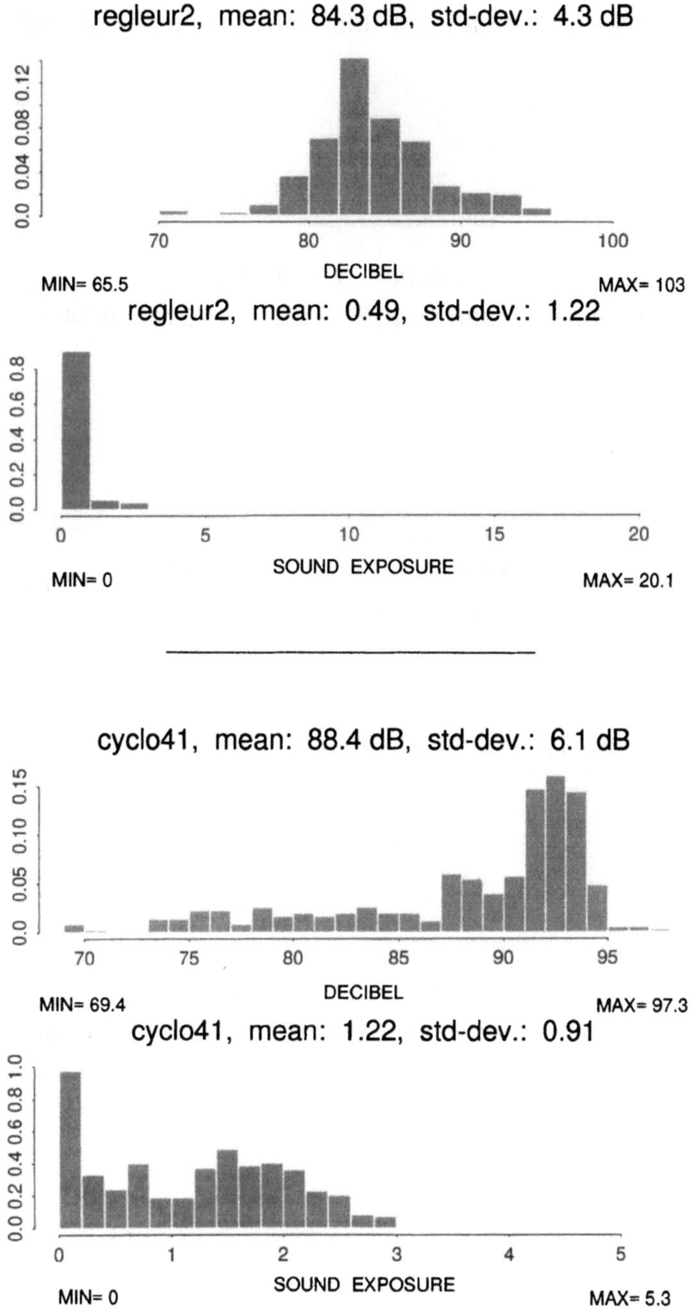

Figure 7. Histograms of regleur2 and cyclo41 1mn integrated data expressed in terms of both a variable directly proportional to the sound exposure and the logarithm of that variable, the decibel scale. The L_{eq} for the regleur2 series (358 values) is 86.9 dB and for the cyclo41 series (348 values) is 90.9 dB.

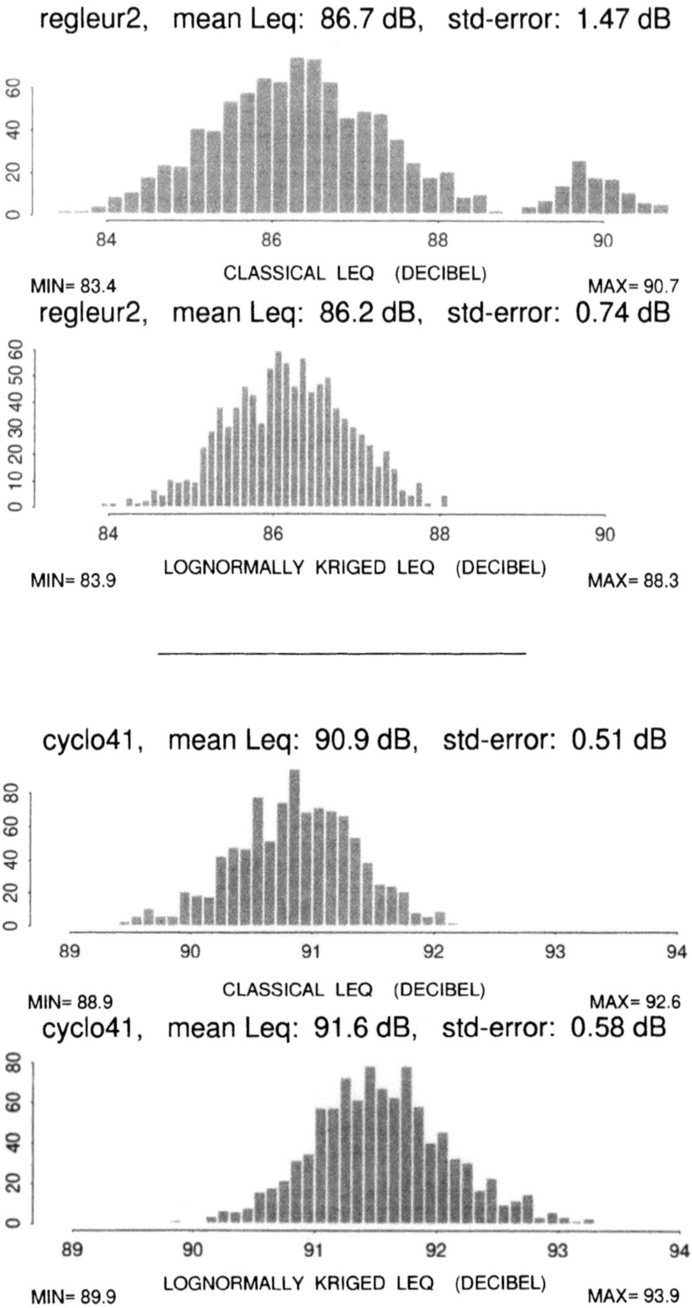

Figure 8. Comparison for regleur2 and cyclo41 1mn data of the classical and the lognormally kriged L_{eq} estimator for 1000 sets of 36 randomly selected samples.

treme value (in terms of sound exposure), is not present on the histogram of the lognormal L_{eq}, because the latter estimator is robust to extreme values in the right tail.

For **cyclo41** (lower graphics on Figure 8) we see that both histograms are fairly symmetrical (with a stronger standard error for the lognormal estimator). The striking feature is the bias for the lognormal estimator. The mean of the classical L_{eq} estimates using 36 samples is equal to the L_{eq} of the 348 values of the series. In comparison, the histogram of the lognormal L_{eq} estimates is shifted to the right by one dB (the mean by .7 dB, the minimum by 1 dB and the maximum by 1.3 dB).

Lognormal estimators are thus only superior in certain situations and cannot be recommended for general use in the statistical treatment of industrial noise data. Geostatistics provides other non-linear estimators which adapt better to an arbitrary distribution of values [18, 11]. However, in present day hearing conservation programs, often less than 30 samples are used in industry to compute 8h-shift L_{eq}values, so that it is difficult to try assessing the shape of an underlying distribution. This situation is likely to change in a near future with the availability of cheap and light personal sound exposure meters, like the one used for measuring the **regleur2** and **cyclo41** series.

4. Conclusion

Despite the publication, ten years ago, of a pioneering and often cited article by Préat [16], geostatistics has not yet received the attention it deserves in the world of industrial hygiene.

A first conclusion of the present paper is that classical statistics based on the independence hypothesis, which are recommended in standards about occupational exposure evaluation, tend, sometimes severely, to underestimate the risk of exceeding the limit values as well as the associated health risks for the workers. Experience shows that data are rarely without any autocorrelation, as required by the classical tests.

The second conclusion of the paper is that the lognormal model, which is still very present in the industrial hygiene literature, can often be inadequate and give poor results even if autocorrelation is taken into account.

Acknowledgement

We wish to thank M. Charreton, A. Good and C. Vallet from the Laboratoire de Chimie du Val Rosay (C.R.A.M. “Rhone-Alpes”) for kindly providing the trichloroethylene data set.

References

1. AFNOR (1987) Méthode de mesurage des niveaux sonores en milieu de travail en vue de l'évaluation du niveau d'exposition sonore quotidienne des travailleurs, *Norme Française*, **NF S 31-084**, Paris.
2. AFNOR (1995) Workplace atmospheres - Guidance for the assessment of exposures by inhalation to chemical agents for comparison with limit values and measurement strategy, *Norme Française*, **NF EN 689**, Paris.
3. DIN (1990) Mittelung von Schallpegeln, *Deutsche Norm*, **DIN 45 641**, Berlin.
4. Francis, M., Selvin, S., Spear, R. and Rappaport, S. (1989) The effect of autocorrelation on the estimation of workers' daily exposures, *American Industrial Hygiene Association Journal*, **50**, 37–43.
5. Guichard, J.C. (1993) L'exposition aux agents chimiques - l'expérience de l'INERIS, *Les Notes Scientifiques et Techniques de l'INRS*, **104**, 93–102.
6. ISO (1995) Acoustics - Guidelines for the measurement and assessment of exposure to noise in a working environment, *International Organization for Standardization*, **ISO/DIS 9612.2**, Geneva.
7. Journel, A.G. and Huijbregts, C.J. (1978) *Mining Geostatistics*, Academic Press, London.
8. Kumagai, S. and Matsunaga, I. (1994) Approaches for estimating the distribution of short-term exposure concentrations for different averaging times, *Annals of Occupational Hygiene*, **38**, 815–825.
9. Kumagai, S., Matsunaga, I. and Kusaka, Y. (1993) Autocorrelation of short-term and daily average exposure levels in workplaces, *American Industrial Hygiene Association Journal*, **54**, 341–350.
10. Lajaunie, C. (1993) Les bases de la modélisation géostatistique de l'effet de support, *Les Notes Scientifiques et Techniques de l'INRS*, **104**, 15–28.
11. Lajaunie, C. (1993) *L'estimation Géostatistique Non Linéaire*. Lecture notes **C–152**, Centre de Géostatistique, Ecole des Mines de Paris, Fontainebleau.
12. Larsen, R.I. (1969) A new mathematical model of air pollutant concentration averaging time and frequency, *Journal of the Air Pollution Control Association*, **19**, 24–30.
13. Malchaire, J. (1994) *Programmes de Conservation de l'Audition*, Masson, Paris.
14. Matheron, G. (1970) *The Theory of Regionalized Variables and its Applications*, Ecole des Mines de Paris, Fontainebleau.
15. Matheron, G. (1973) Effet proportionnel et lognormalité ou: le retour du serpent de mer, Publication **N-374**, Centre de Géostatistique, Ecole des Mines de Paris, Fontainebleau.
16. Préat, B. (1987) Application of geostatistical methods for estimation of the dispersion variance of occupational exposures, *American Industrial Hygiene Association Journal*, **48**, 877–884.
17. Rivoirard, J. (1990) A review of lognormal estimators for in situ reserves, *Mathematical Geology*, **22**, 213–221.
18. Rivoirard, J. (1994) *Introduction to Disjunctive Kriging and Non-Linear Geostatistics*. Oxford University Press, Oxford.
19. Spear, R.C., Selvin, S. and Francis M. (1986) The influence of averaging time on the distribution of exposures, *American Industrial Hygiene Association Journal*, **47**, 365–368.
20. Thiéry, L., Louit, P., Lovat, G., Lucarelli, D., Raymond, F., Servant, J.P. and Signorelli, C. (1994) *Exposition des Travailleurs au Bruit: Méthode de Mesurage*, Editions INRS, **ED 772**, Paris.
21. Thiéry, L., Wackernagel, H. and Lajaunie, C. (1996) Geostatistical analysis of time series of short LAeq values, *Proceedings of INTER-NOISE96*, Liverpool (UK), 2065–2068.
22. Wackernagel, H. (1995) *Multivariate Geostatistics*, Springer-Verlag, Berlin.

CONDITIONAL SIMULATION AND ESTIMATION OF GAUSS-MARKOV RANDOM FIELDS USING THE BAYESIAN NEAREST NEIGHBOR METHOD

Amvrossios C. Bagtzoglou[1, 2] and Rachid Ababou[3]

[1] *Center for Nuclear Waste Regulatory Analyses, Southwest Research Institute, 6220 Culebra Road, San Antonio, Texas 78238-5166, USA, abagtzog@swri.edu*

[2] Now at: *Department of Civil Engineering & Engineering Mechanics, Columbia University, S.W. Mudd Building, 500 West 120th Street, Mail Code 4709, New York, NY 10027, USA, abagtzog@civil.columbia.edu*

[3] *Institut de Mecanique des Fluides de Toulouse, Avenue du Professeur Camille Soula, 31400 Toulouse, FRANCE, ababou@imft.enseeiht.fr*

Abstract

In this paper a specialized method for generating Markovian random fields, with or without conditioning, is presented. Here, the prior fields are assumed to be stationary second-order Gauss-Markov random fields in N-dimensional (N-D) euclidian space. The unconditional Markov fields are generated as numerical solutions of white-noise driven PDEs, for the solution of which finite difference or finite volume discretization on a regular grid yields an algorithm commonly known as the Nearest Neighbor Method (NNM). In addition, starting with the continuous space PDE-driven field, we develop a generalization of the NNM algorithm for the generation and optimal interpolation (estimation) of conditional random fields in the Bayesian sense. This generalized NNM algorithm is called Bayesian-NNM, or B-NNM, and is essentially similar to the classical NNM-type algorithms. The B-NNM method can be used for conditional simulation of random fields, as well as the optimal estimation of spatial fields assuming Markovian prior statistics, based on a collection of measurements or data. These data may be defined either at points or, more importantly, on continuous subdomains of various sizes and shapes, such as lineaments, faults, layers, boreholes, etc. In this paper, we also present exact closed form relations for linear estimation of 1-D Markovian fields, and this without recourse to space discretization. The complete analytical results are used for testing the multi-dimensional B-NNM code in the 1-D case.

1. Introduction and Overview

Geologic media are known to be heterogeneous when examined in detail but their detailed structure is rarely observed on a sufficiently large domain in the field. Nonetheless, a stochastic representation of spatial variability can be used for representing, based on available information, the unobserved fluctuations of material properties. More precisely, the medium is represented as a statistical continuum, with fluctuations that are random but, in a certain statistical sense, continuous. The statistical continuum approach has been used in a number of hydrogeologic studies, and there is

A. Soares et al. (eds.), geoENV I–Geostatistics for Environmental Applications, 477–491.

today an increasing amount of experimental results and field data that supports the usefulness of the stochastic postulate (Gelhar, 1986; King and Smith, 1988; Baker, 1984).

We examine here a more specialized, but powerful, method for generating random fields with or without conditioning. Here, the prior fields are assumed to be stationary second-order Gauss-Markov random fields in N-dimensional euclidian space. Briefly, the unconditional Markov fields are generated as the numerical solutions of white noise-driven Partial Differential Equations (PDEs). A finite difference (FD) or finite volume discretization of this equation on a regular grid yields an algorithm commonly known as the Nearest Neighbor Method (NNM) (Whittle, 1962; King and Smith, 1988; Smith and Freeze, 1979 a,b; Smith and Schwartz, 1980). However, it is important to keep in mind that the random field model at hand is the continuous-space, PDE-driven process, rather than the discrete NNM model itself. Thus, it will be shown that, contrary to early applications of the method, certain strict restrictions need to be imposed on the parameters of the NNM model in order to obtain physically realizable and accurate approximations of the random fields. In addition, starting with the continuous space PDE-driven field, we develop a generalization of the NNM algorithm for the generation and optimal interpolation (estimation) of conditional random fields in the Bayesian sense. This generalized NNM algorithm is called Bayesian-NNM, or B-NNM and is similar to the classical NNM-type algorithms. The B-NNM method can be used for conditional simulation of random fields, as well as the optimal estimation of spatial fields assuming Markovian prior statistics, based on a collection of measurements or data. These data may be defined either at points or on continuous subdomains of various sizes and shapes, such as lineaments, faults, layers, boreholes, etc. While we only illustrate 1-D point conditioning in this paper, the more general case of subdomain conditioning will be important in future geological applications. When no data are provided the conditional B-NNM generator reduces to the unconditional NNM generator. We have also developed exact closed form relations for linear estimation of 1-D Markovian fields, and this without recourse to space discretization. The complete analytical results are used for testing purposes in the 1-D case.

2. Markov Fields and the Klein-Gordon Equation

2.1. ONE-DIMENSIONAL TIME

A discrete-time Markov process, $F(t_n)$, is such that the probability density function of the future state, $F(t_{n+1})$, depends only on the present state, $F(t_n)$, and not on the past states, $\{F(t_{n-1})...F(t_1)\}$:

$$P_{F(t)}\left(t_{n+1} \mid t_n, t_{n-1}...t_1\right) = P_{F(t)}\left(t_{n+1} \mid t_n\right) \tag{1}$$

For instance, the velocity $v(t)$ of a particle with mass m in a liquid follows a continuous Gauss-Markov process that is stationary:

$$\frac{dv}{dt} + \alpha v = \varepsilon(t) \tag{2}$$

where $\alpha = \beta/m$, β is a friction coefficient, and $\varepsilon(t)$ is a random (white) noise of intensity σ_0^2 (due to molecular shocks on the particle). Parameter α is the inverse of a correlation time $\alpha = 1/\tau$, with τ serving as an indicator of the *memory* of the stochastic process. The process $v(t)$ is entirely characterized as follows:

$$v(t) = v(0)\ \exp(-t/\tau) + \int_0^t \exp\left[\frac{-(t-s)}{\tau}\right] \varepsilon(s)\ ds \tag{3}$$

Given that $v(0) = 0$ or $t \to \infty$, this process satisfies:

$$\begin{aligned} E\ [v(t)] &= 0 \\ R_{vv}(t,s) &= \sigma_v^2 \exp\left(\frac{|t-s|}{\tau}\right) \\ \sigma_v^2 &= \sigma_0^2 \end{aligned} \tag{4}$$

A Green's function representation of (2) suggests a direct method for generating $v(t)$ based only on the white-noise process $\varepsilon(t)$. It is worth noting the following: (i) as the correlation time $\tau = 1/\alpha$ goes to zero, $v(t)$ tends to a white noise of vanishing intensity (stationary); and (ii) as the correlation time $\tau = 1/\alpha$ goes to infinity, $v(t)$ tends to a non-stationary Wiener process, that is the integral of a white noise. The latter case is the most interesting; in case of zero-friction, the velocity process is a Wiener process with an infinite correlation scale in time. This process is non-stationary:

$$\frac{dW}{dt} = \varepsilon(t) \tag{5}$$

$$\begin{aligned} E\left[W(t)\right] &= 0 \\ \sigma_w^2(t) &= \sigma_0^2\ t \\ W(t) &= W(0) + \int_0^t \varepsilon(s)\ ds \end{aligned} \tag{6}$$

2.2. ONE-DIMENSIONAL SPACE

A natural way to generalize the Markov property in the non-oriented $\mathbb{R}^1$ space is as follows. For a discrete-space process $F(X_i)$, the Markov property corresponds to the independence of the future state, $F(X_{i+1})$, and the past state, $F(X_{i-1})$, given the present state, $F(X_i)$:

$$P_{F,F}\left(X_{i-1}, X_{i+1} \mid X_i\right) = P_F\left(X_{i-1} \mid X_i\right) \cdot P_F\left(X_{i+1} \mid X_i\right) \tag{7}$$

Similar to the time Gauss-Markov process, one obtains a 1-D space Gauss-Markov process that is stationary and satisfies:

$$\frac{d^2F}{dx^2} - \alpha^2 F(x) = \sigma_o U(x) \tag{8}$$

where $U(x)$ is a white noise with unit intensity, and α is inversely proportional to the integral correlation scale λ of the process $F(x)$:

$$\begin{aligned} R_{uu}(\xi) &= \delta(\xi) \\ \alpha &= 2/\lambda \end{aligned} \tag{9}$$

The integral correlation scale is defined by a one-sided integral as:

$$\lambda = \int_0^\infty P_{FF}(\xi)\, d\xi \tag{10}$$

Note that the right hand side (RHS) of (8) can be replaced by:

$$\sigma_o \frac{dW}{dx} \tag{11}$$

where $W(x)$ is a Wiener process defined in (6). The covariance structure of a zero-mean process satisfying (8) is as follows:

$$\begin{aligned} E\left[F(x)\right] &= 0 \\ R_{FF}(\xi) &= \sigma_F^2 \left(1 + \alpha|\xi|\right) \exp\left(-\alpha|\xi|\right) \\ \sigma_F^2 &= \frac{1}{4\alpha^3}\, \sigma_0^2 \end{aligned} \tag{12}$$

Note that the spatial Gauss-Markov process defined in (8) has the following characteristics: (i) it is stationary, and (ii) imposes a covariance structure that is dependent on dimensionality.

We may now investigate a Green's function representation of the process satisfying (8). Remember that for the time process, this was given by (2) for a given initial condition. Now we need boundary conditions (BCs) rather than an initial condition. In order to obtain an infinite domain representation, the boundaries should be taken at infinity ($x = \pm\infty$). By specifying deterministic homogeneous BCs:

$$F\ (\pm\infty) = 0 \tag{13}$$

we obtain the Green's function representation:

$$F(x) = \left(-\frac{\sigma_0}{2\alpha}\right) \int_{-\infty}^{+\infty} \exp\left(-\alpha|x-\xi|\right)\ U(\xi)\ d\xi \tag{14}$$

As a result, we have obtained an explicit representation for the zero-mean, stationary Markov process characterized by the covariance function (12). Note again that $U(\xi) = \dfrac{dW}{d\xi}$ is in the integral of (14), so this integral exists only because the Wiener process $W(\xi)$ exists.

2.3. MULTI-DIMENSIONAL SPACE

The generalization of the Markov property to the continuous 2-D and 3-D spaces is not a trivial matter. For a discrete multi-dimensional space, the independence property similar to that enunciated in Section 2.2 can be extended quite easily. However, in the continuous space, a more subtle definition is needed. See, for example, the definition of a pseudo-Markov random field in Adler (1981).

The probability of events $\{F(\underline{X}) = F_0(\underline{X})\}$ in the interior of a closed domain (the past) is independent of events $\{F(\underline{X}) = F_1(\underline{X})\}$ in its exterior (the future), given $F(\underline{X})$ in an open neighborhood surrounding the domain boundary (the present). By analogy with the 1-D spatial Gauss-Markov process analyzed above, it turns out that a stationary Gauss-Markov process in either 1-D, 2-D, or 3-D is obtained from the solution of the following PDE (Vanmarcke, 1988; Whittle, 1962):

$$\left(\nabla^2 - \alpha^2\right) F(\underline{X}) = \sigma_0\ U(\underline{X}) \tag{15}$$

The 1-D case has been analyzed above. In 2-D, the Green's function of (15) with a zero RHS (Klein-Gordon equation) involves Hankel functions. The covariance function of the resulting process involves a Bessel function (Vanmarcke, 1988). In 3-D, the Green's function takes a simpler form, when subject to the homogeneous deterministic BCs $F(\underline{X}) = 0$ as X_1, X_2, X_3 or $r = \sqrt{X_1^2 + X_2^2 + X_3^2}$ goes to infinity:

$$G\left(\underline{X},\underline{X}'\right) = \frac{1}{4\pi}\ \frac{\exp(-\alpha|\underline{r} - \underline{r}'|)}{|\underline{r} - \underline{r}'|} \tag{16}$$

The resulting 3-D Gauss-Markov process $F(\underline{X})$ has a Green's function representation, called a fundamental solution to (15):

$$F(\underline{X}) = \frac{\sigma_0}{4\pi} \int\int\int_{-\infty}^{+\infty} \frac{\exp\left(-\alpha\left|\underline{r} - \underline{r}'\right|\right)}{\left|\underline{r} - \underline{r}'\right|} U\left(\underline{X}'\right) d\underline{X}' \tag{17}$$

where

$$\left|\underline{r} - \underline{r}'\right| = \sqrt{\left(X_1 - X_1'\right)^2 + \left(X_2 - X_2'\right)^2 + \left(X_3 - X_3'\right)^2} \tag{18}$$

and

$$d\underline{X}' = dx_1'\ dx_2'\ dx_3' \tag{19}$$

It is not difficult to check that $F(\underline{X})$, as given by (17) is indeed a zero-mean 3-D Gauss-Markov process having the following characteristics (Vanmarcke, 1988):

$$\begin{aligned} R_{FF}(\xi) &= \sigma_F^2 \exp\left(-\alpha\left|\xi\right|\right) \\ \sigma_F^2 &= \frac{\pi}{\alpha}\sigma_o^2 \\ \lambda^3 &= \frac{\pi}{\alpha^3} \end{aligned} \tag{20}$$

where λ is the correlation scale (one-sided) defined as:

$$\lambda^3 = \int\int\int_0^{+\infty} \rho_{FF}(\xi)\, d\xi = \frac{1}{2^3}\int\int\int_{-\infty}^{+\infty} \rho_{FF}(\xi)\, d\xi \tag{21}$$

Once again, it is important to keep in mind the dependence of the covariance model on the problem dimensionality. The discretization of the white noise field $U(\underline{X})$ into the white lattice process $\hat{U}$ was designed so that the variance σ_F^2 be preserved. However, the correlation structure of $F(\underline{X})$ will be preserved only if $\alpha\Delta = \frac{\Delta}{\lambda} << 1$, where Δ is the grid size. The effect of truncating the infinite sum at some value $|i,j,k| \leq n$ can be limited by ensuring that the relative error in the imposed variance, defined as:

$$\varepsilon^2 = \frac{\sigma_F^2 - \sigma_{F^*}^2}{\sigma_F^2} \approx \frac{\pi \exp(-2m\alpha\Delta)}{(2\pi\alpha\Delta)^3} \tag{22}$$

is very small, (i.e., $\varepsilon^2 << 1$). Thus, the two requirements are:

Resolution: $\frac{\Delta}{\lambda} << 1$ for an adequate resolution (spatial integral discretization)

Accuracy: $\varepsilon^2 << 1 \Rightarrow n >> \left(\frac{\Delta}{\lambda}\right)^{-1}$ for an adequate accuracy (cut-off of infinite sum over space at distance equal to n, the lattice spacing).

3. Consistent Discretization of the Klein-Gordon Equation

3.1. ONE-DIMENSIONAL CASE

We examine again the 1-D spatial Gauss-Markov process of equation (8) with the RHS given by (11). Discretization of this PDE leads to:

$$\begin{aligned} \frac{d^2F}{dx^2} &= \frac{F_{i+1} - 2F_i + F_{i-1}}{\Delta^2} + O(\Delta^2) \\ U(x) &= \frac{dW}{dx} = \frac{W_{i+1/2} - W_{i-1/2}}{\Delta} + O(\Delta^2) \\ &= \frac{1}{\Delta}\int_{x_{i-1/2}}^{x_{i+1/2}} U(x)\, dx + O(\Delta^2) \end{aligned} \tag{23}$$

which leads to the FD system:

$$\frac{F_{i+1} - 2F_i + F_{i-1}}{\Delta^2} - \alpha^2 F_i = \sigma_o \frac{1}{\Delta}\int_{x_{i-1/2}}^{x_{i+1/2}} U(x)dx + O(\Delta^2) \tag{24}$$

Observe that $\overline{U}(x_i) = \frac{1}{\Delta}\int_{x_{i-1/2}}^{x_{i+1/2}} U(x)\, dx$ is a white sequence with variance $\sigma_{\overline{u}}^{\,2} = \frac{1}{\Delta}$. For convenience, we define $\hat{U}(x_i)$ the white sequence with unit variance, such that:

$$R_{\hat{u}\hat{u}}(i\Delta) = 0 \text{ for } i \neq 0 \tag{25}$$

$$R_{\hat{u}\hat{u}}(i\Delta) = 1 \text{ for } i = 0 \tag{26}$$

The FD system to be solved can be expressed, using $\overline{U}(x_i) \equiv \frac{1}{\Delta^{1/2}}\hat{U}(x_i)$, as:

$$F_i = a\,(F_{i+1} + F_{i-1}) + b\,\hat{U}(x_i) \tag{27}$$

with

$$a = \frac{1}{2 + (\alpha\Delta)^2}$$
$$b = \frac{\sigma_o \, \Delta^{3/2}}{2 + (\alpha\Delta)^2} = \frac{2(\alpha\Delta)^{3/2}\sigma_F}{2 + (\alpha\Delta)^2} \tag{28}$$

where α and σ_o are known in terms of the *true* $F(\underline{X})$ process:

$$\alpha = \frac{2}{\lambda_F}$$
$$\sigma_o^2 = 4 \, \alpha^3 \, \sigma_F^2 \tag{29}$$

Equation (27) constitutes a system of simultaneous algebraic equations, which is solved numerically for all F_i. The resolution requirement is that:

$$\frac{\Delta}{\lambda} = \frac{1}{2} \, \alpha\Delta << 1 \tag{30}$$

whence the constraint on the parameter a is $1/6 << a < 1/2$. Note that $a >> 1/6$ is an approximate constraint for accuracy, while $a < 1/2$ is a constraint imposed for physical realizability of stationary fields in the continuous space. Also, note that the main problem is that of BCs. In practice, we should solve (8) for BCs obeying (13) on a domain larger than the domain of interest by several correlation scales.

3.2. MULTI-DIMENSIONAL CASE

In 3-D, we obtain after a similar discretization of the governing equation (15):

$$F_{ijk} = a \left(F_{i-1,j,k} + F_{i,j-1,k} + F_{i,j,k-1} + F_{i+1,j,k} + F_{i,j+1,k} + F_{i,j,k+1}\right) + b \, \hat{U}_{i,j,k} \tag{31}$$

where $\hat{U}(i,j,k)$ is a white (uncorrelated) sequence on the lattice, with unit variance. The parameters a and b are now given by:

$$a = \frac{1}{6 + (\alpha \, \Delta)^2}$$
$$b = \frac{\sigma_o \, \Delta^{1/2}}{6 + (\alpha \, \Delta)^2} \tag{32}$$

We have also shown that α and σ_o are related to the true (continuous) process $F(\underline{X})$ according to (20). Obviously we have $0 < a < 1/6$. However, the resolution requirement

$\Delta/\lambda << a < 1/6$ yields $1/8 << a < 1/6$. It is worth noting how strict these constraints are, which is one of the major limitations of this method. On the other hand, however, one should also take note of the very close similarity of (31) to the structure of the system of algebraic equations appearing in any FD-based flow and/or transport problem. As such, the method discussed herein is an extremely efficient algorithmic platform to perform successive realization generation and flow and transport simulations in a Monte Carlo fashion.

4. Bayesian Conditioning of Multi-Dimensional Gauss-Markov Fields

In the probabilistic approach to subsurface flow and contaminant migration, the hydraulic properties of the (naturally heterogeneous) geologic medium are represented as spatially correlated random fields with known *prior* statistics (*posterior* statistics are discussed below). For gaussian fields, or their one-to-one transforms, the priors can be reduced to the first and second order moments: mean and 2-point covariance function. The covariance yields two kinds of information: the single-point variance, and the 2-point correlation function. In hydrogeology, the type of information used for determining these prior statistics may consist of sample statistics obtained directly from site-specific data, statistical information inferred from other geologic sites, and various qualitative or indirect information (geological facies, geophysical logs, etc).

In addition, once prior statistics have been identified, the random field may be further conditioned, that is constrained to match a given set of observations within the specific site at hand. Bayesian conditioning of a stationary prior field $Y(x)$ leads to a non-stationary posterior field, or conditional field, $Y^*(x)$. Denote Z the vector of observations and Y the state vector corresponding to field $Y(x_i)$ at a set of locations x_i. In the case of a gaussian prior, it can be shown that the conditional random field $\hat{Y}(x)$ is a linear combination of the random prior at point x, the random prior at all data points, and the data values themselves. The posterior mean, that is the mean of the conditional field, is denoted $<\hat{Y}>$ or $<Y|Z>$. It can be shown that the posterior mean is the best linear unbiased *estimator* of Y, in the Bayesian sense. Furthermore, in the case of a gaussian prior, it can be shown that $<\hat{Y}>$ is a linear combination of the prior mean vector $<Y>$ and the data vector Z. The resulting estimator $<\hat{Y}>$ is equivalent to the field produced by simple kriging. The reader is referred to Journel and Huijbregts (1978) and Delhomme (1979) for the geostatistical viewpoint on conditional estimation and to Ababou et al. (1994), and references therein, for the Bayesian estimation viewpoint. Both approaches can be viewed as statistical solutions to the problem of optimal spatial interpolation, or optimal estimation of spatial fields.

4.1. POINT CONDITIONING OF 1-D GAUSS-MARKOV FIELDS: ANALYTICAL METHOD AND VERIFICATION

We exploit here the special properties of Markov fields (more precisely, 2nd-order Markov fields) in 1-D space. Using the past-future analogy, this property can be simply stated as follows: the *future* depends only on the *present* and not on the *past*. In other words, if two point measurements are made, defining an interval, the random field values outside that interval depend only on the two point measurements, and not on the random field values inside that interval. The reverse holds true also. That is, the values inside and outside the domain — conditioned in the two point measurements — are mutually independent.

This property allows us to condition the PDE-driven Markov field by solving a 2-point boundary value problem. Moreover, any number of conditioning points may be included. Consider for instance the case of M conditioning points, which define a total of M + 1 intervals:

$$
\begin{aligned}
&(1) : [-\infty,\ x_1] \\
&(2) : [x_1,\ x_2] \\
&\ \cdot \qquad\ \cdot \\
&\ \cdot \qquad\ \cdot \\
&\ \cdot \qquad\ \cdot \\
&(M) : [x_{M-1},\ x_M] \\
&(M+1) : [x_M,\ +\infty]
\end{aligned}
\tag{33}
$$

The conditional Markov field can be generated on $[-\infty,\ +\infty]$ by solving (independently) the 2-point, white noise-driven, boundary value problems:

$$
\begin{gathered}
\frac{d^2Y}{dx^2} - \alpha^2\, Y = f(x) \\
Y(x_i) = Y_i\ ;\ Y(x_{i+1}) = Y_{i+1}
\end{gathered}
\tag{34}
$$

on each interval $[x_i,\ x_{i+1}]$. The BCs at $x_o = -\infty$ and $x_{M+1} = +\infty$ must be zero in order to ensure convergence of the conditional statistics to the unconditional ones at infinite distances from the conditioning points. Thus, we can use $Y_o = 0$, $Y_{M+1} = 0$ for external, far field, BCs. On the other hand, the values $\{Y_1, Y_2, \ldots\ldots Y_M\}$ represent the M measurements (*internal* BCs). The solution of the above 2-point boundary value problem can be obtained by the Green's function approach. The full Green's function for the 1-D operator is developed on a finite interval [-R,+R], with *homogeneous* Dirichlet conditions $Y(+R) = Y(-R) = 0$. We denote this Green's function by $G_R(x,y)$. The final result is:

$$
G_R(x,y) = -\frac{1}{2\alpha}\,\frac{ch\{\alpha\,(2R - |x-y|)\} - ch\{\alpha|x+y|\}}{sh\{2\alpha R\}}
\tag{35}
$$

Incidentally, as $R \to \infty$, we recover the infinite domain (free space) Green's function:

$$G(x,y) = -\frac{1}{2\alpha}\exp(-\alpha|x-y|) \tag{36}$$

which corresponds to the unconditional field.

For each interval $[x_i, x_{i+1}]$ equating $x_{i+1} = 2R_i$ and using a simple transformation of coordinates, the Green's function is found to be:

$$G_{Ri}(x,y) = -\frac{1}{2\alpha}\frac{ch\{\alpha(x_{i+1}-x_i-|x-y|)\} - ch\{\alpha(x+y-x_i-x_{i+1}\}}{sh\{\alpha(x_{i+1}-x_i)\}} \tag{37}$$

where x and y are both in $[x_i, x_{i+1}]$. Therefore, the solution $Y_A(x)$ satisfying the *homogeneous* conditions $Y_A(x_i) = 0$ and $Y_A(x_{i+1}) = 0$ is given by:

$$Y_A(x) = \int_{x_i}^{x_{i+1}} G_{Ri}(x,y)\, f(y)\, dy, \qquad x_i \leq x \leq x_{i+1} \tag{38}$$

This solution needs to be modified in order to obtain the correct BCs, which are usually non-zero. Thus, let:

$$Y(x) = Y_A(x) + Y_B(x) \tag{39}$$

where: (i) Y satisfies the original PDE, which is inhomogeneous and has *inhomogeneous* BCs; (ii) Y_A satisfies the same inhomogeneous equation, but with *homogeneous* conditions $Y_A = 0$; and (iii) Y_B satisfies the homogeneous equations with zero RHS, but with the correct *inhomogeneous* conditions $Y_B(x_i) = Y_i$, $Y_B(x_{i+1}) = Y_{i+1}$. It is easily seen that $Y = Y_A + Y_B$ is indeed the solution of (34) as stated above. To obtain Y_B, we solve:

$$\frac{d^2Y_B}{dx^2} - \alpha^2\, Y_B = 0 \qquad Y_B(x_i) = Y_i\ ;\ Y_B(x_{i+1}) = Y_{i+1} \tag{40}$$

This yields a solution of the form:

$$Y_B(x) = a_i\exp(\alpha x) + b_i\exp(-\alpha x) \qquad (x_i \leq x \leq x_{i+1}) \tag{41}$$

where the coefficients (a_i, b_i) are obtained by substituting the above BCs. This gives:

$$a_i = \frac{Y_{i+1}\exp(-\alpha x_i) - Y_i \exp(-\alpha x_{i+1})}{2\, sh\{\alpha\,(x_{i+1} - x_i)\}}$$
$$b_i = \frac{-\,Y_{i+1}\exp(\alpha x_i) - Y_i \exp(\alpha x_{i+1})}{2\, sh\{\alpha\,(x_{i+1} - x_i)\}} \tag{42}$$

Equations (37) to (42) give a complete characterization of the conditional random field $Y(x)$. The random part $Y_A(x)$ is a zero-mean stochastic integral of white noise, while the remaining part $Y_B(x)$ is deterministic and depends on the measurements. It follows that $Y_B(x)$ represents the posterior mean of the conditional random field $Y(x)$. Thus, we have:

$$< Y(x) > \; = Y_B(x) \tag{43}$$

or:

$$< Y(x) > \; = \frac{Y_{i+1}\, sh\{\alpha(x-x_i)\} - Y_i\, sh\{\alpha(x-x_{i+1})\}}{sh\{\alpha(x_{i+1} - x_i)\}} \tag{44}$$

within each measurement interval $[x_i,\ x_{i+1}]$. Note that (44) gives the kriged Markov field in closed form, that is kriging can be performed analytically in this case. The posterior variance, or estimation variance, can also be developed analytically as follows. We have by construction:

$$Var[Y(x)] = Var[Y_A(x)] \tag{45}$$

Upon substituting the stochastic white noise integral for Y_A, we obtain for $x_i \leq x \leq x_{i+1}$

$$Var[Y(x)] = \sigma_F^2 \int_{x_i}^{x_{i+1}} G_{Ri}(x,y)^2\, dy \tag{46}$$

Substituting the previous expression for G_{Ri}, and integrating, yields:

$$Var[Y(x)] = \frac{1}{4\alpha^3}\,\frac{2\alpha(x_{i+1}-x_i) + sh\{2\alpha(x_{i+1}-x_i)\} - sh\{2\alpha(x-x_i)\} - sh\{2\alpha(x_{i+1}-x)\}}{2sh^2\{\alpha(x_{i+1}-x_i)\}} \tag{47}$$

This expression gives the posterior variance, or kriging estimation variance, for each conditioning interval $x \in [x_i,\ x_{i+1}]$. As a matter of verification, note that we recover the correct prior variance:

$$\sigma_Y^2 = \frac{1}{4\alpha^3} \tag{48}$$

by letting $x_i \to -\infty$ and $x_{i+1} \to +\infty$. This expression coincides with that obtained directly from the Green's function representation of the unconditional Markov field.

4.2. RESULTS AND DISCUSSION

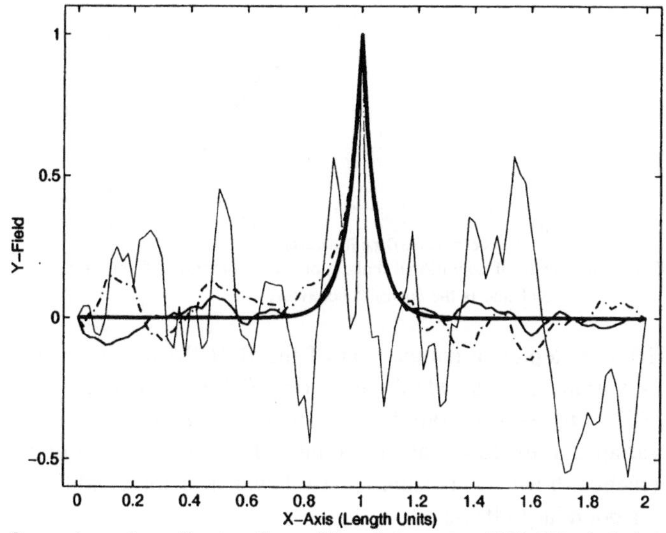

Figure 1. Comparison of a realization (fine solid) and the mean of 200 (thin dash-dot) and 1000 (thin solid) B-NNM realizations to equation (44) (thick solid).

An application and verification of the B-NNM algorithm is presented, together with analytical results, in Figure 1. The 1-D field has a $\lambda = 0.1$ (corresponding to five grid points) and is conditioned by imposing $Y(x=1) = 1.0$ for each of the Monte Carlo realizations. The substantial variability inherent in any particular realization is apparent. One should also note the relatively slow rate of convergence of the mean of the conditional realizations towards the *true* mean given by (44). A detailed analysis of convergence, depicted in Figure 2, indicated that the root mean square error (RMSE) for the mean of the conditional realizations approaches the *true* mean according to $(nmc)^{-1/2}$, where *nmc* is the number of Monte Carlo realizations involved.

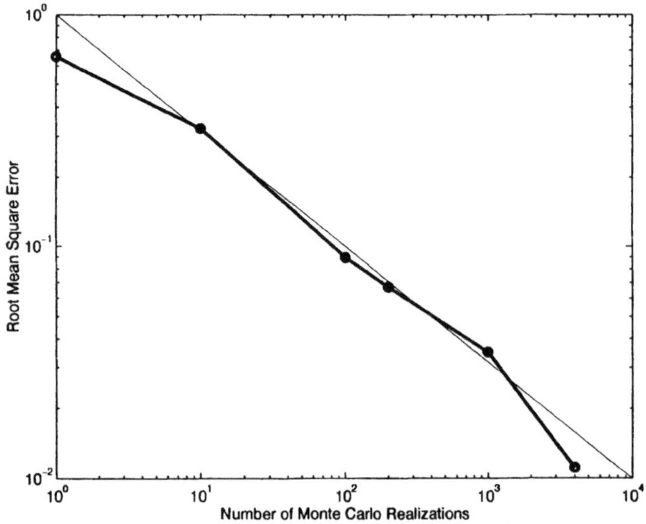

Figure 2. RMSE for the mean of conditional realizations, as a function of *nmc*. Fine solid line is the $(nmc)^{-1/2}$ behavior.

Finally, Figure 3 depicts a comparison of the B-NNM results in terms of the kriging estimation variance, or standard deviation (STD), against the closed form solution given by (47). One should note that the kriging estimation variance converges towards the *true* variance at the same rate as the mean far from the conditioning points. The rate of convergence, however, is significantly lower near the conditioning points, possibly because of boundary effects.

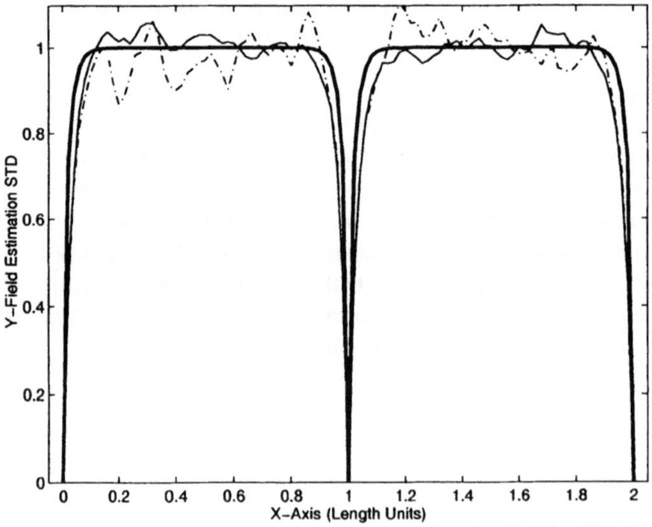

Figure 3. Comparison of kriging estimation STD, as obtained by B-NNM for 200 (thin dash-dot) and 1000 (thin solid) realizations, to equation (47) (thick solid).

5. Conclusions

In this paper, a specialized method for generating conditional random fields has been presented and tested for a 1-D problem against closed form solutions. The method presented here requires very strict resolution and accuracy constraints that need to be imposed for the generated fields to honor the desired covariance structures, which depend on the dimensionality of the problem. On the other hand, the method leads to solution of systems of algebraic equations that are very similar to those appearing in any FD-based flow and/or transport problem. As such, the method presented here is a promising algorithmic platform for conducting environmental assessments in a Monte Carlo fashion and deserves further study.

6. Acknowledgements

This paper was prepared to document work performed by the Center for Nuclear Waste Regulatory Analyses (CNWRA) for the Nuclear Regulatory Commission (NRC) under Contract No. NRC-02-93-005. The activities reported here were performed on behalf of the NRC Office of Nuclear Material Safety and Safeguards, Division of Waste Management, and the NRC Office of Nuclear Regulatory Research, Division of Regulatory Applications. The paper is an independent product of the CNWRA and does not necessarily reflect the views or regulatory position of the NRC. Comments by two anonymous peer-reviewers and the Editor's office helped improve the quality of this paper.

7. References

Ababou, R., Bagtzoglou, A.C., and Wood, E.F. (1994) On the condition number of covariance matrices in kriging, estimation, and simulation of random fields, *Math. Geol.* **26(1)**, 99-133.

Adler, R. (1981) *The Geometry of Random Fields*, Wiley, New York, NY.

Baker, R. (1984) Modeling soil variability as a random field, *Math. Geol.* **16(5)**, 435-448.

Delhomme, J.P. (1979) Spatial variability and uncertainty in groundwater flow parameters: A geostatistical approach, *Water Resour. Res.* **15(2)**, 269-280.

Gelhar, L.W. (1986) Stochastic subsurface hydrology from theory to applications, *Water Res. Res.* **22(9)**, 135-145.

Journel, A.G., and Huijbregts, Ch.J. (1978) *Mining Geostatistics*, Academic Press, New York, NY.

King, P.R., and Smith, P.J. (1988) Generation of correlated properties in heterogeneous porous media, *Math. Geol.* **20(7)**, 863-877.

Smith, L., and Freeze, R.A. (1979a) Stochastic analysis of steady state groundwater flow in a bounded domain: 1. One-dimensional simulations, *Water Resour. Res.* **15(3)**, 521-528.

Smith, L., and Freeze, R.A. (1979b) Stochastic analysis of steady state groundwater flow in a bounded domain: 2. Two-dimensional simulations, *Water Resour. Res.* **15(6)**, 1543-1559.

Smith, L., and Schwartz, F.W. (1980) Mass transport: 1. A stochastic analysis of macroscopic dispersion, *Water Resour. Res.* **16(2)**, 303-313.

Vanmarcke, E. (1988), *Random Fields: Analysis and Synthesis*, MIT Press, Cambridge, Mass.

Whittle, P. (1962) Topographic correlation, power-law covariance functions and diffusion, *Biometrika* **49**, 305-314.

CLUSTERING OF SPATIAL DATA BY THE EM ALGORITHM

C. AMBROISE , M. DANG AND G. GOVAERT

Université de Technologie de Compiègne
URA CNRS 817
BP 529 F-60205 Compiègne cedex - France

Abstract.

A clustering algorithm for spatial data is presented. It seeks a fuzzy partition which is optimal according to a criterion interpretable as a penalized likelihood. We propose to penalize the energy function exhibited by Hathaway (1986) with a term taking into account spatial contiguity constraints. The structure of the EM algorithm may be used to maximize the proposed criterion. The Maximization step is then unchanged and the Expectation step becomes iterative. The efficiency of the new clustering algorithm has been tested with biological images and compared with other clustering techniques.

1. Introduction

When classical clustering techniques are used for partitioning spatial data, the resulting classes will often be geographically very mixed. To avoid this phenomenon, the spatial information of the data has to be taken into account. Very different solutions to this problem have been proposed in the literature.

A natural approach consists in using the geographical coordinates of the individuals, more or less heavily weighted, as an additional pairs of variates - see Berry (1966) or Jain and Farrokhnia (1991).

Another approach groups individuals which are both similar and contiguous (Legendre 1987, Openshaw 1977). This implies the definition of a neighborhood concept. Defining neighborhood relationships is equivalent to building a graph where each element is represented by a node and each

A. Soares et al. (eds.), geoENV I–Geostatistics for Environmental Applications, 493–504.

neighborhood relationship is an edge. Clustering with spatial contiguity constraints can be described as the succession of two steps:

1. The definition of a neighborhood graph. This can be done with standard algorithms such as a Delaunay triangulation (Green and Sibson 1977) or a Gabriel graph (Gabriel and Sokal 1969).
2. Running of a clustering algorithm while respecting the constraints. Many classical clustering algorithms may be modified to take into account the constraints which are summarized by the graph. In Lebart (1978) a classical hierarchical clustering algorithm is adapted.

These procedures produce classes made of adjacent sites. They may separate into different classes individuals which are very similar, if they are geographically far apart.

Oliver and Webster (1989) propose to run clustering algorithms based on a modified dissimilarity matrix. This modified matrix is a combination of the matrix of geographical distances and the dissimilarity matrix computed from the non geographical variables. This kind of procedure seems to work well but has no statistical justification.

Other spatially constrained clustering methods and approaches have been developed in the field of unsupervised image segmentation (Ripley 1988). The specificity of these methods is that they deal with pixels and their regular grid structure. An image may well be considered as a regular lattice of pixels. In this case, the computation of the neighborhood graph is immediate. The most common choices for the neighborhood graph are the 4 neighbor graph (horizontal and vertical adjacencies only) and the 8 neighbor (including diagonals).

There are numerous unsupervised image segmentation algorithms. In this paper, we consider only statistically based methods. In this framework, the Bayesian approach proposes solutions which may be separated into two families (Masson and Pieczinsky 1993):

1. The local methods make assumptions about the pixel or about small groups of adjacent pixels called "context" (Masson and Pieczinsky 1993).
2. The global methods make assumptions about the whole image and generally use a Markov random field model (Geman and Geman 1984, Besag 1974).

Notice that the classes obtained with these methods differ from those obtained with hard contiguity constrained algorithms: a class does not necessarily form a single patch on the image. Pixels of the same class may be present in different parts of the image. Thus the statistical models used in unsupervised segmentation algorithms take the contiguity constraints into account but do not impose "one region classes".

We develop in this paper a new statistical method for spatial clustering which is based on the EM algorithm (Dempster *et al.* 1977) and takes into account the spatial constraints without requiring a partition made of "one region classes". Next section describes the principle of the proposed algorithm. Section 3 relates this method to statistical image segmentation techniques based on Markov Random Fields. Section 4 illustrates the method's performance on the segmentation of a biological gray-level image. The concluding section evocates aspects of this approach that are currently under investigation or seem to be worth of further study.

2. Introducing Spatial Constraints in the EM Algorithm

2.1. MIXTURE ESTIMATION BY THE EM ALGORITHM

In cluster analysis based on Gaussian mixture models (Celeux and Govaert 1995), data are $\mathbb{R}^d$-valued vectors $\mathbf{x}_1, \ldots \mathbf{x}_n$ assumed to be an identically independently distributed (i.i.d.) sample from a mixture of K normal distributions:

$$f(\mathbf{x}_i|\Phi) = \sum_{k=1}^{K} p_k f_k(\mathbf{x}_i|\boldsymbol{\mu}_k, \Sigma_k), \tag{1}$$

where the p_k are the mixing proportions (for $k = 1, \ldots, K, 0 < p_k < 1$, and $\sum_k p_k = 1$), and $f_k(\mathbf{x}|\boldsymbol{\mu}_k, \Sigma_k)$ denotes the density of a Gaussian distribution with mean vector $\boldsymbol{\mu}_k$ and variance matrix $\boldsymbol{\Sigma}_k$. This model also assumes that the unobserved vector of labels, $\mathbf{z} = (z_1, \ldots, z_n)$, is an i.i.d. sample of the multinomial distribution:

$$\text{for } 1 \le i \le n, 1 \le k \le K, \quad P(Z_i = k) = p_k.$$

The EM algorithm is often used to estimate the unknown parameters of the mixture (Dempster *et al.* 1977). It produces a set of parameters that maximizes locally the log-likelihood of the sample, defined as

$$L(\Phi) = \sum_{i=1}^{n} \log f(\mathbf{x}_i).$$

The principle of the EM algorithm consists in building a sequence of estimates $\Phi^0, \Phi^1, \ldots, \Phi^m$, over which the log-likelihood monotonically increases (for all m, $L(\Phi^{m+1}) \ge L(\Phi^m)$). The basic idea in EM is to choose Φ^{m+1} from Φ^m that globally maximizes the expectation defined as

$$\begin{aligned} Q(\Phi|\Phi^m) &\triangleq \sum_{\mathbf{z}} P(\mathbf{z}|\mathbf{x}; \Phi^m) \log P(\mathbf{x}, \mathbf{z}; \Phi) \\ &= \sum_{i=1}^{N} \sum_{k=1}^{K} \log(p_k f_k(\mathbf{x}_i)) P(Z_i = k|\mathbf{x}_i; \Phi^m) \quad \left(\begin{array}{c}\text{for}\\ \text{mixture}\\ \text{models}\end{array}\right) \end{aligned} \tag{2}$$

Thus, starting from an arbitrary value Φ^0, iteration $(m+1)$ of the EM algorithm can be divided in 2 steps:

- **E-step** (*Expectation*): computation of the components of $Q(\Phi|\Phi^m)$ that do not depend on Φ;
- **M-step** (*Maximization*): search $\Phi^{m+1} = \arg\max_\Phi Q(\Phi|\Phi^m)$.

In the case of mixture models, the E-step computes

$$\left[t_{ik}^{m+1} \triangleq P(Z_i = k|\mathbf{x}_i; \Phi^m) = \frac{p_k^m f_k(\mathbf{x}_i|\boldsymbol{\mu}_k^m, \Sigma_k^m)}{f(\mathbf{x}_i|\Phi^m)} \right]_{\substack{1 \leq i \leq n \\ 1 \leq k \leq K}} \tag{3}$$

For a gaussian mixture, the M-step yields, for $k = 1, \ldots, K$

$$\boldsymbol{\mu}_k^{m+1} = \frac{1}{n_k} \sum_{i=1}^{n} t_{ik}^{m+1} \mathbf{x}_i \tag{4}$$

$$\Sigma_k^{m+1} = \frac{1}{n_k} \sum_{k=1}^{K} \sum_{i=1}^{n} t_{ik}^{m+1} (\mathbf{x}_i - \boldsymbol{\mu}_k^{m+1})(\mathbf{x}_i - \boldsymbol{\mu}_k^{m+1})^t \tag{5}$$

$$p_k^{m+1} = \frac{n_k}{n} \tag{6}$$

where $n_k = \sum_{i=1}^{n} t_{ik}^{m+1}$.

2.2 THE EM ALGORITHM AS A FUZZY CLUSTERING METHOD

As Hathaway (1986) highlighted it, the EM algorithm in the case of mixture models is formally equivalent to an alternate optimization of function

$$\mathrm{D}(\mathbf{c}, \Phi) \triangleq \sum_{k=1}^{K} \sum_{i=1}^{n} c_{ik} \log(p_k f_k(\mathbf{x}_i|\boldsymbol{\mu}_k, \Sigma_k)) - \sum_{k=1}^{K} \sum_{i=1}^{n} c_{ik} \log(c_{ik}) \tag{7}$$

where $\mathbf{c} = (c_{ik})_{i=1,n \quad k=1,K}$ defines a fuzzy classification, c_{ik} representing the grade of membership of $\mathbf{x}_i$ to class k ($0 \leq c_{ik} \leq 1$, $\sum_{k=1}^{K} c_{ik} = 1$, $\sum_{i=1}^{n} c_{ik} > 0$, $1 \leq i \leq n, 1 \leq k \leq K$).

To see this, consider the following grouped coordinate ascent on function (7): starting from a given set of initial parameters, Φ^0, criterion $\mathrm{D}(\mathbf{c}, \Phi)$ is alternatively optimized over the possible values of the classification matrix $\mathbf{c}$ with fixed mixture parameters, then over the possible values of mixture parameters Φ with a fixed classification matrix. Thus, iteration $(m+1)$ of this alternate optimization algorithm consists of two steps:

"E"-step: The classification matrix is updated in order to maximize criterion

$$\mathbf{c}^{m+1} = \arg\max_{\mathbf{c}} \mathrm{D}(\mathbf{c}, \Phi^m).$$

Let us write the Lagrangian of $\mathrm{D}(\mathbf{c}, \Phi^m)$ that takes into account the constraints $\sum_{k=1}^{K} c_{ik} = 1, \forall i$:

$$\mathcal{D}(\mathbf{c}) = \mathrm{D}(\mathbf{c}, \Phi^m) + \sum_{i=1}^{n} \lambda_i \left(\left(\sum_{k=1}^{K} c_{ik} \right) - 1 \right), \tag{8}$$

where the λ_i are the Lagrange coefficients corresponding to the constraints.

The necessary conditions of optimality yield:

$$\begin{cases} \frac{\partial \mathcal{D}}{\partial c_{ik}} = \log\left(p_k^m f_k(\mathbf{x}_i | \boldsymbol{\mu}_k^m, \Sigma_k^m)\right) - 1 - \log c_{ik} + \lambda_i = 0 \\ \sum_{k=1}^{K} c_{ik} = 1 \end{cases}$$

which can also be written as:

$$\begin{cases} c_{ik} = \exp\left\{-1 + \lambda_i + \log\left(p_k^m f_k(\mathbf{x}_i | \boldsymbol{\mu}_k^m, \Sigma_k^m)\right)\right\} \\ \sum_{k=1}^{K} \exp\left\{-1 + \lambda_i + \log\left(p_k^m f_k(\mathbf{x}_i | \boldsymbol{\mu}_k^m, \Sigma_k^m)\right)\right\} = 1. \end{cases}$$

The following values are obtained for the c_{ik}^{m+1}:

$$c_{ik}^{m+1} = \frac{p_k^m f_k(\mathbf{x}_i | \boldsymbol{\mu}_k^m, \Sigma_k^m)}{f(\mathbf{x}_i | \Phi^m)}. \tag{9}$$

which is clearly identical to EM's E-step for mixture models (3).

"M"-step: the parameters are reestimated according to

$$\Phi^{m+1} = \arg\max_{\Phi} \mathrm{D}(\mathbf{c}^{m+1}, \Phi).$$

However, combining (2) with (3), and (7) with (9), it can be seen that

$$\mathrm{D}(\mathbf{c}^{m+1}, \Phi) = Q(\Phi | \Phi^m) - \underbrace{\sum_{k=1}^{K} \sum_{i=1}^{n} c_{ik}^{m+1} \log(c_{ik}^{m+1})}_{\text{independent of } \Phi}$$

so that the same Φ^{m+1} is obtained as in the M-step of the EM algorithm.

Since this scheme yields the same calculations as EM applied to mixtures, the latter can be viewed as an alternate optimization method of criterion $\mathrm{D}(\mathbf{c}, \Phi)$. The criteria themselves are related by $L(\Phi^m) = \mathrm{D}(\mathbf{c}^{m+1}, \Phi^m)$.

This interpretation makes clearer the relationships between EM and other clustering techniques, which often optimize criteria very similar to $D(\mathbf{c}, \Phi)$ (Celeux and Govaert 1995).

2.3 PENALIZATION OF HATHAWAY'S CRITERION

We propose to regularize Hathaway's criterion (7) with a term which takes into account the spatial information relative to the data. Let $\mathbf{V}$ be the "neighborhood matrix":

$$v_{ij} = \begin{cases} \alpha > 0 & \text{if } \mathbf{x}_i \text{ and } \mathbf{x}_j \text{ are neighbors} \\ 0 & \text{if } \mathbf{x}_i \text{ and } \mathbf{x}_j \text{ are not neighbors.} \end{cases}$$

To compute the matrix $\mathbf{V}$, one can use the graph structure described in the introduction, or a function of the spatial distance between the elements of the data set.

The regularizing term we propose is the following:

$$G(\mathbf{c}) = \frac{1}{2} \sum_{k=1}^{K} \sum_{i=1}^{n} \sum_{j=1}^{n} c_{ik}.c_{jk}.v_{ij}. \tag{10}$$

The more the classes contain adjacent elements, the greater this term is. Notice that $G(\mathbf{c})$ is only function of the classification matrix. The new criterion we consider is then:

$$U(\mathbf{c}, \Phi) = D(\mathbf{c}, \Phi) + \beta.G(\mathbf{c}) \tag{11}$$

where the coefficient $\beta \geq 0$ gives more or less weight to the spatial homogeneity term relatively to $D(\mathbf{c}, \Phi)$.

A modified version of the EM algorithm can be used to maximize this criterion. We named this version *Neighborhood EM algorithm* (NEM). Since the new regularizing term does not contain any parameter of the mixture, the Maximization step remains unchanged. The Estimation step (maximization of the criterion relatively to the classification matrix with fixed mixture parameters) changes:

1. Initialization: a neighborhood matrix $\mathbf{V}$ is computed according to the spatial relationship; arbitrary initial values are chosen for the parameters of the mixtures $\Phi^{(0)}$ as well as for the classification matrix, $\mathbf{c}^{(0)}$, (n lines, K columns).
2. At each iteration the followings steps are realized until convergence:

   (a) **E-step:**

   $$\mathbf{c}^{m+1} = \arg\max_{\mathbf{c}} U(\mathbf{c}, \Phi^m).$$

The necessary conditions of optimality take the following form:

$$\begin{cases} \frac{\partial U}{\partial c_{ik}} = \log\left(p_k^m f_k(\mathbf{x}_i|\boldsymbol{\mu}_k^m, \Sigma_k^m)\right) + 1 - \log c_{ik} + \lambda_i + \beta \sum_{j=1}^n c_{jk} v_{ij} \\ \sum_{k=1}^K c_{ik} = 1 \end{cases}$$

which may be written as

$$\begin{cases} c_{ik} = \exp\left\{\log(p_k^m f_k(\mathbf{x}_i|\boldsymbol{\mu}_k^m, \Sigma_k^m)) + 1 + \lambda_i + \beta \sum_{j=1}^n c_{jk} v_{ij}\right\} \\ \sum_{k=1}^K \exp\left\{\log(p_k^m f_k(\mathbf{x}_i|\boldsymbol{\mu}_k^m, \Sigma_k^m)) + 1 + \lambda_i + \beta \sum_{j=1}^n c_{jk} v_{ij}\right\} = 1. \end{cases}$$

Finally we get the following equation:

$$c_{ik}^{m+1} = \frac{p_k^m f_k(\mathbf{x}_i|\boldsymbol{\mu}_k^m, \Sigma_k^m) \cdot \exp\{\beta \sum_{j=1}^n c_{jk}^{m+1} v_{ij}\}}{\sum_{\ell=1}^K p_\ell^m f_\ell(\mathbf{x}_i|\boldsymbol{\mu}_\ell^m, \Sigma_\ell^m) \cdot \exp\{\beta \sum_{j=1}^n c_{j\ell}^{m+1} v_{ij}\}} \tag{12}$$

which suggests an iterative computing algorithm of the form $\mathbf{c} = g(\tilde{\mathbf{c}})$, where $\tilde{\mathbf{c}}$ is the old classification matrix. From a practical point of view, a few iterations produce a reasonable new classification matrix $\mathbf{c}$, which can be used for the next *M-step*.

(b) **M-step:**

$$\begin{aligned} \Phi^{m+1} &= \arg\max_{\Phi} U(\mathbf{c}^{m+1}, \Phi) \\ &= \arg\max_{\Phi} D(\mathbf{c}^{m+1}, \Phi). \end{aligned}$$

Thus, to compute the parameters of the mixture, one can use the same formulae as in the M-step of the EM algorithm.

Empirically the NEM algorithm converges with all tested problems, but no proof of local convergence is available yet.

At each iteration the spatial information modifies the partition according to the importance of the β coefficient. The Estimation step smoothes the map of the labels.

The preceding algorithm solves two different tasks which are highly dependent: it estimates the parameters of the mixture and finds a partition. The two tasks are performed at the same time.

2.4 HARD PARTITION

The NEM algorithm may be easily adapted to seek a hard partition ($c_{ik} \in \{0, 1\}$):

- The easiest way is to consider the fuzzy partition obtained at the convergence, and to assign each individual i to the most probable class according to the a posteriori probabilities, i.e. the class k that maximizes c_{ik}.
- As in the CEM algorithm (Celeux and Govaert 1992), it is possible to add an intermediate classification step between the E and the M step.

3. A Bayesian Interpretation

Notice that it is possible to have a Bayesian interpretation of the NEM algorithm. Maximizing the criterion $U(\mathbf{c}, \Phi)$ is equivalent to maximizing

$$\exp\{U(\mathbf{c}, \Phi)\} = \exp\{D(\mathbf{c}, \Phi)\} \cdot \exp\{\beta \cdot G(\mathbf{c})\}. \tag{13}$$

We may write

$$\exp\{\beta \cdot G(\mathbf{c})\} \propto P(\mathbf{c}) \tag{14}$$

where $P(\mathbf{c})$ is a Gibbs distribution with energy function $-\beta \cdot G(\mathbf{c})$. On the other hand, the expression $\exp\{D(\mathbf{c}, \Phi)\}$ may be interpreted as $P_\Phi(\mathbf{x}|\mathbf{c})$, the conditional density of the sample $\mathbf{x}$, knowing the classification matrix $\mathbf{c}$, with parameters Φ characterizing the Gaussian mixture. Thus,

$$\begin{aligned}\exp\{U(\mathbf{c}, \Phi)\} &\propto P(\mathbf{c}) \cdot P_\Phi(\mathbf{x}|\mathbf{c}) \\ &\propto P_\Phi(\mathbf{c}|\mathbf{x})\end{aligned}$$

where $P_\Phi(\mathbf{c}|\mathbf{x})$ is the posterior distribution of the classification matrix $\mathbf{c}$.

Maximizing $U(\mathbf{c}, \Phi)$ is equivalent to finding the matrix $\mathbf{c}$ that maximizes the posterior distribution $P_\Phi(\mathbf{c}|\mathbf{x})$, which is a Gibbs distribution with energy function $-U$. Thus, the NEM algorithm may be interpreted as an algorithm that searches the maximum a posteriori (MAP) estimate of the classification matrix $\mathbf{c}$.

This Bayesian interpretation assumes that there are two random fields: $\mathbf{X} = \{\mathbf{X}_s, s \in S\}$, which is the observed random field and $\mathbf{C} = \{\mathbf{C}_s, s \in S\}$, which is the random field corresponding to the classification matrix (S is the set of sites). The $\mathbf{X}_s$ take their values in $\mathbb{R}^d$ and the $\mathbf{C}_s = (C_{s1}, \cdots, C_{sK})$ in a subset of $[0,1]^K$. Both fields $\mathbf{X}$ and $\mathbf{C}$ follow Gibbs Distributions.

This approach may be used in image segmentation and provides an alternative to the existing unsupervised fuzzy segmentation algorithms (Kent and Mardia 1991, Caillol *et al.* 1993).

In the case of a hard classification matrix, i.e. the $C_s = (C_{s1}, \cdots, C_{sK})$ are random variables which take their values in $\{0,1\}^K$, and with the following neighborhood matrix:

$$v_{ij} = \begin{cases} 1 & \text{if } \mathbf{x}_i \text{ and } \mathbf{x}_j \text{ are neighbors} \\ 0 & \text{if } \mathbf{x}_i \text{ and } \mathbf{x}_j \text{ are not neighbors,} \end{cases}$$

the a priori distribution of the classification matrix is a Markov Random Field model very frequently used in image segmentation (Strauss 1977, Besag 1986):

$$\begin{aligned} P(\mathbf{C} = \mathbf{c}) &= \frac{1}{Z}\exp\{\beta G(\mathbf{c})\} \\ &= \frac{1}{Z}\exp\{\beta \sum_{i<j} v_{ij}(\mathbf{c}_i \cdot \mathbf{c}_j)\} \\ &= \frac{1}{Z}\exp\{\beta(\#\text{neighbors belonging to the same class})\} \end{aligned}$$

where Z is a normalizing constant.

4. An Application to Biological Images

A sample of living cells is laid on a nutritive substance. After a few days new living cells appear and form a thin but visible layer around the original sample. Biologists are interested in determining the surface of the new layer. We have a collection of images of size 512 by 512 [1] of such experimental results and present here the analysis of a representative image.

We aim to distinguish three different kinds of patterns, in order to determine automatically the size of the area covered by the new cells. A "good" segmentation from the biologist point of view should separate the image in three different areas representing:

- the original sample;
- the nutritive substance;
- the new cells.

We have tested three different algorithms:

- the classical EM algorithm, which does not take into account the spatial information;
- the Gibbsian EM algorithm (Chalmond 1989); this unsupervised segmentation method applies the principle of EM to a Markov random field model for the labels $\mathbf{Z}$, and an independant distribution of data $\mathbf{X}$ conditionally on $\mathbf{Z}$; here, the computation of $Q(\Phi|\Phi^m)$ in the E-step requires to simulate the posterior distribution of $\mathbf{Z}$ conditionnally on $\mathbf{x}$: this simulation is achieved by a Gibbs sampler (Geman and Geman 1984);

[1] We are indebted to Eugenio Grapa, from the Université de Technologie de Compiègne, Laboratoire de biologie cellulaire expérimentale, for the biological pictures

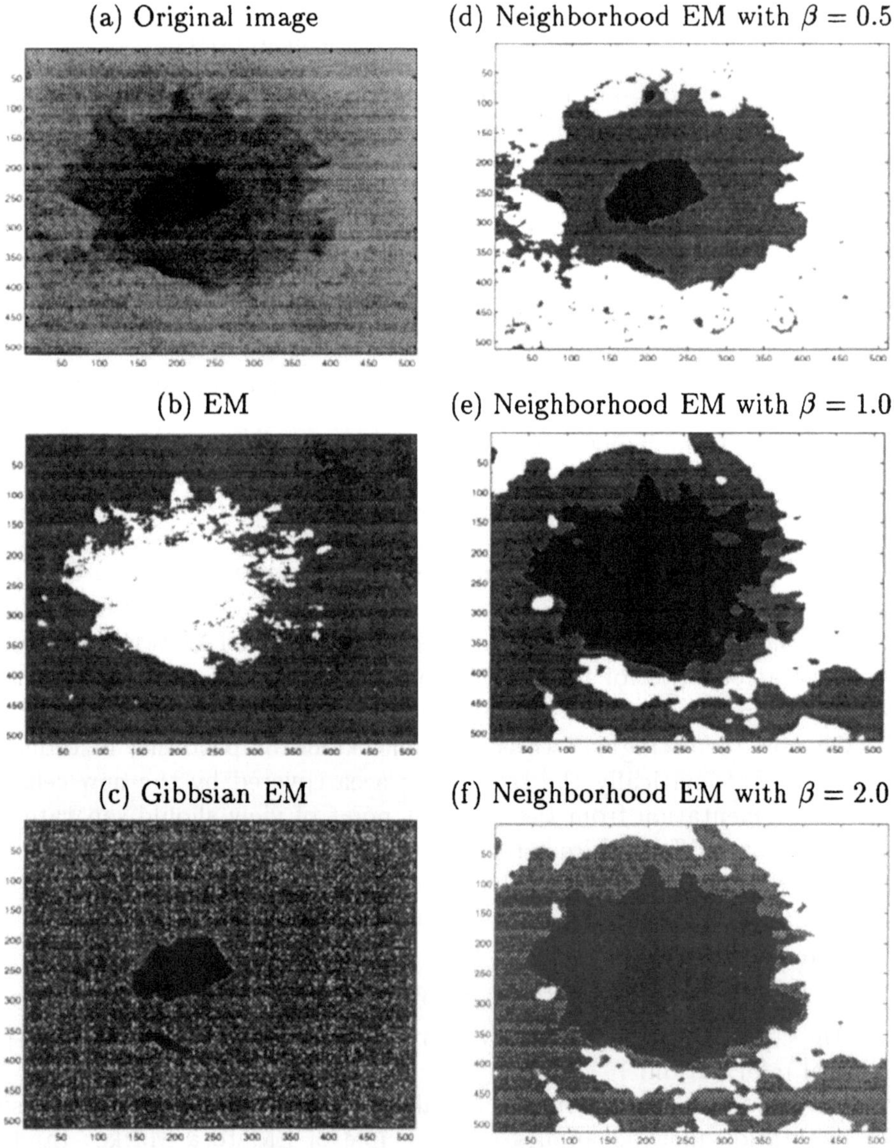

Figure 1. Segmentation of a biological image by different algorithms

– the proposed version of the EM algorithm, which maximizes the previously defined penalized likelihood.

In these experiments, the algorithms were initialized by a "histogram" method, often used in gray level image segmentation. This method consists in clustering the pixels by thresholding the histogram of the gray levels. The thresholds are computed in order to have K classes of equal size. The clustering obtained in this way serves as an initial partition to start the algorithms.

On the original image (Figure 1(a)), the human vision distinguishes clearly the three classes. In fact, each class is far from having an uniform gray level: some pixels representing the new cells have exactly the same color as the nutritive substance pixels. While the human eyes automatically make the necessary adjustments, the unsupervised segmentation of this kind of image is more difficult than one could think at first.

The EM algorithm isolates the nutritive substance, but tends to empty the class corresponding to the original sample (Figure 1(b)) .

The Gibbsian EM algorithm does not make any distinction between the new cells and the nutritive substance (Figure 1(c)).

The NEM algorithm was tested with different values for the parameter β (Figure 1(d)-(f)). When the parameter is very small, the same results are obtained as with the EM algorithm. This seems to be logical, because if we set β to zero, NEM becomes the classical EM algorithm. When $\beta = 0.5$, the result is very satisfying and allows the automatic computation of the surface covered by the new cells. If we try to run NEM with greater values of the parameter, the spatial information becomes preponderant and the segmentation does not make a lot of sense.

5. Discussion

The preceding example showed the practical efficiency of the proposed algorithm on this particular problem. We tested the NEM algorithm on other segmentation tasks and got regularly good results once the β parameter was tuned.

The algorithm was also applied to cluster a data set consisting of socio-economical indices on a set of adjacent counties. The first results are encouraging. Further experiments should test the performance of the algorithm on other spatial data which have an irregular lattice structure.

The main difficulty in applying the NEM algorithm consists in deciding on the value of β. In this problem we validate the value using our knowledge of what the "good" solution should look like. It would be easier to have an automatic estimation of this parameter. To this purpose, the estimation methods that were developed in the framework of image segmentation could be a source of inspiration (Chalmond 1989, Younes 1989, Pieczynski 1994). However, since most of these methods were specifically designed for a regular grid structure of the sites, they would need to be adapted to work on irregularly distributed spatial data.

Theoretical results on the convergence properties of NEM have been recently established. They will be published in a forthcoming paper.

References

Berry, B.J.L. (1966). Essay on commodity flows and the spatial structure of the indian economy. Research paper 111. University of Chicago, departement of geography.

Besag, J.E. (1974). Spatial interaction and the statistical analysis of lattice systems. *Journal of the Royal Statistical Society* **35**, 192–236.

Besag, J.E. (1986). Spatial analysis of dirty pictures. *Journal of the Royal Statistical Society* **48**, 259–302.

Caillol, H., A. Hillion and W. Pieczinsky (1993). Fuzzy random fields and unsupervised image segmentation. *IEEE transactions on geoscience and remote sensing* **31**(4), 801–810.

Celeux, G. and G. Govaert (1992). A classification EM algorithm for clustering and two stochastic versions. *Computational statistics and data analysis* **14**, 315–332.

Celeux, G. and G. Govaert (1995). Gaussian parsimonious clustering models. *Pattern Recognition* **28**, 781–793.

Chalmond, B. (1989). An iterative gibbsian technique for reconstruction of m-ary images. *Pattern Recognition* **22**(6), 747–761.

Dempster, Laird and Rubin (1977). Maximum likelihood from incomplete data via the EM algorithm. *Journal of the Royal Statistical Society* **39**, 1–38.

Gabriel, K.R. and R.R Sokal (1969). A new statistical approach to geographic variation analysis. *Syst. Zool.* **18**, 259–278.

Geman, S. and D. Geman (1984). Stochastic relaxation, gibbs distributions, and the baysian restoration of images. In: *IEEE Transactions on Pattern Analysis and Machine Intelligence*. Vol. PAMI-6. pp. 721–741.

Green, P.J. and R. Sibson (1977). Computing dirichlet tesselations in the plane. *The Computer Journal* **21**, 168–173.

Hathaway, R.J. (1986). Another interpretation of the EM algorithm for mixture distributions. *Journal of Statistics & Probability Letters* **4**, 53–56.

Jain, A.K. and F. Farrokhnia (1991). Unsupervised texture segmentation using gabor filters. *Pattern Recognition* **24**(12), 1167–1186.

Kent, J.T. and K.V. Mardia (1991). Spatial classification using fuzzy membership. *IEER Trans. Patt. Anal. Machine Intell.* **10**(5), 135–143.

Lebart, L. (1978). Programme d'agrégation avec contraintes. *Cahier de l'analyse des données* **3**, 275–287

Legendre, P. (1987). Constrained clustering. *developpments in numerical ecology. NATO ASI Series* **G 14**, 289–307.

Masson, P. and W. Pieczinsky (1993). SEM algorithm and unsupervised statistical segmentation of satellite images. *IEEE transactions on geoscience and remote sensing* **31**(3), 618–633.

Oliver, M.A. and R. Webster (1989). A geostatistical basis for spatial weighting in multivariate classification. *Mathematical Geology* **21**, 15–35.

Openshaw, S. (1977). A geographical solution to scale and aggregation problems in region-building, partitionning and spatial modelling. *Transactions of the institute of british geographers* **2**, 459–472.

Pieczynski, W. (1994). Champs de markov cachés et estimation conditionnelle itérative. *Traitement du Signal* **11**(2), 141–153.

Ripley, B.D. (1988). *Statistical inference for spatial processes.* Cambridge university press.

Strauss, D.J. (1977). Clustering on coloured lattices. *Journal of appl. probab.* **14**, 135–143.

Younes, L. (1989). Parametric inference for imperfectly observed gibbsian fields. *Probability Theory and Related Fields* **82**, 625–645.

Quantitative Geology and Geostatistics

1. F. M. Gradstein, F. P. Agterberg, J. C. Brower and W. S. Schwarzacher: *Quantitative Stratigraphy*. 1985 ISBN 90-277-2116-5
2. G. Matheron and M. Armstrong (Eds.): *Geostatistical Case Studies*. 1987 ISBN 1-55608-019-0
3. *Cancelled*
4. M. Armstrong (Ed.): *Geostatistics*. Proceedings of the 3rd International Geostatistics Congress, held in Avignon, France (1988), 2 volumes. 1989 Set ISBN 0-7923-0204-4
5. A. Soares (Ed.): *Geostatistics Tróia '92*, 2 volumes. 1993 Set ISBN 0-7923-2157-X
6. R. Dimitrakopoulos (Ed.): *Geostatistics for the Next Century*. 1994 ISBN 0-7923-2650-4
7. M. Armstrong and P.A. Dowd (Eds.): *Geostatistical Simulations*. 1994 ISBN 0-7923-2732-2
8. E.Y. Baafi and N.A. Schofield (Eds.): *Geostatistics Wollongong '96*, 2 volumes. 1997 Set ISBN 0-7923-4496-0
9. A. Soares, J. Gómez-Hernandez and R. Froidevaux (Eds.): *geoENV I – Geostatistics for Environmental Applications*. Proceedings of the Geostatistics for Environmental Applications Workshop, Lisbon, Portugal, 18–19 November 1996. 1997 ISBN 0-7923-4590-8

MIX
Papier aus verantwortungsvollen Quellen
Paper from responsible sources
FSC® C105338

If you have any concerns about our products,
you can contact us on
ProductSafety@springernature.com

In case Publisher is established outside the EU,
the EU authorized representative is:
Springer Nature Customer Service Center GmbH
Europaplatz 3, 69115 Heidelberg, Germany

Printed by Libri Plureos GmbH
in Hamburg, Germany